Lecture Notes in Computer Science 13778

Advanced Research in Computing and Software Science
Subline of Lecture Notes in Computer Science

More information about this series at https://link.springer.com/bookseries/558

Kristoffer Arnsfelt Hansen ·
Tracy Xiao Liu · Azarakhsh Malekian (Eds.)

Web and Internet Economics

18th International Conference, WINE 2022
Troy, NY, USA, December 12–15, 2022
Proceedings

 Springer

Editors
Kristoffer Arnsfelt Hansen 🆔
Aarhus University
Aarhus, Denmark

Tracy Xiao Liu
Tsinghua University
Beijing, China

Azarakhsh Malekian 🆔
Rotman School of Management
University of Toronto
Toronto, ON, Canada

ISSN 0302-9743   ISSN 1611-3349   (electronic)
Lecture Notes in Computer Science
ISBN 978-3-031-22831-5   ISBN 978-3-031-22832-2   (eBook)
https://doi.org/10.1007/978-3-031-22832-2

This Springer imprint is published by the registered company Springer Nature Switzerland AG
The registered company address is: Gewerbestrasse 11, 6330 Cham, Switzerland

Preface

This volume contains all regular papers and abstracts presented at the 18th Conference on Web and Internet Economics (WINE 2022) held during December 12–15, 2022, in Troy, NY, USA, at Rensselaer Polytechnic Institute.

Over the last 18 years, the WINE conference series has become a leading interdisciplinary forum for the exchange of ideas and scientific progress across continents on incentives and computation arising in diverse areas, such as theoretical computer science, artificial intelligence, economics, operations research, and applied mathematics. WINE 2022 built on the success of previous editions of WINE (named the Workshop on Internet and Network Economics until 2013) which were held annually from 2005 to 2021.

We continued the successful initiative introduced in 2021 of having a Senior Program Committee, this year composed of 30 researchers from the field. The Program Committee had 63 researchers. The committees reviewed 126 submissions and decided to accept 38 papers. Each paper had at least three single blind reviews, with additional reviews solicited as needed. We are very grateful to all members of the Senior Program Committee and the Program Committee for their insightful reviews and discussions. We thank EasyChair for providing a virtual platform to organize the review process. We also thank Springer for publishing the proceedings and offering support for the Best Paper and Best Student Paper Awards. The Best Paper award was given to "Eliminating Waste in Cadaveric Organ Allocation" by Peng Shi and Junxiong Yin; the Best Student Paper award was given to "Optimal Private Payoff Manipulation against Commitment in Extensive-form Games" by Yurong Chen, Xiaotie Deng, and Yuhao Li.

The program included four invited talks by leading researchers in the field: Vincent Conitzer (Carnegie Mellon University, USA), Robert Kleinberg (Cornell University, USA), Marzena Rostek (University of Wisconsin-Madison, USA), and David Simchi-Levi (Massachusetts Institute of Technology, USA).

We also invited Spotlights beyond WINE talks; the nomination and selection of these talks was decided by Edith Elkind, Michal Feldman, Philipp Strack, and Gabriel Weintraub.

Our special thanks go to the general chairs David Pennock and Lirong Xia and the local organization team.

October 2022

Kristoffer Arnsfelt Hansen
Tracy Xiao Liu
Azarakhsh Malekian

Organization

General Chairs

David Pennock   Rutgers University, USA
Lirong Xia   Rensselaer Polytechnic Institute, USA

Program Committee Chairs

Kristoffer Arnsfelt Hansen   Aarhus University, Denmark
Tracy Xiao Liu   Tsinghua University, China
Azarakhsh Malekian   University of Toronto, Canada

Steering Committee

Xiaotie Deng   Peking University, China
Paul Goldberg   University of Oxford, UK
Christos Papadimitriou   Columbia University, USA
Paul Spirakis   University of Liverpool, UK
Rakesh Vohra   University of Pennsylvania, USA
Andrew Yao   Tsinghua University, China
Yinyu Ye   Stanford University, USA

Senior Program Committee

Itai Ashlagi   Stanford University, USA
Haris Aziz   UNSW Sydney, Australia
Moshe Babaioff   Microsoft Research, Israel
Santiago Balseiro   Columbia University, USA
Omar Besbes   Columbia University, USA
Ozan Candogan   University of Chicago, USA
Jing Chen   Stony Brook University, USA
Xujin Chen   Chinese Academy of Sciences, China
Yi-Chun Chen   National University of Singapore, Singapore
Bolin Ding   Alibaba Group, China
Shahar Dobzinski   Weizmann Institute of Science, Israel
Aris Filos-Ratsikas   University of Liverpool, UK
Hu Fu   Shanghai University of Finance and Economics, China
Nima Haghpanah   Pennsylvania State University, USA
Tobias Harks   University of Augsburg, Germany
Martin Hoefer   Goethe University Frankfurt, Germany
Ayumi Igarashi   National Institute of Informatics, Japan
Yuqing Kong   Peking University, China

Irene Lo	Stanford University, USA
Brendan Lucier	Microsoft Research, USA
Ali Makhdoumi	Duke University, USA
Malesh Pai	Rice University, USA
Qi Qi	Renmin University of China, China
Daniela Saban	Stanford University, USA
Balu Sivan	Google Research, USA
Christos Tzamos	University of Wisconsin-Madison, USA
Carmine Ventre	King's College London, UK
Matt Weinberg	Princeton University, USA
Lirong Xia	Rensselaer Polytechnic Institute, USA
Jun Zhang	Nanjing Audit University, China

Program Committee

Mete Şeref Ahunbay	Technical University of Munich, Germany
Elliot Anshelevich	Rensselaer Polytechnic Institute, USA
Jackie Baek	Massachusetts Institute of Technology, USA
Xiaohui Bei	Nanyang Technological University, Singapore
Hedyeh Beyhaghi	Carnegie Mellon University, USA
Kodric Bojana	Gran Sasso Science Institute, Italy
Modibo Camara	University of Chicago, USA
Zhigang Cao	Beijing Jiaotong University, China
Bhaskar Ray Chaudhury	University of Illinois Urbana-Champaign, USA
Ningyuan Chen	University of Toronto Mississauga, Canada
Yukun Cheng	Suzhou University of Science and Technology, China
Giorgos Christodoulou	Aristotle University of Thessaloniki, Greece
Argyrios Deligkas	Royal Holloway, University of London, UK
Yuan Deng	Google Research, USA
Songzi Du	University of California, San Diego, USA
Zhixuan Fang	Tsinghua University, China
Yiding Feng	Microsoft Research, USA
Diodato Ferraioli	University of Salerno, Italy
Matheus Ferreira	Harvard University, USA
Huiyi Guo	Texas A&M University, USA
Huseyin Gurkan	ESMT Berlin, Germany
Kevin He	University of Pennsylvania, USA
Alexandros Hollender	University of Oxford, UK
Bart de Keijzer	King's College London, UK
Philip Lazos	IOHK, UK
Pascal Lenzner	Hasso Plattner Institute, Germany
Stefanos Leonardos	King's College London, UK
Bo Li	Hong Kong Polytechnic University, China
Jiangtao Li	Singapore Management University, Singapore
Mengling Li	Xiamen University, China
Yingkai Li	Yale University, USA

Thodoris Lykouris	MIT, USA
Will Ma	Columbia University, USA
Jieming Mao	Google Research, USA
Faidra Monachou	Harvard University, USA
Swaprava Nath	Indian Institute of Technology Bombay, India
Dario Paccagnan	Imperial College London, UK
Harry Pei	Northwestern University, USA
Emmanouil Pountourakis	Drexel University, USA
Nidhi Rathi	Aarhus University, Denmark
Scott Rodilitz	UCLA, USA
Kang Rong	Shanghai University of Finance and Economics, China
Daniel Schmand	University of Bremen, Germany
Grant Schoenebeck	University of Michigan, USA
Ariel Schvartzman	DIMACS, USA
Sujoy Sikdar	Binghamton University, USA
Xiang Sun	Wuhan University, China
Biaoshuai Tao	Shanghai Jiao Tong University, China
Nguyễn Kim Thắng	Paris-Saclay University, France
Taiki Todo	Kyushu University, Japan
Cosimo Vinci	University of Salerno, Italy
Alexandros Voudouris	University of Essex, UK
Changjun Wang	Chinese Academy of Sciences, China
Chun Wang	Tsinghua University, China
Zihe Wang	Renmin University of China, China
Zenan Wu	Peking University, China
Yiqing Xing	Johns Hopkins University, USA
Fang-yi Yu	George Mason University, USA
Yuan Yuan	Purdue University, USA
Boyu Zhang	Beijing Normal University, China
Mu Zhang	University of Michigan, USA
Dengji Zhao	ShanghaiTech University, China
Jingtong Zhao	Renmin University of China, China

Additional Reviewers

Hannaneh Akrami	Karl Fehrs
Vincenzo Auletta	Bryce Ferguson
Rangeet Bhattacharyya	Simone Fioravanti
Xiaolin Bu	Miriam Fischer
Linda Cai	Pin Gao
Wenjie Cao	Abheek Ghosh
Xiaomeng Ding	Yue Han
Katharina Eickhoff	Chin-Chia Hsu
Markus Ewert	Chien-Chung Huang
Setareh Farajollahzadeh	Kasper Høgh

Ryota Iijima

Zsuzsanna Jankó

Zhile Jiang

Kai Kang

Tim Koglin

Simon Krogmann

Zhonghong Kuang

Pooja Kulkarni

Rucha Kulkarni

Dominik Kaaser

Alexander Lam

Stefano Leucci

Hannah Li

Junjie Li

Weian Li

Wenhao Li

Zihao Li

Jason Cheuk Nam Liang

Panfeng Liu

Qingsong Liu

Edwin Lock

Xinhang Lu

Themistoklis Melissourgos

Shivika Narang

Rishi Patel

Dominik Peters

Elias Pitschmann

Nicos Protopapas

Pengyu Qian

Kaihua Qin

Rojin Rezvan

David Salas

Marco Schmalhofer

Marc Schroder

Alkmini Sgouritsa

Golnoosh Shahkarami

Garima Shakya

George Skretas

Aikaterini-Panagiota Stouka

Ankang Sun

Hailong Sun

Wei Tang

Sri Aravindakrishnan Thyagarajan

Rohit Vaish

Yllka Velaj

Chenhao Wang

Kangning Wang

Nianyi Wang

Xiaomin Wang

Ziwei Wang

Yi Xiong

Menghan Xu

Xiang Yan

Ruiqi Yang

Shuoguang Yang

Hanrui Zhang

Yuhao Zhang

Tan Zhibin

Yu Zhou

Zhengxing Zou

Song Zuo

Contents

Social Choice

Abstracts

Equilibria in Games

Inefficiency of Pure Nash Equilibria in Series-Parallel Network Congestion Games

Bainian Hao$^{(\boxtimes)}$ and Carla Michini

Department of Industrial and Systems Engineering, University of Wisconsin-Madison,
Madison, WI, USA
{bhao8,michini}@wisc.edu

Abstract. We study the inefficiency of pure Nash equilibria in symmetric unweighted network congestion games defined over series-parallel networks. We introduce a quantity $y(\mathcal{D})$ to upper bound the Price of Anarchy (PoA) for delay functions in class $\mathcal{D}$. When $\mathcal{D}$ is the class of polynomial functions with highest degree p, our upper bound is $2^{p+1} - 1$, which is significantly smaller than the worst-case PoA for general networks. Thus, restricting to symmetric games over series-parallel networks can limit the inefficiency of pure Nash equilibria. We also construct a family of instances with polynomial delay functions that have a PoA in $\Omega(2^p/p)$ when the number of players goes to infinity. Compared with the subclass of extension-parallel networks, whose worst-case PoA is in $\Theta(p/\ln p)$, our results show that the worst-case PoA quickly degrades from sub-linear to exponential when relaxing the network topology. We also consider an alternative measure of the social cost of a strategy profile as the maximum players' cost. We introduce a parameter $z(\mathcal{D})$ and we show that the PoA is at most $y(\mathcal{D})z(\mathcal{D})$, which for polynomial delays of maximum degree p is at most $2^{2p+1} - 2^p$. Compared to the worst-case PoA in general symmetric congestion games, which is in $p^{\Theta(p)}$, our results shows a significant improvement in efficiency. We finally prove that our previous lower bound in $\Omega(2^p/p)$ is still valid for this measure of social cost. This is in stark contrast with the PoA in the subclass of extension-parallel networks, where each pure Nash equilibrium is a social optimum.

Keywords: Congestion games · Series-parallel networks · Price of anarchy

1 Introduction

In a non-cooperative game, rational players act selfishly to maximize their utility. The players influence each other's behaviour, since the quality of each player's strategy depends on the other players' actions. The notion of Nash equilibrium, where no player can improve her cost by unilaterally changing strategy, is the best-known solution concept for predicting a stable outcome of a game. However, since the players act selfishly and independently in a non-cooperative fashion, a Nash equilibrium might be far from minimizing the social cost. The inefficiency

© The Author(s), under exclusive license to Springer Nature Switzerland AG 2022
K. A. Hansen et al. (Eds.): WINE 2022, LNCS 13778, pp. 3–20, 2022.
https://doi.org/10.1007/978-3-031-22832-2_1

of a Nash equilibrium can be measured by comparing its social cost against the minimum social cost that could be achieved. Precisely, the Price of Anarchy (PoA), introduced by Koutsoupias and Papadimitriou [21], is the largest ratio between the cost of a Nash equilibrium and the minimum social cost.

In this paper, we study network congestion games, where each player aims at selecting a shortest path from an origin to a destination, but the cost of each edge is non-decreasing with respect to the total number of players using it. These games are commonly used to model problems in large-scale networks such as routing in communication networks and traffic planning in road networks [19,25] and represent a simple, yet powerful paradigm for selfish resource sharing.

We focus on the inefficiency of *pure* Nash equilibria. Unlike (mixed) Nash equilibria, where each player selects a probability distribution on her strategy set, in a *pure Nash equilibrium* (PNE) each player selects exactly one strategy from her strategy set. Pure Nash equilibria are not guaranteed to exist in general, but congestion games always admit one [26]. We consider two measures of social cost: the *total cost*, which is the sum of all players' costs, and the *maximum cost*, which is the maximum cost of a player in a strategy profile.

Several variants of network congestion games have been studied in the literature, which depend on the combination of a number of parameters. While some parameters seem to only marginally affect the PoA, the impact that graph structure has on the PoA is still not completely understood. Aland et al. [1] leave as an open direction the problem of characterizing "what structures provide immunity against a high PoA and what structures cause it".

The approaches that have been proposed for general network congestion games [1–3,7], later unified in the smoothness framework of Roughgarden [29,30], cannot be used to derive stronger bounds that hold in the presence of special network structures. The two main graph structures for which stronger bounds on the PoA has been provided are parallel-links networks [5,6,15,16,23,32] and extension parallel networks [12]. In this paper, we focus on the larger class of two-terminal series-parallel networks, and we provide upper and lower bounds on the worst-case PoA for (atomic, unweighted, symmetric) network congestion games. These networks can be recognized in linear-time [33] and are relevant in many applications, such as for problems on electric networks, scheduling and compiler optimization. Previous works have highlighted some strong properties of network congestion games defined over series-parallel networks, such as the existence of strong equilibria [20] and optimal tolls [14,24].

First, we consider the total players' cost and arbitrary delay functions. Let $\mathcal{D}$ be a class of nonnegative and non-decreasing functions. We introduce a new parameter $y(\mathcal{D})$ defined as

$$y(\mathcal{D}) = \sup_{d \in \mathcal{D}, \, x \in \mathbb{N}^+} \frac{(x+1)d(x+1) - xd(x)}{d(x)}, \tag{1}$$

which intuitively can be used to upper bound by what percentage the cost of an edge increases when one more player uses the edge. Note that $y(\mathcal{D}) \geq 1$ because $d(x) = (x+1)d(x) - xd(x) \leq (x+1)d(x+1) - xd(x)$. Our main result shows that the worst-case PoA in series-parallel networks is at most $y(\mathcal{D})$.

Theorem 1. *In a symmetric (unweighted) network congestion game on a series-parallel (s,t)-network with delays functions in class $\mathcal{D}$, the PoA w.r.t. the total players' cost is at most $y(\mathcal{D})$.*

The above result has interesting implications when $\mathcal{D}$ is the class of polynomial functions with nonnegative coefficients and highest degree p. We show that in this case $y(\mathcal{D})$ is at most $2^{p+1} - 1$. Our result significantly improves over the worst-case PoA of unweighted congestion games, that is in $\Theta(p/\ln p)^{p+1}$ [1]. We point out that this worst-case PoA is also attained by unweighted *network* congestion games [1], however the construction used in [1] requires asymmetry. In the full version of this paper [17] we consider symmetric congestion games on general networks and we provide a family of instances violating the upper bound of Theorem 1. Moreover, we derive a lower bound on the worst-case PoA in symmetric network congestion games defined over series-parallel networks.

Theorem 2. *The worst-case PoA w.r.t. the total players' cost of a symmetric (unweighted) network congestion game on a series-parallel (s,t)-network, where the delay functions are polynomials with non-negative coefficients and highest degree p, is at least*

$$\frac{1}{1 + l^2 \sqrt[2^p]{r} - rl - \sqrt[2^p]{r} + r},\tag{2}$$

where $r = \left(\frac{2}{2^{p+1}-1}\right)^{\frac{2^p}{2^p-1}}$ and $l = \frac{1}{2}r^{1-\frac{1}{2^p}}$.

We finally prove that our lower bound is in $\Omega\left(\frac{2^p}{p}\right)$, thus also in $\Omega(2^{cp})$ for each $c \in (0,1)$, which almost asymptotically matches the upper bound of $2^{p+1} - 1$. Since the worst-case PoA in extension-parallel networks (a subclass of series-parallel networks) is in $\Theta(p/\ln p)$ [12,13], our result shows that the PoA dramatically increases when relaxing the network topology from extension-parallel to series-parallel.

Next, we consider measuring the social cost of a strategy profile as the maximum players' cost. This variant of the social cost expresses the goal that a central authority might have to maximize fairness by minimizing the cost of the most disadvantaged player. We first consider arbitrary delay functions. To bound the PoA in this setting, introduce a new parameter $z(\mathcal{D})$ defined as

$$z(\mathcal{D}) = \sup_{d \in \mathcal{D},\ x \in \mathbb{N}^+} \frac{d(x+1)}{d(x)}.\tag{3}$$

We first prove that the worst-case PoA in series-parallel networks is at most $y(\mathcal{D})z(\mathcal{D})$.

Theorem 3. *In a symmetric (unweighted) network congestion game on a series-parallel (s,t)-network with delays functions in class $\mathcal{D}$, the PoA w.r.t. the maximum players' cost is at most $z(\mathcal{D})y(\mathcal{D})$.*

When $\mathcal{D}$ is the class of polynomial functions with nonnegative coefficients and maximum degree p we obtain that $z(\mathcal{D})$ is upper bounded by 2^p, thus the PoA is at most $2^{2p+1} - 2^p$. Since the worst-case PoA for general symmetric congestion games and polynomial delays is in $p^{\Theta(p)}$ [7], our result shows a significant drop of the PoA in series-parallel networks.

Finally we show that the lower bound on the PoA w.r.t. the total players' cost also yields a valid lower bound when considering the maximum players' cost. We say that a class of networks $\mathcal{N}$ is *closed under series compositions* if the series composition of two networks G^1 and G^2 in $\mathcal{N}$ still belongs to $\mathcal{N}$.

Theorem 4. *Let $\mathcal{N}$ be a class of networks closed under series compositions and let G be a network in $\mathcal{N}$. Then the worst-case PoA with respect to the maximum social cost of a symmetric (unweighted) network congestion game defined over G is at least the worst-case PoA with respect to the total social cost.*

For series-parallel networks and polynomial delays with nonnegative coefficients and maximum degree p Theorem 4 implies that the worst-case PoA is in $\Omega(2^p/p)$. This is in stark contrast with the result of [10], establishing that the PoA in extension-parallel networks is 1, i.e., any PNE is also a social optimum w.r.t. the maximum players' cost. Thus, relaxing the network topology from extension-parallel to series-parallel dramatically increases the inefficiency of pure Nash equilibria. The reason for this is that the key graph operations that we need to allow are the series compositions, which are forbidden for extension-parallel networks.

1.1 Further Related Work

Total Cost. There is a rich literature concerning the PoA in network congestion games where the social cost is measured based on the players' total cost. Many variants of network congestion games arise from considering different parameters and their combinations. As we shall see, the impact that graph structure has on the inefficiency of pure Nash equilibria varies significantly based on the combination of these parameters.

The first distinction is between atomic and non-atomic congestion games. In *non-atomic* congestion games, the number of players is infinite and each player controls an infinitesimal amount of flow. For these games, Roughgarden [27] proved that the PoA is independent of the network structure and equal to $\rho(\mathcal{D})$, where ρ depends on the class of delay functions $\mathcal{D}$ [31].

For *atomic* games, where each player controls a non-negligible amount of flow, network structure affects the PoA differently, depending on whether all the players have the same effect on congestion. In *weighted* congestion games, where the effect of each player on congestion is proportional to the player's weight, the worst-case PoA is already achieved by very simple networks consisting of only parallel links [4] when $\mathcal{D}$ is the class of polynomial functions with nonnegative coefficients and highest degree p. In contrast, in *unweighted* congestion games the effect of network structure seems significant. For asymmetric congestion games defined over general networks and in the case where $\mathcal{D}$ is the class of polynomial functions with nonnegative coefficients, Christodoulou and Koutsoupias [7] showed that the PoA is in $p^{\Theta(p)}$ (see also [2,3]). Aland et al. [1] later obtained exact values for the worst-case PoA. These exact values admit a lower bound of $\lfloor\phi_p\rfloor^{p+1}$ and an upper bound of ϕ_p^{p+1}, where $\phi_p \in \Theta(p/\ln p)$ is the unique nonnegative real solution to $(x+1)^p = x^{p+1}$. For symmetric congestion games the PoA is again $p^{\Theta(p)}$ [2,3,7]. The worst case PoA drops significantly in the presence of special structure. Lücking et al. [22,23] studied symmetric congestion games on parallel links and proved that the PoA is $4/3$ for linear functions. Later Fotakis [12] extended this result by proving an upper bound of $\rho(\mathcal{D})$ for the larger class of extension parallel networks with delays in class $\mathcal{D}$. Moreover, this upper bound is tight [11,13]. It is known that, for the class of polynomial delays with nonnegative coefficients and highest degree p, $\rho(\mathcal{D}) \in \Theta(p/\ln p)$. This indicates that there is a huge gap between the worst-case PoA in general networks and in extension-parallel networks.

The PoA in symmetric series-parallel network congestion games has been recently investigated only for the specific case of affine delay functions [18], and it has been shown that the worst-case PoA is between $27/19$ and 2 [18], which is strictly worse than the PoA of $4/3$ in extension-parallel networks [12], and strictly better than the PoA of $5/2$ in general networks [8]. One key step to prove the upper bound in [18] consists in using the following inequality introduced in [12]

$$\frac{\text{cost}(f)}{\rho(\mathcal{D})} \le \text{cost}(o) + \Delta(f,o), \tag{4}$$

where $\text{cost}(f)$ and $\text{cost}(o)$ denote the total cost of a PNE flow f and of a social optimum flow o, respectively, and $\Delta(f,o)$, is a quantity that depends on the difference $o - f$. For series-parallel networks with affine delays, Hao and Michini [18] prove that $\Delta(f,o) \le 1/4\,\text{cost}(f)$. This approach cannot be further extended to polynomial delays of maximum degree p, because we would obtain $\Delta(f,o) \le \alpha(p)\,\text{cost}(f)$, where $\alpha(p)$ is a function of p that exceeds $1/\rho(\mathcal{D})$ for large p. Thus, an extension of the approach in [18] would provide an inconsequential bound.

Maximum Cost. The PoA with respect to the maximum players' cost has received less attention. In the non-atomic setting, Roughgarden [28] showed that the PoA is $n - 1$, where n is the number of nodes in the network.

In the atomic setting, Koutsoupias and Papadimitriou [21] first studied weighted congestion games with linear delay functions on m parallel links. For these games, they provided a lower bound of the PoA of $\Omega\left(\frac{\log m}{\log\log m}\right)$ and an upper bound of $O(\sqrt{m\log m})$. Later Czumaj and Vöcking [9] established a tight bound of $\Theta\left(\frac{\log m}{\log\log\log m}\right)$. Christodoulou and Koutsoupias [7] investigated general unweighted congestion games. In the symmetric case, they showed that the PoA is $5/2$ for affine delays and $p^{\Theta(p)}$ for polynomial delays of maximum degree p. In the asymmetric case, for games with N players, they proved that the PoA is in $\Theta(\sqrt{N})$ for affine delays and in $\Omega(N^{\frac{p}{p+1}})$ and $O(N)$ for polynomial delays of maximum degree p.

Epstein et al. [10] characterized efficient network topologies, i.e., graph topologies such that, for any class of non-decreasing delay functions, every PNE is also a social optimum. For unweighted symmetric network congestion games they established that extension-parallel networks are efficient, implying that on these networks the PoA is 1. They also proved that this result is tight, i.e., it does not hold when further relaxing the network topology.

2 Preliminaries

Notation. Let $G = (V, E)$ be an (s, t)-network, i.e., a network with source s and sink t. Directed paths will be simply referred to as paths. A path from node u to node v is called a (u, v)-path. We will only consider *simple* paths, i.e., paths that do not traverse any node multiple times. Paths and cycles of G are regarded as sequences of edges, thus we may for example write $e \in p$ for a path p. An (s, t)-*flow* is an assignment of values to the edges of G such that, at each node u other than s and t, the sum of the values of the edges entering u equals the sum of the values of the edges leaving u. The value of the (s, t)-flow is the sum of the values of the edges entering t. We say a path p is *contained* in an (s, t)-flow f if for all $e \in p$, we have $f_e > 0$. For $n \in \mathbb{N}$, we denote by $[n]$ the set $\{1, \ldots, n\}$.

Network Congestion Games. Let $G = (V, E)$ be an (s, t)-network. We consider a network congestion game on G with N players. The strategy set X^i of player i is the set $\mathcal{P}$ of (s, t)-paths in G. Since all the players have the same origin and destination, their strategy sets all coincide with $\mathcal{P}$ and the game is called *symmetric*. A *state* of the game is a strategy profile $P = (p^1, \ldots, p^N)$ where $p^i \in \mathcal{P}$ is the (s, t)-path chosen by player i, for $i \in [N]$. The set of states of the game is denoted by $X = X^1 \times \cdots \times X^N$. Each state $P = (p^1, \ldots, p^N) \in X$ induces an (s, t)-flow $f = f(P) = \chi^1 + \cdots + \chi^N$ of value N, where χ^i is the incidence vector of p^i for all $i \in [N]$. We say that the (s, t)-paths $p^1, \ldots, p^N$ are a *decomposition* of the (s, t)-flow f if they induce flow f. Note that an (s, t)-flow f of value N can correspond to several states, since there might be multiple decompositions of f into N (s, t)-paths.

For each $e \in E$ we have a nondecreasing delay function $d_e : [N] \to \mathbb{R}_{\geq 0}$. Each player using e incurs a cost equal to $d_e(f_e)$, i.e., the cost of e depends on the total number of players that use e in f. Since d_e is a nondecreasing function, $d_e(j+1) \geq d_e(j)$ for $j \in [N-1]$, which models the effect of congestion. We denote the cost of a path p in G with respect to a flow f by $\text{cost}_f(p) = \sum_{e \in p} d_e(f_e)$. Thus, the cost incurred by player i in state P is $\text{cost}_f(p^i)$. We also define $\text{cost}_f^+(p) = \sum_{e \in p} d_e(f_e + 1)$. Finally, the cost of flow f in G is denoted by $\text{cost}(f) = \sum_{e \in E} f_e d_e(f_e)$. The *total cost* of a state P, denoted by $\text{tot}(P)$, is the sum of all players' costs. Clearly $\text{tot}(P)$ coincides with the cost of the flow $f(P)$:

$$\text{tot}(P) = \sum_{i \in [N]} \text{cost}_{f(P)}(p^i) = \text{cost}(f(P)).$$

We also define the *maximum cost* of P, denoted by $\max(P)$ as the maximum cost of a player in P:

$$\max(P) = \max_{i \in [N]} \text{cost}_{f(P)}(p^i).$$

Pure Nash Equilibria and Social Optima. A *pure Nash equilibrium* (PNE) is a state $(p^1, \ldots, p^i, \ldots, p^N)$ inducing an (s,t)-flow f such that, for each $i \in [N]$ we have

$$\text{cost}_f(p^i) \leq \text{cost}_{\tilde{f}}(\tilde{p}^i) \qquad \forall (p^1, \ldots, \tilde{p}^i, \ldots, p^N) \in X \text{ inducing } (s,t)\text{-flow } \tilde{f} .$$

A PNE represents a stable outcome of the game, since no player $i \in [N]$ can improve her cost if she unilaterally changes strategy by selecting a different (s,t)-path $\tilde{p}^i$. With a slight abuse of terminology, we say that an (s,t)-flow f is a PNE if there exists a PNE $P = (p^1, \ldots, p^N) \in X$ such that $f = f(P)$, i.e., f is the flow induced by P. On the other hand, we are also interested in a *social optimum*. We consider two definitions of social optimum, which depend on whether we measure the cost of a state P according to $\text{tot}(P)$ or $\max(P)$. In the first case, a social optimum is a state that minimizes $\text{tot}(P) = \text{cost}(f(P))$ over all the states $P \in X$. With a slight abuse of terminology, we say that an (s,t)-flow o is a social optimum if o minimizes $\text{cost}(g)$ over all integral (s,t)-flows g of value N. In the second case a social optimum is a state that minimizes $\max(P)$ over all the states $P \in X$. In other words, the social optimum is a state where the maximum player's cost is minimized.

Price of Anarchy. To measure the inefficiency of pure Nash equilibria, we use the definition of (pure) Price of Anarchy. The (pure) *Price of Anarchy* (PoA) is the maximum ratio between the cost of a PNE and the cost of a social optimum. In other words, to compute the PoA we consider the "worst" PNE, i.e., a PNE whose cost is as large as possible. For simplicity, from now on we will refer to the pure PoA as PoA.

We consider two definitions of PoA, which depend on whether we measure the cost of a state P according to $\text{tot}(P)$ or $\max(P)$. In the first case, the PoA is the maximum ratio $\frac{\text{cost}(f)}{\text{cost}(o)}$ such that o is a social optimum flow and f is a PNE flow. In the second case, the PoA is the maximum ratio $\frac{\max(P_f)}{\max(P_o)}$ such that P_o is a social optimum state and P_f is a PNE.

Series-Parallel Networks. An (s,t)-network is series-parallel if it consists of either a single edge (s,t) or of two series-parallel networks composed either in series or in parallel. The *parallel composition* of two networks G_1 and G_2 is an (s,t)-network obtained from the union of G_1 and G_2 by identifying the source of G_1 and the source of G_2 into s, and by identifying the sink of G_1 and the sink of G_2 into t. The *series composition* of G_1 and G_2, denoted by $G_1 \circ G_2$, is an (s,t)-network obtained from the union of G_1 and G_2 by letting s be the source of G_1, t be the sink of G_2, and by identifying the sink of G_1 with the source of G_2. We remark that series-parallel networks are a superclass of parallel-link networks and extension-parallel networks, for which the PoA has been previously studied. An (s,t)-network is extension-parallel if it consists of a single edge (s,t) or of an extension-parallel network and a single edge composed either in series or in parallel.

3   Total Cost

3.1   Upper Bound on the PoA

In this section, we prove the upper bound on the PoA stated in Theorem 1. First, we need to introduce some necessary notation and properties of series-parallel networks. In the following, we denote by f and o a PNE and a social optimum, respectively, of the series-parallel network congestion game. We consider the graph $G(o-f)$ introduced in [12]. Precisely, the node set of $G(o-f)$ is V, and the edge set is $E(o-f) = \{(u,v) : (e = (u,v) \in E$ and $o_e - f_e > 0)$ or $(e = (v,u) \in E$ and $o_e - f_e < 0)\}$. $G(o-f)$ is a collection of simple cycles $\{C_1, \ldots, C_h\}$ such that each C_i carries s_i units of flow. For each $i \in [h]$, define $C_i^+ = \{e = (u,v) \in E : (u,v) \in C_i, o_e > f_e\}$ and $C_i^- = \{e = (u,v) \in E : (v,u) \in C_i, o_e < f_e\}$.

Recall the parameter $y(\mathcal{D})$ we have defined in Sect. 1. In the next four lemmas, we will assume that there exists an index $i \in [h]$ such that C_i^+ is an (s,t)-path, and we will prove that the PoA is at most $y(\mathcal{D})$. Later, we will relax this assumption. Observe that, by definition, C_i^+ is contained in o. In the next lemma, we prove that the cost of C_i^+ with respect to o is at least the average players' cost in the PNE f, that is, $\mathrm{cost}(f)/N$.

Lemma 1. *If C_i^+, $i \in [h]$, is an (s,t)-path, then $\mathrm{cost}_o(C_i^+) \geq \mathrm{cost}(f)/N$.*

Proof. The cost of C_i^+ with respect to flow o satisfies:

$$\mathrm{cost}_o(C_i^+) = \sum_{e \in C_i^+} d_e(o_e) \geq \sum_{e \in C_i^+} d_e(f_e + 1) \geq \frac{\mathrm{cost}(f)}{N}.$$

The first inequality holds since for every $e \in C_i^+$, we have $o_e \geq f_e + 1$. Next we show that the second inequality holds. Denote by P^* the set of N (s,t)-paths in the PNE inducing f. Clearly $\max\{\mathrm{cost}_f(\pi) : \pi \in P^*\} \geq \frac{\mathrm{cost}(f)}{N}$. By contradiction, suppose that $\sum_{e \in C_i^+} d_e(f_e + 1) < \frac{\mathrm{cost}(f)}{N}$. We would obtain that $\max\{\mathrm{cost}_f(\pi) : \pi \in P^*\} > \mathrm{cost}_f^+(C_i^+)$, thus one player would prefer to change her strategy into C_i^+. This contradicts the fact that f is a PNE.   □

In the next lemma, we contemplate adding one unit of flow on an arbitrary (s,t)-path p contained in o, and we lower bound the corresponding increase of the total cost. This will be crucial to derive a lower bound on $\mathrm{cost}_o(p)$ that will be used to relate $\mathrm{cost}(f)$ and $\mathrm{cost}(o)$.

Lemma 2. *Suppose that there exists an index $i \in [h]$, such that C_i^+ is an (s,t)-path. Then every (s,t)-path p contained in o satisfies*

$$\sum_{e \in p} ((o_e + 1)d_e(o_e + 1) - o_e d_e(o_e)) \geq \frac{\mathrm{cost}(f)}{N}.$$

Proof. We will prove this by contradiction. Assume that there is an (s,t)-path p contained in o such that

$$\sum_{e\in p}(o_e+1)d_e(o_e+1) - \sum_{e\in p}o_e d_e(o_e) < \frac{\text{cost}(f)}{N}. \tag{5}$$

We define a new state o' obtained from o by deviating one unit of flow from C_i^+ to p. Let $S = C_i^+ \cap p$. First, the cost difference between o' and o is

$$\text{cost}(o') - \text{cost}(o) = \sum_{e\in C_i^+\backslash S} ((o_e-1)d_e(o_e-1) - o_e d_e(o_e))$$
$$+ \sum_{e\in p\backslash S} ((o_e+1)d_e(o_e+1) - o_e d_e(o_e)).$$

Observe that, since the delay functions are non-decreasing, we have $d_e(o_e-1) \le d_e(o_e)$ for all $e \in C_i^+$, thus

$$\text{cost}(o') - \text{cost}(o) \le \sum_{e\in C_i^+\backslash S} ((o_e-1)d_e(o_e) - o_e d_e(o_e))$$
$$+ \sum_{e\in p\backslash S} ((o_e+1)d_e(o_e+1) - o_e d_e(o_e))$$
$$= - \sum_{e\in C_i^+\backslash S} d_e(o_e) + \sum_{e\in p\backslash S} ((o_e+1)d_e(o_e+1) - o_e d_e(o_e)).$$

Moreover, we have $d_e(o_e+1) \ge d_e(o_e)$ for all $e \in S$, thus

$$0 \le \sum_{e\in S}(o_e+1)(d_e(o_e+1) - d_e(o_e))$$
$$= - \sum_{e\in S} d_e(o_e) + \sum_{e\in S}((o_e+1)d_e(o_e+1) - o_e d_e(o_e)).$$

By summing up these two inequalities we get

$$\text{cost}(o') - \text{cost}(o) \le - \sum_{e\in C_i^+} d_e(o_e) + \sum_{e\in p}((o_e+1)d_e(o_e+1) - o_e d_e(o_e)).$$

By Lemma 1, since C_i^+ is an (s,t)-path, we have $\text{cost}_o(C_i^+) = \sum_{e\in C_i^+} d_e(o_e) \ge \frac{\text{cost}(f)}{N}$. Thus, by (5) we obtain $\text{cost}(o') - \text{cost}(o) < 0$, which contradicts the fact that o is a social optimum. □

By using Lemma 2, we can derive a lower bound on $\text{cost}_o(p)$ similar to the lower bound on $\text{cost}_o(C_i^+)$ stated in Lemma 1, but with an extra factor of $y(\mathcal{D})$.

Lemma 3. *Suppose there exists an index $i \in [h]$ such that C_i^+ is an (s,t)-path, and let P be any decomposition of o. Then for every $p \in P$,*

$$y(\mathcal{D}) \, \text{cost}_o(p) \ge \frac{\text{cost}(f)}{N}.$$

Proof. Since P is a decomposition of o, for each $p \in P$ we have $o_e > 0$ for all $e \in p$. Then we have

$$y(\mathcal{D})\, \text{cost}_o(p) = \sum_{e \in p} y(\mathcal{D}) d_e(o_e) \geq \sum_{e \in p} ((o_e + 1) d_e(o_e + 1) - o_e d_e(o_e)) \geq \frac{\text{cost}(f)}{N},$$

where the first inequality follows the definition of $y(\mathcal{D})$ stated in Eq. (1) and the second inequality follows from Lemma 2. □

Finally, under the assumption that there exists a path C_i^+ from s to t, we are ready to prove that the PoA is at most $y(\mathcal{D})$.

Lemma 4. *If there exists an index $i \in [h]$ such that C_i^+ is an (s,t)-path, then* $\text{cost}(f) \leq y(\mathcal{D})\, \text{cost}(o)$.

Proof. By Lemma 3 we know that given an arbitrary decomposition P of the social optimal flow o, for all $p \in P$, we have $y(\mathcal{D})\, \text{cost}_o(p) \geq \frac{\text{cost}(f)}{N}$. Then we can conclude that:

$$y(\mathcal{D})\, \text{cost}(o) = \sum_{p \in P} y(\mathcal{D})\, \text{cost}_o(p) \geq |P| \frac{\text{cost}(f)}{N} = \text{cost}(f),$$

where the last equality follows from the fact that $|P| = N$. This implies that $\text{cost}(f) \leq y(\mathcal{D})\, \text{cost}(o)$. □

We now relax the assumption that there exists a path C_i^+ from s to t. In order to do this, we will exploit the structure of series-parallel graphs. If G is series-parallel, it is known that for each $i \in [h]$ C_i^+ and C_i^- are two internally disjoint paths in G from a node u_i to a node v_i [12]. For each $i \in [h]$, we identify the pair of nodes u_i, v_i and we define

$$V_i = \{w \in V : \text{there is a } (u_i, v_i)\text{-path containing } w\},$$
$$E_i = \{e \in E : \text{there is a } (u_i, v_i)\text{-path containing } e\},$$

and we let $\mathcal{L} = \{E_1, \ldots, E_h\}$.

Lemma 5. *If G is series-parallel, then $\mathcal{L} = \{E_1, \ldots, E_h\}$ is a laminar family.*

Proof. We prove this lemma by showing that if $E_i \cap E_j \neq \emptyset$ for some i and j in $[h]$, then $E_i \subseteq E_j$ or $E_j \subseteq E_i$. We proceed by induction on $|E|$.

The base case as $|E| = 2$. If the two edges of G are composed in series, then there are no cycles. If they are composed in parallel, then there is only one cycle, i.e., $i = j$, and $E_i = E_j = E$. This implies that the lemma holds for the base case. Now we assume that when $|E| \leq t$, the lemma holds. When $|E| = t + 1$, since G is series-parallel, it can be decomposed either in series or in parallel.

Suppose that G can be decomposed in series into G_1 and G_2. We first show that E_i and E_j are both contained either in the edge set of G_1 or in the edge set G_2. In fact, E_i cannot have edges both in G_1 and in G_2, otherwise C_i^+ and

C_i^- would not be internally disjoint paths. Thus E_i is contained either in the edge set of G_1 or in the edge set G_2. Similarly, E_j is contained either in the edge set of G_1 or in the edge set G_2. Moreover, E_i and E_j cannot belong to different components, otherwise we would have $E_i \cap E_j = \emptyset$. Thus, E_i and E_j both belong to the same component. Assume without loss of generality that this is G_1. Since the number of edges of G_1 is at most t, by the inductive hypothesis we obtain that $E_i \subseteq E_j$ or $E_j \subseteq E_i$, thus the claim is proven in this case.

Now suppose that G can be decomposed in parallel into G_1 and G_2. If E_i and E_j are both contained either in the edge set of G_1 or in the edge set of G_2, then by induction the claim holds. If E_i is contained in the edge set of one component, say G_1, and E_j is contained in the edge set of the other component G_2, then $E_i \cap E_j = \emptyset$, a contradiction. Thus at least one among E_i and E_j has edges both in G_1 and in G_2. Without loss of generality, suppose E_i does. We prove that C_i^+ and C_i^- are (internally disjoint) (s,t)-paths. By contradiction, suppose that C_i^+ and C_i^- are (s_i, t_i)-paths such that $s_i \neq s$ or $t_i \neq t$. Note that s_i and t_i are either both in G_1 or both in G_2. Suppose w.l.o.g. they are both in G_1. Then each (s_i, t_i)-path cannot contain any edge in G_2. Because C_i^+ and C_i^- are (s,t)-paths, by the definition of E_i, we have $E_i = E$. Thus we conclude that $E_j \subseteq E_i$, which proves the claim in this case. $\qquad\square$

By Proposition 1 in [12], if w and w' are two nodes in V_i such that there exist two internally disjoint (w, w')-paths p_1 and p_2, then every (s,t)-path having an edge in common with p_1 contains both w and w' and intersects p_2 only at w and w'. This implies that each (s,t)-path going through u_i also goes through v_i. As a consequence, for each $i \in [h]$ the sub-vectors of f and o that are indexed by the edges of E_i, denoted by $f(E_i)$ and $o(E_i)$, respectively, both define (u_i, v_i)-flows in the subgraph $G_i = (V_i, E_i)$. Define a network congestion game on G_i, where each edge $e \in E_i$ has the same delay d_e as in G, and the number of players N_i is equal to the value of flow $f(E_i)$.

Lemma 6. *If G is series-parallel and E_i is a maximal set in $\mathcal{L}$, then in the network congestion game defined on G_i, $f(E_i)$ and $o(E_i)$ are a PNE flow and a social optimum flow, respectively.*

Proof. Let N_i be the flow value of $f(E_i)$. First we show that $o(E_i)$ also has value N_i. Recall that $G(o - f)$ is a collection of cycles $\{C_1, \dots, C_h\}$ and each C_i carries s_i units of flow. By the definition of $G(o - f)$ we can change f into o as follows: for $j \in [h]$, decrease the flow on C_j^- by s_j and increase the flow on C_j^+ by s_j. By Lemma 5 $\mathcal{L}$ is a laminar family, thus for each $j \in [h]$, the paths C_j^- and C_j^+ are either both in G_i or neither of them in G_i, i.e., either $E_j \subseteq E_i$, or $E_j \cap E_i = \emptyset$. Thus, each step does not change the flow value on G_i. We can conclude that when the procedure ends, the flow value $o(E_i)$ equals the flow value of $f(E_i) = N_i$.

Next, we show that $f(E_i)$ is a PNE flow on G_i. By contradiction, suppose that $f(E_i)$ is not a PNE flow on G_i. This implies that in each decomposition of $f(E_i)$ into N_i (u_i, v_i)-paths there is always one player who can decrease her cost by deviating her strategy to another (u_i, v_i)–path in G_i. This implies that in each decomposition of f into N (s,t)-paths there is always one player that can unilaterally deviate and decrease cost. This contradicts to that f is a PNE flow.

Finally, we show that $o(E_i)$ is a social optimum on G_i. By contradiction, suppose that there is another flow $o'(E_i)$ in G_i of value N_i such that $\text{cost}(o'(E_i)) < \text{cost}(o(E_i))$. Then we can construct a flow o'' such that $o''_e = o_e$ for all $e \in E \setminus E_i$ and $o''_e = o'_e$ for all $e \in E_i$. Then $\text{cost}(o'') < \text{cost}(o)$, contradicting the fact that o is the social optimum. $\qquad\square$

We now consider the graphs G_i, $i \in [h]$, having node set V_i and edge set E_i.

Lemma 7. *If G is series-parallel and E_i is a maximal set in $\mathcal{L}$, then*

$$\text{cost}(f(E_i)) \leq y(\mathcal{D}) \text{cost}(o(E_i)).$$

Proof. According to Lemma 6, the congestion game with N_i players on the two terminal-series parallel graph G_i is such that $f(E_i)$ is a PNE and $o(E_i)$ is a social optimum. Note that u_i and v_i are, respectively, the source and the sink of G_i. Since C_i^+ is a (u_i, v_i)-path, by Lemma 4 we conclude that the lemma holds. $\qquad\square$

We are finally ready to prove Theorem 1, i.e., in a symmetric network congestion game defined over a series-parallel network with delay functions in class $\mathcal{D}$, the PoA is at most $y(\mathcal{D})$.

Proof of Theorem 1. Consider the PNE flow f, the social optimum flow o and the laminar family $\mathcal{L}$ defined previously in this section. We will prove that, since G is series-parallel, then $\text{cost}(f) \leq y(\mathcal{D}) \text{cost}(o)$. Let $E_{C_1}, \ldots, E_{C_l}$ be the maximal sets in $\mathcal{L}$ and denote by $E(\mathcal{L})$ their union. We rewrite $\text{cost}(f)$ as follows.

$$\text{cost}(f) = \sum_{e \notin E(\mathcal{L})} f_e d_e(f_e) + \sum_{e \in E(\mathcal{L})} f_e d_e(f_e).$$

Note that for each edge $e \notin E(\mathcal{L})$ we have $f_e = o_e$. Moreover, $E_{C_1}, \ldots, E_{C_l}$ are a partition of $E(\mathcal{L})$, since they are maximal members of $\mathcal{L}$ that are pairwise disjoint. Thus we can rewrite the above expression as

$$\text{cost}(f) = \sum_{e \notin E(\mathcal{L})} o_e d_e(o_e) + \sum_{i=1}^{l} \sum_{e \in E_{C_i}} f_e d_e(f_e)$$

$$\leq y(\mathcal{D}) \sum_{e \notin E(\mathcal{L})} o_e d_e(o_e) + y(\mathcal{D}) \sum_{i=1}^{l} \sum_{e \in E_{C_i}} o_e d_e(o_e) = y(\mathcal{D}) \text{cost}(o),$$

where the inequality follows from the fact that $y(\mathcal{D}) \geq 1$ and from Lemma 7. $\square$

Let Poly-p be the class of polynomial delay functions with maximum degree p, which are of the form $\sum_{j=0}^{p} a_j x^j$, with $a_j \geq 0$ for $j = 0, \ldots, p$.

Lemma 8. *For the class of polynomial delay functions* Poly-p *it holds that* $y(\text{Poly-}p) \leq 2^{p+1} - 1$.

Proof. By using the definition of $y(\text{Poly-}p)$ in (1) we have that for any $x \in \mathbb{N}^+$

$$y(\text{Poly-}p) = \sup_{a_0,\dots,a_p \in \mathbb{R}_{\geq 0},\ x \in \mathbb{N}^+} \frac{(x+1)\sum_{j=0}^p a_j(x+1)^j - x\sum_{j=0}^p a_j x^j}{\sum_{j=0}^p a_j x^j}$$

$$= \sup_{a_0,\dots,a_p \in \mathbb{R}_{\geq 0},\ x \in \mathbb{N}^+} \frac{\sum_{j=0}^p \left(a_j\left((x+1)^{j+1} - x^{j+1}\right)\right)}{\sum_{j=0}^p a_j x^j}. \tag{6}$$

We now exploit the fact that given two collections of nonnegative real numbers $b_0, \dots, b_p$ and $c_0, \dots, c_p$, we have

$$\frac{\sum_{j=0}^p b_j}{\sum_{j=0}^p c_j} \leq \max_{j=0,\dots,p} \frac{b_j}{c_j}.$$

As a consequence, we can upper bound (6) by

$$\max_{j \in \{0,\dots,p\},\ x \in \mathbb{N}^+} \frac{(x+1)^{j+1} - x^{j+1}}{x^j}. \tag{7}$$

We now upper bound the numerator of the above expression as follows:

$$(x+1)^{j+1} - x^{j+1} = \sum_{k=0}^{j+1} \binom{j+1}{k} x^{j+1-k} - x^{j+1} \leq \sum_{k=1}^{j+1} \binom{j+1}{k} x^j,$$

where the inequality follows from the fact that $j + 1 \geq 1$ and $x \in \mathbb{N}^+$. From (7) we then obtain

$$y(\text{Poly-}p) \leq \max_{j \in \{0,\dots,p\}} \sum_{k=1}^{j+1} \binom{j+1}{k} = \max_{j \in \{0,\dots,p\}} \sum_{k=0}^{j+1} \binom{j+1}{k} - 1 = 2^{p+1} - 1.$$

$\square$

By Theorem 1 and Lemma 8 we obtain that the PoA of series-parallel network congestion games with polynomial delay functions with highest degree is p is at most $2^{p+1} - 1$.

3.2 Lower Bound

In this section, we illustrate how to construct a family of instances that asymptotically achieve the lower bound on the PoA stated in Theorem 2. This construction is an extension to polynomial delays of the construction proposed in [18] for affine delays. Let $\{q_1, \dots, q_N\}$ be an ordered sequence of positive numbers such that $\sum_{i=1}^N q_i = 1$ and $q_{i+1} = \frac{1}{2^p}\sum_{j=1}^i \frac{q_j}{i}$ for $i \in [N-1]$. Let $m \in [N-1]$. We define a new sequence $\{s_1, \dots, s_N\}$ by averaging $\{q_1, \dots, q_m\}$. Precisely,

$s_1 = \cdots = s_m = \frac{\sum_1^m q_i}{m}$ and $s_j = q_j$ for $j \geq m+1$. We construct a series-parallel (s,t)-network G with delays in Poly-p, and an (s,t)-flow f of value N recursively. Let G_m be a single (s,t)-edge with flow f_m of value m and delay equal to $\frac{s_1 x}{m}$. For every $i \in [m, N-1]$, we construct G_{i+1} and f_{i+1} using G_i and f_i as follows: we compose in parallel G_i and a new (s,t)-edge with flow value 1 and delay function $s_{i+1}x^p$ and call the new network $\tilde{G}_i$ and the new (s,t)-flow $\tilde{f}_i$. Next, we compose in series $i+1$ copies of $\tilde{G}_i$ with flow $\tilde{f}_i$ to get G_{i+1} and f_{i+1}. We also divide the delay functions by $i+1$. Then we set $f = f_N$. Finally we compose G_N in parallel with m new (s,t)-edges $e_1, \ldots, e_m$ with delay function $\frac{1}{N}x^p$ to get G. By construction, G is a series-parallel network with polynomial delay functions having non-negative coefficients and maximum degree p.

To prove Theorem 2, we first show that f is a PNE. Then we define a new (s,t)-flow h that is obtained from f by deviating $k \in [m]$ units of flows from the most expensive (s,t)-paths in f to the k parallel (s,t)-edges in G with delay function $\frac{1}{N}x^p$. The parameters r and l in (2) are defined as $r = \frac{m}{N}$, $l = \frac{k}{m}$. The complete proof of Theorem 2 is given in the full version of this paper [17].

We now argue that the worst case PoA is in $\Omega(2^p/p)$. By substituting the expression of l in the denominator of (2), we obtain

$$1 + l^2 \sqrt[2^p]{r} - rl - \sqrt[2^p]{r} + r = 1 - \frac{1}{4}r^{2-\frac{1}{2^p}} + r - r^{\frac{1}{2^p}}. \tag{8}$$

Since $r, l \in [0,1]$, we can upper bound the above expression with

$$1 + r - r^{\frac{1}{2^p}} = 1 + \left(\frac{2}{2^{p+1}-1}\right)^{\frac{2^p}{2^p-1}} - \left(\frac{2}{2^{p+1}-1}\right)^{\frac{1}{2^p-1}}$$

$$\leq 1 + \left(\frac{2}{2^{p+1}-1}\right)^{\frac{2^p}{2^p-1}} - \left(\frac{1}{2^p}\right)^{\frac{1}{2^p-1}} \leq 1 - \left(1 - \frac{1}{2^p}\right)\left(\frac{1}{2^p-1}\right)^{\frac{1}{2^p-1}}.$$

Finally, we have that $\lim_{p \to \infty} \frac{1 - \left(1 - \frac{1}{2^p}\right)\left(\frac{2}{2^{p+1}-1}\right)^{\frac{1}{2^p-1}}}{\frac{p}{2^p}} = 0$, proving that (8) is in $O(p/2^p)$, which implies that when N goes to infinity the PoA is at least in $\Omega(2^p/p)$.

4   Maximum Cost

In this section, we measure the social cost of a state P as the maximum players' cost in P, and we derive an upper bound and a lower bound on the PoA with respect to this notion of cost. Recall that given any state P, $tot(P)$ is the total cost of P and $max(P)$ is the maximum cost of a player in P.

We first prove the upper bound on the PoA stated in Theorem 3.

Proof of Theorem 3. Let P_o be the social optimum with respect to the total cost, and let $P_{\hat{o}}$ be the social optimum with respect to the maximum cost. Let $P_f = \{p_f^1, \ldots, p_f^N\}$ be an arbitrary PNE. We will show that $max(P_f) \leq z(\mathcal{D})y(\mathcal{D})max(P_{\hat{o}})$.

Because P_f is a PNE and $\max(P_f)$ is the cost of a player, we have $\max(P_f) \le \text{cost}_f^+(p_f^i)$ for any $i \in [N]$. Moreover, by (3), we have $\text{cost}_f^+(p_f^i) \le z(\mathcal{D})\, \text{cost}_f(p_f^i)$. In other words, the most expensive path in P_f has cost no greater than $z(\mathcal{D})$ times the cost of any other path in P_f. Thus we can conclude that

$$N \cdot \max(P_f) \le \sum_{i=1}^{N} \text{cost}_f^+(p_f^i) \le z(\mathcal{D}) \sum_{i=1}^{N} \text{cost}_f(p_f^i) = z(\mathcal{D})\, \text{tot}(f),$$

i.e., the most expensive path in P_f has cost no greater than $z(\mathcal{D})$ times the average players' cost in P_f. Moreover,

$$z(\mathcal{D})\, \text{tot}(P_f) \le z(\mathcal{D}) y(\mathcal{D})\, \text{tot}(P_o) \tag{9}$$
$$\le z(\mathcal{D}) y(\mathcal{D})\, \text{tot}(P_{\hat{o}}) \tag{10}$$
$$\le z(\mathcal{D}) y(\mathcal{D})(N \cdot \max(P_{\hat{o}})). \tag{11}$$

Inequality (9) directly follows Theorem 1. Inequality (10) holds since P_o is the social optimum state with respect to the total cost, which implies that $\text{tot}(P_o) \le \text{tot}(P_{\hat{o}})$. Inequality (11) holds because $\max(P_{\hat{o}})$ is the maximum player's cost in $P_{\hat{o}}$. $\qquad\square$

We now consider the class Poly-p of polynomial delays with nonnegative coefficients and maximum degree p, and we prove that $z(\text{Poly-}p)$ is at most 2^p.

Lemma 9. *For the class of polynomial delay functions* Poly-p *it holds that* $z(\text{Poly-}p) \le 2^p$.

Proof. By the definition of $z(\text{Poly-}p)$ in (3) we have that for any $x \in \mathbb{N}^+$

$$z(\text{Poly-}p) = \max_{x \in \mathbb{N}^+} \frac{\sum_{j=0}^{p} a_j (x+1)^j}{\sum_{j=0}^{p} a_j x^j}$$

Note that given two collections of nonnegative real numbers $b_0, \ldots, b_p$ and $c_0, \ldots, c_p$, we have

$$\frac{\sum_{j=0}^{p} b_j}{\sum_{j=0}^{p} c_j} \le \max_{j=0,\ldots,p} \frac{b_j}{c_j}.$$

Thus,

$$z(\text{Poly-}p) = \max_{x \in \mathbb{N}^+} \frac{\sum_{j=0}^{p} a_j (x+1)^j}{\sum_{j=0}^{p} a_j x^j} \le \max_{x \in \mathbb{N}^+} \max_{j=0,\ldots,p} \frac{a_j (x+1)^j}{a_j x^j} \le 2^p.$$

$\qquad\square$

Finally, we prove that, for any class of delay functions, and as long as the network's structure is preserved under series compositions, any lower bound on the PoA with respect to the total social cost is also valid when measuring the social cost in terms of the maximum players' cost.

Proof of Theorem 4. We start with an instance of an atomic, unweighted, symmetric network congestion game on a (s, t)-network G, where P_f is a PNE, P_o is a social optimum with respect to the total players' cost, and the PoA is $\text{cost}(P_f)/\text{cost}(P_o)$. Our goal is to construct a new instance on a network G', and to define a PNE $P_{f'}$ and a social optimum $P_{o'}$ with respect to the maximum players' cost, such that

$$\frac{\max(P_{f'})}{\max(P_{o'})} = \frac{\text{cost}(P_f)}{\text{cost}(P_o)}.$$

We construct G' as follows. First, let $G_1, \ldots, G_N$ be N duplicates of G and let G' be the (s, t)-network obtained by composing in series $G_1, \ldots, G_N$. We remark that any graph structure possessed by G is still valid for G', by our assumption. Let $P_f = \{p_f^1, \ldots, p_f^N\}$ and $P_o = \{p_o^1, \ldots, p_o^N\}$. For each $i \in [N]$ let $P_{f_i} = \{p_{f_i}^1, \ldots, p_{f_i}^N\}$ and $P_{o_i} = \{p_{o_i}^1, \ldots, p_{o_i}^N\}$ be the corresponding duplicates of P_f and P_o in G_i, respectively. For each player $i \in [N]$ we define the strategy $p_{f'}^i$ of player i in $P_{f'}$ by having the player choose the path $p_{f_j}^{j(i)}$ in G_j, where $j(i) = (i + N - 1) \mod N$. For example, the strategy of player 2 in $P_{f'}$ is obtained by composing in series the paths $p_{f_1}^2, p_{f_2}^3, \ldots, p_{f_{N-1}}^N, p_{f_N}^1$. Analogously, we define the strategy $p_{o'}^i$ of player i in $P_{o'}$ by having the player choose the path $p_{o_j}^{j(i)}$ in G_j. It can be checked that $P_{f'} = \{p_{f'}^1, \ldots, p_{f'}^N\}$ is a PNE for the new instance defined on G' (otherwise we would contradict that f is a PNE in the original instance). Similarly, it can be checked that $P_{o'} = \{p_{o'}^1, \ldots, p_{o'}^N\}$ is the social optimum in G' with respect to the total cost (otherwise we would contradict that o is a social optimum in the original instance).

Observe that, since in our construction we are permuting the players' strategies, all the players have the same cost, both in $P_{f'}$ and in $P_{o'}$. Moreover this cost is equal to $\text{tot}(P_f)$ in $P_{f'}$ and to $\text{tot}(P_o)$ in $P_{o'}$. Thus, $\max(P_{f'}) = \text{tot}(P_f)$ and $\max(P_{o'}) = \text{tot}(P_o)$. Now let $\hat{f}$ and $\hat{o}$ be the worst PNE and the social optimum in the new instance. We conclude that

$$\frac{\text{tot}(P_f)}{\text{tot}(P_o)} = \frac{\max(P_{f'})}{\max(P_{o'})} \leq \frac{\max(P_{\hat{f}})}{\max(P_{\hat{o}})},$$

which implies the statement of this theorem. □

5   Conclusion

Our contributions fill a gap in the literature on the PoA of atomic, unweighted, symmetric network congestion games, which tackles either general networks, or very simple network structures, such as parallel-link networks and extension-parallel networks.

In this paper we have focused on symmetric games. The worst-case PoA for unweighted congestion games over general networks [1] is achieved in the asymmetric case. On the other hand, Bhavalkar et al. proved that PoA of symmetric (unweighted) congestion games is as large as in asymmetric ones [4]. What impact does symmetry have in the presence of network structure? Consider the class of polynomial delays Poly-p. If we relax the symmetry assumption, the upper bound of Theorem 1 does not hold. In fact, the PoA in asymmetric congestion games defined over parallel-link networks is as large as in asymmetric congestion games defined over general networks [16]. What happens if we instead stay in the realm of symmetric *network* congestion games, with no assumption on the network structure? In the full version of this paper [17], we provide a construction that violates the upper bound of Theorem 1, even if only by one.

References

1. Aland, S., Dumrauf, D., Gairing, M., Monien, B., Schoppmann, F.: Exact price of anarchy for polynomial congestion games. In: Durand, B., Thomas, W. (eds.) STACS 2006. LNCS, vol. 3884, pp. 218–229. Springer, Heidelberg (2006). https://doi.org/10.1007/11672142_17
2. Awerbuch, B., Azar, Y., Epstein, A.: The price of routing unsplittable flow. In: Proceedings of the Thirty-Seventh Annual ACM Symposium on Theory of Computing, pp. 57–66. STOC 2005, Association for Computing Machinery, New York, NY, USA (2005)
3. Awerbuch, B., Azar, Y., Epstein, A.: The price of routing unsplittable flow. SIAM J. Comput. **42**(1), 160–177 (2013)
4. Bhawalkar, K., Gairing, M., Roughgarden, T.: Weighted congestion games: the price of anarchy, universal worst-case examples, and tightness. ACM Trans. Econ. Comput. **2**(4), 1–23 (2014)
5. Bilò, V., Vinci, C.: On the impact of singleton strategies in congestion games. In: Pruhs, K., Sohler, C. (eds.) 25th Annual European Symposium on Algorithms (ESA 2017). Leibniz International Proceedings in Informatics (LIPIcs), vol. 87, pp. 17:1–17:14. Schloss Dagstuhl-Leibniz-Zentrum fuer Informatik, Dagstuhl, Germany (2017)
6. Caragiannis, I., Flammini, M., Kaklamanis, C., Kanellopoulos, P., Moscardelli, L.: Tight bounds for selfish and greedy load balancing. Algorithmica **61**(3), 606–637 (2011)
7. Christodoulou, G., Koutsoupias, E.: The price of anarchy of finite congestion games. In: Proceedings of the Thirty-Seventh Annual ACM Symposium on Theory of Computing, pp. 67–73. STOC 2005, Association for Computing Machinery, New York, NY, USA (2005)
8. Correa, J., de Jong, J., de Keijzer, B., Uetz, M.: The inefficiency of Nash and subgame perfect equilibria for network routing. Math. Oper. Res. **44**(4), 1286–1303 (2019)
9. Czumaj, A., Vöcking, B.: Tight bounds for worst-case equilibria. ACM Trans. Algorithms **3**(1), 1–7 (2007)
10. Epstein, A., Feldman, M., Mansour, Y.: Efficient graph topologies in network routing games. Games Econ. Behav. **66**(1), 115–125 (2009)
11. Fotakis, D.: Stackelberg strategies for atomic congestion games. In: Arge, L., Hoffmann, M., Welzl, E. (eds.) ESA 2007. LNCS, vol. 4698, pp. 299–310. Springer, Heidelberg (2007). https://doi.org/10.1007/978-3-540-75520-3_28
12. Fotakis, D.: Congestion games with linearly independent paths: convergence time and price of anarchy. In: Monien, B., Schroeder, U.-P. (eds.) SAGT 2008. LNCS, vol. 4997, pp. 33–45. Springer, Heidelberg (2008). https://doi.org/10.1007/978-3-540-79309-0_5

13. Fotakis, D.: Stackelberg strategies for atomic congestion games. Theor. Comp. Sys. **47**(1), 218–249 (2010)
14. Fotakis, D., Spirakis, G.P.: Cost-balancing tolls for atomic network congestion games. Internet Math. **5**(4), 343–363 (2008)
15. Gairing, M., Lücking, T., Mavronicolas, M., Monien, B., Rode, M.: Nash equilibria in discrete routing games with convex latency functions. In: Díaz, J., Karhumäki, J., Lepistö, A., Sannella, D. (eds.) ICALP 2004. LNCS, vol. 3142, pp. 645–657. Springer, Heidelberg (2004). https://doi.org/10.1007/978-3-540-27836-8_55
16. Gairing, M., Schoppmann, F.: Total latency in singleton congestion games. In: Deng, X., Graham, F.C. (eds.) WINE 2007. LNCS, vol. 4858, pp. 381–387. Springer, Heidelberg (2007). https://doi.org/10.1007/978-3-540-77105-0_42
17. Hao, B., Michini, C.: Inefficiency of pure Nash equilibria in series-parallel network congestion games, November 2021. Optimization Online
18. Hao, B., Michini, C.: The price of anarchy in series-parallel network congestion games. Math. Program. 1–31 (2022). https://doi.org/10.1007/s10107-022-01803-w
19. Harker, P.T.: Multiple equilibrium behaviors on networks. Transp. Sci. **22**(1), 39–46 (1988)
20. Holzman, R., Monderer, D.: Strong equilibrium in network congestion games: increasing versus decreasing costs. Int. J. Game Theory **44**(3), 647–666 (2015)
21. Koutsoupias, E., Papadimitriou, C.: Worst-case equilibria. In: Meinel, C., Tison, S. (eds.) STACS 1999. LNCS, vol. 1563, pp. 404–413. Springer, Heidelberg (1999). https://doi.org/10.1007/3-540-49116-3_38
22. Lücking, T., Mavronicolas, M., Monien, B., Rode, M.: A new model for selfish routing. In: Diekert, V., Habib, M. (eds.) STACS 2004. LNCS, vol. 2996, pp. 547–558. Springer, Heidelberg (2004). https://doi.org/10.1007/978-3-540-24749-4_48
23. Lücking, T., Mavronicolas, M., Monien, B., Rode, M.: A new model for selfish routing. Theoret. Comput. Sci. **406**(3), 187–206 (2008)
24. Nickerl, J.: The minimum tollbooth problem in atomic network congestion games with unsplittable flows. Theor. Comp. Sys. **65**(7), 1094–1109 (2021)
25. Orda, A., Rom, R., Shimkin, N.: Competitive routing in multiuser communication networks. IEEE/ACM Trans. Netw. **1**(5), 510–521 (1993)
26. Rosenthal, R.W.: A class of games possessing pure-strategy Nash equilibria. Internat. J. Game Theory **2**, 65–67 (1973)
27. Roughgarden, T.: The price of anarchy is independent of the network topology. J. Comput. Syst. Sci. **67**(2), 341–364 (2003)
28. Roughgarden, T.: The maximum latency of selfish routing. In: Proceedings of the Fifteenth Annual ACM-SIAM Symposium on Discrete Algorithms, pp. 980–981. SODA 2004, Society for Industrial and Applied Mathematics, USA (2004)
29. Roughgarden, T.: Intrinsic robustness of the price of anarchy. In: Proceedings of the Forty-First Annual ACM Symposium on Theory of Computing, pp. 513–522. STOC 2009, Association for Computing Machinery, New York, NY, USA (2009)
30. Roughgarden, T.: Intrinsic robustness of the price of anarchy. J. ACM **62**(5), 1–42 (2015)
31. Roughgarden, T., Tardos, E.: How bad is selfish routing? J. ACM **49**(2), 236–259 (2002)
32. Suri, S., Tóth, C.D., Zhou, Y.: Selfish load balancing and atomic congestion games. In: Proceedings of the Sixteenth Annual ACM Symposium on Parallelism in Algorithms and Architectures, pp. 188–195. SPAA 2004, Association for Computing Machinery, New York, NY, USA (2004)
33. Valdes, J., Tarjan, R., Lawler, E.: The recognition of series parallel digraphs. SIAM J. Comput. **11**(2), 298–316 (1982)

Insightful Mining Equilibria

Mengqian Zhang[1] iD, Yuhao Li[2] iD, Jichen Li[3] iD, Chaozhe Kong[3] iD,
and Xiaotie Deng[3(✉)] iD

[1] Department of Computer Science and Engineering, Shanghai Jiao Tong University,
Shanghai 200240, China
mengqian@sjtu.edu.cn
[2] Columbia University, New York, NY 10027, USA
yuhaoli@cs.columbia.edu
[3] Center on Frontiers of Computing Studies, School of Computer Science,
Peking University, Beijing 100871, China
{limo923,kcz,xiaotie}@pku.edu.cn

Abstract. The selfish mining attack, arguably the most famous game-theoretic attack in blockchain, indicates that the Bitcoin protocol is not incentive-compatible. Most subsequent works mainly focus on strengthening the selfish mining strategy, thus enabling a single strategic agent more likely to deviate. In sharp contrast, little attention has been paid to the resistant behavior against the selfish mining attack, let alone further equilibrium analysis for miners and mining pools in the blockchain as a multi-agent system. In this paper, first, we propose a novel strategy called insightful mining to counteract the selfish mining attack. By infiltrating an undercover miner into the selfish pool, the insightful pool could acquire the number of its hidden blocks. We prove that, with this extra insight, the utility of the insightful pool is strictly greater than the selfish pool's when they have the same mining power. Then we investigate the mining game where all pools can choose to be honest or take the insightful mining strategy. We characterize the Nash equilibrium of such a game and derive three corollaries: (a) each mining game has a *pure* Nash equilibrium; (b) there are at most two insightful pools under some equilibrium no matter how the mining power is distributed; (c) honest mining is a Nash equilibrium if the largest mining pool has a fraction of mining power no more than 1/3. Our work explores, for the first time, the idea of spying in the selfish mining attack, which might shed new light on researchers in the field.

Keywords: Blockchain · Selfish mining · Markov process · Insightful mining · Mining game

This work was supported by Science and Technology Innovation 2030 - "New Generation Artificial Intelligence" Major Project No. 2018AAA0100901. This work has been performed with support from the Algorand Foundation Grants Program.
Y. Li—Supported by NSF grants CCF-1563155, CCF-1703925, IIS-1838154, CCF-2106429 and CCF-2107187.

K. A. Hansen et al. (Eds.): WINE 2022, LNCS 13778, pp. 21–37, 2022.
https://doi.org/10.1007/978-3-031-22832-2_2

1 Introduction

Bitcoin [16], as the pioneering blockchain ecosystem, proposes an electronic payment system without any trusted party. It creatively uses *Proof-of-Work* (PoW) to incentivize all *miners* to solve a cryptopuzzle (also known as *mining*). The winner will gain the record-keeping rights to generate a *block* and be awarded the newly minted tokens. As more and more computational power is invested into mining, it may take sole miner months or even years to find a block [24]. In order to reduce the uncertainty, a group of miners forms a *mining pool* to share their computational resources. Under the leadership of the pool manager, all miners in a pool solve the same puzzle in parallel and share the block rewards. In the Bitcoin system, so long as all participants behave honestly, one's expected revenue will be proportional to its hashing power.

However, in practice, miners are rational and may act strategically. Thus, game theory naturally stands out as a tool for analyzing the robustness of the Bitcoin protocol. The conventional wisdom would expect a proof of the incentive compatibility of the Bitcoin protocol and subsequently the strategyproofness against manipulative miners.

Such a hope was broken by the seminal work [6], which proposed the selfish mining strategy, arguably the most well-known game-theoretic attack in blockchain. It indicates that the Bitcoin mining protocol is not incentive-compatible. The key idea behind the attack is to induce honest miners to waste their mining power. As a result, the selfish pool could obtain more revenue than its fair share.

Pushing this approach to the extreme, [23] expanded the action space of selfish mining, modeled it as a Markov Decision Process (MDP), and pioneered a novel technique to resolve the non-linear objective function of the MDP to get a more powerful selfish mining strategy, for a revenue arbitrarily close to the optimum. A series of works have since been initiated to study the mining strategies of a rational pool under the same assumption that other pools behave honestly [7,11,14,17,18,21].

In sharp contrast, little attention has been paid to the incentive of other pools, which plays an important role in studying the strategic interactions among participants and understanding the stable state of blockchain systems. In this paper, we propose and study the following vital questions.

1. *Can a pool strategically defend against the selfish mining attack?*
2. *Moreover, what equilibrium will the ecosystem of different types of agents eventually reach?*

1.1 Our Contributions

In this work, we propose a strategy called *insightful mining* (Fig. 1). Once detecting a selfish pool, an *insightful pool* that adopts the insightful mining strategy can infiltrate an undercover miner into it to monitor the number of hidden blocks.[1]

---

[1] We discuss this action in more detail in Sect. 3.1.

With this key information, the insightful pool clearly knows the real-time state of the mining competition and thus responds strategically. From a high-level view, when observing that the selfish pool is taking the lead, the insightful pool would behave honestly to end its leading advantage as quickly as possible. On the other hand, when the insightful pool is taking the lead, it will take action similar to selfish mining, regarding the selfish pool and the honest pool as "others". Note that by infiltrating spies, a strategic player can gain more information (e.g., the hash values of hidden blocks) than the length of the private branch. With this information, there are lots of things that a player could do. This paper, however, focuses on the insightful mining strategy, which only utilizes the number of hidden blocks.

Although using very little information, the strategy firmly answers our first research question: A pool can strategically defend against the selfish mining attack with the insightful mining strategy. Specifically, the system consists of three types of players: the honest pool, the selfish pool, and the insightful pool. With different mining strategies, the three players may hold different branches and have asymmetric information during the mining competition. The honest pool, following the protocol, has the public information (i.e., the length of its public branch). The selfish pool keeps a selfish branch and is aware of the length of the public branch and its selfish branch. Owing to the infiltrated spy, the insightful pool learns all information (in particular, the length of the honest branch, the selfish branch, and its insightful branch). We model their interactions as a two-dimensional Markov reward process with an infinite number of states (Table 1 and Fig. 2). We prove that when there is a selfish pool and an insightful pool with the same mining power, the insightful pool will get a strictly greater expected revenue than the selfish pool (Theorem 1). This demonstrates that the extra insight significantly reverses the selfish pool's advantage.

Then we investigate the scene where all n mining pools are strategic. Besides counteracting the selfish mining attack, insightful mining can be adopted directly as a mining strategy. Specifically, insightful mining resembles selfish mining if there is no pool mining selfishly. We study the mining game where each pool plants spies into all other pools and chooses either to follow the Bitcoin protocol or to take the insightful mining strategy. Such a mining game can be formulated as an n-player normal-form game. Note that although there are 2^n pure strategy profiles, the payoff function of each player is explicitly represented (Proposition 1). Our main result is a characterization theorem of the Nash equilibrium in mining games (Theorem 2). Concretely, Theorem 2 derives three corollaries: (a) each mining game has a *pure* Nash equilibrium; (b) there are at most two insightful pools under some equilibrium no matter how the mining power is distributed; (c) honest mining is a Nash equilibrium if the largest mining pool has a fraction of total hashing power no more than $1/3$. These corollaries are surprising. Taking (a) as an example, there is no guarantee of the existence of pure Nash equilibria in general.

Beyond our theoretical results, we also conduct several simulations to understand insightful mining (Sect. 5). First, we visualize the relative revenue of the selfish pool and the insightful pool when they have the same mining power. An interesting observation is that when their hashing power is larger than 1/3, the insightful pool can gain most of the revenue. Besides, we explore the performance of the insightful mining strategy when they have different mining power. Simulation results provide compelling evidence that the insightful pool could still gain more revenue even if it holds less mining power than the selfish pool.

In the end, we discuss the role of the undercover miner in the context of selfish mining and blockchain, which sheds new light on future research directions (Sect. 6).

1.2 Related Work

The classic selfish mining attack was first proposed and mathematically modeled as a Markov reward process in the seminal paper [6]. Observing that the classic selfish mining strategy could be suboptimal for a large parameter space, several works [17,23] further generalized the system as a Markov Decision Process (MDP) to find the optimal selfish mining strategy. Aiming to solve the average-MDP with a non-linear objective function, [23] proposed a binary search procedure by converting the problem into a series of standard MDPs. A recent work [28] developed a more efficient method called Probabilistic Termination Optimization, converting the average-MDP into only one standard MDP.

Studying other agents' incentives against one selfish miner was more challenging due to the tremendous state spaces and complicated Markov reward processes. The work of [15] presented some simulation results on systems involving multiple selfish miners [6] or involving multiple stubborn miners [17]. On the learning side, a recent work [12] proposed a novel framework called SquirRL, which is based on deep reinforcement learning (deep-RL) techniques. Their experiments suggest that adopting selfish mining might not be the optimal choice when facing selfish mining. We *prove* such a result by providing the insightful mining strategy and the dominating theorem (Theorem 1). The strength of SquirRL is a more general strategy space generated by deep-RL. However, we highlight that it cannot cover our insightful mining strategy since our greatest strength comes from our undercover miner's insights (information), which have not been discussed in the broad selfish mining context.

To our best knowledge, the most related work that theoretically studied the equilibria with multiple selfish mining pools is [4]. Due to the analytical challenges of infinite states in the classic selfish mining strategy, they proposed a simplified version called semi-selfish mining, where the strategic mining pool will only keep a private chain of the length of at most two. Such a restriction makes the Markov reward process have finite states (as long as there is a finite number of semi-selfish miners) and simplifies the equilibrium analysis. However, our insightful mining strategy works against the classic selfish mining strategy and may also keep an arbitrary long private chain. While this leads to a 2-dimensional Markov reward process with an infinite number of states, the techniques in the mathematical analysis are sufficient for us to prove the desired dominating theorem (Theorem 1) and equilibrium characterization (Theorem 2).

2 Preliminaries

2.1 Proof of Work

In the context of blockchain, Proof of Work was first introduced in Bitcoin [16]. As mentioned, the security of Bitcoin heavily relies on the Proof-of-Work scheme, which has also been widely adopted by other blockchain systems like Ethereum [2]. The past decade has seen a great amount of research around PoW, with respect to its block rewards design [3], strategic deviation [13], the difficulty adjustment algorithm [10,19], energy costs [8], and so on.

Taking Bitcoin as an example, PoW requires a miner to randomly engage in the hashing function calls to solve a cryptopuzzle. Typically, miners should search for a *nonce* value satisfying that

$$H(previous\ hash;\ address;\ Merkle\ root;\ nonce) \leq D \qquad (1)$$

where $H(\cdot)$ is a commonly known cryptographic hash function (*e.g.*, SHA-256 in Bitcoin); *previous hash* is the hash value of the previous block; *address* is the miner's address to receive potential rewards; *Merkle root* is an integrated hash value of all transactions in the block; and D is the target of the problem and reflects the difficulty of this puzzle.[2] Started from the genesis block, all miners compete to find a feasible solution, thus generating a new block appended to the previous one. In return, they will be awarded the newly minted bitcoins for their efforts in maintaining the blockchain system. The standard Bitcoin protocol treats the longest chain as the main chain. Once encountering two blocks at the same block height, miners randomly choose one to follow according to the uniform tie-breaking rule. Thus, in order to be accepted by more miners, it is suggested to publish the newly generated block immediately. In this paper, the miners who stick to the Bitcoin protocol are referred to be *honest*.

2.2 Mining Pool

With more and more hashing power invested into mining, the chances of finding a block as a sole miner are quite slim. Nowadays, miners tend to participate in organizations called mining pools.

Generally, a mining pool comprises a pool manager and several peer miners. All participants shall cooperate to solve the same puzzle. Specifically, each miner will receive a task like (1) above from the pool manager and a work unit a work unit containing a particular range of nonce. Instead of trying all possible nonce values, the miner only needs to search for the answer from the received work unit. In this way, all miners in the pool work in parallel. Once any miner finds a valid solution, this pool succeeds in this mining competition. Then a new task will be organized and further released to all miners in the pool. Also, participants will share the mining rewards according to the reward allocation protocol like Pay

---

[2] For security, the difficulty of puzzles will be adjusted automatically to ensure that the mean interval of block generation is 10 min.

Per Share (PPS), proportional (PROP), Pay Per Last N Shares (PPLNS) [27], and so on. In expectation, the miners' rewards are proportional to their hashing power. As a result, miners who join the mining pool can significantly reduce the variance of mining rewards. Currently, most of the blocks in Bitcoin are generated by mining pools such as AntPool [1], Poolin [20], F2Pool [9].

2.3   Selfish Mining

It has long been believed that the Bitcoin protocol is incentive-compatible. However, Eyal and Sirer [6] indicate that it is not the case. It describes a well-known attack called selfish mining. A pool could receive higher rewards than its fair share via the selfish mining strategy. This attack ingeniously exploits the conflict-resolution rule of the Bitcoin protocol, in which when encountering a fork, only one chain of blocks will be considered valid. With the selfish mining strategy, the attacker deliberately creates a fork and forces honest miners to waste efforts on a stale branch. Specifically, the selfish pool strategically keeps its newly found block secret rather than publishing it immediately. Afterward, it continues to mine on the head of this private branch. When the honest miners generate a new block, the selfish pool will correspondingly publish one private block at the same height and thus create a fork. Once the selfish pool's leads reduce to two, an honest block will prompt the selfish pool to reveal all its private blocks. As a well-known conclusion, assuming that the honest miners apply the uniform tie-breaking rule, if the fraction of the selfish pool's mining power is greater than 25%, it will always get more benefit than behaving honestly.

3   Insightful Mining Strategy

3.1   Model and Strategy

This paper considers a system of n miners. Each miner i has m_i fraction of total hashing power, such that $\sum_{i=1}^{n} m_i = 1$. Let $\mathcal{H}$, $\mathcal{S}$, $\mathcal{I}$ denote the set of honest miners, selfish miners, and insightful miners, respectively. As the honest miners strictly follow the Bitcoin protocol and do not hide any block information from each other, they are regarded as a whole, referred to as the *honest pool* in the paper. Similarly, all selfish miners who adopt the selfish mining strategy combine together to behave as a single agent, which is called the *selfish pool*. The remaining miners form the *insightful pool* and adopt the insightful strategy stated later. Let α and β denote the fraction of mining power controlled by the selfish pool and the insightful pool, respectively. We have $\alpha = \sum_{i \in \mathcal{S}} m_i$ and $\beta = \sum_{i \in \mathcal{I}} m_i$. Then the total power of the honest pool can be represented as $1 - \alpha - \beta$. Following the previous work [6,23], in this paper, we also assume that the time to broadcast a block is negligible and the transaction fee is negligible. In other words, the pools' revenue mainly comes from block rewards. In addition, the block generation is treated as a randomized model, where a new block is generated in each time slot.

Now we describe the insightful mining strategy. Before getting into the details, we state that the insightful pool could learn how many blocks the selfish pool has been hiding by doing the following. The manager of the insightful pool shall pretend to join the selfish pool as a spy. As a pool member, it will receive a mining task from the manager of the selfish pool. The hash value of the previous block can be parsed from the task. Normally, this hash value corresponds to the last block of the main chain. Once the selfish pool mines a block,[3] its manager will keep the block private and publish a new task based on it. From the spy's perspective, there is no newly published block in the system, but the selfish manager releases a new task based on an unknown block. Then it is reasonable to believe that the selfish manager is hiding blocks. Furthermore, the number of hidden blocks is exactly the number of recently received tasks with unmatched *previous hash*.

By working as a spy,[4] the insightful pool has a clear understanding of the system's situation, *i.e.*, the mining progress of each player. Although all pools are mining after the main chain, the three players may hold different sub-chain (also referred to as *branch*) during the mining competition. Let l_h, l_s, l_i denote the length of honest branch, selfish branch, and insightful branch respectively. In the process of mining, the honest pool only knows the public information l_h. The selfish pool is aware of both l_h and l_s, while the insightful pool can observe all three lengths. Then the three types of players compete to generate blocks based on their own information. Their competition works in rounds. Each round begins with a global consensus on the current longest chain. When the selfish pool and insightful pool reveal all their private blocks, or they have no hidden blocks while the honest pool finds a block (see *Case 1* below), the round ends, leading to a new global consensus. For the first block in a round, there are three possible cases.

Case 1: the honest pool generates the first block. With probability $1 - \alpha - \beta$, the honest pool mines a block and broadcasts it immediately. In this case, the insightful pool accepts this newly generated block and mines after it. According to the selfish mining strategy, the selfish pool will do the same. Consequently, all players reach a consensus in this case and compete for the next block.

Case 2: the selfish pool generates the first block. With probability α, the selfish pool mines a block. Based on the selfish mining strategy, the selfish pool will keep it private, aiming to further extend its lead. After observing this situation through the spy in the selfish pool, the insightful pool behaves honestly until the selfish pool reveals all its hidden blocks. Recall that when facing two branches of the same height, the honest pool chooses one of them uniformly. The insightful pool, however, will deterministically mine on the opposite of the selfish branch.

---

[3] A member of the selfish pool finds an acceptable nonce to the cryptopuzzle and submits it to the manager.

[4] We assume that the mining power of this spy is negligible, as well as its revenue from the selfish pool.

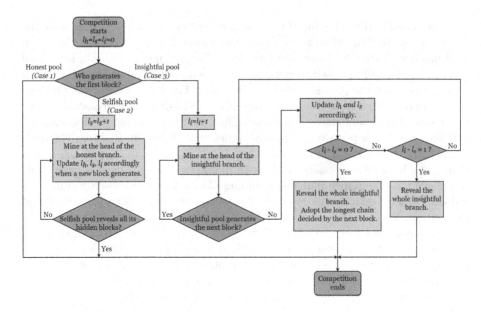

Fig. 1. Flow chart of the insightful mining strategy. l_h, l_s and l_i are the length of the honest branch, selfish branch, and insightful branch, respectively.

The key insight behind this strategy is to prompt[5] the selfish pool to reveal all its hidden blocks and end its leading advantage as quickly as possible.

Case 3: the insightful pool generates the first block. With probability β, the insightful pool mines a block. It hides this block and takes the following actions, which are similar to selfish mining. The insightful pool keeps a watchful eye on how many blocks the selfish pool and the honest pool have mined respectively. In the following competition, when its lead is larger than one (*i.e.*, $l_i - \max\{l_h, l_s\} > 1$), the insightful pool always hides all its mined blocks. Otherwise, it reveals the private branch all at once. Here, the way of releasing blocks is different from selfish mining, which reveals blocks one by one in response to honest behavior.

The above three cases complete the description of our insightful mining strategy. We also show the flow chart of the strategy in Fig. 1. We emphasize that even if there is no selfish pool, the insightful mining could also work as an independent strategy, where *Case 2* never appears.

Remark 1. Note that with different strategies, players in the system have asymmetric information. Each of them can be characterized by the depth of their strategic thought, which forms a hierarchy of levels of iterated rationality.

---

[5] The meaning of "prompt" is that by generating blocks on the opposing branch, the selfish pool will be encouraged to reveal its hidden blocks one by one. Note that in this process, the selfish pool does not know the insightful pool exists, which is critical to the strategy design in game theory.

- Level zero. The honest pool, as the naive level-0 player, truthfully follows the protocol and has the public information (*i.e.*, l_h).
- Level one. The selfish pool, as the level-1 player, acts on the belief that other players are level-0 players. It adopts the selfish mining strategy, keeps a selfish branch, and is aware of l_h and l_s.
- Level two. Due to the infiltrated spy, the insightful pool works as a more sophisticated level-2 player. It observes that the population consists of both level-0 and level-1 players, learns all information (*i.e.*, l_h, l_s, and l_i), and adopts the insightful mining strategy.

Next, we will discuss the revenue of the three types of players with different levels of cognition. The scenario where all players are at the same cognitive level will be explored in Sect. 4.

3.2   Markov Reward Process

To analyze the relative revenue of different players under the insightful mining strategy, we use a two-dimensional state $s = (x, y)$ to reflect the system status and further model the mining events as a Markov Reward Process. The state x denotes the selfish pool's lead over the honest pool, *i.e.*, the number of blocks that the selfish pool has not revealed. Similarly, y is the insightful pool's lead over the selfish pool. Thus, we have $x, y \in \mathbb{N} \cup \{0'\}$ ($0'$ will be explained soon). Here, zero means the selfish pool (corresponding to x) or the insightful pool (corresponding to y) has no hidden blocks. Specifically, it contains two different states, which we use 0 and $0'$ to distinguish. Take x as an example. The state $x = 0$ indicates that the honest pool and the selfish pool are in agreement about a public chain. In other words, their branches are exactly the same. The state $x = 0'$ means that the selfish pool and others (the honest pool or the insightful pool) hold a separate branch of the same length, and the selfish pool has revealed all blocks on its branch. In the state of $0'$, the next block will break the tie and decides the longest chain. For y, the meanings of state 0 and $0'$ are similar to the above, with the insightful pool and others (the selfish pool and the honest pool) as two players.

Let $Pr[s, \tilde{s}]$ denote the probability of changing from state s to state $\tilde{s}$. The vector $r[s, \tilde{s}]$ represents the expected reward obtained from this state transition. It contains three components corresponding to the revenue of the honest pool, the selfish pool, and the insightful pool, respectively. With the help of these notations, Table 1 lists the detailed state transitions and corresponding revenues in the system. Specifically, the item (1) formalizes Case 1 in Sect. 3.1. Items (2)-(9) correspond to Case 2, and Case 3 contains items (10)–(24). The detailed analysis of each transition can be found in [26]. Figure 2 illustrates the overall state transitions in a more intuitive way. We denote the Markov Reward Process of Fig. 2 by $Markov(\alpha, \beta)$.

Recall that a branch will win at the end of one round. It is easy to verify that in our design, each block of the final winning branch will be awarded to some player once and only once.

Table 1. The state transitions and corresponding revenues.

No.	State s	State $\tilde{s}$	$Pr[s,\tilde{s}]$	$r[s,\tilde{s}]$	Conditions
1	$(0,0)$	$(0,0)$	$1-\alpha-\beta$	$(1,0,0)$	
2	$(0,0)$	$(1,0)$	α	$(0,0,0)$	
3	$(1,0)$	$(0',0)$	$1-\alpha-\beta$	$(0,0,0)$	
4	$(0',0)$	$(0,0)$	1	$(\frac{3-3\alpha-\beta}{2},\frac{1+3\alpha-\beta}{2},\beta)$	
5	$(1,0)$	$(1,0')$	β	$(0,0,0)$	
6	$(1,0')$	$(0,0)$	1	$(1-\alpha-\beta,\frac{1+3\alpha-\beta}{2},\frac{1-\alpha+3\beta}{2})$	
7	$(x,0)$	$(x+1,0)$	α	$(0,0,0)$	$\forall x \geq 1$
8	$(2,0)$	$(0,0)$	$1-\alpha$	$(0,2,0)$	
9	$(x,0)$	$(x-1,0)$	$1-\alpha$	$(0,1,0)$	$\forall x \geq 3$
10	$(0,0)$	$(0,1)$	β	$(0,0,0)$	
11	$(0,1)$	$(1,0')$	α	$(0,0,0)$	
12	$(0,1)$	$(0,0')$	$1-\alpha-\beta$	$(0,0,0)$	
13	$(0,0')$	$(0,0)$	1	$(\frac{3-2\alpha-3\beta}{2},\alpha,\frac{1+3\beta}{2})$	
14	$(0,1)$	$(0,2)$	β	$(0,0,0)$	
15	$(0,2)$	$(0,0)$	$1-\beta$	$(0,0,2)$	
16	(x,y)	$(x,y+1)$	β	$(0,0,0)$	$\forall x \in \{0'\}\bigcup \mathbb{N}, y \geq 2$
17	$(0,y)$	$(0,y-1)$	$1-\alpha-\beta$	$(0,0,1)$	$\forall y \geq 3$
18	(x,y)	$(x+1,y-1)$	α	$(0,0,1)$	$\forall x \geq 0, y \geq 3$
19	$(1,y)$	$(0',y)$	$1-\alpha-\beta$	$(0,0,0)$	$\forall y \geq 2$
20	$(2,y)$	$(0,y)$	$1-\alpha-\beta$	$(0,0,0)$	$\forall y \geq 2$
21	(x,y)	$(x-1,y)$	$1-\alpha-\beta$	$(0,0,0)$	$\forall 3 \geq 2, y \geq 2$
22	$(x,2)$	$(0,0)$	α	$(0,0,2)$	$\forall x \geq 1$
23	$(0',2)$	$(0,0)$	$1-\beta$	$(0,0,2)$	
24	$(0',y)$	$(0,y-1)$	$1-\beta$	$(0,0,1)$	$\forall y \geq 3$

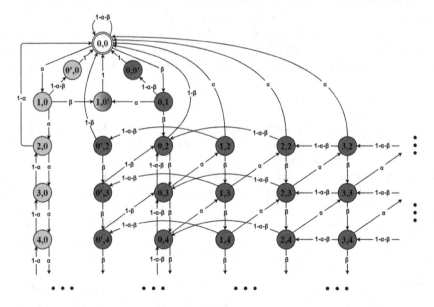

Fig. 2. The Markov Process of the system under the insightful mining strategy.

3.3   The Dominating Theorem

Let $M := \{H, IM, SM\}$. The utility of each $i \in M$ is the relative revenue (denoted by $RREV_i$) defined as follows:

$$
\mathbb{E}\left[\liminf_{T \to \infty} \frac{\sum_{t=1}^{T} r_i[s_t, s_{t+1}] \mid s_0 = (0,0), s_{t+1} \sim Pr[s_t, s_{t+1}]}{\sum_{t=1}^{T} \sum_{j \in M} r_j[s_t, s_{t+1}] \mid s_0 = (0,0), s_{t+1} \sim Pr[s_t, s_{t+1}]}\right].
$$

Note that the transition probability $Pr[s_t, s_{t+1}]$, $r_i[s_t, s_{t+1}]$, and $RREV_i$ ($\forall i \in M$) should depend on the mining power α and β. Here we simplify the notation without ambiguity.

Like previous work [12], here we focus on the scenario where the selfish pool and the insightful pool have the same mining power. The following theorem asserts that, in this case, the expected revenue of the insightful pool is strictly greater than the expected revenue of the selfish pool. The scenario with different pool sizes (*i.e.*, $\alpha \neq \beta$) will be explored in Sect. 5.

Theorem 1. *Let α and β be the fraction of mining power that the selfish pool and the insightful pool control, respectively. When $0 < \alpha = \beta < \frac{1}{2}$, $RREV_{SM}(\alpha, \beta) < RREV_{IM}(\alpha, \beta)$ holds.*

Here, we give the intuition why Theorem 1 holds, and the formal proof can be found in [26]. First, when the selfish pool takes the lead (*Case 2*), the insightful pool cooperates with the honest pool as a whole. However, when the insightful pool is taking the lead (*Case 3*), the selfish pool still competes with the honest pool (*i.e.*, inducing it to waste the mining power on a stale branch), which causes their internal friction. The second intuition is that, when facing two branches with the same length (one is the honest branch and the other is the selfish branch), the insightful pool can clearly know the selfish pool's branch and play against it. Conversely, when confronted with an honest branch and an insightful branch of the same length, the selfish pool will uniformly choose one of them. These two reasons enable the insightful pool to get more revenue than the selfish pool when both have the same mining power.

4   The Mining Game and Equilibria

In this section, we consider the scenario where all n mining pools are strategic and study its Nash equilibrium. Specifically, during the competitive interaction, the honest and selfish pool may realize the existence of insightful mining and learn to do the same, where all players are at the same level of recognition. It is worth noting that insightful mining is a well-defined strategy and can be adopted directly. If there is no selfish pool in the system, insightful mining will look the same as selfish mining. Then we consider the scenario where each pool can choose to follow the Bitcoin protocol truthfully or take the insightful mining strategy. We formally define its strategy space in Sect. 4.1, analyze the utility functions in Sect. 4.2, and characterize the Nash equilibrium in Sect. 4.3.

4.1   Strategy Space

There are n mining pools, and we denote by $[n] := \{1, \cdots, n\}$. The fraction of their hashing power is denoted by $\{m_1, \cdots, m_n\}$ and we have $\sum_{i=1}^{n} m_i = 1$. Each pool i will infiltrate undercover miners into all other pools to monitor their real-time state, namely, whether a certain pool is mining selfishly and, if any, how many blocks are hidden. As a result, each pool i could adopt the insightful mining strategy.

In the mining game, each pool has two strategies: *refined honest mining* and *insightful mining*, denoted by *RHonest* and *Insightful* respectively. The insightful mining strategy is exactly the same as we proposed before, while the refined honest mining is a slightly modified version of the standard mining strategy. Specifically, refined honest mining requires the pool to mine after the longest public chain and to publish its newly-generated block immediately. If someone hides the block, each pool could detect it through the spy therein. Then when facing two branches of the same length, the pool adopting *RHonest* shall clearly follow the honest branch instead of choosing one of them uniformly.

It is important to note that, in this mining game, at most one player is hiding blocks at any time. This is because once an insightful pool mines the first block and hides it, each other pool adopting no matter *RHonest* or *Insightful* will play against it until this mining competition ends. This makes the following analysis of the expected reward function fairly clean and enables us to complete the equilibrium analysis.

4.2   Expected Reward Functions

This section gives the formula of the expected reward function $ER_i(x_1, \cdots, x_n)$ of each pool i under the pure strategy profile $(x_1, \cdots, x_n) \in \{RHonest, Insightful\}^n$.

Proposition 1. *For an n-player mining game $(m_1, \cdots, m_n)$, let $(x_1, \cdots, x_n)$ be a (pure) strategy profile. Let c be a value depending on $(m_1, \cdots, m_n)$ and $(x_1, \cdots, x_n)$.[6] Let $Q \subseteq [n]$ be the set of pools that adopt Insightful strategy. Then we have*

$$ER_i(x_1, \cdots, x_n) = \begin{cases} c \cdot \left(f(m_i) + m_i \cdot \sum_{j \in Q} 2m_j(1 - m_j) \right), i \in Q; \\ c \cdot \left(m_i + m_i \cdot \sum_{j \in Q} 2m_j(1 - m_j) \right), \quad i \notin Q, \end{cases} \quad (2)$$

where $f(y) := y^2 \cdot (2 - 3y)/(1 - 2y)$.

The proof of Proposition 1 can be found in [26].

---

[6] We note that c will not affect the calculation of a pool's relative revenue in the subsequent section.

4.3   Equilibria Characterization

The following theorem characterizes the pure Nash equilibria of the mining game. We refer readers to [26] for the proofs.

Theorem 2. *For an n-player mining game* $(m_1, \cdots, m_n)$ *with* $m_1 \geq \cdots \geq m_n$, *there are three types of pure Nash equilibrium* $(x_1, \cdots, x_n)$, *where*

(1) $(x_1 = \cdots x_n = RHonest)$ *is a Nash equilibrium if and only if* $m_1 \leq 1/3$;
(2) $(x_1 = Insightful, x_2 = \cdots x_n = RHonest)$ *is a Nash equilibrium if and only if* $m_1 \geq 1/3$ *and* $m_2 \leq g(m_1)$;
(3) $(x_1 = x_2 = Insightful, x_3 = \cdots x_n = RHonest)$ *is a Nash equilibrium if and only if* $m_1 \geq 1/3$ *and* $m_2 \geq g(m_1)$,

where $g(y) := \frac{-y^3 + 2y^2 + y - 1}{2y^2 + 4y - 3}$.

Remark 2 (Interpretation of two thresholds in Theorem 2). The analysis of Theorem 2 (1) is to consider the case where one player (say player 1) is deciding to choose *RHonest* or *Insightful* while all other players are adopting *RHonest*. Note that when such a player is adopting *Insightful*, it is the only one that may hide some blocks, and whenever it hides blocks, all other pools will play against it. This case corresponds to the $\gamma = 0$ case of the seminal work [6],[7] where they also got a 1/3 threshold (see Observation 1 in [6]). However, the cases of Theorem 2 (2) and (3) are much more interesting since there exists more than one strategic player with complicated (but explicit) utility functions. For the $g(\cdot)$ function, we note that the threshold $g(m_1) < 1/3$ whenever $m_1 > 1/3$. The interpretation is from the following observation: When player 1 behaves honestly, player 2's relative revenue is exactly proportional to its hashing power (say m_2). But when player 1 adopts *Insightful* ($m_1 \geq 1/3$ by Theorem 2 (1)), the relative revenue of player 2 is lower than m_2 (see proof in [26] for the specific revenue function). Hence, player 2 is more likely to deviate from *RHonest* if someone else (*i.e.*, player 1 here) has been behaving strategically. As a result, the threshold for player 2 to adopt *Insightful* is also lower than (the original) 1/3, and the exact bound is $g(m_1)$.

Theorem 2 has the following three corollaries.

Corollary 1. *Every n-player mining game* $(m_1, \cdots, m_n)$ *has a pure Nash equilibrium.*

Corollary 2. *For an n-player mining game* $(m_1, \cdots, m_n)$, *(RHonest,* $\cdots$, *RHonest) is a Nash equilibrium if* $m_1 \leq 1/3$.

Corollary 3. *For every n-player mining game* $(m_1, \cdots, m_n)$, *there is an equilibrium with at most two insightful pools.*

---

[7] In [6], γ denotes the ratio of honest miners that choose to mine on the private block when facing two branches with the same length.

5   Simulation

This section conducts several simulations to evaluate the effectiveness of the insightful mining strategy. Three agents are considered: the honest pool, the selfish pool, and the insightful pool. Their interactions are simulated as a discrete-time random walk process. In each step, one of the pools generates a block with a probability proportional to its hashing power, and others respond according to their strategies. The simulation ends after 2e9 steps. Then we calculate each pool's relative revenue during the process, which is defined as the proportion of blocks it generates on the main chain to the total number of blocks therein.

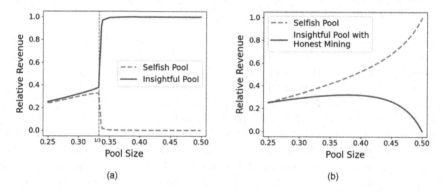

(a)   (b)

Fig. 3. Relative revenue of the selfish pool and the insightful pool with the same mining power. The selfish pool adopts the selfish mining strategy. (a) The insightful pool adopts the insightful mining strategy. (b) The insightful pool mines honestly.

Recall that α and β are the fractions of hashing power of the selfish pool and the insightful pool, respectively. First, we focus on the scenario where the insightful pool and the selfish pool have the same hashing power, *i.e.*, $\alpha = \beta$. Figure 3(a) visualizes the relative revenue of the insightful pool and the selfish pool when their hashing power belongs to $(0.25, 0.5)$. As can be seen, the insightful pool can always gain more revenue than the selfish pool. It is exactly consistent with our theoretical result in Theorem 1. Surprisingly, if their hashing power is larger than $1/3$ (*i.e.*, $\alpha = \beta > 1/3$), the insightful pool can gain most of the revenue. For a clear comparison, we also show their relative revenue under the circumstance that the insightful pool mines honestly in Fig. 3(b). As mentioned in the Introduction, the insightful pool suffers heavy losses in this case, which grow rapidly with the pool size increasing. Comparing Fig. 3(a) and 3(b) shows that the insightful mining strategy dramatically helps the pool turn things around when facing selfish mining.

Then we explore the scenario where $\alpha > \beta$, to consider whether less hashing power can also enable the insightful pool to earn more. Here, two definitions of "more revenue" are studied. One is the aforementioned relative revenue, which corresponds to the dashed line in Fig. 4. It demonstrates the threshold above

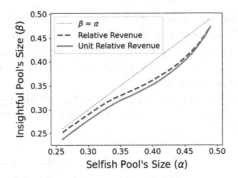

Fig. 4. Threshold of the insightful pool's size, above which it could obtain more relative revenue or unit relative revenue than the selfish pool.

which $RREV_{IM}(\alpha, \beta) > RREV_{SM}(\alpha, \beta)$. The other is the unit relative revenue. The solid line in Fig. 4 represents the corresponding threshold, above which we have $\frac{RREV_{IM}(\alpha,\beta)}{\beta} > \frac{RREV_{SM}(\alpha,\beta)}{\alpha}$. This curve is below the former. Both curves have similar trends, and they are all below the line of $\beta = \alpha$. It provides compelling evidence that with the insightful mining strategy, less computing power can also yield more revenue.

6   Discussion

In blockchain, the action of planting a spy in the pool has been deeply discussed in the context of Block Withholding Attack [5,22]. In such an attack, the attacker infiltrates miners into opponent pools to reduce their revenue. The undercover miner sends only partial solutions (*i.e.*, proofs of work) to the pool manager to share rewards. If it luckily finds a full solution which means a valid block, the undercover miner will discard the full proof of work directly, causing a loss to the victim pool. Our work explores, for the first time, the idea of spying in the selfish mining attack. It will shed new light on the researchers in the field.

Infiltrating spies dramatically expands the action spaces that a pool can take to counteract the selfish mining attack. Besides insightful mining, other strategies are worth exploring. Here, we roughly describe a potential idea. Recalling that the spy can actually extract the hash value of the latest hidden block from the new task issued by the pool manager. With this information, other pools can mine directly behind the latest block, although its full contents are not yet known.[8] By this strategy, all pools could follow the longest chain, which makes selfish mining ineffective. In other words, keeping the block secret for the

---

[8] Such an idea was discussed in [25]. In that context, the strategic miner mines on a newly generated block directly even before it is validated. To avoid potential conflict, the miner can choose to embed no transaction in the block being mined and just try to win the potential block rewards. Our discussion mainly focuses on the role of spies against the selfish mining attack.

selfish pool is equivalent to revealing it honestly, which extremely benefits the blockchain system. Nevertheless, such a strategy might not be the best choice for strategic mining pools. Further research should be undertaken to investigate the optimal mining strategy. It is also worthwhile to extend the action of planting spies to other blockchain scenarios.

Back to our work, insightful mining tells us that insight brings more revenue to a pool. It would be interesting to study the interactions between the insightful mining strategy and other strategies or protocols.

Acknowledgements. We would like to thank anonymous reviewers, Hongyin Chen, Yurong Chen, Zhaohua Chen, Zhijian Duan, Wenhan Huang, Hanyu Li, Kai Li, Qian Wang, and Xiang Yan for helpful comments on an earlier draft.

References

1. Antpool: World leading BTC mining pool (2014). https://antpool.com/. Accessed 20 Oct 2022
2. Buterin, V., et al.: A next-generation smart contract and decentralized application platform. White Paper. **3**(37), 1–2 (2014)
3. Chen, X., Papadimitriou, C., Roughgarden, T.: An axiomatic approach to block rewards. In: Proceedings of the 1st ACM Conference on Advances in Financial Technologies, pp. 124–131 (2019)
4. Cossío, F.J.M., Brigham, E., Sela, B., Katz, J.: Competing (semi-)selfish miners in bitcoin. In: Proceedings of the 1st ACM Conference on Advances in Financial Technologies, AFT 2019, Zurich, Switzerland, 21–23 October 2019, pp. 89–109. ACM (2019)
5. Eyal, I.: The miner's dilemma. In: 2015 IEEE Symposium on Security and Privacy, pp. 89–103. IEEE (2015)
6. Eyal, I., Sirer, E.G.: Majority is not enough: Bitcoin mining is vulnerable. In: Christin, N., Safavi-Naini, R., (eds.) International Conference on Financial Cryptography and Data Security. LNAI, vol. 7151, pp. 436–454. Springer, Cham (2014). https://www.springerprofessional.de/en/majority-is-not-enough-bitcoin-mining-is-vulnerable/4391572
7. Feng, C., Niu, J.: Selfish mining in Ethereum. In: 2019 IEEE 39th International Conference on Distributed Computing Systems (ICDCS), pp. 1306–1316. IEEE (2019)
8. Fiat, A., Karlin, A., Koutsoupias, E., Papadimitriou, C.: Energy equilibria in proof-of-work mining. In: Proceedings of the 2019 ACM Conference on Economics and Computation, pp. 489–502 (2019)
9. Gencer, A.E., Basu, S., Eyal, I., van Renesse, R., Sirer, E.G.: Decentralization in bitcoin and Ethereum networks. In: Meiklejohn, S., Sako, K. (eds.) FC 2018. LNCS, vol. 10957, pp. 439–457. Springer, Heidelberg (2018). https://doi.org/10.1007/978-3-662-58387-6_24
10. Goren, G., Spiegelman, A.: Mind the mining. In: Proceedings of the 2019 ACM Conference on Economics and Computation, pp. 475–487 (2019)
11. Grunspan, C., Perez-Marco, R.: Selfish mining in Ethereum. In: Pardalos, P., Kotsireas, I., Guo, Y., Knottenbelt, W. (eds.) Mathematical Research for Blockchain Economy. SPBE, pp. 65–90. Springer, Cham (2020). https://doi.org/10.1007/978-3-030-53356-4_5

12. Hou, C., et al.: SquirRL: automating attack analysis on blockchain incentive mechanisms with deep reinforcement learning. In: 28th Annual Network and Distributed System Security Symposium, NDSS 2021, virtually, 21–25 February 2021. The Internet Society (2021)

13. Kiayias, A., Koutsoupias, E., Kyropoulou, M., Tselekounis, Y.: Blockchain mining games. In: Proceedings of the 2016 ACM Conference on Economics and Computation, EC 2016, Maastricht, The Netherlands, 24–28 July 2016, pp. 365–382. ACM (2016)

14. Li, Q., Chang, Y., Wu, X., Zhang, G.: A new theoretical framework of pyramid Markov processes for blockchain selfish mining. J. Syst. Sci. Syst. Eng. **30**(6), 667–711 (2021)

15. Liu, H., Ruan, N., Du, R., Jia, W.: On the strategy and behavior of bitcoin mining with n-attackers. In: Proceedings of the 2018 on Asia Conference on Computer and Communications Security, pp. 357–368 (2018)

16. Nakamoto, S.: Bitcoin: A peer-to-peer electronic cash system. Decentralized Bus. Rev. 21260 (2008)

17. Nayak, K., Kumar, S., Miller, A., Shi, E.: Stubborn mining: generalizing selfish mining and combining with an eclipse attack. In: 2016 IEEE European Symposium on Security and Privacy (EuroS&P), pp. 305–320. IEEE (2016)

18. Negy, K.A., Rizun, P.R., Sirer, E.G.: Selfish mining re-examined. In: Bonneau, J., Heninger, N. (eds.) FC 2020. LNCS, vol. 12059, pp. 61–78. Springer, Cham (2020). https://doi.org/10.1007/978-3-030-51280-4_5

19. Noda, S., Okumura, K., Hashimoto, Y.: An economic analysis of difficulty adjustment algorithms in proof-of-work blockchain systems. In: Proceedings of the 21st ACM Conference on Economics and Computation, pp. 611–611 (2020)

20. Poolin: A great bitcoin and multi-cryptocurrency mining pool (2017). https://www.poolin.com/. Accessed 20 Oct 2022

21. Ritz, F., Zugenmaier, A.: The impact of uncle rewards on selfish mining in Ethereum. In: 2018 IEEE European Symposium on Security and Privacy Workshops (EuroS&PW), pp. 50–57. IEEE (2018)

22. Rosenfeld, M.: Analysis of bitcoin pooled mining reward systems. arXiv preprint arXiv:1112.4980 (2011)

23. Sapirshtein, A., Sompolinsky, Y., Zohar, A.: Optimal selfish mining strategies in bitcoin. In: Grossklags, J., Preneel, B. (eds.) FC 2016. LNCS, vol. 9603, pp. 515–532. Springer, Heidelberg (2017). https://doi.org/10.1007/978-3-662-54970-4_30

24. Schrijvers, O., Bonneau, J., Boneh, D., Roughgarden, T.: Incentive compatibility of bitcoin mining pool reward functions. In: Grossklags, J., Preneel, B. (eds.) FC 2016. LNCS, vol. 9603, pp. 477–498. Springer, Heidelberg (2017). https://doi.org/10.1007/978-3-662-54970-4_28

25. Sompolinsky, Y., Zohar, A.: Bitcoin's underlying incentives. Commun. ACM **61**(3), 46–53 (2018)

26. Zhang, M., Li, Y., Li, J., Kong, C., Deng, X.: Insightful mining equilibria. arXiv preprint arXiv:2202.08466 (2022)

27. Zolotavkin, Y., García, J., Rudolph, C.: Incentive compatibility of pay per last n shares in bitcoin mining pools. In: Rass, S., An, B., Kiekintveld, C., Fang, F., Schauer, S. (eds) Decision and Game Theory for Security. GameSec 2017. LNCS, vol. 10575, pp. 21–39. Springer, Cham (2017). https://doi.org/10.1007/978-3-319-68711-7_2

28. Zur, R.B., Eyal, I., Tamar, A.: Efficient MDP analysis for selfish-mining in blockchains. In: Proceedings of the 2nd ACM Conference on Advances in Financial Technologies, pp. 113–131 (2020)

Learning and Online Algorithms

Online Allocation and Display Ads Optimization with Surplus Supply

Melika Abolhassani[1], Hossein Esfandiari[1], Yasamin Nazari[2],
Balasubramanian Sivan[1], Yifeng Teng[1(✉)], and Creighton Thomas[1]

[1] Google, New York, NY, USA
melika.abolhasani@gmail.com,
{esfandiari,balusivan,yifengt,creighton}@google.com
[2] University of Salzburg, Salzburg, Austria
ynazari@cs.sbg.ac.at

Abstract. In this work, we study a scenario where a publisher seeks to maximize its total revenue across two sales channels: guaranteed contracts that promise to deliver a certain number of impressions to the advertisers, and spot demands through an Ad Exchange. On the one hand, if guaranteed contract is not fully delivered, it incurs a penalty for the publisher. On the other hand, the publisher might be able to sell an impression at a high price in the Ad Exchange. How does a publisher maximize its total revenue as a sum of the revenue from the Ad Exchange and the loss from the under-delivery penalty? We study this problem parameterized by *supply factor f*: a notion we introduce that, intuitively, captures the number of times a publisher can satisfy all its guaranteed contracts given its inventory supply. In this work we present a fast simple deterministic algorithm with the optimal competitive ratio. The algorithm and the optimal competitive ratio are a function of the supply factor, penalty, and the distribution of the bids in the Ad Exchange.

Beyond the yield optimization problem, classic online allocation problems such as online bipartite matching of Karp-Vazirani-Vazirani [25] and its vertex-weighted variant of Aggarwal et al. [2] can be studied in the presence of the additional supply guaranteed by the supply factor. We show that a supply factor of f improves the approximation factors from $1 - 1/e$ to $f - fe^{-1/f}$. Our approximation factor is tight and approaches 1 as $f \to \infty$.

Keywords: Online resource allocation · Online advertising · Ad exchange

1 Introduction

An overwhelming majority of publishers on the web monetize their service by displaying ads alongside their content. The revenue stream of such publishers typically comes from two key channels, often referred to as direct sales and indirect sales. In the direct sales channel the publisher strikes several contracts with some major advertisers. The price of such contracts are often negotiated and

© The Author(s), under exclusive license to Springer Nature Switzerland AG 2022
K. A. Hansen et al. (Eds.): WINE 2022, LNCS 13778, pp. 41–59, 2022.
https://doi.org/10.1007/978-3-031-22832-2_3

decided on a per-impression basis before the serving begins. In the indirect sales channel, the ad is selected by seeking, in real-time, bids in an Ad Exchange platform (AdEx for short). In this case an auction is conducted to select the winner and decide how much they pay. A comprehensive yield optimization consists of jointly optimizing the publisher's revenue across both channels. In fact, revenue optimization in this context is significantly important since the display ads industry represents a giant ($> \$50B$) marketplace and is fast growing even at its current mammoth size.

Basic Setting and Preliminaries. We begin by formally describing our setting. The joint yield optimization problem can be modeled as an online edge-weighted and vertex-capacitated bipartite matching problem. There is a set A of offline vertices that correspond to the advertisers with contracts (direct sales), and there is an additional special offline vertex a_d representing AdEx (indirect sales). Advertiser $a \in A$ has capacity n_a and we have $n_{a_d} = \infty$. The capacity n_a represents the number of impressions demanded[1] by contractual advertiser a. Let $N = \sum_{a \in A} n_a$. There is a penalty c that the publisher pays an advertiser for every undelivered impression[2]: i.e., if at the end of the algorithm we assign $k_a < n_a$ impressions to $a \in A$, the publisher pays $c(n_a - k_a)$ to a (there is no benefit to the publisher for delivering beyond n_a impressions). The publisher is not obligated to deliver any impression to AdEx, and thus doesn't incur any penalty from a_d.

Arrival Model. Advertisers are represented as *offline vertices*. Users/queries, arrive *online* in an adversarial manner, and they constitute the online vertex set. When an online vertex (query) arrives, the set of its incident edges to offline vertices (representing the offline nodes that are eligible to be assigned this query) becomes known to the algorithm. *Every* arriving query has an edge to the AdEx node a_d, i.e., every query can be sent to an exchange seeking a bid. All edges incident on any node $a \in A$ have the same weight[3] and the edges incident on the AdEx node a_d could have an arbitrary weight depending on the highest bid from the Exchange. AdEx is modeled by the publicly known distribution D of highest bids in the exchange: i.e., regardless of the query that arrives, when it is assigned to a_d, the publisher accrues a profit that is equal to a draw from

---

[1] We use the terms demand and capacity interchangeably. Technically, offline nodes do not have any capacities, they just have demands. However, for a node that demands n_a impressions, assigning more than n_a impressions is always suboptimal, so essentially n_a can be interpreted as a capacity as well. The AdEx offline node is an exception where the capacity is infinite, in the sense that it is always profitable to assign an additional online node to AdEx.

[2] We later discuss relaxing the penalty c to depend on the advertiser a.

[3] Unweighted edges for contractual advertisers is fine because these contracts are mostly based on the number of impressions delivered. In a few cases the contracts are based on the number of clicks or conversions, in which case the edges will be weighted based on the probability of click or conversion. Contracts based on impressions form such a large majority, that having unweighted edges, is almost wlog.

D. The publisher's basic problem is to decide, on a per-query basis, whether to assign the query to a contract advertiser (if so, whom) or to AdEx.

Objective. Publisher's goal is to maximize its overall revenue. Publishers typically have pre-negotiated prices p_a for each contractual advertiser a. The total revenue of the publisher will be the sum of three parts (i) the revenue from AdEx (i.e., the sum of edge weights of queries assigned to AdEx), (ii) the revenue from contracts: $\sum_{a \in A} n_a \cdot p_a$, and (iii) the revenue loss due to under-delivery, i.e., the negative of the penalty paid. Note that (ii) is a constant, and is unaffected by the allocation algorithm. Thus, while computing competitive ratio, we compute it w.r.t. the sum of (i) and (iii).

Supply Factor. An important concept that we introduce is what we call a *supply factor* of an instance, which captures the (potentially fractional) number of times that a publisher will be able to satisfy their contractual advertisers' demands. Formally, let a complete matching be defined as one where all contractual advertisers' demands n_a are fully satisfied, i.e., all the offline vertices are fully saturated. The supply factor of an instance is defined as the largest positive real number f s.t., there exists an *offline solution* with f complete matchings, s.t., these f matchings are vertex disjoint on the online vertices (the offline vertices are clearly not vertex disjoint across these f matchings; rather, the number of edges incident on any offline vertex $a \in A$, summed over all these f matchings, is $f \cdot n_a$). If there are many such matchings, we pick one to be the supply-factor-determining-offline-solution. In this work, we assume that the number of arriving online queries is exactly $fN = f \sum_{a \in A} n_a$. The algorithm designer is aware of f, the n_a's, and the highest bid distribution from AdEx.

There are several important practical aspects of the yield optimization problem that previous work do not capture that we aim to address:

1. The first aspect is that publishers typically have more inventory than they are able to sell via the direct sales channel (contracts), and indeed that is the main reason that most publishers are selling through the indirect sales channel of AdEx as well. Most previous works on joint yield optimization either address the objectives of the two channels separately (bi-criteria objective), or study them in the absence of supply factor/penalties/AdEx bid distribution. Studying the yield optimization problem with a single unified objective (AdEx revenue - penalty) in the presence of supply factor and AdEx bid distribution surfaces the nature of the optimal tradeoff between the supply factor and how on-track a contract is towards hitting its goals. Clearly, when a contract is lagging behind, we should allocate a query to AdEx only when the AdEx bid is high enough. But how does this "high enough" vary as we increase/decrease the publisher's supply, captured by the supply factor f? This is explicitly answered in our work. Similarly the dependence on the penalty and AdEx distribution are also explicitly revealed.

2. Even in classic online allocation problems like the online bipartite matching of Karp et al. [25] and the online vertex-weighted bipartite matching of Aggarwal

et al. [2], it is interesting to inquire what happens to the competitive ratio when there is a supply factor $f \geq 1$.

3. Prior works mostly studied the problem in a fully stochastic model or a fully adversarial model. In reality, while user browsing patterns might have significant variations across days, in response to events, state-of-mind etc. (and hence an adversarial arrival of queries is reasonable), advertiser bidding/spending patterns are far more predictable because advertisers have daily and hourly spending budgets. We incorporate this in our model by having a distribution D over the highest bids from AdEx, even though query arrival is adversarial. The inclusion of AdEx bid distribution, not only represents reality better, but also leads to a crisp algorithm that sheds ample light on the role of the distribution in the joint yield optimization problem. Further modeling bids in AdEx via a known distribution is standard in literature (e.g. [10, 27]).

1.1 Our Results

One of our contributions, as just discussed, is to present an economical model that crisply captures the reality of display ads monetization. Our main result is a fast simple deterministic algorithm that obtains the optimal competitive ratio as n_a values grow large. The algorithm is as follows: let $0 = r_1 < \cdots < r_d$ be the points in the support of the distribution D of highest bid in AdEx (highest bid is often referred to as reward for short). As a pre-processing step, compute d thresholds $s_1 < \cdots < s_d$ as a function of f (we define $s_0 = 0$ and $s_{d+1} = 1$), c and the AdEx bid distribution. Let the satisfaction-ratio $SR(a)$ of a contractual advertiser a be the ratio of the number of impressions delivered to the contract thus far, to the number of impressions n_a requested by the contract. For each arriving query, the algorithm picks the contract with the lowest satisfaction ratio, call it s. Find u such that $s \in [s_{u-1}, s_u)$. Assign the query to AdEx if the highest bid r in the exchange exceeds r_{d+1-u}. And if not, assign the query to the contract with the lowest satisfaction ratio. Algorithm 1 summarizes this. We highlight a few important aspects of this algorithm.

1. Once the pre-processing step is over (which is a one-time computation), the algorithm is very simple to implement in real time while serving queries, even in a distributed fashion. Each relevant advertiser a for the current query (i.e., each offline node a with a matching edge to the current online node) just responds with its satisfaction ratio $SR(a)$. From there on, the algorithm simply computes the smallest satisfaction ratio, does a lookup over the thresholds that are pre-computed, and decides the allocation based on how big the AdEx bid is.

2. The algorithm is quite intuitive. As the satisfaction ratio of the most needy contract gets lower, the AdEx bid has to be correspondingly higher to merit snatching this impression from the contract. This tradeoff happens to take such a simple symmetric form, where one looks for the mirror image in $\vec{r}$, namely r_{d+1-u}, of the index u to which the satisfaction ratio gets mapped

---

ALGORITHM 1: Optimal algorithm for general AdEx distribution

Input: AdEx distribution D with support $0 = r_1 < ... < r_d$, penalty c, and supply factor f.

Preprocessing: Compute thresholds $s_1, ..., s_d$ (we discuss how in Optimization Problem 1).

for *each query arriving online* **do**

    Let r be the highest AdEx bid for this query.

    Let a be the advertiser, with an edge to this query, and with the lowest satisfaction ratio $SR(a)$.

    if $SR(a) = 1$ **then**

       | Assign the impression to AdEx.

    end

    else

       Find u such that $SR(a) \in [s_{u-1}, s_u)$.

       if $r \le r_{d+1-u}$ **then**

          | Assign the impression to advertiser a.

       end

       else

          | Assign the impression to AdEx.

       end

    end

end

---

is quite surprising. Importantly, the supply factor and penalty are used only in the pre-processing step to compute the thresholds, and don't appear in serving time at all.

3. The algorithm need not fully know the highest bid r from AdEx. It just needs to be able to compare the highest bid against a reserve price of r_{d+1-u}. Further, extending the algorithm to deal with multiple Ad Exchanges is simple: broadcast the same reserve to all exchanges, and pick the highest bidding exchange that clears the reserve (we just need to know which exchange is the highest bidder, and whether they clear the reserve, not the exact value of the bid). If no exchange clears the reserve, allocate to the advertiser a with the lowest $SR(a)$.

4. While the algorithm is intuitive in hindsight, it is far from obvious that it obtains the optimal competitive ratio.

As mentioned earlier, apart from analyzing the joint yield optimization problem, we also show the benefits of surplus supply in classic online algorithmic problems. For the seminal online bipartite matching problem of [25], we can show that the same RANKING algorithm of [25] with a supply factor of f yields a tight competitive ratio of $f - fe^{-1/f}$, which increases with f and approaches 1 as $f \to \infty$. Likewise for the vertex-weighted generalization of this problem studied by [2], the same generalized vertex-weighted RANKING algorithm of [2] (a.k.a PERTURBED GREEDY) yields a competitive ratio of $f - fe^{-1/f}$.

Overview of Analysis Techniques. We use a max-min approach to analyze the performance of our algorithm. Given the thresholds $s_1 < \cdots < s_d$, our algorithm is completely defined. Therefore the adversary can compute the instance that minimizes the optimal objective of our algorithm given the thresholds, and the algorithm can optimize the thresholds $s_1 < \ldots s_d$ knowing the best response of the adversary. The minimization problem of the adversary can be captured by a succinct LP, and we reason about the structure of the optimal solution to this LP. This sets up the maximization problem of the algorithm, which turns out to be a non-linear, non-convex optimization problem. Nevertheless, we develop a simple poly-time dynamic programming algorithm that obtains the optimal solution (optimal thresholds $s_1, \ldots, s_d$) up to a small additive error. For tightness, we construct an example which is a modified version of the "upper triangular graph" of Karp et al. [25], and show that no algorithm can obtain an objective value larger than the objective value achieved as the optimal solution to the max-min problem described above. This establishes that the class of threshold-based algorithms is optimal. To act as a warm up to ease into the general distribution section, we begin with the special case of distributons with support size two. In this case, the maximization problem of the algorithm in the max-min problem above is a single-variable concave maximization problem, and already yields clear insights on how the optimal threshold computed by the algorithm depends on the supply factor f and the penalty c.

Extensions. A natural question to ask is what happens if the publishers have different under-delivery penalties c_a for different advertisers. To show a proof of concept extension of our results to this setting, we consider the simpler setting of our problem where the AdEx rewards are equal to r for every query (i.e., a deterministic distribution D), and show how the technique and results extend to handle different c_a's. We conjecture that the same approach extends to the general AdEx distributions as well, and leave it as an open problem. In a different direction, in this work, we focus on a deterministic algorithm because of its many virtues when deployed in a production system: the ability to replay and hence debug easily, ex-post fairness, etc. While we show that it achieves the optimal competitive ratio (i.e., even randomized algorithms cannot improve further), this necessarily requires n_a values being large (for a deterministic algorithm to be optimal, large budgets are necessary even for the much simpler B-matching problem [23]). In practice, however, large budget assumption essentially always holds, as advertiser contractual demands are much larger than the edge weight of 1. Nevertheless, one could ask whether one could use randomized algorithms to remove the dependence of n_a's being large. Again, as a proof of concept extension of our results, we show that for the special case where AdEx rewards equal to r for every query, randomized algorithms can get the same competitive ratio as deterministic ones for any value of n_a, not just large ones.

Related Notions of Surplus Supply. The first related concept is the *bid-to-budget ratio* notion used in several works (e.g. [11, 14]). On the surface level, it might appear that the notion of supply factor is just like the "large budgets" assumption, where it is assumed that the budget (in our case the number of impressions n_a demanded by each advertiser a) is much larger than the bid (i.e., the value of an edge). However these two concepts are quite different. In particular, even with the large budgets assumption, without a supply factor larger than 1, any algorithm will be very conservative and will essentially always allocate to the contracts (assuming the penalty is larger than the AdEx reward). The supply factor is a property of the entire setup of the publisher: the demands of the contracts and the nature of traffic (set of online nodes arriving, i.e., users/queries that visit their website).

Several work consider other notions of surplus supply. Karande et al. [24] show a competitive ratio of $1 - O(1/\sqrt{f})$ for the online matching problem when there are f *edge-disjoint matchings*. Note that, while related, this assumption is not equivalent to our definition of supply factor f that asks for f vertex disjoint matchings (disjoint on the online side). On the negative side, Cohen and Wajc [9] prove that no online matching algorithm can achieve a competitive ratio better than $1 - O(1/\sqrt{d})$ for d-regular graphs. This result does not violate our stated bounds since f-regular graphs only have f edge disjoint matchings, and this does not imply a supply factor of f. Naor and Wajc [31] consider assumptions on vertex degrees that can be seen as a *per vertex* surplus supply notion.

Comparison to Closely Related Work. In terms of works that consider joint optimization across the two channels, the closest to ours is that of Dvořák and Henzinger [15], who also consider the objective of maximizing revenue across two channels: the fundamental differences are (a) the absence of a supply factor in their work, (b) they model adversarially both the arrivals and the AdEx bids, and (c) they achieve separate approximation factors for each channel as opposed to our approximating the joint unified objective. Equally close is the work of Balseiro et al. [5], who study the same problem, with the differences being (a) the absence of a supply factor, (b) they model stochastically both the arrivals and AdEx bids.

Another closely related work is by Devanur and Jain [12] in which they consider the adwords problem with concave returns in the objective: while their model can capture penalties, it does not handle the AdEx reward distribution. Our model takes the reward distribution and penalties into account simultaneously. Additionally, the supply factor notion is absent in [12].

There are some results on bi-objective online allocation which targets two orthogonal objective functions, e.g. revenue and customer satisfaction. Korula et al. studies this problem targeting a weighted objective and a cardinality objective [26]. Aggarwal et al. studies this problem with a budgeted allocation objective and a cardinality objective [1]. Esfandiari et al. improves and extended these results to a more general setting [16].

A number of works consider the optimization problem without the presence of AdEx. Feldman et al. [20] study the problem with worst case arrivals and achieve a $1 - 1/e$ competitive ratio as the n_a's grow large. Feldman et al. [19] study the general packing LPs in a random permutation arrival model and show how to achieve $1 - \epsilon$ approximation as the n_a's grow large, and Devanur and

Hayes [11] study the related Adwords problem in the same random permutation model to achieve a $1 - \epsilon$ approximation.

Agrawal et al. [4] show how to attain $1 - \epsilon$ for general packing LPs with better convergence rates on how fast n_a's need to go to ∞. Devanur et al. [14] consider general packing and covering LPs in an i.i.d. model with unknown distribution and achieve even better convergence rates. Agrawal and Devanur [3] study online stochastic convex programming. Mirrokni et al. [30] study the Adwords problem and design algorithms that simultaneously perform well for both stochastic and adversarial settings, and Balseiro et al. [6] do this for generalized allocation problems with non-linear objectives using dual mirror descent. Variants of display ads problem without the large capacity assumption (and without the supply factor assumption) are also considered recently by [7,18,21,32]. We refer the reader to Choi et al. [8] for a literature review on the display ads market as it is too vast to cover in entirety here. The differentiating factors of all these works from ours is that even if these works were to add an AdEx node with infinite capacity, (a) they do not consider the supply factor, (b) and they do not have a unified objective.

Another related work by Esfandiari et al. [17] considers the allocation problem in a mixed setting, where a fraction of queries arriving are adversarial, and a fraction are stochastic. They then characterize their competitive ratio, by this *prediction fraction*. The setting we consider is different, as we allow fully adversarial queries. We only assume a known AdEx distribution, which we argued is often more predictable than the user traffic.

Karp et al. [25] wrote the seminal paper on online bipartite matching, and Aggarwal et al. [2] consider the generalization of it to vertex weighted settings. Mehta et al. [29] introduced the influential Adwords problem and gave a $1 - 1/e$ approximation for it, with a recent breakthrough result by Huang et al. [22] showing how to beat a $1/2$ approximation for this problem even with small budgets. Devanur et al. [13] give a randomized primal dual algorithm that gives a unified analysis of [2,25,29]. We refer the reader to [28] for a survey on the online matching literature.

2 Optimal Algorithm for Binary Ad Exchange Distribution

In this section, we consider a special case where the highest AdEx bid (referred to as AdEx reward often) of each query is drawn from a distribution D of support size two. We consider the general distribution in Sect. 3. We first provide an algorithm, and later show that this algorithm is optimal. Formally we consider the following setting:

Definition 1 (Binary reward distribution with parameters q and r). *We consider the setting where AdEx reward distribution D is 0 with probability q, and is r with probability $1 - q$.*

Without loss of generality we assume that the two support points are 0 and r, rather than r_1 and r_2 for $0 < r_1 < r_2$. This is because, in the latter case, we can subtract r_1 from each support point, and also from the penalty, and it yields the distribution in the format we need. Also, without loss of generality we assume that the support point r in the distribution is such that $r < c$ where c is the penalty. Note that if $r \geq c$, then clearly whenever the AdEx reward is r (i.e., non-zero), an optimal algorithm can always allocate the query to AdEx, so there is nothing to study here.

2.1 An Optimal Algorithm

Now we propose a simple greedy algorithm (we basically specialize Algorithm 1 for binary distributions), analyze its performance and establish its optimality. The analysis can be extended to the more general distributions of AdEx rewards, but with more involved techniques.

Algorithm 2 is our algorithm for binary reward distributions. Here, we compute an appropriate threshold s as a pre-processing step. At arrival of a query, let a be the available advertiser (i.e., an advertiser with an edge to the incoming vertex) with the lowest satisfaction ratio $SR(a)$. The algorithm allocates the impression to AdEx if and only if $SR(a) \geq s$ and the query has non-zero AdEx reward of r. I.e., the algorithm first greedily allocates queries to available advertisers that are furthest from being satisfied, no matter how large the AdEx weight of arriving queries. However, when the advertisers are satisfied to some extent (i.e., their $SR \geq s$), satisfying contracts becomes less of a priority, and AdEx is preferred when it offers non-zero reward.

ALGORITHM 2: Optimal algorithm for binary AdEx bid distribution

---

Input: Binary AdEx distribution with parameter q and r, penalty c, and supply factor f.

Preprocessing: Set the threshold $s = \max\left(0, 1 + fq\ln(1 - \frac{r}{c})\right)$ (see Proposition 2).

for *each query arriving online* **do**
    Let a be a matching advertiser with the lowest satisfaction ratio.
    if $SR(a) = 1$ **then**
       | Assign the impression to AdEx.
    end
    else if $SR(a) \geq s$ *and AdEx reward is* r **then**
       | Assign the impression to AdEx.
    end
    else
       | Assign the impression to a.
    end
end

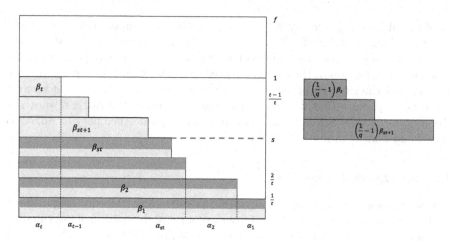

Fig. 1. Analysis of Algorithm 2: The figure can be viewed from the POV of a single advertiser, as well as from the POV of all advertisers. For a single advertiser a, the demand n_a is divided into t intervals. Red colored rectangles represent queries with AdEx reward r, and green rectangles represent queries with AdEx reward 0. The threshold is represented by s. The figure on the LHS is for allocation to the contract advertiser a, while the figure on the RHS is for allocation to AdEx. In the LHS figure, each horizontal rectangle of height $1/t$ represents a value of β_j. When smallest SR is below the threshold, all the queries (regardless of AdEx reward) are allocated to the contract, so each height $1/t$ rectangle below s is constituted by both red and green queries. Above the threshold s, all rectangles are only colored green since only queries with AdEx reward 0 are allocated to a contract. In the RHS figure, since AdEx gets allocated only queries of non-zero value, all rectangles are red. Also, for each query with AdEx reward 0 allocated to an advertiser with SR greater than s, in expectation $(1/q - 1)$ queries get allocated to AdEx (because in expectation for every query with AdEx reward 0, $1/q - 1$ queries have reward r). The α's in the bottom of the figure come into picture when all the advertisers are taken together. We will soon show that $\alpha_j - t(\beta_j - \beta_{j+1})$. Thus the bottom right corner piece rectangle in the LHS figure represents α_1/t etc. (Color figure online)

Before proving the competitive ratio, we set some notation that we use in our analysis throughout the paper. These concepts are also demonstrated in Fig. 1. Let t be a sufficiently large integer used to discretize the total demand of each advertiser into equal intervals of length $1/t$. The right picture to have in mind is $n_a \gg t \gg 1$. We call any given advertiser a to be of *type j*, if at the *end of the algorithm*, $SR(a) \in (\frac{j-1}{t}, \frac{j}{t}]$. For type 1 alone we let the SR interval be closed on both sides, namely $[0, \frac{1}{t}]$. Let A_j be the set[4] of advertisers of type j; and let $\alpha_j = \mathbb{E}[\sum_{a \in A_j} n_a]$ be the total demand of advertisers in A_j. For simplicity we assume that an advertiser $a \in A_j$ gets allocated exactly $\frac{j}{t}n_a$ impressions: this leads to an additive error $O(\frac{1}{t})$ in analysis, which is negligible

---

[4] Note that A_j is a random set depending on the realization of AdEx rewards over all queries.

when $t \to \infty$. Let β_j be the expected total number (across all advertisers) of allocated impressions s.t., at the time of allocation the assigned advertiser had satisfaction ratio $\in [\frac{j-1}{t}, \frac{j}{t})$. Finally, let $N = \sum_{a \in A} n_a$ be the total demand of all advertisers.

By definition of α, β we get the following (see also Fig. 1):

$$\beta_j = \sum_{a \in \cup_{\ell \geq j} A_\ell} \frac{1}{t} n_a = \frac{1}{t}(N - \sum_{\ell < j} \alpha_\ell). \tag{1}$$

Thus

$$\alpha_j = t(\beta_j - \beta_{j+1}). \tag{2}$$

Lemma 1. *Based on definition of α, β as described, for any $j \leq t - 1$,*

$$\sum_{\ell \leq j} f\alpha_\ell \leq \begin{cases} \sum_{\ell \leq st} \beta_\ell + \sum_{st < \ell \leq j} \frac{1}{q} \beta_\ell, & \text{if } j \geq st; \\ \sum_{\ell \leq j} \beta_\ell, & \text{if } j < st. \end{cases} \tag{3}$$

Proof. The RHS represents the set of queries that, when they arrived, the most deserving (lowest SR) contractual advertiser that was eligible was of type at most j. To see this note that when the lowest SR is $\frac{j}{t} < s$, every arriving query is allocated to the contract (hence the second line of RHS). When the lowest SR is at least s, only a q fraction of the considered queries are allocated to the contract—thus the considered queries = allocated queries / q, which is the first line of RHS.

The LHS represents the number of queries that were allocated to an advertiser of type at most j in the supply-factor-determining-offline-solution.

It is immediate that LHS is at most RHS because every query counted in the LHS will count for RHS when it arrives.

Notice that the total expected reward of the algorithm can be divided into the following parts:

- The baseline penalty is if no impression is allocated to contracts, the total such penalty is $-Nc$. The total AdEx reward that may be obtained by assigning everything to AdEx is $Nf(1-q)r$. The next points capture the change to the objective when we move away from this extreme solution of giving everything to AdEx.
- Any impression that is allocated to an advertiser with satisfaction ratio $\frac{j}{t} < s$ (which is the set of impressions counted in β_j for $j \leq st$), with probability $(1-q)$, loses a reward of r from AdEx. Thus in expectation each impression has reward $c - (1-q)r$ added to the objective;
- Each time an impression is allocated to an advertiser with satisfaction ratio $\frac{j}{t} \geq s$ (which is the set of impressions counted in β_j for $j > st$), the impression always has reward 0 for AdEx, but adds c to the objective.

Therefore the expected total reward ALG of the algorithm is

$$\text{ALG} = Nf(1-q)r - Nc + \sum_{j \le st}(c-(1-q)r)\beta_j + \sum_{st<j\le t} c\beta_j. \tag{4}$$

We can add (2) and (3) as constraints, to get a linear program that lower bounds the reward of the algorithm as follows:

$$\text{minimize} \quad Nf(1-q)r - Nc + \sum_{j \le st}(c-(1-q)r)\beta_j + \sum_{st<j\le t} c\beta_j \tag{5}$$

$$\text{s.t.} \quad ft\beta_1 - ft\beta_{j+1} \le \sum_{\ell \le j}\beta_\ell, \qquad\qquad \forall j, 1 \le j \le st;$$

$$ft\beta_1 - ft\beta_{j+1} \le \sum_{\ell \le st}\beta_\ell + \sum_{st<\ell\le j}\frac{1}{q}\beta_\ell, \qquad \forall j, st < j \le t;$$

$$\beta_1 = \frac{N}{t}; \; \beta_j \ge 0, \forall j, 1 \le j \le t.$$

The constraints are explained immediately by expanding and doing a telescopic summation using (2) and (3). We set $\beta_1 = N/t$ because in all but pathological instances we have that every advertiser ends up with at least $/1t$ fraction of their demand satisfied (note that t is large, just that $n_a \gg t \gg 1$). Even in the pathological instances where this is not true, i.e., only $\beta_1 < N/t$ holds, by setting $\beta_1 = N/t$, there is just a $O(1/t)$ additive error we have introduced. Namely, when proving optimality of our algorithm, we will just have proved it up to additive $O(1/t)$ terms. We first observe that the optimal solution of the LP (5) is achieved when all non-trivial constraints are tight.

Proposition 1. *By setting β values as follows we get an optimal solution to the linear program (5):*

$$\beta_j^* = \begin{cases} \frac{N}{t}\left(1 - \frac{1}{tf}\right)^{j-1}, & \text{if } j \le st+1; \\ \frac{N}{t}\left(1 - \frac{1}{tf}\right)^{st}\left(1 - \frac{1/q}{tf}\right)^{j-st-1}, & \text{if } j > st+1. \end{cases}$$

We can use the above observations on structure of ALG to compute the appropriate threshold in the following proposition:

Proposition 2. *The objective of the algorithm is maximized when the threshold is set to $s = \max\left(0, 1 + fq\ln(1 - \frac{r}{c})\right)$.*

Proof. Using Proposition 1 we have,

$$\text{ALG} \geq Nf(1-q)r - Nc + \sum_{j \leq st}(c - (1-q)r)\beta_j^* + \sum_{st < j \leq t} c\beta_j^*$$

$$= Nf(1-q)r - Nc + (c - (1-q)r)Nf\left(1 - \left(1 - \frac{1}{tf}\right)^{st}\right)$$

$$+ cqfN\left(1 - \frac{1}{tf}\right)^{st}\left(1 - \left(1 - \frac{1/q}{tf}\right)^{t-st}\right)$$

$$= Nf(1-q)r - Nc + (c - (1-q)r)Nf(1 - e^{-\frac{s}{f}}) + cqfNe^{-\frac{s}{f}}(1 - e^{-\frac{1-s}{qf}})$$

$$= Nc(f-1) + (1-q)(r-c)fNe^{-x} - qfNce^{\frac{1-q}{q}x - \frac{1}{qf}},$$

where $x = \frac{s}{f} \in [0, \frac{1}{f}]$. Then to maximize the reward, we consider the following expression in the right hand side:

$$RHS(x) = Nc(f-1) + (1-q)(r-c)fNe^{-x} - qfNce^{\frac{1-q}{q}x - \frac{1}{qf}}. \tag{6}$$

We have the unique zero point of $RHS'(x)$ is $x^* = q(\ln(1 - r/c) + \frac{1}{qf}) < 1/f$. This means that $RHS(x)$ is maximized either when $s^* = fx^*$ or $s^* = 0$.

Useful Insights. Interesting insights already flow out of this binary support distribution case. It shows that the optimal threshold s^* that we set is an affine function of the supply factor f. The higher the supply factor, lower the threshold we set (note that the coefficient of f in s^*, namely $q\ln(1 - r/c)$ is negative). Also, the dependence on the penalty c and AdEx reward r are quite non-trivial and intriguing. The binary support is often a good first-order approximation of reality when we bucket bids into "high" and "low" types.

2.2   Optimality of Algorithm 2

We now prove the optimality of the algorithm in the previous section by showing an example for which no algorithm can perform better. Consider a binary distribution with parameter q and r as defined earlier. We use a modification of the "upper triangular graph" instance of [25] as follows:

Example 1. Suppose that there are m advertisers, and each advertiser demands n impressions. There are $fmn = fN$ queries arriving in m groups $G_1, \cdots, G_m$, with queries in group G_i have an edge to the same $m-i+1$ advertisers determined as follows: consider a random permutation $\pi : [m] \rightarrow [m]$, then the queries in group G_i are available to advertisers j with $\pi(j) \geq i$.

At a high-level, in this instance, all advertisers are available to the first group of queries arriving. Then with each group one random advertiser is removed from the set of available advertisers to the group. We next argue that Algorithm 2 is optimal for this instance by showing that any online algorithm will not lead to a better reward.

Theorem 1. *For Example 1, the competitive ratio of any randomized online algorithm matches the competitive ratio obtained by Algorithm 2 up to a small additive factor $O\left(\frac{1}{m}\right)$.*

Proof. First we have the following observation about deterministic algorithms. By Yao's minimax principle, we only need to consider the performance of any deterministic algorithm over the randomness of the instance.

Fix any deterministic algorithm. Let q_{ij1} be the fraction of queries in G_i with AdEx reward 0 that is allocated to advertiser $\pi^{-1}(j)$, and q_{ij2} be the fraction of queries in G_i with AdEx reward r that is allocated to advertiser $\pi^{-1}(j)$. Then for $u = 1$ and 2,

$$\mathbb{E}_\pi[q_{iju}] \leq \begin{cases} \frac{1}{m-i+1}, & \text{if } j \geq i; \\ 0, & \text{if } j < i. \end{cases}$$

Also later we use $\mathbb{E}_\pi[q_{iju}] = \mathbb{E}_\pi[q_{imu}]$. This is because for each i, there are $m - i + 1$ random advertisers that have an edge connected to impressions in G_i. If $j \geq i$, then $\pi^{-1}(j)$ is a uniformly at random advertiser among this group of $m - i + 1$ advertisers. Thus $\mathbb{E}_\pi[q_{iju}] \leq \frac{1}{m-i+1}$ and for any $j, j' \geq i$ it holds $\mathbb{E}_\pi[q_{iju}] = \mathbb{E}_\pi[q_{ij'u}]$. If $j < i$, then advertiser $\pi^{-1}(j)$ does not have an edge to impressions in G_i. Then the expected reward we get from the algorithm, using the same reasoning from the previous section, is

$$-Nc + fN(1-q)r + \sum_{i=1}^{m} \sum_{j=i}^{m} \left(\frac{fN}{m} q\mathbb{E}_\pi[q_{ij1}]c + \frac{fN}{m}(1-q)\mathbb{E}_\pi[q_{ij2}](c-r) \right).$$

Here the first term and the second term are the total reward from not allocating anything to the contract advertisers, while the third term is the total reward gain from the allocation of the algorithm: there are in expectation $\frac{fN}{m}q$ queries with AdEx reward 0 (or $\frac{fN}{m}(1-q)$ with reward r) from group G_i and $\mathbb{E}_\pi[q_{ij1}]$ (or $\mathbb{E}_\pi[q_{ij2}]$) fraction of them are allocated to advertiser $\pi^{-1}(j)$, with each impression contributing to a reward gain c (or $c - r$) compared to being allocated to AdEx.

As we discussed $\mathbb{E}_\pi[q_{iju}] = \mathbb{E}_\pi[q_{imu}]$ for any $j \geq i, u = 1, 2$. Hence we can simplify the overall expectation for all $j \geq i$:

$$-cN + fN(1-q)r + \sum_{i=1}^{m}(m-i+1)\frac{fN}{m} \left(q\mathbb{E}_\pi[q_{im1}]c + (1-q)\mathbb{E}_\pi[q_{im2}](c-r) \right).$$

Then the reward of the algorithm is upper bounded by the solution of the following linear program, where y_{iu} variables represent the expected value $\mathbb{E}_\pi[q_{imu}]$.

$$\text{maximize} \quad -cN + fN(1-q)r + \frac{fN}{m}\sum_{i=1}^{m}(m-i+1)\left(qy_{i1}c + (1-q)y_{i2}(c-r)\right)$$

$$s.t. \quad \sum_{i=1}^{m}\left(\frac{fN}{m}qy_{i1} + \frac{fN}{m}(1-q)y_{i2}\right) \leq \frac{N}{m};$$

$$0 \leq y_{i1}, y_{i2} \leq \frac{1}{m-i+1}, \forall i, 1 \leq i \leq m. \tag{7}$$

Here the left hand side of the first constraint is the total expected number of allocated impressions to advertiser $\pi^{-1}(m)$, which is at most $n = \frac{N}{m}$.

Next, we show a structure on any optimal solution to this LP, that captures a threshold based behavior that we can be related to the algorithm we presented in the previous section. The proof is omitted.

Lemma 2. *For an optimal solution* $\mathbf{y}$ *to the above LP, there exists thresholds* $1 \le z_2 \le z_1 \le m$, *such that,* $y_{iu} = \frac{1}{m-i+1}$ *for* $i < z_u$, *and* $y_{iu} = 0$ *for* $i > z_u$ *for* $u = 1, 2$.

From the above lemma, we know that the optimal strategy for Example 1 has the following form: for queries in group $G_1, \cdots, G_{z_2}$, all impressions are allocated uniformly to all available advertisers; for queries in group $G_{z_2+1}, \cdots, G_{z_1}$, only queries with AdEx reward 0 are allocated uniformly to all available advertisers; for queries in group $G_{z_1+1}, \cdots, G_m$, no impression is allocated a contract.

By setting the $\mathbf{y}$ values, as determined by Lemma 2, we can simplify the objective function of linear program (7) with threshold z_1 and z_2 and bound the reward ALG obtained from an online algorithm as follows: the objective is

$$-cN + (1-q)r + \left(\sum_{i=1}^{z_2} (qc + (1-q)(c-r)) + \sum_{i=z_2+1}^{z_1} qc \right)$$

$$= -cN + (1-q)r + z_1 qc + z_2(1-q)(c-r)z_2.$$

Then we get,

$$\text{ALG} \le \max_{z_1, z_2} \quad -cN + (1-q)fNr + z_1 qc + z_2(1-q)(c-r)z_2 \qquad (8)$$

$$\text{s.t.} \sum_{i=1}^{z_2} f \cdot \frac{1}{m-i+1} + \sum_{i=z_2+1}^{z_1} f \cdot \frac{1}{m-i+1} q = 1.$$

When m is large enough, the constraint can be replaced by

$$f \ln \frac{m}{m-z_2} + fq \ln \frac{m-z_2}{m-z_1} = 1.$$

Let $x = \frac{m}{m-z_2} \in [0, \frac{1}{f}]$. We can express z_1 and z_2 by x as $z_1 = m(1 - e^{\frac{x(1-q)}{q} - \frac{1}{fq}})$ and $z_2 = m(1 - e^{-x})$. Apply these to (8) we have ALG upper bounded by

$$\max_{x \in [0, \frac{1}{f}]} -cN + (1-q)fNr + m(1 - e^{\frac{x(1-q)}{q} - \frac{1}{fq}})qc + m(1 - e^{-x})(1-q)(c-r)$$

$$= \max_{x \in [0, \frac{1}{f}]} Nc(f-1) + (1-q)(r-c)fNe^{-x} - qfNce^{\frac{1-q}{q}x - \frac{1}{qf}}.$$

Notice that the optimization problem here is identical to the optimization problem (6) that we described in the analysis of Algorithm 2. Thus the upper bound of the performance of any online algorithm for this instance matches the lower bound of the performance of Algorithm 2 for any underlying graph. As the optimal offline allocation has the same expected reward for any instance, we prove the optimality of Algorithm 2.

Understanding the Optimality of Algorithm 2. Algorithm 2 is optimal in the following sense. Although due to the complexity of the problem we did not give a closed-form competitive ratio of the algorithm, we are able to provide a lower bound of the net reward of the algorithm (characterized by (6)) on any instance with the same supply factor f, total demand of advertisers N, binary bid distribution parameterized by q and r, and penalty c, but with possibly different underlying online advertiser-impression bipartite graph. On the other hand, we show that for the same set of parameters, there exists an underlying graph such that no online algorithm can obtain reward more than (6) (plus some negligible terms). Thus the optimality of the algorithm can be viewed as a worst-case optimality over all possible online bipartite graphs.

3   General Ad Exchange Distribution

In this section, we briefly discuss how to generalize the setting with a binary AdEx distribution to the setting with arbitrary binary distribution, omitting the details. For a general AdEx reward distribution, we use a similar max-min approach as in Sect. 2. Although the max-min problem of the algorithm becomes multi-variate, non-linear and non-convex, for general AdEx distributions we can still establish that the non-linear mathematical programs obtained in the maximization problem of the algorithm and in the hard example are identical.

Suppose that each query has an AdEx reward drawn from a discrete distribution D with a fixed support size d:[5]

Definition 2 (AdEx distribution with parameters $(r_i, q_i)_{i \in [d]}$). *Consider an AdEx distribution D with support size d, rewards $0 = r_1 \leq r_2 \leq \dots \leq r_d \leq c$, where probability of that the reward is $r \leq r_i$ is q_i. Also we set $q_0 = 0, q_d = 1$.*

Our algorithm is presented in Algorithm 1 (see Sect. 1). For any query that arrives, if a is the advertiser the lowest satisfaction ratio, and $SR(a) \in [s_{u-1}, s_u)$, then the impression is allocated to a if and only if its AdEx reward $r \leq r_{d+1-u}$. Here we define $s_0 = 0$ for completeness. We propose Algorithm 1, a threshold-based algorithm in which a set of thresholds $s_1, \dots, s_d$ are chosen based on an optimization problem that takes D, f, c into account. For t being a large enough integer, as a generalization of Proposition 1 define β_j^* for every $j \in [t]$ as follows:

$$
\beta_j^* = \begin{cases}
\frac{N}{t}\left(1 - \frac{1/q_d}{tf}\right)^{j-1}, & \text{if } j \leq s_1 t + 1; \\[2ex]
\frac{N}{t}\left(1 - \frac{1/q_d}{tf}\right)^{s_1 t - s_0 t}\left(1 - \frac{1/q_{d-1}}{tf}\right)^{j - s_1 t - 1}, & \text{if } s_1 t + 1 < j \leq s_2 t + 1; \\[2ex]
\dots & \\[1ex]
\frac{N}{t}\left(1 - \frac{1/q_d}{tf}\right)^{s_1 t - s_0 t}\dots\left(1 - \frac{1/q_{d+2-u}}{tf}\right)^{s_{u-1}t - s_{u-2}t}\left(1 - \frac{1/q_{d+1-u}}{tf}\right)^{j - s_{u-1}t - 1}, & \\
& \text{if } s_{u-1} t + 1 < j \leq s_u t + 1; \\[2ex]
\dots & \\[1ex]
\frac{N}{t}\left(1 - \frac{1/q_d}{tf}\right)^{s_1 t - s_0 t}\dots\left(1 - \frac{1/q_{d+2-u}}{tf}\right)^{s_{d-1}t - s_{d-2}t}\left(1 - \frac{1/q_1}{tf}\right)^{j - s_{d-1}t - 1}, & \\
& \text{if } s_{d-1} t + 1 < j \leq s_d t = t.
\end{cases}
$$

---

[5] The assumption on a fixed support, can be relaxed using a standard discretization approach at a small cost in the competitive ratio that depends on this discretization.

We have the following optimization problem that decides the value of $s_1, \cdots, s_d$:

Optimization Problem 1. Given an AdEx distribution D with parameters $(r_i, q_i)_{i \in [d]}$, find $0 \leq s_1 \leq s_2 \leq \cdots \leq s_d = 1$ that maximizes the following objective such that β_j^* values satisfy the above constraints:

$$-cN + \sum_{u=1}^{d} fN(q_u - q_{u-1})r_u + \sum_{u=1}^{d} \sum_{j=s_{u-1}t+1}^{s_u t} \beta_j^*(c - \mathbb{E}_D[r|r \leq r_{d+1-u}]).$$

Theorem 2. *For any $f \geq 1$, and AdEx distribution with parameters $(r_i, q_i)_{i \in [d]}$, Algorithm 1 with thresholds determined by Optimization Problem 1 is optimal.*

References

1. Aggarwal, G., Cai, Y., Mehta, A., Pierrakos, G.: Biobjective online bipartite matching. In: Liu, T.-Y., Qi, Q., Ye, Y. (eds.) WINE 2014. LNCS, vol. 8877, pp. 218–231. Springer, Cham (2014). https://doi.org/10.1007/978-3-319-13129-0_16
2. Aggarwal, G., Goel, G., Karande, C., Mehta, A.: Online vertex-weighted bipartite matching and single-bid budgeted allocations. In: Randall, D. (ed.) Proceedings of the Twenty-Second Annual ACM-SIAM Symposium on Discrete Algorithms, SODA 2011, San Francisco, California, USA, 23–25 January 2011, pp. 1253–1264. SIAM (2011)
3. Agrawal, S., Devanur, N.R.: Fast algorithms for online stochastic convex programming. In: Indyk, P. (ed.) Proceedings of the Twenty-Sixth Annual ACM-SIAM Symposium on Discrete Algorithms, SODA 2015, San Diego, CA, USA, 4–6 January 2015, pp. 1405–1424. SIAM (2015)
4. Agrawal, S., Wang, Z., Ye, Y.: A dynamic near-optimal algorithm for online linear programming. Oper. Res. **62**(4), 876–890 (2014)
5. Balseiro, S.R., Feldman, J., Mirrokni, V.S., Muthukrishnan, S.: Yield optimization of display advertising with ad exchange. Manag. Sci. **60**(12), 2886–2907 (2014)
6. Balseiro, S.R., Lu, H., Mirrokni, V.S.: Dual mirror descent for online allocation problems. In: Proceedings of the 37th International Conference on Machine Learning, ICML 2020, July 13–18 2020, Virtual Event. Proceedings of Machine Learning Research, vol. 119, pp. 613–628. PMLR (2020)
7. Blanc, G., Charikar, M.: Multiway online correlated selection. In: 2021 IEEE 62nd Annual Symposium on Foundations of Computer Science (FOCS), pp. 1277–1284. IEEE (2022)
8. Choi, H., Mela, C.F., Balseiro, S.R., Leary, A.: Online display advertising markets: a literature review and future directions. Inf. Syst. Res. **31**(2), 556–575 (2020)
9. Cohen, I.R., Wajc, D.: Randomized online matching in regular graphs. In: Proceedings of the Twenty-Ninth Annual ACM-SIAM Symposium on Discrete Algorithms, pp. 960–979. SIAM (2018)
10. Derakhshan, M., Golrezaei, N., Leme, R.P.: Lp-based approximation for personalized reserve prices. In: Karlin, A., Immorlica, N., Johari, R. (eds.) Proceedings of the 2019 ACM Conference on Economics and Computation, EC 2019, Phoenix, AZ, USA, 24–28 June 2019, p. 589. ACM (2019)

11. Devanur, N.R., Hayes, T.P.: The adwords problem: online keyword matching with budgeted bidders under random permutations. In: Chuang, J., Fortnow, L., Pu, P. (eds.) Proceedings 10th ACM Conference on Electronic Commerce (EC-2009), Stanford, California, USA, 6–10 July 2009, pp. 71–78. ACM (2009)
12. Devanur, N.R., Jain, K.: Online matching with concave returns. In: Proceedings of the Forty-Fourth Annual ACM Symposium on Theory of Computing, pp. 137–144 (2012)
13. Devanur, N.R., Jain, K., Kleinberg, R.D.: Randomized primal-dual analysis of RANKING for online bipartite matching. In: Proceedings of the Twenty-Fourth Annual ACM-SIAM Symposium on Discrete Algorithms, SODA 2013, New Orleans, Louisiana, USA, 6–8 January 2013, pp. 101–107. SIAM (2013)
14. Devanur, N.R., Jain, K., Sivan, B., Wilkens, C.A.: Near optimal online algorithms and fast approximation algorithms for resource allocation problems. J. ACM **66**(1), 7:1-7:41 (2019)
15. Dvoák, W., Henzinger, M.: Online ad assignment with an ad exchange. In: Bampis, E., Svensson, O. (eds.) WAOA 2014. LNCS, vol. 8952, pp. 156–167. Springer, Cham (2015). https://doi.org/10.1007/978-3-319-18263-6_14
16. Esfandiari, H., Korula, N., Mirrokni, V.: Bi-objective online matching and submodular allocations. Adv. Neural. Inf. Process. Syst. **29**, 2739–2747 (2016)
17. Esfandiari, H., Korula, N., Mirrokni, V.: Allocation with traffic spikes: mixing adversarial and stochastic models. ACM Trans. Econ. Comput. (TEAC) **6**(3–4), 1–23 (2018)
18. Fahrbach, M., Huang, Z., Tao, R., Zadimoghaddam, M.: Edge-weighted online bipartite matching. In: 2020 IEEE 61st Annual Symposium on Foundations of Computer Science (FOCS), pp. 412–423. IEEE (2020)
19. Feldman, J., Henzinger, M., Korula, N., Mirrokni, V.S., Stein, C.: Online stochastic packing applied to display ad allocation. In: de Berg, M., Meyer, U. (eds.) ESA 2010. LNCS, vol. 6346, pp. 182–194. Springer, Heidelberg (2010). https://doi.org/10.1007/978-3-642-15775-2_16
20. Feldman, J., Korula, N., Mirrokni, V., Muthukrishnan, S., Pál, M.: Online ad assignment with free disposal. In: Leonardi, S. (ed.) WINE 2009. LNCS, vol. 5929, pp. 374–385. Springer, Heidelberg (2009). https://doi.org/10.1007/978-3-642-10841-9_34
21. Gao, R., He, Z., Huang, Z., Nie, Z., Yuan, B., Zhong, Y.: Improved online correlated selection. In: 2021 IEEE 62nd Annual Symposium on Foundations of Computer Science (FOCS), pp. 1265–1276. IEEE (2022)
22. Huang, Z., Zhang, Q., Zhang, Y.: Adwords in a panorama. In: 61st IEEE Annual Symposium on Foundations of Computer Science, FOCS 2020, Durham, NC, USA, 16–19 November 2020, pp. 1416–1426. IEEE (2020)
23. Kalyanasundaram, B., Pruhs, K.: An optimal deterministic algorithm for online b-matching. Theor. Comput. Sci. **233**(1–2), 319–325 (2000)
24. Karande, C., Mehta, A., Tripathi, P.: Online bipartite matching with unknown distributions. In: Proceedings of the Forty-Third Annual ACM Symposium on Theory of Computing, pp. 587–596 (2011)
25. Karp, R.M., Vazirani, U.V., Vazirani, V.V.: An optimal algorithm for on-line bipartite matching. In: Proceedings of the Twenty-Second Annual ACM Symposium on Theory of Computing, pp. 352–358 (1990)
26. Korula, N., Mirrokni, V.S., Zadimoghaddam, M.: Bicriteria online matching: maximizing weight and cardinality. In: Chen, Y., Immorlica, N. (eds.) WINE 2013. LNCS, vol. 8289, pp. 305–318. Springer, Heidelberg (2013). https://doi.org/10.1007/978-3-642-45046-4_25

27. Leme, R.P., Pál, M., Vassilvitskii, S.: A field guide to personalized reserve prices. In: Bourdeau, J., Hendler, J., Nkambou, R., Horrocks, I., Zhao, B.Y. (eds.) Proceedings of the 25th International Conference on World Wide Web, WWW 2016, Montreal, Canada, 11–15 April 2016, pp. 1093–1102. ACM (2016)
28. Mehta, A.: Online matching and ad allocation. Found. Trends Theor. Comput. Sci. 8(4), 265–368 (2013)
29. Mehta, A., Saberi, A., Vazirani, U., Vazirani, V.: Adwords and generalized online matching. Journal of the ACM (JACM) 54(5), 22-es (2007)
30. Mirrokni, V.S., Gharan, S.O., Zadimoghaddam, M.: Simultaneous approximations for adversarial and stochastic online budgeted allocation. In: Proceedings of the Twenty-Third Annual ACM-SIAM Symposium on Discrete Algorithms, SODA 2012, Kyoto, Japan, January 17–19, 2012. pp. 1690–1701. SIAM (2012)
31. Naor, J., Wajc, D.: Near-optimum online ad allocation for targeted advertising. ACM Transactions on Economics and Computation (TEAC) 6(3–4), 1–20 (2018)
32. Shin, Y., An, H.C.: Making three out of two: Three-way online correlated selection. arXiv preprint arXiv:2107.02605 (2021)

Online Ad Allocation in Bounded-Degree Graphs

Susanne Albers and Sebastian Schubert[✉]

Department of Computer Science, Technical University of Munich, Boltzmannstr. 3, 85748 Garching, Germany
albers@in.tum.de, sebastian.schubert@tum.de

Abstract. We study the AdWords problem defined by Mehta, Saberi, Vazirani and Vazirani [10]. A search engine company has a set of advertisers who wish to link ads to user queries and issue respective bids. The goal is to assign advertisers to queries so as to maximize the total revenue accrued. The problem can be formulated as a matching problem in a bipartite graph G. We assume that G is a (k, d)-graph, introduced by Naor and Wajc [11]. Such graphs model natural properties on the degrees of advertisers and queries.

As a main result we present a deterministic online algorithm that achieves an optimal competitive ratio. The competitiveness tends to 1, for arbitrary $k \geq d$, using the standard small-bids assumption where the advertisers' bids are small compared to their budgets. Hence, remarkably, nearly optimal ad allocations can be computed deterministically based on structural properties of the input. So far competitive ratios close to 1, for the AdWords problem, were only known in probabilistic input models.

Keywords: Online ad allocation · Targeted advertising · Deterministic algorithm · AdWords problem

1 Introduction

Ad allocation in Internet advertising and in particular the AdWords problem have received considerable research interest over the past years, see e.g. [3–7, 10, 11] and references therein. The worldwide digital ad spending has reached nearly half a trillion US\$ in 2021 [1]. Search ads are the dominant advertising format with a share of over 40%. Scientifically, ad allocation incorporates challenging assignment problems that are relevant in more general auction design.

We investigate the AdWords problem, which was introduced by Mehta et al. [10]. A search engine company, such as Google, Yahoo or Microsoft Bing, has a set of advertisers who wish to link ads to users of the search engine as they enter search queries. The information on the query keywords allows for highly targeted ad assignments. Formally, the setting can be modeled as an online matching problem in a bipartite graph $G = (A, Q, E)$. The vertex set A represents the set of advertisers. These vertices are known in advance. Each

© The Author(s), under exclusive license to Springer Nature Switzerland AG 2022
K. A. Hansen et al. (Eds.): WINE 2022, LNCS 13778, pp. 60–77, 2022.
https://doi.org/10.1007/978-3-031-22832-2_4

advertiser $a \in A$ has a daily budget $B(a)$ he can spend. Furthermore, he specifies a set $K(a)$ of relevant keywords. There is a bid $w_{a,k}$ for each $k \in K(a)$. In manual bidding, these bids are determined by the advertiser; in automated bidding they are set by the advertising platform. The vertices of Q are the individual queries that are entered by the users. These vertices arrive online, one by one. Whenever a new query $i \in Q$ with a keyword k arrives, the advertising platform checks which advertisers $a \in A$ are interested in presenting an ad to the query, i.e. for which advertisers $k \in K(a)$. For any such advertiser a, the graph G contains an edge $(a, i) \in E$ of weight $w_{a,i} = w_{a,k}$. The advertising platform must decide immediately which advertiser a to assign to query i, for a revenue of $w_{a,i}$. For simplicity it is assumed that at most one advertiser is assigned to a query. The goal is to maximize the total revenue accrued by the platform at the end of the day.

We study the AdWords problem in bipartite graphs $G = (A, Q, E)$ that satisfy natural degree properties. Such properties were first identified and formalized by Naor and Wajc [11]. Obviously, each advertiser $a \in A$ has a high degree: During a day a great number of queries are entered at the search engine, and the advertiser sets up a campaign that reaches a reasonable fraction of the user population. However, his budget is not high enough to present an ad to all interesting queries, i.e. $\sum_{(a,i)\in E} w_{a,i} >> B(a)$. On the other hand, each query in Q has a relatively small degree: It is of interest to a smaller number of advertisers, within the huge pool A of advertisers a search engine company has. As an example, consider a typical query that may address a particular sports discipline (or travel to a specific location, or a given consumer product). The number of advertisers that sell suitable equipment (or tickets/services, or the requested product) is small, compared to the total number $|A|$ of advertisers. The set of advertisers covers essentially all aspects of daily and commercial life.

We assume that G is a bipartite (k, d)-graph, as defined by Naor and Wajc [11] for the AdWords problem. Here k and d are positive integers. (1) Each advertiser $a \in A$ has a total bid volume (total weighted degree) satisfying $\sum_{(a,i)\in E} w_{a,i} \geq k \cdot B(a)$. Since bids are small compared to the budget, i.e. $w_{a,i} << B(a)$, the last inequality implies a very high degree for a. (2) Each query $i \in Q$ has a degree $d(i) \leq d$. This captures the fact that a limited number of advertisers can offer a suitable ad. In practice $k \geq d$. We assume that $d \geq 2$. If $d = 1$, then a GREEDY algorithm constructs an optimal assignment.

The AdWords problem generalizes the classical online bipartite matching problem, introduced by Karp et al. [8], where each vertex in A can be matched only once and the goal is to maximize the cardinality of the constructed matching.

We analyze the performance of online algorithms for the AdWords problem using competitive analysis. Given an input graph G, let $\text{ALG}(G)$ be the total revenue (weight) of the assignment constructed by an online algorithm ALG. Let $\text{OPT}(G)$ be the corresponding value of an optimal offline algorithm OPT. Algorithm ALG is c-competitive if $\text{ALG}(G) \geq c \cdot \text{OPT}(G)$ holds, for all G. In our analyses we will focus on bipartite (k, d)-graphs G.

Related Work. As mentioned above, the AdWords problem was formally defined by Mehta et al. [10]. They presented a deterministic online algorithm that assigns an incoming query i to the advertiser a who maximizes $w_{a,i} \cdot e^{-(1-T(a))}$, where $T(a)$ is the fraction of the advertiser's budget that has been spent so far. The algorithm achieves a competitive ratio of $1 - 1/e \approx 0.632$, using the standard *small-bids assumption* where the bids are small compared to the advertisers' budgets. More precisely, let $R_{\max} = \max_{(a,i) \in E} \frac{w_{a,i}}{B(a)}$ be the maximum ratio between the bid of any advertiser and his budget. With the small-bids assumption, $R_{\max}$ tends to 0. Mehta et al. also showed that no randomized online algorithm can obtain a competitive ratio greater than $1 - 1/e$. Hence their algorithm achieves an optimal competitiveness.

Buchbinder et al. [4] developed a primal-dual algorithm for AdWords that attains a competitive ratio of $(1-1/c)(1-R_{\max})$, where $c = (1+R_{\max})^{1/R_{\max}}$. As $R_{\max} \to 0$, the competitiveness tends to $1-1/e$. Buchbinder et al. also examined a setting where the degree of each incoming query is upper bounded by d and gave an algorithm with a competitive ratio of nearly $1 - (1 - 1/d)^d$. Azar et al. [3] showed that this ratio is best possible, also for randomized algorithms. The expression $1 - (1 - 1/d)^d$ is always greater than $1 - 1/e$ but approaches the latter value as d increases.

The class of (k, d)-graphs was defined by Naor and Wajc [11], who studied online bipartite matching and the AdWords problem. They devised deterministic online algorithms that achieve a competitive ratio of $1 - (1 - 1/d)^k$. This ratio holds for online bipartite matching and the vertex-weighted extension where all edges incident to a vertex $a \in A$ have the same weight. Furthermore, the ratio holds for AdWords with equal bids per advertiser, i.e. each advertiser has the same bid to all his relevant keywords. The ratio of $1 - (1 - 1/d)^k$ is best possible for online bipartite matching and the vertex-weighted extension if $k \geq d$. For AdWords with arbitrary bids, Naor and Wajc gave an algorithm with a competitive ratio of $(1 - R_{\max})(1 - (1 - 1/d)^k)$. They also showed an upper bound of $(1 - R_{\max})(1 - (1 - 1/d)^{k/R_{\max}})$ on the best possible competitiveness if $k \geq d$. For increasing k/d, the expression $1 - (1 - 1/d)^k$ tends to 1. For $k \approx d$ increasing, it tends again to $1 - 1/e$.

In our recent work [2] we studied the online b-matching problem in (k, d)-graphs. Each vertex a on the left-hand side of the bipartite graph is a server with a capacity of b_a, meaning that it may be matched with up to b_a incoming requests that arrive as right-hand side vertices. The goal is to maximize the cardinality of the constructed matching. We developed deterministic online algorithms that achieves an optimal competitiveness.

For the Adwords problem without the small-bids assumption, a Greedy algorithm attains a competitive ratio of $1/2$ [9]. This ratio was recently improved to 0.506 [7].

All of the above results hold in the adversarial input model where the query vertices Q and their arrival order are determined by an adversary. The AdWords problem has also been examined in stochastic input models. In the random-order model, a random permutation of the vertices of Q arrives. Alternatively, the vertices of Q are drawn i.i.d. from a known or unknown distribution. For AdWords in the random-arrival order, Devanur and Hayes [5] developed a $(1-\varepsilon)$-competitive algorithm, for any $\varepsilon > 0$. The algorithm needs to know (approximately) the number $|Q|$ of queries. Furthermore, the maximum bid of any advertiser must be upper bounded by roughly $\varepsilon^3/|A|^2$ times the revenue of an optimal solution. For AdWords in the unknown i.i.d. model, Devanur et al. [6] gave an algorithm with a competitiveness of $1 - O(\sqrt{R_{\max}})$.

Our Contributions. We investigate the AdWords problem in (k, d)-graphs. For each advertiser $a \in A$, arbitrary bids for the relevant keywords in $K(a)$ are allowed.

In Sect. 2 we present a deterministic online algorithm, called ALLOCATION. At the heart of the algorithm is a continuous function V with two arguments. The first argument considers budget that has already been spent by a given advertiser. The second argument keeps track of the total bid volume that has been issued by him so far. For technical reasons, the arguments have to be scaled down. The ad assignment works as follows. At any time while queries of Q are processed and for each advertiser $a \in A$, ALLOCATION maintains a value V_a. Whenever a new query $i \in Q$ arrives, bids are issued by the advertisers a who find the corresponding query interesting and their total issued bid volume increases. This results in updates of V_a. The query is assigned to the advertiser a who maximizes the increase in V_a, multiplied by $B(a)$.

We formulate and analyze ALLOCATION as a primal-dual algorithm. We prove that it achieves a competitive ratio of $(1 - R_{\max})c^*$, where

$$c^* := 1 - \frac{1}{b} \left(\sum_{i=1}^{b} i \binom{kb}{b-i} \frac{1}{(d-1)^{b-i}} \right) \left(1 - \frac{1}{d} \right)^{kb}$$

and $b := \lfloor 1/R_{\max} \rfloor$. The ratio of c^* is best possible. No deterministic online algorithm can achieve a higher competitive ratio, even if all advertisers issue uniform bids equal to 1 for all their relevant keywords. This follows from our work [2] on the b-matching problem, which can be viewed as a special AdWords problem in which all advertisers have bids equal to 1 for all their keywords. The competitive factor of c^* is complex but exact in all terms. In [2] we also show that c^* tends to 1 as b tends to infinity, for arbitrary $k \geq d$. Consequently, the competitive ratio of ALLOCATION tends to 1, for all $k \geq d$, with the small-bids assumption as $R_{\max}$ tends to 0. The ALLOCATION algorithm generalizes our algorithm for the b-matching problem [2] but the extension is non-trivial, for bids of arbitrary value.

A major technical contribution of this paper is the construction of the function V that guides the assignment decisions. Section 3 is devoted to it. A crucial aspect is that V must be continuous in both arguments because bids of arbitrary value must be handled, in terms of budgets spent and bid volumes issued. On a high-level, V is a linear combination of data points with integral arguments. In fact, V induces a grid of parallelograms whose vertices are function values with integral arguments. Bidding operations and ad assignments have the property that the new function values are still in the same parallelogram on in a neighboring one to the right or above. Even this seemingly simple "interpolation" of the function V is hard to analyze. We show how to define V and prove properties that are essential to establish an optimal competitiveness in the primal-dual analysis of our algorithm.

A strength of our results is that ALLOCATION achieves a competitiveness arbitrarily close of 1, for all $k \geq d$, using the small-bids assumption. Hence nearly optimal allocations can be computed deterministically and online based on natural structural properties of the input. Recall that without degree constraints, the best competitive ratio is $1 - 1/e \approx 0.632$, again with the small-bids assumption. So far, competitive factors close to 1 for AdWords were only known in probabilistic input models, based on small bids. Our results improve upon the previous best factor of $(1 - R_{\max})(1 - (1 - 1/d)^k)$ by Naor and Wajc [11] and settle open questions raised by them.

2   An Algorithm for AdWords

In this section, we present the online algorithm ALLOCATION for the AdWords problem. It makes use of an upper bound on $R_{\max}$. Every bid of advertiser a is upper bounded by $\max_{k \in K(a)} w_{a,k}$. Hence, by slightly abusing notation we now set $R_{\max} := \max_{a \in A, k \in K(a)} \frac{w_{a,k}}{B(a)}$. The latter value can be computed easily, given $B(a)$ and the bids for the keywords in $K(a)$, for any $a \in A$, because this information is available in advance. In automated bidding one can assume that bids are updated once a day.

Let $b := \lfloor 1/R_{\max} \rfloor$. We will show that ALLOCATION achieves a competitive ratio of $(1 - R_{\max})c^*$, where again

$$c^* := 1 - \frac{1}{b} \left(\sum_{i=1}^{b} i \binom{kb}{b-i} \frac{1}{(d-1)^{b-i}} \right) \left(1 - \frac{1}{d} \right)^{kb}.$$

Our algorithm uses the standard convention that allows more queries to be assigned to an advertiser a than he can pay for with his budget $B(a)$. However, the actual revenue generated by a is upper bounded by $B(a)$. This is reasonable, as advertisers typically do not mind if their ads are shown to more users for free. At the end of this section, we discuss the necessary changes to ALLOCATION if this is not permitted. It results in a slightly worse competitiveness, which still tends to 1 under the small-bids assumption.

ALLOCATION is a generalization of the primal-dual algorithm WEIGHTEDASSIGNMENT [2]. It maintains a carefully chosen value for each advertiser a, which depends on the sum of all the bids for queries assigned to a (in the following load l_a) and the sum of all the bids made by a (in the following degree δ_a), so far. The primal and dual (fractional) linear program modelling the AdWords problem are given below. The primal program uses a variable $m(a, i)$ for each edge $(a, i) \in E$. It denotes what fraction of query i is assigned to advertiser a.

$$\textbf{P: max} \sum_{(a,i)\in E} m(a, i) \cdot w_{a,i}$$

$$\text{s.t.} \sum_{i:(a,i)\in E} m(a, i) \cdot w_{a,i} \leq B(a), \ (\forall a \in A)$$

$$\sum_{a:(a,i)\in E} m(a, i) \leq 1, \ (\forall i \in Q)$$

$$m(a, i) \geq 0, \ (\forall (a, i) \in E)$$

$$\textbf{D: min} \sum_{a\in A} B(a) \cdot x(a) + \sum_{i\in Q} y(i)$$

$$\text{s.t.} \ w_{a,i} \cdot x(a) + y(i) \geq w_{a,i}, \ (\forall (a, i) \in E)$$

$$x(a), y(i) \geq 0, \ (\forall a \in A, \forall i \in Q)$$

The algorithm ALLOCATION is detailed below. It uses the value $w_a := B(a)/b$, which can be seen as the maximum bid that the advertiser a may make. For now, consider the function V as a black box. We only state the properties of V crucial for the analysis in this section. A detailed definition of V and the proofs for the stated properties are given in Sect. 3.

Algorithm 1: ALLOCATION

---

1 Initialize $x(a) = 0$, $y(i) = 0$ and $m(a, i) = 0$, $\forall a \in A$ and $\forall i \in Q$;
2 **while** *a new query $i \in Q$ arrives* **do**
3   Let $N(i)$ denote the set of neighbors a of i with remaining budget;
4   **if** $N(i) = \emptyset$ **then**
5 | Do not assign i;
6   **else**
7   Assign i to $a :=$
             $$\arg\max \left\{ B(a') \cdot \left(V\left(\tfrac{l_{a'}}{w_{a'}}, \tfrac{\delta_{a'}+w_{a',i}}{w_{a'}}\right) - V\left(\tfrac{l_{a'}}{w_{a'}}, \tfrac{\delta_{a'}}{w_{a'}}\right) \right) : a' \in N(i) \right\};$$
8   Update $m(a, i) \leftarrow 1$;
9   Set $x(a) \leftarrow V\left(\tfrac{l_a + w_{a,i}}{w_a}, \tfrac{\delta_a + w_{a,i}}{w_a}\right)$;
10   **forall** $a' \neq a \in N(i)$ **do**
11 | Set $x(a') \leftarrow V\left(\tfrac{l_{a'}}{w_{a'}}, \tfrac{\delta_{a'}+w_{a',i}}{w_{a'}}\right)$;
12   **end**
13   **end**
14 **end**

As we shall see, the function V has three parameters: b, k and d, which are all positive integers. We have $b = \lfloor 1/R_{\max} \rfloor$, while k and d denote the parameters of the underlying (k, d)-graph. The crucial properties of V are then as follows.

(a) It holds that $V(0,0) = 0$.

(b) For all $l \in [0, \infty)$, $\delta \in [l, \infty)$ and $w \in [0, 1]$, it holds that

$$V(l + w, \delta + w) - V(l, \delta) + (d - 1) \cdot \left(V(l, \delta + w) - V(l, \delta) \right) \leq \frac{w}{b \cdot c^*}.$$

(c) It holds that $V(l, \delta) = 1$, if $l \geq b$ or $\delta \geq k \cdot b$.

With these properties, we can show the competitiveness of ALLOCATION in three steps.

1. Let ΔP and ΔD denote the increase in the value of the primal and dual solution of any iteration, respectively. We show

$$\frac{\Delta P}{\Delta D} \geq c^*.$$

2. The algorithm produces a feasible dual solution.
3. The algorithm produces an almost feasible primal solution.

Proof of 1. Note that the dual variables are only updated by ALLOCATION if a query i is assigned to an advertiser a. In this case, the increase in the primal solution is equal to $\Delta P = w_{a,i}$, since $m(a, i)$ is set to one. Afterwards, the dual variables are updated in Lines 9 and 11. Property a) implies that $x(a) = V(l_a/w_a, \delta_a/w_a)$ is true initially. It is then easy to verify that this invariant is maintained until a has no budget left. This means that we have a total increase in the value of the dual solution of

$$\Delta D = B(a) \cdot \left(V\left(\frac{l_a + w_{a,i}}{w_a}, \frac{\delta_a + w_{a,i}}{w_a} \right) - V\left(\frac{l_a}{w_a}, \frac{\delta_a}{w_a} \right) \right)$$
$$+ \sum_{a' \in N(i) \setminus \{a\}} B(a') \cdot \left(V\left(\frac{l_{a'}}{w_{a'}}, \frac{\delta_{a'} + w_{a',i}}{w_{a'}} \right) - V\left(\frac{l_{a'}}{w_{a'}}, \frac{\delta_{a'}}{w_{a'}} \right) \right).$$

Recall that $|N(i)| \leq d$. Since the algorithm chose a in Line 7, we can upper bound ΔD by

$$\Delta D \leq B(a) \cdot \left(V\left(\frac{l_a + w_{a,i}}{w_a}, \frac{\delta_a + w_{a,i}}{w_a} \right) - V\left(\frac{l_a}{w_a}, \frac{\delta_a}{w_a} \right) \right.$$
$$\left. + (d - 1) \cdot \left(V\left(\frac{l_a}{w_a}, \frac{\delta_a + w_{a,i}}{w_a} \right) - V\left(\frac{l_a}{w_a}, \frac{\delta_a}{w_a} \right) \right) \right).$$

We now use Property (b), applied with $l = l_a/w_a$, $\delta = \delta_a/w_a$ and $w = w_{a,i}/w_a$. Note that $w \in [0, 1]$ is always true, since w_a is an upper bound for $w_{a,i}$. We obtain

$$\Delta D \leq B(a) \cdot \frac{w_{a,i}}{w_a} \frac{1}{b \cdot c^*} = \frac{w_{a,i}}{c^*},$$

where the last step is true because $B(a)/w_a = b$, by definition. This finishes the proof.   □

Proof of 2. Observe that $x(a) = 1$ for all advertisers $a \in A$ by the end of the algorithm implies dual feasibility. As mentioned before, ALLOCATION maintains that $x(a) = V(l_a/w_a, \delta_a/w_a)$, as long as a still has remaining budget. Hence, if an advertiser a still has remaining budget by the end of the algorithm, we have $\delta_a \geq k \cdot B(a)$, by definition of a (k, d)-graph. This implies $\delta_a/w_a \geq k \cdot b$. Property (c) then immediately yields $x(a) = 1$.

Now, consider the case that a has no budget remaining in the end. Let δ'_a denote the value of δ_a when a had no budget remaining for the first time. When the last query was assigned to a, $x(a)$ was updated to $V(l_a/w_a, \delta'_a/w_a)$. It holds that $l_a \geq B(a)$ and thus $l_a/w_a \geq b$. Property (c) again implies $x(a) = 1$.   □

Proof of 3. The algorithm stops assigning queries to an advertiser a once a has no budget left. Hence, we have

$$\sum_{i \in Q} m(a, i) \cdot w_{a,i} \leq B(a) + \max_{i \in Q} w_{a,i}.$$

In these cases, the actual amount of money paid by the advertiser is $B(a)$. Let ALG denote the total revenue generated in the assignment created by ALLOCATION. It follows that

$$\text{ALG} \geq P \cdot \min_{a \in A} \left\{ \frac{B(a)}{B(a) + \max_{i \in Q} w_{a,i}} \right\} = P \cdot \frac{1}{1 + R_{\max}} \geq P \cdot (1 - R_{\max}).$$

□

Theorem 1. *Let $b = \lfloor 1/R_{\max} \rfloor$. ALLOCATION achieves a competitive ratio of $(1 - R_{\max})c^*$ for the AdWords problem on (k, d)-graphs.*

Proof. Let ALG and OPT denote the total amount of revenue accrued in the assignment created by ALLOCATION and the optimal offline algorithm, respectively. Step 2 and weak duality implies OPT $\leq D$. Step 3 gives ALG $\geq P \cdot (1 - R_{\max})$. Applying step 1 at every iteration of the algorithm further yields $P/D \geq c^*$. Overall, we can conclude

$$\frac{\text{ALG}}{\text{OPT}} \geq (1 - R_{\max})\frac{P}{D} \geq (1 - R_{\max})c^*.$$

□

As mentioned at the start of this section, ALLOCATION is allowed to assign more queries to an advertiser than he can pay for with his budget. If this is not permitted, we have to change Line 3 such that the set $N(i)$ also excludes advertisers a where $w_{a,i}$ exceeds the remaining budget of a. Furthermore, we redefine l_a and δ_a such that they now exclude bids that were ignored for this reason. We denote the resulting algorithm by ALLOCATIONR. At every step of this algorithm, it still holds that $\Delta P/\Delta D \geq c^*$. Moreover, ALLOCATIONR now always produces a feasible primal solution. However, it may create an infeasible dual solution, since we cannot guarantee $\delta_a \geq k \cdot B(s)$ anymore. If $l_a \leq B(a) - w_a$ by the end of the algorithm, then a never made a bid bigger than his remaining budget. Hence, we have $\delta_a \geq k \cdot B(s)$ and thus $x(a) = 1$ in this case. However, if $l_a > B(a) - w_a$, then bids of a may be neglected by the algorithm. In the worst case, $x(a)$ may not be increased any further once we have $l_a = \delta_a = B(a) - w_a + \varepsilon$. This results in a lower bound of $x(a) > V(b-1, b-1)$ for all $a \in A$. It is possible to show that

$$\alpha := V(b-1, b-1) = 1 - \left(\frac{1}{b \cdot c^*}\right)\left(1 - \frac{1}{d}\right)^{(k-1)b+1} < (1 - R_{\max}).$$

Nonetheless, observe that α tends to 1 as b tends to infinity, which is equivalent to $R_{\max}$ tending to 0.

Again, let ALGR and OPT denote the total amount of revenue generated in the assignment created by ALLOCATIONR and the optimal offline algorithm, respectively. If we were to multiply the value of each dual variable by $(1/\alpha)$, we would obtain a feasible dual solution. This implies $\text{OPT} \leq (1/\alpha)D$, by weak duality. Moreover, we have $\text{ALGR} = P$ and $P/D \geq c^*$. Overall, we obtain

$$\frac{\text{ALGR}}{\text{OPT}} \geq \alpha \cdot \frac{P}{D} \geq \alpha \cdot c^*.$$

Theorem 2. ALLOCATIONR *achieves a competitive ratio of $\alpha \cdot c^*$ for the AdWords problem on (k, d)-graphs.*

3   The Function V

In this section, we give the definition of $V(l, \delta)$, for all $l \in [0, \infty)$ and $\delta \in [l, \infty)$. Moreover, we show that it attains the desired properties. We start by adopting the definition of V from [2]. There, $V(l, \delta)$ (with three parameters b, k and d) is defined for $l \in \{0, 1, \ldots, b\}$ and $\delta \in \{l, l+1, \ldots, kb\}$. This definition satisfies

$$p(l, \delta) + (d-1) \cdot q(l, \delta) = \frac{1}{b \cdot c^*}, \tag{1}$$

for all $l \in \{0, 1, \ldots, b-1\}$ and $\delta \in \{l, l+1, \ldots, kb-1\}$, where

$$p(l, \delta) := V(l+1, \delta+1) - V(l, \delta),$$
$$q(l, \delta) := V(l, \delta+1) - V(l, \delta).$$

Moreover, it holds that $V(0, 0) = 0$ and $V(l, \delta) = 1$, if $l = b$ or $\delta = kb$. The exact definition of $V(l, \delta)$ is complex and not needed for the extension in the following. Nevertheless, we want to mention it here for the sake of completeness. It holds that

$$V(l, l) := \frac{1}{b \cdot c^*}\left(\sum_{i=1}^{b-l} i \binom{kb-l}{b-l-i} \frac{1}{(d-1)^{b-l-i}}\right)\left(1 - \frac{1}{d}\right)^{kb-l} + 1 - \frac{b-l}{b \cdot c^*}$$

and

$$V(l, \delta) := \sum_{i=l}^{b-1}(-1)^{i-l} \frac{1}{(d-1)^{i-l}}\binom{\delta-l}{i-l}\left(\frac{d}{d-1}\right)^{\delta-i}\left(V(l, l) + \frac{b-i}{b \cdot c^*} - 1\right)$$

$$+ 1 - \frac{b-l}{b \cdot c^*}.$$

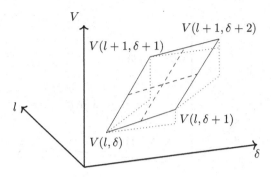

Fig. 1. Definition of V on the continuous domain inside a parallelogram. V is defined by a linear combination of the values of V at the corners. The dashed lines indicate the linear functions that are fit in (2) and (3).

We start extending this definition by setting $V(l,\delta) := 1$, if $l \geq b$ or $\delta \geq kb$. By that, V already satisfies Properties a) and c) from Sect. 2. For Property b) to hold, we need to fill the remaining gaps carefully. It is done by a technique that is reminiscent of linear interpolation. The details are given in the following. Subsequently, we show that the resulting function V has the desired property in Lemma 1.

Consider the parallelogram with corners (l,δ), $(l,\delta+1)$, $(l+1,\delta+1)$ and $(l+1,\delta+2)$, where $l \in \{0,1,\ldots,b-1\}$ and $\delta \in \{l,l+1,\ldots,kb-2\}$. We extend V along the edges of said parallelogram by fitting linear functions between the values of V at these corners. This means, we define

$$V(l,\delta+x) := V(l,\delta) + x \cdot (V(l,\delta+1) - V(l,\delta)) = V(l,\delta) + x \cdot q(l,\delta),$$

and

$$V(l+y,\delta+y) := V(l,\delta) + y \cdot (V(l+1,\delta+1) - V(l,\delta)) = V(l,\delta) + y \cdot p(l,\delta),$$

for all $x,y \in [0,1]$. We use the same idea to define V for all the points inside this parallelogram, i.e. we fit a linear function between $V(l+x,\delta+x)$ and $V(l+x,\delta+x+1)$, for all $x \in [0,1]$ (see Fig. 1). This results in

$$\begin{aligned}
V(l+x,\delta+x+y) &:= V(l+x,\delta+x) + y\big(V(l+x,\delta+x+1) - V(l+x,\delta+x)\big) \\
&= V(l,\delta) + xp(l,\delta) \\
&\quad + y\big(V(l,\delta+1) + xp(l,\delta+1) - V(l,\delta) - xp(l,\delta)\big) \\
&= V(l,\delta) + xp(l,\delta) + y\big(q(l,\delta) + xp(l,\delta+1) - xp(l,\delta)\big) \\
&= V(l,\delta) + yq(l,\delta) + x\big((1-y)p(l,\delta) + yp(l,\delta+1)\big).
\end{aligned}$$

$$(2)$$

Alternatively, we could have also decided to fit a linear function between $V(l,\delta+y)$ and $V(l+1,\delta+y+1)$. This would result in

$$V(l+x,\delta+x+y) := V(l,\delta) + xp(l,\delta) + y\big((1-x)q(l,\delta) + xq(l+1,\delta+1)\big).$$

$$(3)$$

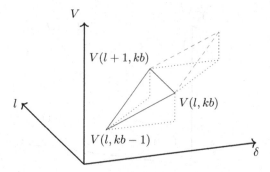

Fig. 2. The definition of V on the continuous domain inside a triangle. The triangle is extended to a parallelogram using dummy definitions, indicated in red. Then, V is again defined by a linear combination of the values at the corners. (Color figure online)

We want to note that these two definitions are identical, since

$$y \cdot q(l, \delta) + x \cdot \big((1 - y)p(l, \delta) + yp(l, \delta + 1)\big)$$
$$= y \cdot q(l, \delta) + x \cdot (l, \delta) - xy \cdot \big(p(l, \delta) - p(l, \delta + 1)\big)$$
$$= y \cdot q(l, \delta) + x \cdot p(l, \delta) - xy \cdot \big(q(l, \delta) - q(l + 1, \delta + 1)\big)$$
$$= x \cdot p(l, \delta) + y \cdot \big((1 - x)q(l, \delta) + xq(l + 1, \delta + 1)\big),$$

where the second equality is true because

$$p(l, \delta) + q(l + 1, \delta + 1) = V(l, \delta + 2) - V(l, \delta) = q(l, \delta) + p(l, \delta + 1).$$

Interestingly, Definition (2) reveals the gradient of the linear function that is fit between $V(l, \delta + y)$ and $V(l + 1, \delta + y + 1)$. It is the convex combination of $p(l, \delta)$ and $p(l, \delta + 1)$ with coefficients $(1 - y)$ and y. Similarly, Definition (3) yields that the gradient of the line between $V(l + x, \delta + x)$ and $V(l + x, \delta + x + 1)$ is the convex combination of $q(l, \delta)$ and $q(l + 1, \delta + 1)$ with coefficients $(1 - x)$ and x. For this reason, we will use Definition (2) to quantify the difference in V if both l and δ are updated. On the other hand, Definition (3) will be useful for determining the difference in V if only δ changes and l remains the same.

Note that we have not defined V inside the triangles with corners $(l, kb - 1)$, (l, kb) and $(l + 1, kb)$, for $l \in \{0, 1, \ldots, b - 1\}$. There, we want to extend V by fitting a plane through the values of V at the three corners. For this, we introduce the dummy definitions $p(l, kb) := p(l, kb - 1)$ and $q(l + 1, kb) := q(l, kb - 1)$. This extends the triangle to a parallelogram, such that the dummy value $V(l + 1, kb + 1)$ is embedded in the desired plane (see Fig. 2). Hence, we can use the definitions above for the points within the triangle.

We want to show that these dummy definitions also satisfy (1), since this property is crucial for the subsequent proofs. Since it holds that $V(l, kb) = 1$, we have $p(l, kb - 1) = 1 - V(l, kb - 1) = q(l, kb - 1)$, for all $l \in \{0, \ldots, b - 1\}$. Plugging this into (1) yields $p(l, kb - 1) = q(l, kb - 1) = 1/(bc^*d)$. This implies that

$$p(l, kb) + (d-1) \cdot q(l, kb) = \frac{1}{bc^*d} + (d-1) \cdot \frac{1}{bc^*d} = \frac{1}{b \cdot c^*},$$

where $l \in \{0, 1, \ldots, b-1\}$. Here, we additionally define $q(0, kb) := 1/(bc^*d)$ for the sake of completeness.

Lemma 1. *For all $l \in [0, \infty)$, $\delta \in [l, \infty)$ and $w \in [0, 1]$, it holds that*

$$V(l+w, \delta+w) - V(l, \delta) + (d-1) \cdot \big(V(l, \delta+w) - V(l, \delta)\big) \leq \frac{w}{b \cdot c^*}. \quad (4)$$

Before we can prove Lemma 1, we need the following monotonicity lemmas.

Lemma 2. *For all $l \in \{0, 1, \ldots, b-1\}$ and $\delta \in \{l+1, l+2, \ldots, kb\}$, it holds that*

$$p(l, \delta) \leq p(l, \delta-1).$$

Proof. We want to start by mentioning the connection

$$p(l, \delta) \gtreqless p(l', \delta') \iff q(l, \delta) \gtreqless q(l', \delta'), \quad (5)$$

which holds for all $l, l' \in \{0, 1, \ldots, b-1\}$ and $\delta, \delta' \in \{l, l+1, \ldots, kb\}$. It is a direct implication of (1), since

$$p(l, \delta) + (d-1) \cdot q(l, \delta) = \frac{1}{b \cdot c^*} = p(l', \delta') + (d-1) \cdot q(l, \delta')$$

$$\Rightarrow \quad p(l, \delta) = p(l', \delta') + (d-1) \cdot (q(l', \delta') - q(l, \delta)).$$

Recall that we defined $p(l, kb) := p(l, kb-1)$, for all $l \in \{0, 1, \ldots, b-1\}$, which already implies the lemma for $\delta = kb$. For all $\delta < kb$, we will prove the lemma by showing $q(l, \delta) \geq q(l, \delta-1)$, which is identical to $p(l, \delta) \leq p(l, \delta-1)$ by (5).

For $l = b-1$, one can show by induction over δ that

$$V(b-1, \delta) = 1 - \frac{1}{bc^*d} \sum_{i=0}^{kb-\delta-1} \left(\frac{d-1}{d}\right)^i.$$

Hence, it holds that

$$q(b-1, \delta) = \frac{1}{bc^*d} \left(\frac{d-1}{d}\right)^{kb-\delta-1} \geq \frac{1}{bc^*d} \left(\frac{d-1}{d}\right)^{kb-\delta} = q(b-1, \delta-1).$$

So far we have shown that $q(b-1, \delta) \geq q(b-1, \delta-1)$ holds for all δ. We will finish the proof inductively. Consider a load level $l \leq b-2$. For all $\delta \in \{l+1, l+2, \ldots, kb-2\}$, the induction hypothesis states that $q(l+1, \delta+1) \geq q(l+1, \delta)$. Now, consider $q(l, \delta-1)$ and suppose for the sake of contradiction that $q(l, \delta-1) < q(l+1, \delta)$. Plugging this into

$$p(l, \delta-1) + q(l+1, \delta) = V(l+1, \delta+1) - V(l, \delta-1) = q(l, \delta-1) + p(l, \delta) \quad (6)$$

yields $p(l, \delta) > p(l, \delta - 1)$. By (5), it follows that $q(l, \delta) < q(l, \delta - 1)$. Together with our assumption and the induction hypothesis, we conclude that $q(l, \delta) < q(l + 1, \delta + 1)$. To summarize, we showed that $q(l, \delta - 1) < q(l + 1, \delta)$ implies $q(l, \delta) < q(l+1, \delta+1)$. Thus, we can apply these arguments repeatedly to obtain $q(l, kb - 2) < q(l + 1, kb - 1)$. However, we can show that this is false, yielding a contradiction. Therefore, our assumption does not hold, implying $q(l, \delta - 1) \geq q(l + 1, \delta)$. Plugging this into (6) and then using (5) gives $q(l, \delta) \geq q(l, \delta - 1)$.

We finish the proof by justifying our claim. We show that $q(l, kb - 2) = q(l+1, kb-1)$ holds if $l \leq b-2$. As mentioned before, $V(l+1, kb) = V(l+2, kb) = 1$ implies that $p(l+1, kb-1) = q(l+1, kb-1)$. Using this in (1) gives $q(l+1, kb-1) = 1/(bc^*d)$. We can use the exact same arguments to show $q(l, kb - 1) = 1/(bc^*d)$, as also $V(l, kb) = 1$. This means that $V(l + 1, kb - 1) = V(l, kb - 1)$ holds. This further implies

$$p(l, kb-2) = V(l+1, kb-1) - V(l, kb-2) = V(l, kb-1) - V(l, kb-2) = q(l, kb-2).$$

Plugging this into (1) again yields $q(l, kb - 2) = 1/(bc^*d)$, finishing the proof. $\square$

Lemma 3. *For all* $l \in \{1, 2, \ldots, b\}$ *and* $\delta \in \{l, l + 1, \ldots, kb\}$, *it holds that*

$$q(l, \delta) \leq q(l - 1, \delta - 1).$$

Proof. First, recall that for $\delta = kb$, we defined $q(l, kb) = q(l - 1, kb - 1)$. Hence, this lemma also trivially holds for $\delta = kb$. For $\delta < kb$, consider the parallelogram with lower left corner $V(l - 1, \delta - 1)$. Lemma 2 implies that $p(l - 1, \delta - 1) \geq p(l - 1, \delta)$. Since $q(l - 1, \delta - 1) + p(l - 1, \delta) = p(l - 1, \delta - 1) + q(l, \delta)$, we get $q(l, \delta) \leq q(l - 1, \delta - 1)$. $\square$

Proof of Lemma 1. At first, we assume that $l + w \leq b$ and $\delta + w \leq kb$. This means that $(l + w, \delta + w)$, $(l, \delta + w)$ and (l, δ) are all contained in one of the previously mentioned parallelograms. Thus, the value of V at these points is defined using (2) or (3). We do a case distinction over the different possible locations of $(l + w, \delta + w)$ and $(l, \delta + w)$ with respect to (l, δ). Since $w \leq 1$, $(l + w, \delta + w)$ is either in the same parallelogram as or in the one to the top of the parallelogram containing (l, δ). Similarly, $(l, \delta + w)$ is either in the same parallelogram as or in the one to the right of (l, δ). Thus, we consider the following four cases.

Case 1. Both $(l + w, \delta + w)$ and $(l, \delta + w)$ are in the same parallelogram as (l, δ). Let (l', δ') be the lower left corner of this parallelogram. Further, define $y := l - l'$, $x := \delta - y - \delta'$, $p = V(l+w, \delta+w) - V(l, \delta)$ and $q = V(l, \delta+w) - V(l, \delta)$ (see Fig. 3).

Using Definition (2), we obtain

$$p = w \cdot ((1-x)p(l', \delta') + xp(l', \delta'+1)) = w \cdot p(l', \delta') + wx \cdot (p(l', \delta'+1) - p(l', \delta')).$$

Lemma 2 implies that $p(l', \delta + 1) \leq p(l', \delta')$. Hence, we get $p \leq w \cdot p(l', \delta')$. Similarly, by using Definition (3), we have

$$q = w \cdot ((1-y)q(l', \delta') + yq(l'+1, \delta'+1)) = w \cdot q(l', \delta') + wy \cdot (q(l'+1, \delta'+1) - q(l', \delta')),$$

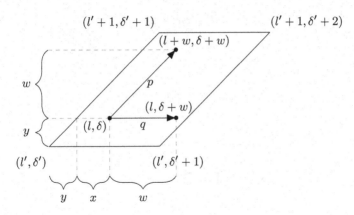

Fig. 3. Case 1. All three points are in the same parallelogram.

which we can upper bound by $w \cdot q(l', \delta')$ with Lemma 3. Overall, we have

$$b \cdot (p + (d-1) \cdot q) \leq w \cdot b \cdot (p(l', \delta') + (d-1) \cdot q(l', \delta')) = \frac{w}{c^*},$$

where the last equality is true according to (1).

Case 2. $(l+w, \delta+w)$ is in the parallelogram to the top of the one containing both (l, δ) and $(l, \delta+w)$. Let (l', δ') be the lower left corner of the bottom parallelogram. We define again $y := l - l'$, $x := \delta - y - \delta'$ and $q = V(l, \delta+w) - V(l, \delta)$. Moreover, we set $w_1 := 1 - y$ and $w_2 := w - w_1$. It holds that $w_2 > 0$. Let $p_1 = V(l+w_1, \delta+w_1) - V(l, \delta)$ and $p_2 = V(l+w, \delta+w) - V(l+w_1, \delta+w_1)$ (see Fig. 4).

We obtain

$$p_1 = w_1 \cdot ((1-x)p(l', \delta') + xp(l', \delta'+1)),$$
$$p_2 = w_2 \cdot ((1-x)p(l'+1, \delta'+1) + xp(l'+1, \delta'+2)),$$
$$q = w \cdot ((1-y)q(l', \delta') + yq(l'+1, \delta'+1)).$$

Similarly to Case 1, we can upper bound p_1 and p_2 with the help of Lemma 2 by

$$p_1 \leq w_1 \cdot p(l', \delta') \quad \text{and} \quad p_2 \leq w_2 \cdot p(l'+1, \delta'+1).$$

In order to upper bound q, we use $w = w_1 + w_2$ and $y = 1 - w_1$

$$
\begin{aligned}
q &= (w_1 + w_2) \cdot q(l', \delta') \\
&\quad + (w_1 + w_2 - w_1^2 - w_1 w_2) \cdot (q(l'+1, \delta'+1) - q(l', \delta')) \\
&= w_1 \cdot q(l', \delta') + w_2 \cdot q(l'+1, \delta'+1) \\
&\quad + (w_1 - w_1^2 - w_1 w_2) \cdot (q(l'+1, \delta'+1) - q(l', \delta')) \\
&= w_1 \cdot q(l', \delta') + w_2 \cdot q(l'+1, \delta'+1) \\
&\quad + w_1(1 - (w_1 + w_2)) \cdot (q(l'+1, \delta'+1) - q(l', \delta')) \\
&\leq w_1 \cdot q(l', \delta') + w_2 \cdot q(l'+1, \delta'+1),
\end{aligned}
$$

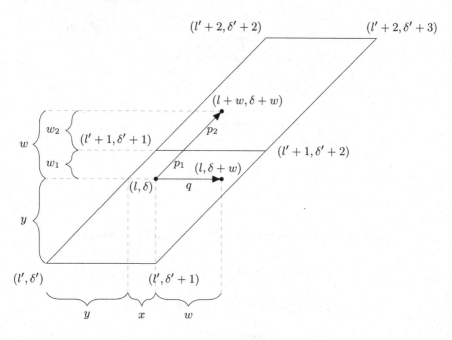

Fig. 4. Case 2. The points are contained by two parallelograms on top of each other.

where the last inequality follows from Lemma 3 and $w_1(1 - (w_1 + w_2)) \geq 0$ as $w_1 + w_2 = w \leq 1$. Thus, we overall get

$$
\begin{aligned}
b \cdot (p_1 + p_2 + (d-1) \cdot q) &\leq w_1 \cdot b \cdot (p(l', \delta') + (d-1) \cdot q(l', \delta')) \\
&\quad + w_2 \cdot b \cdot (p(l'+1, \delta'+1) + (d-1) \cdot q(l'+1, \delta'+1)) \\
&= \frac{w_1 + w_2}{c^*} = \frac{w}{c^*}.
\end{aligned}
$$

Case 3. $(l, \delta + w)$ is in the parallelogram to the right of the one containing both (l, δ) and $(l + w, \delta + w)$. This case is analogous to case 2.

Case 4. The three points are in three different parallelograms. Let (l', δ') be the lower left corner of the parallelogram containing (l, δ). Let $y := l - l'$, $x := \delta - y - \delta'$, $w_{l1} := 1 - y$ and $w_{l2} := w - w_{l1}$. Moreover, we also have here $w_{\delta 1} = 1 - x$ and $w_{\delta 2} = w - w_{\delta 1}$. It holds that $w_{l2} > 0$ and $w_{\delta 2} > 0$. Let $p_1 = V(l + w_{l1}, \delta + w_{l1}) - V(l, \delta)$, $p_2 = V(l + w, \delta + w) - V(l + w_{l1}, \delta + w_{l1})$, $q_1 = V(l, \delta + w_{\delta 1}) - V(l, \delta)$ and $q_2 = V(l, \delta + w) - V(l, \delta + w_{\delta 1})$ (see Fig. 5).

It holds that

$$
\begin{aligned}
p_1 &= w_{l1} \cdot ((1-x)p(l', \delta') + xp(l', \delta'+1)), \\
p_2 &= w_{l2} \cdot ((1-x)p(l'+1, \delta'+1) + xp(l'+1, \delta'+2)), \\
q_1 &= w_{\delta 1} \cdot ((1-y)q(l', \delta') + yq(l'+1, \delta'+1)), \\
q_2 &= w_{\delta 2} \cdot ((1-y)q(l', \delta'+1) + yq(l'+1, \delta'+2)).
\end{aligned}
$$

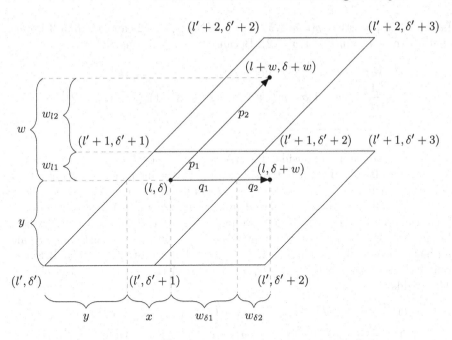

Fig. 5. Case 4. All three points are in different parallelograms.

First, we substitute $w_{l1} = 1 - y$ and $w_{\delta 1} = 1 - x$ to obtain

$$p_1 = (1 - x)(1 - y)p(l', \delta') + x(1 - y)p(l', \delta' + 1)),$$
$$q_1 = (1 - x)(1 - y)q(l', \delta') + (1 - x)yq(l' + 1, \delta' + 1)).$$

Next, we use $w_{l2} = y + w - 1$ and $w_{\delta 2} = x + w - 1$

$$p_2 = (1 - x)yp(l' + 1, \delta' + 1) + xyp(l' + 1, \delta' + 2))$$
$$+ (w - 1) \cdot ((1 - x)p(l' + 1, \delta' + 1) + xp(l' + 1, \delta' + 2)),$$
$$q_2 = x(1 - y)q(l' + 1, \delta' + 1) + xyq(l' + 1, \delta' + 2))$$
$$+ (w - 1) \cdot ((1 - y)q(l' + 1, \delta' + 1) + yq(l' + 1, \delta' + 2)).$$

Observe that the factors (excluding those with $(w - 1)$) before corresponding p and q values are identical. Moreover, they sum up to one, i.e.

$$xy + x(1 - y) + (1 - x)y + (1 - x)(1 - y) = 1.$$

Hence, we get

$$b \cdot \big(p_1 + p_2 + (d - 1) \cdot (q_1 + q_2)\big)$$
$$= \frac{1}{c^*} + (w - 1)b \cdot \big[(1 - x)p(l' + 1, \delta' + 1) + xp(l' + 1, \delta' + 2)$$
$$+ (d - 1) \cdot ((1 - y)q(l' + 1, \delta' + 1) + yq(l' + 1, \delta' + 2))\big].$$

Lemma 2 implies that $p(l' + 1, \delta' + 1) \geq p(l' + 1, \delta' + 2)$ and Lemma 3 implies that $q(l', \delta' + 1) \geq q(l' + 1, \delta' + 2)$. Because $w - 1 \leq 0$, this yields

$$b \cdot \big(p_1 + p_2 + (d - 1) \cdot (q_1 + q_2)\big)$$
$$\leq \frac{1}{c^*} + (w - 1)b \cdot \big[p(l' + 1, \delta' + 2) + (d - 1)q(l' + 1, \delta' + 2)\big]$$
$$= \frac{1}{c^*} + \frac{w - 1}{c^*} = \frac{w}{c^*}.$$

At last, we want to consider the case $l + w > b$ or $\delta + w > kb$. If $l > b$ or $\delta > kb$, the left side of (4) is 0, trivially satisfying the inequality. Hence, we focus on the case where $l \in (b - w, b]$ and $\delta \leq kb$ or $l \leq b$ and $\delta \in (kb - w, kb]$. Then not all the relevant values of V are defined using (2) and (3). However, once we leave the area of these definitions, we know that V is defined as 1 everywhere. Let $w_l = \min\{w, b - l\}$ and $w_\delta = \min\{w, kb - \delta\}$. Note that at least one of w_l and w_δ has to be strictly less than w. If $w_\delta \leq w_l$, it holds that $\delta + w_\delta = kb$. It follows that $V(l + w, \delta + w) = 1 = V(l + w_\delta, \delta + w_\delta)$ and $V(l, \delta + w) = 1 = V(l, \delta + w_\delta)$. This yields

$$b \cdot \big(V(l + w, \delta + w) - V(l, \delta) + (d - 1) \cdot (V(l, \delta + w) - V(l, \delta))\big)$$
$$= b \cdot \big(V(l + w_\delta, \delta + w_\delta) - V(l, \delta) + (d - 1) \cdot \big(V(l, \delta + w_\delta) - V(l, \delta)\big)\big)$$
$$\leq \frac{w_\delta}{c^*} \leq \frac{w}{c^*}.$$

The first inequality is true since $l + w_\delta \leq b$ and $\delta + w_\delta = kb$, meaning that we can use the calculations of the case distinctions.

On the other hand, if $w_l < w_\delta$, it holds that $l + w_l = b$. This means that $V(l + w, \delta + w) = 1 = V(l + w_l, \delta + w_l)$ and $V(l, \delta + w) = V(l, \delta + w_\delta)$. Moreover, let $m := \lfloor w_\delta / w_l \rfloor$. We partition the increase $V(l, \delta + w_\delta) - V(l, \delta)$ into $(m + 1)$ parts, i.e. m parts of length w_l and one part of length $w_\delta \bmod w_l$. Then, the idea is to add the necessary terms such that we can apply the results from the case distinctions above to all $m + 1$ parts. More, precisely

$$b \cdot \big(V(l + w, \delta + w) - V(l, \delta) + (d - 1) \cdot (V(l, \delta + w) - V(l, \delta))\big)$$
$$\leq b \cdot \sum_{i=1}^{m} \Big(V(l + w_l, \delta + i w_l) - V(l, \delta + (i - 1)w_l)$$
$$+ (d - 1) \cdot \big(V(l, \delta + i w_l) - V(l, \delta + (i - 1)w_l)\big)\Big)$$
$$+ b \cdot \Big(V(l + (w_\delta \bmod w_l), \delta + w_\delta) - V(l, \delta + m w_l)$$
$$+ (d - 1) \cdot \big(V(l, \delta + w_\delta) - V(l, \delta + m w_l)\big)\Big)$$
$$\leq \sum_{i=1}^{m} \frac{w_l}{c} + \frac{w_\delta \bmod w_l}{c} = \frac{w_\delta}{c} \leq \frac{w}{c}.$$

For the first inequality to hold, we have to show that $V(l + w_l, \delta + i \cdot w_l) - V(l, \delta + (i - 1) \cdot w_l)$, for $2 \leq i \leq m$, as well as $V(l + (w_\delta \bmod w_l), \delta + w_\delta) - V(l, \delta + m \cdot w_l)$

are non-negative. Using Definition (2), we can see that such a term could only be negative if there was a pair of l' and δ' such that $p(l', \delta') < 0$. However, by repeatedly applying Lemma 2, we would obtain $0 > p(l', \delta') \geq p(l', kb)$. As shown previously, we have $p(l', kb) = 1/(bc^*d) > 0$, which is a contradiction. □

References

1. https://www.emarketer.com/content/worldwide-digital-ad-spending-year-end-update
2. Albers, S., Schubert, S.: Tight bounds for online matching in bounded-degree graphs with vertex capacities. In: Proceedings of 30th Annual European Symposium on Algorithms (ESA). LIPIcs, Leibniz International Proceedings in Informatics (2022). http://arxiv.org/abs/2206.15336
3. Azar, Y., Cohen, I., Roytman, A.: Online lower bounds via duality. In: Proceedings of 28th Annual ACM-SIAM Symposium on Discrete Algorithms (SODA), pp. 1038–1050. SIAM (2017)
4. Buchbinder, N., Jain, K., Naor, J.S.: Online primal-dual algorithms for maximizing ad-auctions revenue. In: Arge, L., Hoffmann, M., Welzl, E. (eds.) ESA 2007. LNCS, vol. 4698, pp. 253–264. Springer, Heidelberg (2007). https://doi.org/10.1007/978-3-540-75520-3_24
5. Devanur, N., Hayes, T.: The adwords problem: online keyword matching with budgeted bidders under random permutations. In: Proceedings of 10th ACM Conference on Electronic Commerce (EC), pp. 71–78. ACM (2009)
6. Devanur, N., Sivan, B., Azar, Y.: Asymptotically optimal algorithm for stochastic adwords. In: Proceedings of 13th ACM Conference on Electronic Commerce (EC), pp. 388–404. ACM (2012)
7. Huang, Z., Zhang, Q., Zhang, Y.: Adwords in a panorama. In: Proceedings of the 61st IEEE Annual Symposium on Foundations of Computer Science (FOCS), pp. 1416–1426 (2020)
8. Karp, R., Vazirani, U., Vazirani, V.: An optimal algorithm for on-line bipartite matching. In: Proceedings of 22nd Annual ACM Symposium on Theory of Computing (STOC), pp. 352–358 (1990)
9. Lehmann, B., Lehmann, D., Nisan, N.: Combinatorial auctions with decreasing marginal utilities. Games Econ. Behav. **55**(2), 270–296 (2006)
10. Mehta, A., Saberi, A., Vazirani, U., Vazirani, V.: Adwords and generalized online matching. J. ACM **54**(5), 22 (2007)
11. Naor, J., Wajc, D.: Near-optimum online ad allocation for targeted advertising. ACM Trans. Economics and Comput. **6**(3–4), 16:1–16:20 (2018)

Online Team Formation Under Different Synergies

Matthew Eichhorn[1]($\boxtimes$), Siddhartha Banerjee[1], and David Kempe[2]

[1] Cornell University, Ithaca, NY 14850, USA
{meichhorn,sbanerjee}@cornell.edu
[2] University of Southern California, Los Angeles, CA 90089, USA

Abstract. Team formation is ubiquitous in many sectors: education, labor markets, sports, etc. A team's success depends on its members' latent types, which are not directly observable but can be (partially) inferred from past performances. From the viewpoint of a principal trying to select teams, this leads to a natural exploration-exploitation trade-off: retain successful teams that are discovered early, or reassign agents to learn more about their types? We study a natural model for online team formation, where a principal repeatedly partitions a group of agents into teams. Agents have binary latent types, each team comprises two members, and a team's performance is a symmetric function of its members' types. Over multiple rounds, the principal selects matchings over agents and incurs regret equal to the deficit in the number of successful teams versus the optimal matching for the given function. Our work provides a complete characterization of the regret landscape for all symmetric functions of two binary inputs. In particular, we develop team-selection policies that, despite being agnostic of model parameters, achieve optimal or near-optimal regret against an adaptive adversary.

Keywords: Online team formation · Regret · Combinatorial bandits

1 Introduction

An instructor teaching a large online course wants to pair up students for assignments. The instructor knows that a team performs well as long as at least one of its members has some past experience with coding, but unfortunately, there is no available information on the students' prior experience. However, the course staff can observe the *performance* of each team on assignments, and so, over multiple assignments, would like to reshuffle teams to try and quickly maximize the overall number of successful teams. How well can one do in such a situation?

Team formation is ubiquitous across many domains: homework groups in large courses, workers assigned to projects on online labor platforms, police officers paired up for patrols, athletes assigned to teams, etc. Such teams must often be formed without prior information on each individual's latent skills or

Extended version with proofs can be found at https://arxiv.org/abs/2210.05795.

K. A. Hansen et al. (Eds.): WINE 2022, LNCS 13778, pp. 78–95, 2022.
https://doi.org/10.1007/978-3-031-22832-2_5

personality traits, albeit with knowledge of how these latent traits affect team performance. The lack of information necessitates a natural trade-off: a principal must decide whether to exploit successful teams located early or reassign teammates to gain insight into the abilities of other individuals. The latter choice may temporarily reduce the overall rate of success.

To study this problem, we consider a setting (described in detail in Sect. 2) where agents have binary latent types, each team comprises two members, and the performance of each team is given by the same *synergy function*, i.e., some given symmetric function of its members' types. Over multiple rounds, the principal selects matchings over agents, with the goal of minimizing the cumulative *regret*, i.e., the difference between the number of successful teams in a round versus the number of successful teams under an optimal matching. Our main results concern the special case of symmetric Boolean synergy functions—in particular, we study the functions EQ and XOR (in Sect. 3), OR (in Sect. 4) and AND (in Sect. 5). While this may at first appear to be a limited class of synergy functions, in Sect. 2.3, we argue that these four functions are in a sense the *atomic primitives* for this problem; our results for these four settings are sufficient to handle *arbitrary symmetric synergy functions*.

The above model was first introduced by Johari et al. [12], who considered the case where agent types are i.i.d. Bernoulli(p) (for known p) and provide asymptotically optimal regret guarantees under AND (and preliminary results for OR). As with any bandit setting, it is natural to ask whether one can go beyond a stochastic model to admit *adversarial* inputs. In particular, the strongest adversary one can consider here is an *adaptive adversary*, which observes the choice of teams in each round, and only then fixes the latent types of agents. In most bandit settings, such an adversary is too strong to get any meaningful guarantees; among other things, adaptivity precludes the use of randomization as an algorithmic tool, and typically results in every policy being as bad as any other. Nevertheless, in this work, we provide a *near-complete characterization of the regret landscape for team formation under an adaptive adversary*. In particular, in a setting with n agents of which k have type '1', we present algorithms that are agnostic of the parameter k, and yet when faced with an adaptive adversary, achieve optimal regret for EQ and XOR, and near-optimal regret bounds under OR and AND (and therefore, using our reduction in Sect. 2.3, achieve near-optimal regret for any symmetric function).

While our results are specific to particulars of the model, they exhibit several noteworthy features. First, despite the adversary being fully adaptive, our regret bounds differ only by a small constant factor from prior results for AND under i.i.d. Bernoulli types [12]; such a small gap between stochastic and adversarial bandit models is uncommon and surprising. Next, our bounds under different synergy functions highlight the critical role of these functions in determining the regret landscape. Additionally, our algorithms expose a sharp contrast between learning and regret minimization in our setting: while the rate of learning increases with more exploration, minimizing regret benefits from maximal exploitation. Finally, to deal with adaptive adversaries in our model, we use techniques from extremal graph theory that are atypical in regret minimization; we hope that these ideas prove useful in other complex bandit settings.

1.1 Related Work

Regret minimization in team formation, although reminiscent of *combinatorial bandits/semi-bandits* [4–6,8,16], poses fundamentally new challenges arising from different synergy functions. In particular, a crucial aspect of bandit models is that rewards and/or feedback are linear functions of individual arms' latent types. Some models allow rewards/feedback to be given by a non-linear *link* function of the sum of arm rewards [7,10], but typically require the link function to be well-approximated by a linear function [17]. In contrast, our team synergy functions are *non-linear*, and moreover, are not well-approximated by any non-linear function of the sums of the agents' types.

One way to go beyond semi-bandit models and incorporate pairwise interactions is by assuming that the resulting reward matrix is low-rank [13,20,24]. The critical property here is that under perfect feedback, one can learn all agent types via a few 'orthogonal' explorations; this is true in our setting under the XOR function (Sect. 3), but not for other Boolean functions. Another approach for handling complex rewards/feedback is via a Bayesian heuristic such as Thompson sampling or information-directed sampling [9,14,19,23]. While such approaches achieve near-optimal regret in many settings, the challenge in our setting is in updating priors over agents' types given team scores. We hope that the new approaches we introduce could, in the future, be combined with low-rank decomposition and sampling approaches to handle more complex scenarios such as shifting types and corrupted feedback.

In addition to the bandit literature, there is a parallel stream on learning for team formation. Rajkumar et al. [18] consider the problem of learning to partition workers into teams, where team compatibility depends on individual types. Kleinberg and Raghu [15] consider the use of individual scores to estimate team scores and use these to approximately determine the best team from a pool of agents. Singla et al. [21] present algorithms for learning individual types to form a single team under an online budgeted learning setting. These works concentrate on pure learning. In contrast, our focus is on minimizing regret. Finally, there is a line of work on strategic behavior in teams, studying how to incentivize workers to exert effort [2,3], and how to use signaling to influence team formation [11]. While our work eschews strategic considerations, it suggests extensions that combine learning by the principal with strategic actions by agents.

2 Model

2.1 Agents, Types, and Teams

We consider n *agents* who must be paired by a principal into *teams of two* over a number of rounds; throughout, we assume that n is even. Each agent has an unknown latent *type* $\theta_i \in \{0,1\}$. These types can represent any dichoto-mous attribute: "left-brain" vs. "right-brain" (Sect. 3), "low-skill" vs. "high-skill" (Sects. 4 and 5), etc. We let k denote the number of agents with type 1, and assume that k is fixed *a priori* but unknown.

In each round t, the principal selects a matching M_t, with each edge $(i,j) \in M_t$ representing a *team*. We use the terms "edge" and "team" interchange-ably. The *success* of a team $(i,j) \in M_t$ is $f(\theta_i, \theta_j)$, where $f : \{0,1\}^2 \to \mathbb{R}$ is some known symmetric function of the agents' types. In Sects. 3-5, we restrict our focus to Boolean functions, interpreting $f(\theta_i, \theta_j) = 1$ as a success and $f(\theta_i, \theta_j) = 0$ as a failure. The algorithm observes the success of each team, and may use this to select the matchings in subsequent rounds; however, the algorithm can-not directly observe agents' types. For any matching M, we define its *score* as $S(M) := \sum_{(i,j) \in M} f(\theta_i, \theta_j)$—in the special case of Boolean functions, this is the number of successful teams.

A convenient way to view the Boolean setting is as constructing an edge-labeled *exploration graph* $G(V, E_1, E_2, \dots)$, where nodes in V are agents, and the edge set $E_t := \bigcup_{t' \le t} M_{t'}$ represents all pairings played up to round t. Upon being played for the first time, an edge is assigned a label $\{0,1\}$ corresponding to the success value of its team. *Known 0-agents* and *known 1-agents* are those whose types can be inferred from the edge labels. The remaining agents are *unknown*. The *unresolved subgraph* is the induced subgraph on the unknown agents.

2.2 Adversarial Types and Regret

The principal makes decisions facing an *adaptive adversary*, who knows k (unlike the principal, who only knows n) and, in each round, is free to assign agent types *after seeing the matching chosen by the principal*, as long as (1) the assignment is consistent with prior observations (i.e., with the exploration graph), and (2) the number of 1-agents is k. Note that this is the strongest notion of an adversary we can consider in this setting; in particular, since the adversary is fully adaptive and knows the matching *before* making decisions, randomizing does not help, and so it is without loss of generality to consider only deterministic algorithms.

We evaluate the performance of algorithms in terms of additive *regret* against such an adversary. Formally, let M^* be any matching maximizing $S(M^*)$—note that for any Boolean team success function, $S(M^*)$ is a fixed function of n and k. In round t, an algorithm incurs regret $r_t := S(M^*) - S(M_t)$, and its total regret is the sum of its per-round regret over an a priori infinite time horizon. Note, however, that after a finite number of rounds, a naïve algorithm that enumerates all matchings can determine, and henceforth play, M^*; thus, the optimal regret is always finite. Moreover, the "effective" horizon (i.e., the time until the algorithm learns M^*) of our algorithms is small.

2.3   Symmetry Synergy Functions and Atomic Primitives

In subsequent sections, we consider the problem of minimizing regret under four Boolean synergy functions $f\colon \{0,1\}^2 \to \{0,1\}$: EQ, XOR, OR, and AND. Interestingly, the algorithms for these four settings suffice to handle any symmetric synergy function $f\colon \{0,1\}^2 \to \mathbb{R}$. We argue this below for synergy functions that take at most two values; We handle the case of synergy functions f taking three different values at the end of Sect. 3.1.

Lemma 1. *Fix some $\ell \leq u$, let $f\colon \{0,1\}^2 \to \{\ell, u\}$ be any symmetric synergy function, and let $r^f(n,k)$ denote the optimal regret with n agents, of which k have type 1.*
    Then, $r^f(n,k) = (u-\ell) \cdot r^g(n,k)$ for one of $g \in \{$EQ,XOR,AND,OR$\}$.

Proof. First, note that without loss of generality, we may assume that $f(0,0) \leq f(1,1)$. Otherwise, we can swap the labels of the agent types without altering the problem. Note that this immediately allows us to reduce team formation under the Boolean NAND and NOR function to the same problem under AND and OR, respectively. Next, note that if $f(0,0) = f(1,0) = f(1,1)$, then the problem is trivial, as all matchings have the same score. Otherwise, we may apply the affine transformation $f \mapsto \frac{1}{u-\ell} \cdot f - \frac{\ell}{u-\ell}$ to the output to recover a Boolean function:

– When $f(0,1) < f(0,0) = f(1,1)$, we recover the EQ function.
– When $f(0,0) = f(1,1) < f(0,1)$, we recover the XOR function.
– When $f(0,0) = f(0,1) < f(1,1)$, we recover the AND function.
– When $f(0,0) < f(0,1) = f(1,1)$, we recover the OR function.

The structure of the problem remains unchanged since total regret is linear in the number of each type of team played over the course of the algorithm. The regret simply scales by a factor of $u - \ell$.   ∎

3   Uniform and Diverse Teams

We first focus on forming teams that promote uniformity (captured by the Boolean EQ function) or diversity (captured by the XOR function). In addition, we also show that the algorithm for EQ minimizes regret under any general symmetric synergy function taking three different values.

3.1   Uniformity (EQ)

We first consider the *equality* (or EQ) synergy function, $f^{\mathsf{EQ}}(\theta_i, \theta_j) = \overline{\theta_i \oplus \theta_j}$. Here, an optimal matching M^* includes as few $(0,1)$-teams as possible, and thus $S(M^*) = \frac{n}{2} - (k \bmod 2)$. If k (and thus $n-k$) is even, then all agents can be paired in successful teams; else, any optimal matching must include one unsuccessful team with different types. For this setting, Theorem 1 shows that a simple policy (Algorithm 1) achieves optimal regret for *all* parameters n and k.

Algorithm 1. FORM UNIFORM TEAMS

Round 1: Play an arbitrary matching.
Round 2: Swap unsuccessful teams in pairs as $\{(a,b),(c,d)\} \rightarrow \{(a,c),(b,d)\}$. Repeat remaining teams (including one unsuccessful team when k is odd).
Round 3: If $\{(a,b),(c,d)\}, \{(a,c),(b,d)\}$ are both unsuccessful, play $\{(a,d),(b,c)\}$. Repeat remaining teams.

Theorem 1. *Define* $r^{EQ}(n,k) := 2 \cdot \big(\min(k,n-k) - (k \bmod 2) \big)$. *Then,*

1. *Algorithm 1 learns an optimal matching by round 3, and incurs regret at most* $r^{EQ}(n,k)$.
2. *Any algorithm incurs regret at least* $r^{EQ}(n,k)$ *in the worst case.*

Proof. For the upper bound on the regret, note that every unsuccessful team includes a 0-agent and a 1-agent. Thus, there is a re-pairing of any two unsuccessful teams that gives rise to two successful teams. If the re-pairing in round 2 is unsuccessful, the only other re-pairing, selected in round 3, must be successful. There will be $k \bmod 2$ unsuccessful teams in round 3, making it an optimal matching. At most $\min(k,n-k)$ $(0,1)$-teams can be chosen in each of rounds 1–2, implying that the maximum regret in each of these rounds is $\min(k,n-k) - (k \bmod 2)$. Since Algorithm 1 incurs regret only in rounds 1–2, its total regret is at most $r^{EQ}(n,k)$.

For the converse (Claim 2), we argue that against *any* algorithm, the adversary can always induce regret $\min(k,n-k) - (k \bmod 2)$ in each of rounds 1–2. Note that after round 2, the exploration graph is a union of two (not necessarily disjoint) matchings, and hence consists of a disjoint union of even-length cycles and isolated (duplicated) edges; this is independent of the algorithm, as it holds for any pair of perfect matchings. Since the graph is bipartite, the adversary can assign types such that no pair of the minority type is adjacent in the graph by starting with the labeling according to the bipartition, then arbitrarily relabeling a subset of the minority side to make the labeling consistent with k. ∎

A similar argument allows us to complete our treatment of general (symmetric) synergy functions from Sect. 2.3.

Corollary 1. *For any symmetric synergy function* $f : \{0,1\}^2 \rightarrow \mathbb{R}$ *such that* $f(0,0) \neq f(0,1) \neq f(1,1)$, *there is a regret-minimizing algorithm that locates an optimal matching within two rounds.*

Proof. By applying an affine transformation to the outputs as in Sect. 2.3, we may assume without loss of generality that $f(0,0) = 0$, and $f(1,1) = 1$. There are three cases to consider:

- $f(0,1) = \frac{1}{2}$: The problem is trivial, since all matchings have the same score.
- $f(0,1) > \frac{1}{2}$: The optimal matching includes as many 1–0 agent teams as possible. After the first (arbitrary) matching, every agent is either part of a known 1–0 team or has a known identity (as a member of a 0–0 or 1–1 team). Thus, one can always select an optimal matching in the second round.

- $f(0,1) < \frac{1}{2}$: The optimal matching includes as many 1–1 agent teams as possible, just as in the EQ setting. Note that the three distinct values of f allow us to distinguish between $(0,0)$, $(0,1)$, and $(1,1)$ teams. The same adversarial policy ensures that all 0–1 teams remain sub-optimally paired in round 2, so we exactly recover the EQ setting.

∎

3.2 Diversity (XOR)

Next we consider the XOR success function, $f^{\mathsf{XOR}}(\theta_i, \theta_j) = \theta_i \oplus \theta_j$, which promotes diverse teams. Now $S(M^*) = \min(k, n-k)$, since any optimal matching M^* includes as many $(0,1)$-teams as possible. Define $x^+ := \max(0, x)$; we again show that a simple policy (Algorithm 2) has optimal regret for all n, k.

Algorithm 2. FORM DIVERSE TEAMS

Round 1: Play an arbitrary matching; let $\{(1,2),\ldots,(\ell-1,\ell)\}$ denote unsuccessful teams.
Round 2: Replay all successful teams, and construct a single cycle over all unsuccessful teams (i.e., play teams $\{(\ell,1),(2,3),\ldots,(\ell-2,\ell-1)\}$).
Round 3: Play any inferred optimal matching (see Theorem 2).

Theorem 2. *Define $r^{\mathsf{XOR}}(n,k) := 2 \cdot \big(\min(k, n-k) - 1 - (k \bmod 2) \big)^+$. Then,*

1. *Algorithm 2 learns an optimal matching after round 2, and incurs regret at most $r^{\mathsf{XOR}}(n,k)$.*
2. *Any algorithm incurs regret at least $r^{\mathsf{XOR}}(n,k)$ in the worst case.*

Proof. For the achievability in Claim 1, note that each edge $(i, (i+1) \bmod \ell)$ of the cycle constructed in the algorithm has the following property: if the edge is successful in round 2, then its endpoints have opposite types; otherwise, they have the same type. By following edges around the cycle, the algorithm can therefore construct the sets $S^=$ of agents with the same type as agent 1, and $S^{\neq}$ of agents with the opposite type. Subsequently, it is optimal to match $\min(|S^=|, |S^{\neq}|)$ teams of (known) opposite-type agents, and match the extraneous agents into unsuccessful teams.

Among agents $1, \ldots, \ell$, there are $k - \frac{n-\ell}{2}$ 1-agents and $n - k - \frac{n-\ell}{2}$ 0-agents; thus, the round 1 regret is $r_1 := \min(k, n-k) - \frac{n-\ell}{2}$ (note that When k is odd, one team must be successful in round 1). Since no regret is incurred after round 2, the adversary must maximize the regret in round 2 conditioned on the choice of ℓ. This is achieved by assigning type 1 to agents $1, \ldots, k - \frac{n-\ell}{2}$, and type 0 to agents $k - \frac{n-\ell}{2} + 1, \ldots, \ell$. Since $(\ell, 1)$ and $(k - \frac{n-\ell}{2}, k - \frac{n-\ell}{2} + 1)$ are the only successful teams (as long as agents 1 to ℓ do not all have the same type), the regret in round 2 is $(r_1 - 2)^+$. The total regret $\left(2\min(k, n - k) - n + \ell - 2\right)^+$ is monotone increasing in ℓ, with the maximum attained at $\ell = n - 2(k \bmod 2)$. Substituting, we get the upper bound.

For the converse (Claim 2), we describe a policy for the adversary that ensures regret at least $r^{\mathsf{XOR}}(n, k)$. In round 1, the adversary reveals $k \bmod 2$ successful teams, resulting in regret $\min(k, n - k) - (k \bmod 2)$. In round 2, the exploration graph must consist of a disjoint union of even-length cycles (including isolated duplicated edges).

First, when k is odd, consider the component containing the one revealed successful team from round 1. If the component has just two agents (i.e., the algorithm repeats the team), then we again get one successful team. Otherwise, if the team is part of a longer cycle, the adversary puts an odd number of adjacent 0s and an odd number of adjacent 1s in the cycle, such that the previously successful team is (0, 1). Since the edge is not repeated, and only one other (0, 1)-team is created, the algorithm gets at most one successful team in this cycle. The remaining cycles contain an even number of 1-agents, so we appeal to below.

When k is even, the adversary fills cycles with 0-agents until they are exhausted, then labels all remaining agents as 1-agents. At most one cycle contains both agent types. Placing the 0-agents contiguously in this cycle ensures only two adjacent successful teams. Since all cycle lengths are even, as is $n - k$, these successful teams will be an even number of edges apart; in particular, the adversary can ensure that they are both edges from round 2, making the assignment consistent with round 1. In total, the algorithm obtains at most $2 + (k \bmod 2)$ successful teams in round 2, giving total regret at least $r^{\mathsf{XOR}}(n, k)$. ∎

4 The Strongest Link Setting (OR)

We next consider the Boolean OR synergy function, that is, $f^{\mathsf{OR}}(\theta_i, \theta_j) = \theta_i + \theta_j$. Adopting the terminology of Johari et al. [12], we refer to this setting as the *strongest link* model: interpreting 0/1-agents as having low/high skill, a team is successful when it has at least one high-skill member.

Observe that under OR, we have $S(M^*) = \min(k, n/2)$, since any optimal matching M^* includes a maximal set of $(0, 1)$-teams. Define $\alpha := \frac{n-k}{n}$ to be the *fraction of low-skill agents*; our regret bounds in this setting are more conveniently phrased in terms of α. In particular, our first result establishes the following lower bounds on the regret incurred by *any* algorithm.

Theorem 3. *For the strongest link setting, any algorithm incurs regret at least* $L^{\mathsf{OR}}(\alpha) \cdot n$ *in the worst-case, where*

$$L^{\mathsf{OR}}(\alpha) = \begin{cases} \frac{13\alpha}{17} & 0 \le \alpha \le \frac{1}{2} \\ \frac{6-9\alpha}{4} & \frac{1}{2} < \alpha \le \frac{6}{11} \\ \frac{3-4\alpha}{3} & \frac{6}{11} < \alpha \le \frac{3}{5} \\ \frac{1-\alpha}{2} & \frac{3}{5} < \alpha \le 1 \end{cases}$$

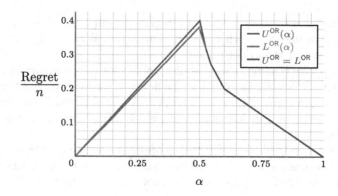

Fig. 1. *Our regret bounds (Thoerems 3 to 5) under the Strongest Link model, as functions of $\alpha := \frac{n-k}{n}$, the fraction of low-skill agents. The bounds match for $\frac{10}{19} \le \alpha \le 1$.*

Proof Sketch. The bounds are established using a common underlying adversarial strategy that forces any algorithm to incur an unavoidable regret. The structure of the strategy is as follows: The adversary first reveals a chosen fraction of 0-agents in round 1. Subsequently, when the algorithm explores an unresolved agent, the adversary reveals it to be a 0-agent whenever possible. This leads to a tension between inducing high regret in the first round (by revealing 0-edges), and leaving more unresolved 0-agents for later rounds. $L^{OR}(\alpha)$ is obtained by choosing the fraction of initially revealed 0-agents to maximize regret under this tension. A full proof is presented in the extended version. ∎

The lower bound in Theorem 3 is plotted in Fig. 1, and notably varies greatly with α. Nevertheless, we provide a policy (Algorithm 3) that manages to achieve *nearly matching regret across all* α, while being agnostic of k (and thus α). Both bounds are plotted in Fig. 1; despite the functions being piecewise linear, they match exactly for $\alpha \ge \frac{10}{19}$, and $U^{OR}(\alpha) - L^{OR}(\alpha) < 0.018$ for all α.

4.1 The MAXEXPLOIT WITH 4-CLIQUES Algorithm

To simplify our analysis, we introduce some terminology: we say that two unknown 0-agents become *discovered* when they are paired to form an unsuccessful team. An unknown agent is *explored* when its type is revealed by pairing it with a known 0-agent. Our policy for this setting, MAXEXPLOIT WITH 4-CLIQUES, is given in Algorithm 3. The algorithm exploits the inferred types of agents to the greatest possible extent; a maximal number of known 1–0 agent teams are played in each round. Exploration is only done using known 0-agents that cannot be included in such a pair. If only two agents in a 4-cycle are explored, we treat the other two agents as unknown, even if their type is deducible.

First, to see that the algorithm terminates, note that in each iteration of the loop, at least one known 0-agent is used for exploration, revealing the type of another agent. Thus, the algorithm makes progress and eventually terminates.

Algorithm 3. MAXEXPLOIT WITH 4-CLIQUES
___
Round 1: Select an arbitrary matching.
while #{known 0-agents} > #{known 1-agents} **and** #{unknown agents} > 0
do
   Pair each known 1-agent with a known 0-agent.
   Use extra known 0-agents to explore both members of successful teams.
   (In round 3, explore all members of 4-cycles whenever possible[3].)
   Round 2: Re-pair remaining unknown successful teams into 4-cycles.
   (If number of remaining unknown successful teams is odd, repeat one team.)
   Round 3: In each 4-cycle with undiscovered agents, re-pair to form a 4-clique.
   Round 4+: Re-play the matching from round 1 on unexplored successful teams.
___

Next, note that unknown agents are always in successful teams throughout the algorithm (as both members of an unsuccessful team can be deduced as 0-agents). Upon termination, the algorithm can play an optimal matching: either all agents are known, or there are enough known 1-agents to match all known 0-agents, and the other successful teams of unknown agents can be safely replayed.

Let d_t be the number of 0-agents discovered in round t by pairing two unknown agents, and e_t the number of 0-agents revealed by exploration with a known 0-agent. We define

$$\Delta_t := \#\{\text{known 0-agents after round } t\} - \#\{\text{known 1-agents after round } t\}$$

The following lemma studies how Δ_t evolves over rounds t.

Lemma 2. $\Delta_1 = d_1$, and $2e_t = \Delta_t \leq \Delta_{t-1}$, for all $t \geq 2$.

Proof. In round 1, the algorithm discovers d_1 0-agents, and no 1-agent (since there is no exploration); hence $\Delta_1 = d_1$. Consider the 4-cycle and 4-clique edges played in rounds 2–3. If such an edge comprises two 0-agents, then the other two agents in its cycle or clique must be 1-agents. In particular, the addition of d_t known 0-agents in these rounds is exactly counterbalanced by the deduction of their neighboring d_t 1-agents, so discovery does not contribute to $\Delta_{t+1} - \Delta_t$.

Next, consider any round $t \geq 2$. The algorithm first pairs all known 1-agents with known 0-agents, so exactly Δ_{t-1} agents are used for exploration. Each exploration must discover either a 0-agent or a 1-agent, so $\Delta_t = \Delta_{t-1} + e_t - (\Delta_{t-1} - e_t) = 2e_t$. Since members of successful teams are explored in pairs, at most half of all explorations can reveal 0-agents. Thus, $e_t \leq \frac{\Delta_{t-1}}{2}$. ∎

For the subsequent analysis, there are two distinct regimes depending on the *fraction* of low-skill agents α. When most agents are low-skill ($\alpha > \frac{1}{2}$), the optimal configuration includes some $(0,0)$-teams, but no $(1,1)$-teams, and r_t equals the number of $(1,1)$-teams in M_t. On the other hand, when most agents are high-skill ($\alpha \leq \frac{1}{2}$), the optimal configuration consists entirely of successful teams, and an algorithm's round-t regret r_t is the number of $(0,0)$-teams in M_t. Consequently, the analysis in each regime is very different.

4.2   Majority High-Skill Regime ($\alpha \leq \frac{1}{2}$)

We begin the analysis by focusing on the case when $\alpha \leq \frac{1}{2}$. Recall that the total regret in this regime equals the total number of $(0,0)$ teams the algorithm plays.

Theorem 4. *For* $\alpha \leq \frac{1}{2}$, *Algorithm 3 has regret at most* $\frac{4}{5} \cdot \alpha n$.

Proof. First, note that in the regime $\alpha \leq \frac{1}{2}$, the algorithm never pairs two *known* 0-agents; known 0-agents are paired with known 1-agents or used for exploration. Hence, the number of $(0,0)$-teams selected, and thus the regret, in round t is $e_t + \frac{d_t}{2}$. (Note that $e_1 = 0$.)

After round 3, by Lemma 2, there are $2e_3$ more known 0-agents than 1-agents. The unresolved agents are contained in 4-cliques of successful teams, which must each contain at least three 1-agents. Thus, exploring any 0-agent means that the algorithm can deduce three 1-agents. After e_3 such explorations, the algorithm locates $3e_3$ 1-agents, terminating the loop. The regret incurred in rounds 4 and later is thus at most e_3, giving total regret at most $\frac{d_1}{2} + \frac{d_2}{2} + \frac{d_3}{2} + e_2 + 2e_3$.

We can now bound the regret incurred by Algorithm 3 by formulating the adversary's problem of choosing the worst-case number of revealed zeros in each round as an LP with variables $\{d_1, d_2, d_3, e_2, e_3\}$. Applying Lemma 2 to rounds 2 and 3, we obtain that $e_2 \leq \frac{d_1}{2}$ and $e_3 \leq e_2$. In addition, $d_1 + d_2 + d_3 + e_2 + 2e_3 \leq \alpha n$ ensures that the number of 0-agents revealed by the adversary is at most the total number of 0-agents. Put together, we get the following LP:

$$
\begin{aligned}
\text{Maximize:} \quad & \frac{d_1}{2} + \frac{d_2}{2} + \frac{d_3}{2} + e_2 + 2e_3 \\
\text{Subject to:} \quad & e_2 \leq \frac{d_1}{2} \\
& e_3 \leq e_2 \\
& d_1 + d_2 + d_3 + e_2 + 2e_3 \leq \alpha n \\
& d_1, d_2, d_3, e_2, e_3 \geq 0
\end{aligned}
$$

Solving, we get $(d_1, d_2, d_3, e_2, e_3) = (\frac{2\alpha n}{5}, 0, 0, \frac{\alpha n}{5}, \frac{\alpha n}{5})$ as the adversary's best strategy, with regret at most $\frac{4}{5}\alpha n$. ∎

4.3   Majority Low-Skill Regime ($\alpha > \frac{1}{2}$)

A different, more involved, analysis shows that Algorithm 3 is also near-optimal when $\alpha > \frac{1}{2}$.

Theorem 5. *For* $\alpha > \frac{1}{2}$, *Algorithm 3 learns an optimal matching after incurring regret at most* $U^{OR}(\alpha) \cdot n$, *where*

$$
U^{OR}(\alpha) = \begin{cases}
\frac{10-16\alpha}{5} & \frac{1}{2} \leq \alpha < \frac{10}{19}, \\
\frac{6-9\alpha}{4} & \frac{10}{19} \leq \alpha < \frac{6}{11}, \\
\frac{3-4\alpha}{3} & \frac{6}{11} \leq \alpha < \frac{3}{5}, \\
\frac{1-\alpha}{2} & \frac{3}{5} \leq \alpha \leq 1.
\end{cases}
$$

Note that $\lim_{\alpha \downarrow \frac{1}{2}} U^{OR}(\alpha) = \frac{2}{5}$, which matches $\lim_{\alpha \uparrow \frac{1}{2}} U^{OR}(\alpha)$ from Theorem 4. Before proceeding, we define $s_t^{00}, s_t^{01}, s_t^{11}$ to be the number of $(0,0), (0,1)$, and $(1,1)$-teams the algorithm plays in round t, respectively. Since there are $(1-\alpha)n$ 1-agents in total, $s_t^{01} = (1-\alpha)n - 2s_t^{11}$; in turn, since there are αn 0-agents, $s_t^{00} = \frac{1}{2} \cdot (\alpha n - s_t^{01}) = s_t^{11} + (\alpha - \frac{1}{2}) \cdot n > s_t^{11}$. We now prove Theorem 5 via a series of lemmas.

Lemma 3. *The adversary has a best response to Algorithm 3 with the following properties:*

1. *It never reveals pairs of unknown agents as $(0,0)$-teams after round 1.*
2. *It never reveals any $(1,1)$-team until all $(0,1)$-teams have been revealed.*

Proof. For the first claim, suppose that the adversary reveals a $(0,0)$-team among the re-paired teams in round $t = 2$ or $t = 3$. The 4-cycle or 4-clique containing this $(0,0)$-team contains two 0-agents and two 1-agents, so two $(0,1)$-teams were selected in each of the first $t-1$ rounds. The adversary can force the same regret, and provide the same information, by relabeling these agents so a $(0,0)$-team and a $(1,1)$-team are revealed in round 1, the two 1-agents are explored in round 2, and (if $t = 3$) the $(0,1)$-teams are repeated in round 3. By repeating this relabeling, we arrive at an adversary strategy of the same regret, of the claimed form.

For the second claim, recall that the algorithm's regret in the regime $\alpha > \frac{1}{2}$ is exactly the number of $(1,1)$-teams it plays. We will describe a scheme charging $(1,1)$-teams played in round t to 0-agents explored in round t.

Consider some round $t \geq 2$ in which s_t^{11} $(1,1)$-teams are played. Since $s_t^{00} > s_t^{11}$, and all $(0,0)$-teams result from exploration[1] by the first claim, we can charge one distinct explored 0-agent for each such $(1,1)$-team. Thus, the number of *uncharged* explored 0-agents in round t is exactly $s_t^{00} - s_t^{11} = (\alpha - \frac{1}{2}) \cdot n$, independent of t and s_t^{11}.

The total number of 0-agents explored in rounds $t \geq 2$ is exactly s_1^{01}. If the algorithm runs for T rounds, exactly $(T-1) \cdot (\alpha - \frac{1}{2}) \cdot n$ explored 0-agents remain uncharged. Thus, the number of *charged* 0-agents, which equals the regret incurred after round 1, is $s_1^{01} - (T-1) \cdot (\alpha - \frac{1}{2}) \cdot n$. Conditioned on $s_1^{00}, s_1^{01}, s_1^{11}$, the regret is therefore maximized by minimizing T; that is, the adversary wants the algorithm to finish in as few rounds as possible. To minimize the number of rounds T, the adversary should maximize the number of 0-agents available for exploration. The adversary accomplishes this by having the algorithm explore $(0,1)$-teams before any $(1,1)$-teams. ∎

We now focus, without loss of generality, on such an adversary. This lets us bound the regret in terms of $(s_1^{00}, s_1^{01}, s_1^{11})$.

---

[1] Except in the last round, where known $(0, 0)$-teams may be played; however, no $(1, 1)$-teams are played in this round.

Lemma 4. *Conditioned on $s_1^{00}, s_1^{01}, s_1^{11}$, Algorithm 3 has regret at most*

$$\left\lfloor \frac{s_1^{01} + s_1^{11}}{s_1^{00}} \right\rfloor s_1^{11} + \min(s_1^{11}, (s_1^{01} + s_1^{11}) \mod s_1^{00}).$$

Proof. From Lemma 3 we know that only $(0, 1)$-teams are explored before any $(1, 1)$-team is explored. Each explored $(0, 1)$-team results in pairing a 0-agent with a newly discovered 1-agent, and also adds a known 0-agent. Thus as long as the algorithm explores only $(0, 1)$-teams, the number of pairs of 0-agents available for exploration stays constant at s_1^{00}. Therefore, the total number of rounds of exploration until all agents in successful teams are explored is $\left\lceil \frac{s_1^{01} + s_1^{11}}{s_1^{00}} \right\rceil$. One subtlety here is that the exploration of $(1, 1)$-teams decreases the available 0-agents. However, because $s_1^{11} < s_1^{00}$, the first round in which a $(1, 1)$-team can be explored is one round before the last; in this case, the last round only explores $(1, 1)$-teams, and the number of 0-agents available for this exploration exceeds the number of 1-agents to be explored. Thus, the bound on the number of rounds of exploration does hold. In the last round of exploration, no $(1, 1)$-teams can be played, so no regret is incurred. We therefore focus on the first $T = \left\lfloor \frac{s_1^{01} + s_1^{11}}{s_1^{00}} \right\rfloor$ rounds of exploration. Again because $s_1^{11} < s_1^{00}$, none of the first $T - 1$ rounds of exploration explore any $(1, 1)$-team; thus, each of these rounds, as well as the very first round of the algorithm, incurs a regret of s_1^{11}.

In round T, the total regret is the number of $(1, 1)$-teams explored in round $T + 1$ (because these teams are still played in round T). This number is either $(s_1^{01} + s_1^{11}) \mod s_1^{00}$ (if *only* $(1, 1)$-teams are explored in round $T+1$, then it is the total number of explored teams in round $T + 1$), or s_1^{11} (if some $(0, 1)$-teams are explored in round $T+1$, then *all* $(1, 1)$-teams are explored in round $T+1$). Thus, the regret in the Tth round of exploration is the minimum of the two terms. We thus obtain the total regret of the algorithm as: $\left\lfloor \frac{s_1^{01} + s_1^{11}}{s_1^{00}} \right\rfloor \cdot s_1^{11} + \min(s_1^{11}, (s_1^{01} + s_1^{11}) \mod s_1^{00})$. ∎

s_t^{00} turns out to be further constrained, as follows:

Lemma 5. *In Algorithm 3, if the adversary reveals 0-agents only by exploration in rounds $t \geq 2$, then $s_1^{00} > \frac{\alpha}{5} n$.*

Proof. The number of 0-agents discovered in round 1 is $2s_1^{00}$. In rounds 2 and 3 combined, the algorithm discovers an additional $e_2 + e_3$ 0-agents. The number of 1-agents discovered in round 2 is $\Delta_1 - e_2 = 2s_1^{00} - e_2$, and in round 3, it is $\Delta_2 - e_3 = 2e_2 - e_3$, by Lemma 2. Thus, the number of unknown 0-agents after round 3 is $\alpha n - 2s_1^{00} - e_2 - e_3$, and the number of unknown 1-agents is $(1 - \alpha) \cdot n - 2s_1^{00} - e_2 + e_3$. But notice also that after round 3, all unknown agents form 4-cliques of successful edges, which can contain at most one 0-agent each. Therefore, there must be at least three times as many remaining 1-agents as 0-agents, so $(1 - \alpha) \cdot n - 2s_1^{00} - e_2 + e_3 \geq 3(\alpha n - 2s_1^{00} - e_2 - e_3)$. Rearranging, we obtain that $4s_1^{00} \geq (4\alpha - 1) \cdot n - 2e_2 - 4e_3$. By Lemma 2, we get that $e_3 \leq e_2 \leq s_1^{00}$, so the previous inequality in particular implies that $4s_1^{00} \geq (4\alpha - 1) \cdot n - 6s_1^{00}$, or $s_1^{00} \geq \frac{4\alpha - 1}{10} \cdot n > \frac{\alpha}{5} \cdot n$, because $\alpha > \frac{1}{2}$. ∎

We obtain the piecewise-linear bound in Theorem 5 by maximizing the bound in Lemma 4 subject to the constraint in Lemma 5 (and using $s_t^{01} = \alpha n - 2s_t^{00}$, $s_t^{11} = s_t^{00} - (\alpha - \frac{1}{2})n$). Details of this calculation can be found in the extended version.

5   The Weakest Link Setting (AND)

Finally, we consider the Boolean AND synergy function. If, as before, we interpret 0/1-agents as having low/high skill, then (in the terminology of Johari et al. [12]), this corresponds to a *weakest link* model: the difficulty of the task ensures any team with a low-skill member is unsuccessful. To simplify the analysis, we assume throughout that k is even.

Theorem 6. *For the weakest link model, any algorithm incurs regret at least* $L^{AND}(n, k) := n - k$.

Proof Sketch. The total regret for an algorithm equals half the number of $(0, 1)$-teams selected over the duration of the algorithm, since the optimal solution would re-pair these agents into $(0, 0)$-teams and $(1, 1)$-teams. We consider a myopic greedy adversary which reveals as few 1-agents as possible in each round, and argue that such an adversary can ensure that each 0-agent is paired with at least two 1-agents during the execution of any algorithm. The proof is presented via a series of lemmas in the extended version.   ∎

5.1   The RING FACTORIZATION WITH REPAIRS Algorithm

The fact that the regret of an algorithm is half the number of $(0, 1)$-teams selected suggests that we want the algorithm's chosen matchings to quickly locate (and pair) all of the 1-agents, while minimizing the number of times each 0-agent is paired with a 1-agent. Playing matchings according to a *1-factorization* (that is, a partition of the complete graph K_n into perfect matchings) ensures that no team is ever repeated. This intuition is used in the EXPONENTIAL CLIQUES algorithms of Johari et al. [12], who show that when each agent has independent Bernoulli(k/n) type, this algorithm has expected regret $\frac{3}{4}(n - k) + o(n)$, which is asymptotically optimal. Against an adaptive adversary, however, an arbitrary 1-factorization is not enough to get good regret; for example, a 1-factorization that first builds the Turán graph $T(n, \frac{n}{k})$ [1,22] has regret $\frac{1}{2}k(n - k)$. Similarly, the performance of EXPONENTIAL CLIQUES in the worst case is also much worse.

Lemma 6. EXPONENTIAL CLIQUES *incurs regret* $2(n - k - 1)$ *against an adaptive adversary.*

Proof. Consider an instance on $n = 2^j + 2$ agents with $k = 2$ having high skill. An adaptive adversary can ensure that the two 1-agents comprise the last unexplored team. Over its first $2^j - 1$ rounds, Exponential Cliques builds a 2^j-clique in the explored subgraph while repeating the remaining team $2^j - 1$ times.

Subsequently, it must spend 2^j additional rounds exploring all teams comprising a member of this repeated edge and a member of the clique, resulting in regret $2(2^j - 1) = 2(n - k - 1)$. ∎

Our main algorithm for this setting, RING FACTORIZATION WITH REPAIRS, leverages a particular 1-factorization, which we call the RING FACTORIZATION. We organize the agents into two nested rings and choose matchings so that closer agent pairs under this ring geometry are matched earlier. In the first round, agents in corresponding positions in the rings are matched. Over the next four rounds, matchings are chosen to pair each agent with the four agents in adjacent positions in the rings, and this process repeats at greater distances. The structure and order of the four matchings chosen in each "phase" are critical. A formal description of this 1-factorization is given in the extended version. We visualize the first 5 matchings in the constructions for $n = 10$ and $n = 12$ in Fig. 2.

$n = 10 \ (m = 5):$

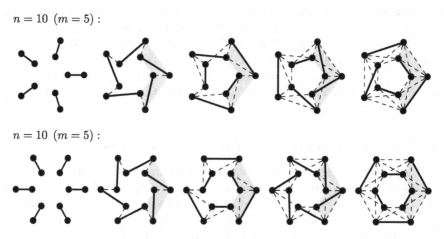

$n = 10 \ (m = 5):$

Fig. 2. *The first five rounds (i.e., Phases 0 and 1) of* RING FACTORIZATION *on 10 (top) and 12 (bottom) agents. The last four matchings illustrate the general matching sequence for cycles in intermediate phases; the blue highlighted section of each matching is repeated based upon the size of the cycle.*

Theorem 7. RING FACTORIZATION WITH REPAIRS *(Algorithm 4) locates an optimal matching after incurring regret at most* $U^{AND}(n, k) := n - k + \left\lfloor \frac{\min(k, n-k)}{4} \right\rfloor$.

Proof Sketch. As mentioned before, the double-ring structure of our factorization defines a notion of distance between agents (namely, the difference between their column indices modulo m). By selecting matchings according to this factorization, each agent is paired up with other agents in non-decreasing order of distance. Consider this pairing from the perspective of a 0-agent x. We will show

---

Algorithm 4. RING FACTORIZATION WITH REPAIRS (Sketch)

---

while #{unknown agents} > 0 **do**

  Select a matching via RING FACTORIZATION for n.

  if a $(1,1)$-team is revealed **then**

    Perform a case-specific "repair" step (possibly over multiple rounds) that partitions agents into known $(0,0)$-teams, known $(1,1)$-teams, and an intermediary stage of the RING FACTORIZATION construction of size $n' < n$. (See Appendix B in the extended version)

    Play known $(0,0)$- and $(1,1)$-teams, and continue playing matchings according to RING FACTORIZATION on the remaining n' agents.

---

that roughly speaking (with some exceptions which require technical work and slightly weaken the bound), x will be paired with at most two 1-agents before being identified as a 0-agent. If each 0-agent is paired with at most two 1-agents before discovery, then we get an overall regret bound of $n - k$. Consider three 1-agents $\{y_1, y_2, y_3\}$, all located in different columns from x. Then, two of these 1-agents—say, y_1 and y_2—lie on the same side of x. Thus, y_1 and y_2 are strictly closer to each other than the further of the two (say, y_1) is to x. In particular, y_1 and y_2 were paired before y_1 is paired with x, and so must have been revealed as 1-agents. Thus, y_1 will never be paired with x. Since this holds for every triple of 1-agents, x cannot be paired with three 1-agents.

While the above argument encapsulates the main intuition, the technical challenge is removing the assumption that y_1, y_2, y_3 were all in different columns from x. "Repairing" the cycle to account for the case of a 1-agent in the same column as x largely accounts for the additional term in the regret bound. The full analysis of these "repair" steps is intricate; see Appendix B in the extended version for details.   ■

The bounds of Theorems 6 and 7 are off by an additive term $\lfloor \min(k, n - k)/4 \rfloor$. The lower bound is simpler, and it is tempting to think that it may be tight; unfortunately, this is not true in general; in Appendix C of the extended version, we show that any algorithm on the instance with $n = 10$, $k = 4$ must incur regret at least 7 (which coincidentally matches the upper bound in Theorem 7, though it is unclear if this extends to larger settings). Closing this gap is an interesting and challenging direction for future work.

6 Conclusion

Our work provides near-optimal regret guarantees for learning an optimal matching among agents under any symmetric function of two binary variables. While our results are specific to each function, they exhibit several noteworthy common features. First, although we consider an adaptive adversary, it is not hard to see that the regret bounds with i.i.d. Bernoulli types can only improve by a small constant factor; such a small gap between stochastic and adversarial models is uncommon. Next, for all our settings, minimizing regret turns out to require maximal exploitation (in contrast to quickly learning all agent types, which would benefit from more exploration). Finally, the problems appear to get harder for $k = \frac{n}{2}$, and also handling the weakest link setting (i.e., the Boolean AND function) is more challenging than other synergy functions. These phenomena hint at underlying information-theoretic origins, and formalizing these may help in reasoning about more complex models.

Our work raises three natural future directions:

1. It would be desirable to close the gaps between our bounds for the OR and AND settings. In each case, however, our results suggest that the optimal procedures may depend heavily on number-theoretic properties of n and k which can expose a further level of complication.
2. We consider only perfect feedback, which in itself presented interesting challenges, but may be unrealistic in real-world settings. Our results likely extend to some noisy feedback models by repeatedly playing a team and averaging their scores. However, quantifying the relationship between the amount of noise and the expected additional regret is an open problem.
3. The restriction to teams of size 2, and binary agent types, are the main restrictions of our model. For a more general theory of team formation, it is desirable to consider larger teams and other synergy functions (in particular, threshold functions); doing so is a rich and challenging open direction.

Acknowledgments. We would like to thank anonymous reviewers for useful feedback. Part of the work was done when SB and ME were visiting the Simons Institute for the Theory of Computing for the semester on Data-Driven Decision Processes; they also acknowledge support from the NSF under grants ECCS-1847393 and CNS-195599, and the ARO MURI grant W911NF1910217. DK acknowledges support from ARO MURI grant ARO W911NF1810208.

References

1. Aigner, M.: Turán's graph theorem. Am. Math. Mon. **102**(9), 808–816 (1995)
2. Babaioff, M., Feldman, M., Nisan, N.: Combinatorial agency. In: Proceedings of 7th ACM Conference on Electronic Commerce, pp. 18–28 (2006)
3. Carlier, G., Ekeland, I.: Matching for teams. Econ. Theory **42**, 397–418 (2010)
4. Cesa-Bianchi, N., Lugosi, G.: Combinatorial bandits. J. Comput. Syst. Sci. **78**(5), 1404–1422 (2012)
5. Chen, W., Wang, Y., Yuan, Y.: General framework and applications. In: ICML, Combinatorial Multi-armed Bandit (2013)
6. Combes, R., Sadegh Talebi, M., Proutiere, A., Lelarge, M.: Combinatorial bandits revisited. In: Proceedings of 29th Advances in Neural Information Processing Systems, pp. 2116–2124 (2015)
7. Devanur, N.R., Jain, K.: Online matching with concave returns. In: Proceedings of the Forty-Fourth Annual ACM Symposium on Theory of Computing, pp. 137–144 (2012)
8. Gai, Y., Krishnamachari, B., Jain, R.: Combinatorial network optimization with unknown variables: multi-armed bandits with linear rewards and individual observations. IEEE/ACM Trans. Networking **20**(5), 1466–1478 (2012)

9. Gopalan, A., Mannor, S., Mansour, Y.: Thompson sampling for complex online problems. In: International Conference on Machine Learning, pp. 100–108. PMLR (2014)
10. Han, Y., Wang, Y., Chen, X.: Adversarial combinatorial bandits with general nonlinear reward functions. In: International Conference on Machine Learning, pp. 4030–4039 (2021)
11. Hssaine, C., Banerjee, S.: Information signal design for incentivizing team formation. In: Proceedings of 14th Conference on Web and Internet Economics (WINE) (2018)
12. Johari, R., Kamble, V., Krishnaswamy, A.K., Li, H.: Exploration vs. exploitation in team formation. In: Proceedings of 14th Conference on Web and Internet Economics (WINE) (2018)
13. Katariya, S., Kveton, B., Szepesvari, C., Vernade, C., Wen, Z.: Stochastic rank-1 bandits. In: Artificial Intelligence and Statistics, pp. 392–401. PMLR (2017)
14. Kirschner, J., Lattimore, T., Krause, A.: Information directed sampling for linear partial monitoring. In: Conference on Learning Theory, pp. 2328–2369. PMLR (2020)
15. Kleinberg, J., Raghu, M.: Team performance with test scores. In: Proceedings of 16th ACM Conference on Economics and Computation (2015)
16. Branislav Kveton, Zheng Wen, Azin Ashkan, and Csaba Szepesvári. Tight regret bounds for stochastic combinatorial semi-bandits. In Proc. 18th Intl. Conf. on Artificial Intelligence and Statistics, 2015
17. Merlis, N., Mannor, S.: Tight lower bounds for combinatorial multi-armed bandits. In: Conference on Learning Theory, pp. 2830–2857. PMLR (2020)
18. Rajkumar, A., Mukherjee, K., Tulabandhula, T.: Learning to partition using score based compatibilities. In: Proceedings of 16th International Conference on Autonomous Agents and Multiagent Systems, pp. 574–582 (2017)
19. Russo, D., Van Roy, B.: Learning to optimize via information-directed sampling. Advances in Neural Information Processing Systems, 27 (2014)
20. Sentenac, F., Yi, J., Calauzenes, C., Perchet, V., Vojnovic, M.: Pure exploration and regret minimization in matching bandits. In: ICML21, pp. 9434–9442 (2021)
21. Singla, A., Horvitz, E., Kohli, P., Krause, A.: Learning to hire teams. In: Third AAAI Conference on Human Computation and Crowdsourcing (2015)
22. Turán, P.: Egy gráfelméleti szélsoértékfeladatról (on an extremal problem in graph theory). Mat. Fiz. Lapok **48**(3), 436–453 (1941)
23. Wang, S., Chen, W.: Thompson sampling for combinatorial semi-bandits. In: International Conference on Machine Learning, pp. 5114–5122. PMLR (2018)
24. Zimmert, J., Seldin, Y.: Factored bandits. Advances in Neural Information Processing Systems 31 (2018)

Stability of Decentralized Queueing Networks Beyond Complete Bipartite Cases

Hu Fu[1(✉)], Qun Hu[1], and Jia'nan Lin[2]

[1] ITCS, Shanghai University of Finance and Economics, Shanghai, China
`fuhu@mail.shufe.edu.cn, 2019212804@sufe.edu.cn`
[2] Rensselaer Polytechnic Institute, Troy, NY, USA
`linj21@rpi.edu`

Abstract. Gaitonde and Tardos [3,4] recently studied a model of queueing networks where queues compete for servers and re-send returned packets in future rounds. They quantify the amount of additional processing power that guarantees a decentralized system's stability, both when the queues adapt their strategies from round to round using no-regret learning algorithms, and when they are patient and evaluate the utility of a strategy over long periods of time.

In this paper, we generalize Gaitonde and Tardos's model and consider scenarios where not all servers can serve all queues (i.e., the underlying graph is an incomplete bipartite graphs) and, further, when packets need to go through more than one server before their completions (i.e., when the underlying graph is a DAG). For the bipartite case, we obtain bounds comparable to those by Gaitonde and Tardos, with the factor slightly worse in the patient queueing model. For the more general multi-layer systems, we show that straightforward generalizations of the queues' utilities and servers' priority rules in [3] may lead to unbounded gaps between centralized and decentralized systems when the queues use no regret strategies. We give new utilities and service priority rules that are aware of the queue lengths, and show that these suffice to restore the bounded gap between centralized and decentralized systems.

Keywords: Queueing networks · Price of anarchy · No-regret learning dynamics

1 Introduction

A recurrent theme in algorithmic game theory is to analyze systems operated by decentralized, strategic agents, in comparison with those run by centralized authorities. Since Koutsoupias and Papadimitriou [6] introduced the concept of *Price of Anarchy*, it has been applied and studied in various games such as routing in congestion games [8], network resource allocation [5], auctions [2],

---

Supported by the Fundamental Research Funds for the Central Universities of China. Part of the work was done when the third author was visiting Shanghai University of Finance and Economics.

K. A. Hansen et al. (Eds.): WINE 2022, LNCS 13778, pp. 96–114, 2022.
https://doi.org/10.1007/978-3-031-22832-2_6

among many other settings. Recently, Gaitonde and Tardos [3,4] introduced a routing game in queueing systems, where queues compete for servers each round, and packets not processed successfully in one round go back to their queues and have to be re-sent in the future. Unlike most games previously studied, in such systems, the strategies and outcomes of one round have carryover effect in future rounds, introducing intricate dependencies among the rounds. Gaitonde and Tardos developed bicriteria bounds that quantify the loss of efficiency due to decentralized strategic behaviors in such systems in two settings: in [3], the queues evaluate the utility of their strategies from round to round, and adopt no-regret learning algorithms in their routing decisions; in [4], the queues are "patient", and fix their strategies over long periods of time over which they evaluate their performances.

In both [3] and [4], all servers can process requests from all queues, and a packet leaves the system once it is processed by a server. These are simplifying modelling assumptions: in many queueing systems, each queue's packets may only be processed by certain servers, and a packet may need to go through more than one server before leaving the system. In this work, we model such added complexities by seeing the queues and servers as nodes of a directed acyclic graph (DAG). A queue can send requests to a server only if it has an outgoing edge to the server. Packets arrive at given rates to nodes with no incoming edges, and leave the system when they are successfully processed by servers with no outgoing edges; nodes with both incoming and outgoing edges are both servers and queues—after it successfully processes a packet, the packet joins its queue and waits to be sent to the next server. The case considered by Gaitonde and Tardos [3] corresponds to complete bipartite graphs. We examine whether and how their results generalize to more general settings.

Our Results. We first characterize networks that can be stable under a centralized policy, where stability roughly means that the number of packets accumulated in the system is bounded. As in [3], the main lesson of the characterization is that it is without loss of generality for a centralized policy to fix for each queue a distribution and sample a server from this distribution at each time step, independently of the history and all other happenings in the system. For bipartite graphs (Theorem 2) our proof takes a perspective arguably simpler than that in [3], and this perspective is instrumental in showing the conditions for general DAGs, which are considerably more involved.

We then consider decentralized systems where queues use no-regret learning strategies. For general bipartite graphs, we show that the bound in [3] generalizes with minor modification. We inherit much of the proof framework of [3], including a potential function argument and various apparatus for analyzing the random processes, although in the key step of the argument where one uses the no regret property to bound the number of "old" packets processed over a time window, our proof has to take into account the underlying graph structure, and makes a connection with the *dual* form of the conditions for centralized stability. The eventual stability conditions we give (Theorem 3) when queues use no-regret learning strategies is also expressed as a scaled dual form of the centralized stability conditions. As a consequence, the main bicriteria comparison result in [3] extends to general bipartite graphs: a decentralized system is stable if it can be made stable under a centralized policy with the arrival rates doubled. Interestingly, the dual variables in our decentralized stability condition take values from a smaller range ($\{0, 1\}$) than in the centralized stability condition (where they may be any nonnegative numbers). For complete bipartite graphs, it can be shown that even for the centralized stability condition, the dual variables need only take $0, 1$ values. In this sense, our results suggest that the gap between the two conditions tends to be smaller for incomplete bipartite graphs.

Networks with more than one layer of servers are even more interesting. A major conclusion reached in [3] is that a server's rule of priority for packets simultaneously sent to it is crucial for the system's stability. In the complete bipartite graphs, it was shown that, if the servers pick a packet uniformly at random, then no said bicriteria bound can be given; in contrast, the bicriteria result was obtained when servers are assumed to prioritize older packets. Another important factor in the model is the queues' utilities: it was assumed in [3] that a queue collects utility of 1 if its packet is successfully cleared by a server, and 0 otherwise. Our results for general bipartite graphs inherit both these modelling assumptions. However, for graphs of even three layers, we give an example showing that no finite bound of the bicriteria form can be obtained if one directly extends the utility and the priority rule from [3]. Intuitively, in order for the system not to lose too much efficiency, information on the underlying graph is important when there are multiple layers: a server with strong processing capacity may be poorly connected in the next layer, and myopic strategies easily send too many packets to such a server. Therefore, the queues' utilities need to incorporate more information for their strategies to better align with the system's stability; on the other hand, if they are fed with too much global information, the difference could blur between centrally controlled systems and decentralized ones. A natural question to raise is whether it is possible to incentivize the queues using only local information so that their selfish behaviors do not hurt the system efficiency too much. We answer this question in the affirmative, showing that the *lengths* of queues in the neighboring nodes provide just this information. We propose a new service priority rule, under which the servers prioritize packets from the *longest* queues. We also propose new utility functions for queues, with which a queue of length L_i, when it sends a packet to a server j whose own queue is of length L_j, obtains utility $L_i - L_j$ if the packet is successfully processed by j. In particular, with this new utility function, a queue never sends its packets to a server whose current queue is longer than itself. We show that when the new service priority rule and utilities are adopted, the bicriteria result is restored: a queueing system is stable with queues that use no-regret strategies as long as it is stable under a centralized policy even when the packet arrival rates are doubled.

Lastly, we extend the model with patient queues to bipartite graphs. Gaitonde and Tardos [4] showed for complete bipartite graphs that, when queues are patient, with appropriately defined long-term utilities, a Nash equilibrium always exists, and a system is stable under any Nash equilibrium as long as it is stable under a centralized policy even with $\frac{e}{e-1}$ times the original arrival rates. To this end, they developed elaborate tools for computing the long-term utilities given the queues' strategies. These tools generalize straightforwardly in general bipartite graphs, but the delicate deformation argument in the proof of their bicriteria result does not easily generalize. Our proof again makes use of the dual form of the condition for centralized stability, which provides a matching between the fastest growing queues in an equilibrium and servers.

In the full version of this paper, we also consider two other variants of the problem: in one model, whether a server can process a packet is not determined by which queue the packet is from, but is an intrinsic property of the packet; in the other one, the arrival of packets at each queue is not from a Bernoulli distribution, but is controlled by an adversarial, as in the model of Borodin et al. [1]. In both variants, we show that the bicriteria results persist when queues use no-regret strategies. Lastly, we give a tighter bicriteria result for the model in [3], where the underlying graph is a complete bipartite graph. We show that a queueing system is stable with queues that play no-regret strategies as long as it is stable under a centralized policy even when the k-th largest packet arrival rate is increased by a factor $\frac{2k-1}{k}$ for each k.

Further Related Works. We refer to Gaitonde and Tardos [3,4] for pointers to related works in algorithmic game theory and no regret learning. Sentenac et al. [9] considered the same model as in [3] but when queues use *cooperative* learning. When incentives are removed from the problem, they show that the queues can essentially learn the necessary system parameters and reach a stable outcome as long as the system is stable under a centralized policy.

2 Preliminaries

2.1 Queue-G Model

A *Queue-G Model* is a $G = (V, E, \boldsymbol{\lambda}, \boldsymbol{\mu})$, where $(V = S_1 \cup S_2 \cup S_3, E)$ constitutes a directed acyclic graph, and $\boldsymbol{\lambda}$ and $\boldsymbol{\mu}$ are the arrival and processing rates on the nodes. A node i with no incoming edge is a *source*, and has an *arrival rate* λ_i. For each i, $\lambda_i \in (0, 1)$. S_1 denotes the set of sources. All the other nodes are *servers*, and each server j has a *processing rate* μ_j. A server with no outgoing edge is a *terminal*. S_3 denotes the set of terminals. The set of non-terminal server nodes is $S_2 := V - S_1 - S_3$. An edge $(i, j) \in E$ means that node i can send packets to node j. For $i \in S_1 \cup S_2$, we denote by $N^{\text{out}}(i) := \{j \in V : (i, j) \in E\}$ the set of out-neighbors of i, and for a server i, we denote by $N^{\text{in}}(i) := \{j \in V : (j, i) \in E\}$ the set of in-neighbors of i.

Let Q_t^i denote the number of packets at node i at the beginning of time step t. For all $i \in V$, $Q_0^i = 0$. At each time step t, the following events happen, in two phases:

(I) Packet sending: each node i with $Q_t^i > 0$ chooses a server j from $N^{\text{out}}(i)$ and sends to j the oldest packet (with the earliest timestamp) in i's queue. In a centralized system, a central authority dictates for each node if and where to send its packet at each time step; in a decentralized system, each node strategizes over this decision.

(II) Packet arrival and processing: at each source $i \in S_1$, a packet with timestamp t arrives with probability λ_i; each server $i \in S_2 \cup S_3$, if it receives
any packet in phase (I), chooses one such packet according to some *service
priority rule* to process, and succeeds with probability μ_i. The arrivals of
packets at each source and the successes of their processing at each server
are all mutually independent events. A packet cleared by server $j \in S_2$ joins
the queue of server j; a packet cleared by a server in S_3 leaves the system.
A packet not chosen by or not successfully processed by a server goes back
to the node that sends it. It follows that any $i \in S_3$ has $Q_t^i = 0$ at any time
step t.

Gaitonde and Tardos [3] considered a special case of the Queue-G Model,
where there are no non-terminal servers and every source can send packets to
every server, i.e., $S_2 = \emptyset$ and $E = S_1 \times S_3$, and the service priority rule at each
server is to choose the oldest packet (breaking ties arbitrarily).[1]

We refer to this special case as the *Queue-CB Model* ("CB" for complete
bipartite).

If we only have $S_2 = \emptyset$ (and allow any $E \subseteq S_1 \times S_3$), we have the *Queue-B
Model*.

2.2 Stability and No Regret Learning

Let $Q_t := \sum_{i \in V} Q_t^i$ be the total number of packets in the queueing system at
the start of time step t. We inherit from [3] the notion of stability:

Definition 1. *Under some scheduling policy (either with a central authority or
with queues strategizing), a queueing system is* strongly stable *if for any $a > 0$,
there is a constant C_a only related to a, such that $\mathbb{E}[(Q_t)^a] \leq C_a$ for all t. A
queueing system is* **almost surely strongly stable** *if with probability 1, the
following event happens: for any $a > 0$, $Q_t = o(t^a)$.*

Gaitonde and Tardos [3] showed that if a queueing system is strongly stable,
then it is almost surely strongly stable. We therefore focus on showing strong
stability, and often refer to a strongly stable system simply as stable.

The following theorem by Pemantle and Rosenthal [7], also used in [3], is the
workhorse for showing stability.

Theorem 1 ([7]). *Let $X_0, \cdots, X_n$ be nonnegative random variables. If there
are constants $b, c, d > 0$ and $p > 2$ such that $X_0 \leq b$ and, for all n,*

$$\mathbb{E}(|X_{n+1} - X_n|^p \mid X_0, \cdots, X_n) \leq d; \tag{1}$$
$$X_n > b \quad \Rightarrow \quad \mathbb{E}(X_{n+1} - X_n \mid X_0, \cdots, X_n) \leq -c, \tag{2}$$

*then for any $a \in (0, p-1)$, there is $C = C(p, a, b, c, d)$ such that $\mathbb{E}(X_n)^a < C$
for all n.*

---

[1] For ease of presentation, we made minor changes from Gaitonde and Tardos [3]'s
model, in the order of packet sending and packet arrival. It is easy to see that
the difference is negligible for the analysis of the system's stability, which is an
asymptotic quality, to be defined below.

We refer to (2) as the *negative drift condition*, and (1) as the *bounded jump condition*.

We now introduce utilities of queues, as defined in [3]. The utility of a queue at time step t is the number of packets cleared from this queue at time step t. Let $a_i(t)$ denote the server that node i chooses at time step t. (A node i may not choose any server, in which case we let $a_i(t) = 0$ and we set $\mu_0 = 0$.) Let $\mathcal{F}_t$ denote the history of the system up to the beginning of time step t. We use $u_t^i(a_i(t), a_{-i}(t)|\mathcal{F}_t)$ to denote the utility of node i when node i chooses server $a_i(t)$ and the other nodes choose $a_{-i}(t)$, given history $\mathcal{F}_t$. We should specify the content of a history: $\mathcal{F}_t$ only includes information on which packets, up to time step t, were cleared and the age of the currently oldest packet in each node, but does not include the queue size Q_t^i. This makes sure that, for the k-th packet in node i that is cleared at time step t, the time difference between its arrival and that of the $(k+1)$-st packet is independent of the history $\mathcal{F}_{t'}$ for all $t' < t$, and obeys the geometric distribution with parameter λ_i.

Lastly, we define the *regret* of a node i up to time w as the difference between its utility in a real sample path and what it could have achieved by always playing a best fixed action in hindsight.

Definition 2. *For a time window from time step $t_0 - w$ to $t_0 - 1$, the* regret *of queue i for actions $a_i(t_0 - w), \ldots, a_i(t_0 - 1)$ is*

$$\mathrm{Reg}_i(w, t_0) := \max_{j : j \in N^{\mathrm{out}}(i)} \sum_{t=t_0-w}^{t_0-1} u_t^i(j, a_{-i}(t)|\mathcal{F}_t) - \sum_{t=t_0-w}^{t_0-1} u_t^i(a_i(t), a_{-i}(t)|\mathcal{F}_t).$$

Note that, in this definition, the utility obtained by playing the best fixed strategy is evaluated using "real" histories ($\mathcal{F}_t$'s) observed under the actual actions taken by the node. It does not use counterfactual histories generated by playing the fixed strategy. We often drop the parameter t_0 when it is clear from the context.

Definition 3. *Given fixed $\delta \in (0, 1)$, queue i's scheduling policy is* no regret *if, for any time window from time step $t_0 - w$ to $t_0 - 1$, with probability at least $1 - \delta$, $\mathrm{Reg}_i(w, t_0) \leq \varphi_\delta(w)$, where $\varphi_\delta(w) = o(w)$ may depend only on δ and the number of nodes in the queueing system.*

3   Bipartite Queueing Systems

In this section we derive necessary and sufficient conditions for the existence of a centralized policy that stabilizes a queueing system in a bipartite graph (a Queue-B model). We then give a sufficient condition that guarantees the stability of such systems when all queues adopt no-regret strategies. For the special case when the underlying graph is complete bipartite, our conditions degenerate to the ones given by Gaitonde and Tardos [3].

3.1 Stability Conditions Under Centralized Policies

A Queue-B model as defined in Sect. 2 simply consists of n queues on one side and m servers on the other. Server j is able to clear a packet from queue i if and only if there is an edge between the two. It is easy to see that a centralized policy never benefits from sending packets from two queues to a same server in a single time step, as the server picks up only one of them. Therefore, with loss of generality, the routing dictated by a centralized policy at any step gives a matching of the queues to the servers. (Some queues may be asked not to send their packets, and some servers may be allowed to be idle for that round.) It is less clear whether a centralized policy benefits from making intricate use of the history when it decides on the matching at each step. It turns out, for the system to be stable (Definition 1), it is without loss of generality to consider history oblivious centralized policy, which samples a matching from a fixed distribution over matchings from step to step. A Queue-B model can be stable under any centralized policy if and only if it can be stable under such a policy. This is the essence of the following theorem.

Recall that a fractional matching matrix $P \in [0,1]^{n \times m}$ is such that $\sum_j P_{ij} \le 1$ for all $i \in [n]$ and $\sum_i P_{ij} \le 1$ for all $j \in [m]$.

Theorem 2. *Given a Queue-B model with n queues and m servers, with arrival rates $\boldsymbol{\lambda} = (\lambda_1, \cdots, \lambda_n)$ and processing rates $\boldsymbol{\mu} = (\mu_1, \ldots, \mu_m)$, there is a centralized policy under which the system is stable if and only if there exists a fractional matching matrix $P \in [0,1]^{n \times m}$, such that $P\boldsymbol{\mu} \succ \boldsymbol{\lambda}$, where $\succ$ denotes element-wise greater than.*

Sufficiency of the condition is a consequence of Birkhoff-von Neumann theorem. The argument of necessity makes use of the observation that, conditioning on any event in the system, the expected routing decision made by a centralized policy is expressible as a fractional matching matrix. One may as well condition on the event that all queues have arrivals considerably larger than the expectations, which occurs with constant probability. This part of the argument is arguably simpler than the proof in [3], and makes possible the more involved proof for more general graphs (Theorem 4). The proofs missing due to lack of space are deferred to the full version of this paper.

Before moving on to decentralized Queue-B models, we derive a dual form of the conditions in Theorem 2. The dual form plays a crucial role in our analysis of the systems' stability under no-regret policies.

Lemma 1. *Given a Queue-B model with arrival rates $\boldsymbol{\lambda}$ and processing rates $\boldsymbol{\mu}$, the following two conditions are equivalent:*

(1) There is a fractional matching matrix P such that $P\boldsymbol{\mu} \succ \boldsymbol{\lambda}$.
(2) For any $\boldsymbol{\alpha} \in \mathbb{R}_+^n$, there is a matching matrix $M \in \{0,1\}^{n \times m}$, such that $\boldsymbol{\alpha}^\top M \boldsymbol{\mu} > \boldsymbol{\alpha}^\top \boldsymbol{\lambda}$.

The lemma is an application of Farkas' lemma. The proof of this lemma is deferred to the full version. It is worth pointing out that, when the underlying graph is a complete bipartite graph, it suffices to have the condition (2) satisfied for only $\boldsymbol{\alpha} \in \{0,1\}^n$. This difference plays a role in the contrast between complete and incomplete bipartite graphs when the system is decentralized, as we explain in the next section.

3.2   Stability Conditions Under Decentralized, No-Regret Policies

In this section we give conditions under which, in a queueing system on an incomplete bipartite graph (the Queue-B model), if all queues use no-regret strategies, the system is stable. Our conditions are most easily comparable with the dual form of centralized stability conditions stated in Lemma 1. When the underlying graph is a complete bipartite graph, the conditions are identical to those by Gaitonde and Tardos [3], as we discuss below. The technique in this part is largely inherited from [3], although our proof reveals an interesting connection between the dual form of stability conditions and key steps in the proof. The sufficient condition is the following:

Assumption 1. *There is a constant $\beta > 0$, such that for any $\boldsymbol{\alpha} = (\alpha_1, ..., \alpha_n) \in \{0,1\}^n$, there is a matching matrix M, such that $\frac{1}{2}(1-\beta)\boldsymbol{\alpha}^\top M \boldsymbol{\mu} > \boldsymbol{\alpha}^\top \boldsymbol{\lambda}$.*

A quick comparison between this and dual condition in Lemma 1 suggests that, if one has a Queue-B model which can be made stable by a centralized policy, then, doubling its processing capabilities guarantees its stability when the queues use no-regret strategies. Note though that the range of $\boldsymbol{\alpha}$ is much smaller in Assumption 1 ($\{0,1\}^n$) than in Lemma 1 ($\mathbb{R}_+^n$). For complete bipartite graphs, this difference vanishes (see remark following Lemma 1), but in general bipartite graphs, this difference is real. This suggests that in incomplete bipartite graphs, the gap between centralized and decentralized systems tends to be smaller than in complete bipartite graphs.

Theorem 3. *If a Queue-B model queueing system satisfies Assumption 1, and queues use no-regret learning strategies with $\delta = \frac{\beta}{128n}$, then the system is strongly stable.*

Following Gaitonde and Tardos [3], we introduce a potential function with the intention to apply Theorem 1 to its square root. The *age* of a packet that arrives in the system at time t_1 is defined to be $t_2 - t_1$ at time t_2. Let T_t^i be the age of the oldest packet in queue i at time step t, and let $\boldsymbol{T}$ be the vector $(T_t^1, \cdots, T_t^n)$. Note that Q_t^i, the length of the queue, is at most T_t^i. For a positive integer $\tau > 0$, define

$$\Phi_\tau(\mathbf{T}_t) := \sum_{i:T_t^i \geq \tau} \lambda_i (T_t^i - \tau).$$

The potential function Φ is defined as

$$\Phi(\mathbf{T}_t) := \sum_{\tau=1}^{\infty} \Phi_\tau(\mathbf{T}_t) = \sum_{\tau=1}^{\infty} \sum_{i:T_t^i \geq \tau} \lambda_i (T_t^i - \tau) = \frac{1}{2} \sum_{i=1}^{n} \lambda_i T_t^i (T_t^i - 1).$$

We analyze the system by dividing the time steps into windows of length w each, for some large enough w. Let $Z_\ell := \sqrt{\Phi(\mathbf{T}_{\ell \cdot w})}$ be the square root of the potential function at the beginning of the ℓ-th window. The main work lies in showing that $(Z_\ell)_\ell$ satisfies the conditions of Theorem 1, which implies $\mathbf{E}[Z_\ell^a]$ is bounded for any $a > 0$. This in turn implies that $\mathbf{E}[(\sum_i T_t^i)^a]$ is bounded, and so is $\mathbf{E}[(Q_t)^a]$.

Lemma 2. *[Negative drift condition.] Denote by $\lambda_{(n)}$ the minimum element of λ.*
Let $b = \frac{w}{\sqrt{2\lambda_{(n)}}} \max\left(\frac{8}{\beta}\left(\sum_{i=1}^n \lambda_i\right), 16n^2\right)$, $c = -\frac{\sqrt{2\lambda_{(n)}}\beta w}{64}$. Then $Z_0 = 0 \le b$ and, for all ℓ,

$$Z_\ell > b \quad \Rightarrow \mathbf{E}\left[Z_{\ell+1} - Z_\ell \mid Z_0, \cdots, Z_\ell\right] \le -c.$$

Lemma 3. *[Bounded jump condition.] For each even integer $p \ge 2$, there is a constant d_p, such that for all ℓ*

$$\mathbf{E}\left[|Z_{\ell+1} - Z_\ell|^p \mid Z_0, \cdots, Z_\ell\right] \le d_p.$$

Lemma 3 is identical to the corresponding part in Gaitonde and Tardos [3], and we omit its proof. The main difference between our proof and [3] is in the proof of the negative drift condition (Lemma 2). We present here the key steps of our proof, and the rest is deferred to full version.

Following Gaitonde and Tardos [3], for a given $\tau > 0$, we say a packet is τ-*old* if its age is at least τ at time step $\ell \cdot w$, i.e., if its arrival time is no later than $\ell w - \tau$. Let J_τ be the set of queues which have τ-old packets at time step $(\ell+1) \cdot w$. For a queue i, if by time step $(\ell+1) \cdot w$, it still has packets that arrived before time step $\ell \cdot w$, let $\tau_i = \max_{\tau > 0 : J_\tau \ni i} \tau$ be the age of the oldest packet in queue i; otherwise, set $\tau_i = 0$. Let N_τ^i be the number of τ-old packets cleared from queue i during the time window from time step $\ell \cdot w$ to $(\ell+1) \cdot w$. Similarly, for a server j, let L_τ^j be the number of τ-old packets cleared by server j during this time window. Next, define $N_\tau = \sum_{i \in [n]} N_\tau^i = \sum_{j \in [m]} L_\tau^j$ as the number of τ-old packets cleared during this time window. Lastly, let C_t^j be the indicator variable for server j succeeding in processing a packet if it picks one up.

Lemma 4. *For any $\tau > 0$ and $\epsilon > 0$, if $\sum_{t=\ell \cdot w}^{(\ell+1) \cdot w - 1} C_t^j \ge (1-\epsilon)\mu_j w$ for each j, then $N_\tau \ge \frac{1-\epsilon}{1-\beta} \sum_{i \in J_\tau} \lambda_i w - \sum_{i=1}^n \mathrm{Reg}_i(w, (\ell+1) \cdot w)$.*

Proof. Any queue $i \in J_\tau$ has a τ-old packet throughout the time window. For a server j which can serve queue i, consider the counterfactual utility i may gain during this time window by sending a request to j at each step. Let X_{jt}^τ be the indicator variable for the event that *some queue* (which may not be i) sends a τ-old packet to server j at time step t. Then at any time step t when $X_{jt}^\tau = 0$, server i's packet would have been picked up by server j had i sent a request, because no other packet sent to j is τ-old, so the packet from i has priority. Recall that C_t^j is the indicator variable for server j succeeding in processing a packet if it picks one up. So queue i would have gained utility 1 at time t by sending a request to j if $C_t^j = 1$ and $X_{jt}^\tau = 0$. Over the time window, queue i's counterfactual utility could have been $\sum_{t=\ell\cdot w}^{(\ell+1)\cdot w-1}(1 - X_{jt}^\tau)$. Note that queue i's actual utility is N_τ^i, so by definition of regret, we have

$$N_\tau^i \geq \sum_{t=\ell\cdot w}^{(\ell+1)\cdot w-1} C_t^j(1 - X_{jt}^\tau) - \text{Reg}_i(w, (\ell+1)\cdot w).$$

On the other hand, whenever $X_{jt}^\tau S_t^j = 1$, server j successfully clears a τ-old packet. Therefore, $L_\tau^j = \sum_{t=\ell\cdot w}^{(\ell+1)\cdot w-1} C_t^j X_{jt}^\tau$. Then, for a pair of queue $i \in J_\tau$ and server j that can serve i, we have

$$N_\tau^i + L_\tau^j \geq \sum_{t=\ell\cdot w}^{(\ell+1)\cdot w-1} C_t^j - \text{Reg}_i(w, (\ell+1)\cdot w). \tag{3}$$

Now we are ready to apply Assumption 1. Let $\boldsymbol{\alpha}$ be the indicator vector for the set $J_\tau \subseteq [n]$, i.e., $\alpha_i = 1$ if $i \in J_\tau$, and $\alpha_i = 0$ otherwise. By Assumption 1, we can find a matching matrix M_τ such that $\frac{1}{2}(1 - \beta)\boldsymbol{\alpha}^\top M_\tau \boldsymbol{\mu} > \boldsymbol{\alpha}^\top \boldsymbol{\lambda}$. Let U_τ be the edge set such that $(i,j) \in U_\tau \Leftrightarrow M_\tau(i,j) = 1$. Then,

$$\frac{1}{2}(1 - \beta) \sum_{(i,j)\in U_\tau} \mu_j > \sum_{i\in J_\tau} \lambda_i. \tag{4}$$

Now, we are ready to give a lower bound for N_τ:

$$2N_\tau = \sum_{i=1}^n N_\tau^i + \sum_{j=1}^m L_\tau^j \geq \sum_{(i,j)\in U_\tau} (N_\tau^i + L_\tau^j)$$

$$\geq \sum_{(i,j)\in U_\tau} \left(\sum_{t=\ell\cdot w}^{(\ell+1)\cdot w-1} C_t^j - \text{Reg}_i(w, (\ell+1)\cdot w) \right) \tag{5}$$

$$\geq \sum_{(i,j)\in U_\tau} (1 - \epsilon)\mu_j w - \sum_{i=1}^n \text{Reg}_i(w, (\ell+1)\cdot w) \tag{6}$$

$$\geq \frac{2(1 - \epsilon)}{1 - \beta} \sum_{i\in J_\tau} \lambda_i w - \sum_{i=1}^n \text{Reg}_i(w, (\ell+1)\cdot w), \tag{7}$$

where the second inequality uses (3), and the last inequality uses (4).

We sketch the rest of the proof, and all details are deferred to the full version. When the servers' realized processing capacities are close to their expectations (as in the condition of Lemma 4) and when the queues' regret are small (which should happen with high probability by assumption), the lower bound given by Lemma 4 on N_τ implies a lower bound on the decrease in the potential function due to packet clearing (Lemma 5). We can further bound the increase in the potential function due to aging over the time interval. (Lemma 6). We define an event A, specified in the full version of this paper, which happens with high probability, and under which all of these events (of concentration and no regret) happen.

Recall that τ_i is the age of the oldest packet in queue i at time step $(\ell+1) \cdot w$, where the age is measured by time step $\ell \cdot w$. Let $\boldsymbol{\tau} = \{\tau_1, \cdots, \tau_n\}$.

Lemma 5. *Under* event A, $\Phi(\mathbf{T}_{\ell \cdot \mathbf{w}}) - \Phi(\boldsymbol{\tau}) \geq \frac{1-2\epsilon}{1-\beta} \sum_{i=1}^{n} \lambda_i \tau_i w$.

Lemma 6. *Under* event A, $\Phi(\mathbf{T}_{(\ell+1) \cdot \mathbf{w}}) - \Phi(\boldsymbol{\tau}) \leq \sum_{i=1}^{n} \lambda_i \tau_i w + \frac{1}{2} \sum_{i=1}^{n} \lambda_i w^2$.

With a small probability, event A does not happen, and it is relatively straightforward to upper bound the increase in the potential in this case. (Most pessimistically, no packets is cleared during the time window and T_t^i in each queue grows by w.)

Lemma 7. *If event A does not happen,* $\Phi(\mathbf{T}_{(\ell+1) \cdot \mathbf{w}}) - \Phi(\mathbf{T}_{\ell \cdot \mathbf{w}}) \leq \sum_{i=1}^{n} \lambda_i T_{\ell \cdot w}^i w + \frac{1}{2} \sum_{i=1}^{n} \lambda_i w^2$.

Lemma 2 follows from combining Lemma 5, 6 and 7.

4 Queueing Systems with Multiple Layers

In this section we study queueing systems where packets or tasks may need to go through more than one servers before their completions. After a packet is successfully processed by an intermediate server, it immediately joins the queue forming at their server, waiting to be sent to the next server. In Sect. 4.1, we give sufficient and necessary conditions for such a queueing system to be stable under a centralized policy. In Sect. 4.2, we show that, when one extends the utility and service priority rules from Gaitonde and Tardos [3]'s model to such networks, it is impossible to obtain a PoA result comparable to Theorem 3. In Sect. 4.2, we introduce new utilities and service priority rules that are aware of local queue lengths, and show that they suffice to restore conditions for stability under decentralized, no-regret strategies.

4.1 Stability Under Centralized Policies

As we reasoned for the bipartite case, it never benefits a central planner to send packets from more than one queues to the same server in a single time step, therefore, it is without loss of generality to consider policies under which, at each time step, the edges along which packets are sent from a set of vertex-disjoint paths. (Note that one such path need not start from a source or end at a terminal.) In general, at each step this set of paths may be sampled from a distribution that depends on the history. As in the bipartite case, the following characterization of stable systems shows it without loss of generality to let this distribution be the same from step to step, regardless of what happened in the past. The proof though is considerably more involved than in the bipartite case.[2]

Theorem 4. *Given a Queue-G model* $(V, E, \boldsymbol{\lambda}, \boldsymbol{\mu})$, *the following statements are equivalent.*

1. *There exists a centralized policy under which the system is stable.*
2. *The following linear system is feasible:*

$$\lambda_i < \sum_j z_{ij} \mu_j, \qquad\qquad \forall i \in S_1; \qquad (8)$$

$$\mu_i \sum_j z_{ji} < \sum_j z_{ij} \mu_j, \qquad \forall i \in S_2 \text{ with } \prod_{j \in N^{\mathrm{in}}(i)} z_{ji} > 0; \qquad (9)$$

$$\sum_j z_{ij} \leq 1, \qquad\qquad \forall i \in S_1 \cup S_2; \qquad (10)$$

$$\sum_j z_{ji} \leq 1, \qquad\qquad \forall i \in S_2 \cup S_3; \qquad (11)$$

$$z_{ij} = 0, \qquad\qquad \forall (i,j) \notin E; \qquad (12)$$

$$z_{ij} \geq 0, \qquad\qquad \forall (i,j) \in E. \qquad (13)$$

3. *The following linear system in* $(f_{i\pi})_{i \in S_1, \pi \in \Pi}$ *is feasible, where* Π *is the set of paths from a node in* S_1 *to a node in* S_3:

$$\sum_{i \in S_1} \sum_{y \in N^{\mathrm{out}}(x)} \sum_{\pi \in \Pi : \pi \ni (x,y)} \frac{f_{i\pi}}{\mu_y} \leq 1, \forall x \in S_1 \cup S_2; \qquad (14)$$

$$\sum_{i \in S_1} \sum_{y \in N^{\mathrm{in}}(x)} \sum_{\pi \in \Pi : \pi \ni (y,x)} f_{i\pi} \leq \mu_x, \forall x \in S_2 \cup S_3; \qquad (15)$$

$$\sum_{\pi \in \Pi} f_{i\pi} > \lambda_i, \forall i \in S_1; \qquad (16)$$

$$\sum_{i \in S_1} \sum_{\pi \in \Pi : \exists y, (y,x) \in \pi} f_{i\pi} = \sum_{i \in S_1} \sum_{y \in N^{\mathrm{out}}(x)} \sum_{\pi \in \Pi : \pi \ni (x,y)} f_{i\pi}, \forall x \in S_2; \quad (17)$$

$$f_{i\pi} = 0, \forall i \in S_1, \pi \in \Pi \text{ s.t. } i \text{ not on } \pi. \qquad (18)$$

We relegate the proof of the theorem to the full version. The fact that constraints (8)–(13) being feasible implies the stability of a centralized policy is a relatively straightforward consequence of a generalization of Birkhoff-von Neumann theorem. Proving the other direction is considerably more involved than for the bipartite case, and it is for this purpose that we introduce the third condition in Theorem 4. We show that the feasibility of constraints (14)–(18) implies the feasibility of constraints (8)–(13), then we show that a system for which constraints (14)–(18) are not feasible cannot be stable.

---

[2] It is relatively easy to extend the argument in Theorem 2 to show the necessity of the conditions in Theorem 4, except for the strictness of the signs in (8) and (9).

Again we give a dual form of conditions for centralized stability, which are central to our analysis of the systems' stability under no-regret policies. Its proof can be found in the full version of this paper.

Definition 4. *A vertex-disjoint path (one such path need not start from a source or end at a terminal) is a collection of edges. Any two edges in the path can't have the same head or the same tail.*

Lemma 8. *Given a Queue-G model $(V, E, \boldsymbol{\lambda}, \boldsymbol{\mu})$ with n sources and m servers, the following two conditions are equivalent:*

(1) The linear system of the second statement in Theorem 4 is feasible.
(2) For any $\boldsymbol{\alpha} \in \{\mathbb{R}_+^{n+m} \mid \alpha_i = 0 \text{ if } i \in S_3\}$, there is a vertex-disjoint path set U, such that $\sum_{(i,j)\in U}(\alpha_i - \alpha_j)\mu_j > \sum_{i\in S_1} \alpha_i\lambda_i$.

4.2   Decentralized Multi-Layer Networks

System Failure with Myopic Queues. In a queueing system with multiple layers, (i.e., when $S_2 \neq \emptyset$), a natural extension of the utility in bipartite systems as defined in Sect. 2 is to let a queue earn utility 1 at a time step if one of its packet is successfully processed by the server it is sent to. The hope is that when all queues focus on getting their packets processed by the *next* server, the system runs relatively efficiently. Unfortunately, as the following example shows, when the queues run no-regret strategies on such utilities, they may be too short-sighted for the decentralized system to have performance comparable to a centralized one, even if one increases the processing capacities by any constant factor.

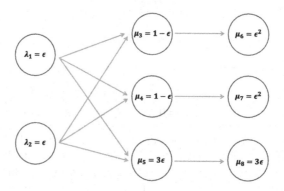

Fig. 1. A queueing system with two layers of servers. A centralized policy sending packets from both sources to server 5 makes the system stable, but the two sources may find it a no-regret strategy to send requests to servers 3 and 4, respectively.

Example 1. The system shown in Fig. 1 is stable under a centralized policy. One feasible solution to the linear system given in Theorem 4 is $z_{15} = z_{25} = 0.4$, $z_{58} = 1$, with all other coordinates of $\mathbf{z}$ set to 0. It is not difficult to see that, if both queue 1 and queue 2 send their requests to server 3 and server 4, respectively, they play no-regret strategies, but the system is unstable because packets accumulate at servers 3 and 4. The phenomenon persists even when the processing capacities are increased by a factor of $\frac{1}{\epsilon}$.

Stability with Queue Length Aware Utilities. Example 1 suggests that the instantaneous, local feedback is not enough to align the queues' interests with the system's efficiency. In this section we show that, when we incorporate one other piece of local information, the *queue lengths*, into the queues' utilities and the service priority rule, we can recover the bicriteria results we showed for single-layer systems in Sect. 3.

Recall that Q_t^i denotes the length of queue i at time t. Our new utility for queue $i \in S_1 \cup S_2$ for sending a request to server j at time t is

$$u_t^i(j, a_{-i}(t) \mid \mathcal{F}_t) = \begin{cases} Q_t^i - Q_t^j, & \text{if the packet sent to } j \text{ is successfully processed;} \\ 0, & \text{otherwise.} \end{cases}$$

Note that this utility function immediately implies that it is never in a queue's interest to send a request to a server with a queue longer than itself. Also recall that $Q_t^j = 0$ for any $j \in S_3$ at any time t. The history $\mathcal{F}_t$ now includes information on which packets have been cleared and the queue size Q_t^i.[3]

We also change the servers' priority rules to preferring requests from longer queues. With the new utilities and service priority rules, the sufficient condition we obtain for decentralized stability is:

Assumption 2. *There is a $\beta > 0$ such that for any $\boldsymbol{\alpha} \in \{\mathbb{R}_+^{n+m} \mid \alpha_i = 0 \text{ if } i \in S_3\}$, there is a vertex-disjoint path set U such that*

$$\frac{1}{2}(1 - \beta) \sum_{(i,j) \in U} (\alpha_i - \alpha_j)\mu_j > \sum_{i \in S_1} \alpha_i \lambda_i$$

Theorem 5. *If a Queue-G model queueing system satisfies Assumption 2 holds, and nodes use no-regret learning strategies with $\delta = \frac{\beta \mu_{(m)}}{96(n+m)^2}$, then the queueing system is strongly stable.*

A quick comparison between Assumption 2 and the dual form of condition for centralized stability in Lemma 8 shows that, a queueing system is guaranteed

---

[3] This is different from the setup in Sect. 2. We now no longer have the independence between the time interval between packet arrivals and histories prior to the their clearing. As will be clear in the proof, this independence is no longer needed in the proof. The introduction of queue lengths makes the change in the potential more directly connected with the queues' utilities.

to be stable with queues using no-regret learning strategies as long as the system can be stable under a centralized policy with twice as many packet arrivals.

As in the proof for Theorem 3, we introduce a similar potential function

$$\Phi\left(\mathbf{Q}_t\right) := \frac{1}{2} \sum_{i \in S_1 \cup S_2} Q_t^i(Q_t^i - 1), \tag{19}$$

and define Z_ℓ as its square root at the beginning of the ℓ-th window of length w. There is change in the proofs for both the negative drift condition and the bounded jump condition. We detail here the main different steps in Lemma 9 for the negative drift condition, and relegate the rest to the full version.

Lemma 9. *[Negative drift condition.]* Let $b = \frac{8\sqrt{2(n+m)}}{\beta\mu_{(m)}}(\sum_{i \in S_1} \lambda_i^2 w^2 + \sum_{i \in S_2 \cup S_3} \mu_j w^2 + w)$, $c = -\frac{\sqrt{2}\mu_{(m)}\beta w}{128(n+m)}$. Then $Z_0 = 0 \le b$ and, for all ℓ,

$$Z_\ell > b \quad \Rightarrow \mathbf{E}\left[Z_{\ell+1} - Z_\ell \mid Z_0, \cdots, Z_\ell\right] \le -c.$$

Lemma 10. *[Bounded jump condition.]* *For each even integer $p \ge 2$, there is a constant d_p, such that for all ℓ,*

$$\mathbf{E}\left[|Z_{\ell+1} - Z_\ell|^p \mid Z_0, \cdots, Z_\ell\right] \le d_p.$$

At time step $(\ell+1) \cdot w$, let τ_i be the number of unprocessed packets at node i which arrived before time step $\ell \cdot w$. Note the difference from the definition of τ_i in the proof of Theorem 3—there, τ_i is the *age* of the oldest packet in queue i at time ℓw. For any node i and $t \in [\ell \cdot w, (\ell+1) \cdot w]$, $\tau_i \le Q_t^i \le \tau_i + w$. Recall that C_t^j is the indicator variable for server j succeeding in processing a packet if it picks one up, and u_t^i is the utility of queue i at time step t. For a server j, define its *contribution* v_t^j to be u_t^i if j successfully clears a packet from queue i at time step t, and is 0 if it fails to do so. Then, at any time step t, $\sum_{i \in S_1 \cup S_2} u_t^i = \sum_{j \in S_2 \cup S_3} v_t^j$. A key observation is that, with the new utility functions, when a packet is cleared, the decrease in potential is exactly equal to the increase in the corresponding queue's utility. We can therefore calculate the decrease in potential function by tracking the sum of utilities of all nodes. The following key lemma lower bounds the total utility over a time window as a function of the number of old packets (τ_i's), assuming the realized processing capacities of all servers are around their expectations. Its use of vertex-disjoint paths paves the way for applying Assumption 2. The following lemma gives the lower bound of the utility of queues since queues use no regret learning strategies: for a pair of a long queue and a server, either the server clears many packets from long queues, generating a certain amount of utility or queues will generate a certain amount of utility since its utility will no less than the utility if it always sends packets to this server and its packets have priority since its queue length is large.

Lemma 11. *For any $\epsilon > 0$, if $\sum_{t=\ell \cdot w}^{(\ell+1) \cdot w - 1} C_t^j \ge (1 - \epsilon)\mu_j w$ for each $j \in S_2 \cup S_3$, then for any set U of vertex-disjoint paths, $\sum_{i \in S_1 \cup S_2} \sum_{t=\ell \cdot w}^{(\ell+1) \cdot w - 1} u_t^i \ge \frac{1}{2}\sum_{(i,j) \in U}(\tau_i - \tau_j - w)(1 - \epsilon)\mu_j w - \sum_{i \in S_1 \cup S_2} \operatorname{Reg}_i(w, (\ell+1) \cdot w)$.*

We sketch the rest of the proof, and relegate details to the full version.

When the servers' realized processing capacities are close to their expectations (as in the condition of Lemma 11) and when the queues' regrets are small (which should happen with high probability by assumption), the lower bound given by Lemma 11 on the utility of queues implies a lower bound on the decrease in the potential function due to packet processing (Lemma 12). When the packet arrivals in the sources are close to their expectations, we can further bound the increase in the potential function due to packet arrival (Lemma 13). We define an event B, specified in the full version of this paper, which happens with high probability, and under which all of these events (of concentration and no regret) happen. Roughly speaking, when B happens, the increase in the potential due to packet arrivals is at least offset by the decrease due to packet clearing when we use Assumption 2 to generate an appropriate U in Lemma 11 and relate the term $\frac{1}{2}\sum_{(i,j)\in U}(\tau_i - \tau_j)\mu_j w$ to $\sum_{i\in S_1}\tau_i\lambda_i$. With a small probability, event B does not happen, and it is relatively straightforward to upper bound the increase in the potential in this case.

To put things formally, recall that τ_i is the length of packets in queue i at time step $(\ell+1)\cdot w$, which arrived in queue i before time step $\ell\cdot w$; let $\boldsymbol{\tau} = \{\tau_1, \cdots, \tau_n\}$. Given τ_i for each node i, by Assumption 2, let U^* be the vertex-disjoint path such that

$$\frac{1}{2}(1-\beta) \sum_{(i,j)\in U^*} (\tau_i - \tau_j)\mu_j > \sum_{i\in S_1} \tau_i\lambda_i.$$

Lemma 12. *Under* event B, $\Phi(\mathbf{Q}_{\ell\cdot\mathbf{w}}) - \Phi(\boldsymbol{\tau}) \geq \frac{\beta}{4}\sum_{(i,j)\in U^*}(\tau_i - \tau_j)\mu_j w + (1 + \frac{\beta}{8})\sum_{i\in S_1}\lambda_i\tau_i w - \sum_{i\in S_2\cup S_3}\mu_i w^2 - w.$

Lemma 13. *Under* event B, $\Phi(\mathbf{Q}_{(\ell+1)\cdot\mathbf{w}}) - \Phi(\boldsymbol{\tau}) \leq (1 + \frac{\beta}{8})\sum_{i\in S_1}\lambda_i\tau_i w + \sum_{i\in S_1}\lambda_i^2 w^2.$

With a small probability, event B does not happen, and it is relatively straightforward to upper bound the increase in the potential in this case. (Most pessimistically, no packets is cleared during the time window and for each time step, there is a packet arriving at each source node, then the queue length of each source node grows by w.)

Lemma 14. *If event* B *does not happen, then* $\Phi(\mathbf{Q}_{(\ell+1)\cdot\mathbf{w}}) - \Phi(\mathbf{Q}_{\ell\cdot\mathbf{w}}) \leq \sum_{i\in S_1} Q_{\ell\cdot w}^i w + w^2.$

Lemma 9 follows from combining Lemma 12, 13 and 14.

5 Patient Queueing Model

In this section we extend a model introduced in [4], where queues do not vary their routing policies from step to step, but evaluate the utility of a fixed routing policy over a long period of time. On complete bipartite graphs, Gaitonde and

Tardos [4] showed that Nash equilibria always exist in the resulting game, and that a system is stable under *any* Nash as long as it can be stable under a central policy with $\frac{e}{e-1}$ times as much arrival rates. We obtain a similar result for incomplete bipartite graphs, but the factor in our bicriteria result worsens to 2. We leave for future work to decide whether this factor can be improved.

Below we first describe the model in more detail, before presenting our result.

5.1   Model Description

A bipartite *Patient Queueing Model* has the same packet arrival, routing and processing procedures as in a Queue-B model described in Sect. 2; the servers' priority rule is to pick the oldest packet. The main difference is that the queues are "patient": each queue fixes a routing strategy in the form of a distribution over the servers it can reach, and evaluates its utility/cost over a long time period. Formally, the strategy space of queue i is the simplex over the servers i can send requests to: $\Delta_i := \Delta(N^{\text{out}}(i))$. By adopting a strategy $p_i \in \Delta_i$, queue i in each round samples a server according to the distribution given by p_i, independently of all history and other happenings in the system, and sends a request to the sampled server if its queue is non-empty. Let $\boldsymbol{p}_{-i}$ be the strategies of the queues other than i, then the *cost* of queue i for using strategy p_i is $c_i(p_i, \boldsymbol{p}_{-i}) := \lim_{t \to \infty} \frac{T_t^i}{t}$, where T_t^i is the age of the oldest packet in queue i at time step t. Each queue aims to minimize its cost. A strategy profile $\mathbf{p}$ is a *Nash equilibrium* if for each queue i, $p_i \in \operatorname{argmin}_{p' \in \Delta_i} c_i(p', \boldsymbol{p}_{-i})$.

Gaitonde and Tardos [4] considered a special case of the Patient queueing Model, where the underlying bipartite graph is complete, i.e., $E = S_1 \times S_3$. They gave an algorithm that, given a strategy profile $\mathbf{p}$, computes an $r_i(\mathbf{p})$ for each queue i, with $r_i(\mathbf{p})$ equal to $c_i(\mathbf{p})$ almost surely. This algorithm played a crucial role in the derivation of their main result. The algorithm extends directly to our more general setting, and is also a key step in our result in this section. We present this algorithm next.

5.2   Gaitonde and Tardos's Algorithm for Computing Costs

Algorithm 1 is a straightforward generalization (to general bipartite graphs) of Gaitonde and Tardo's [4] algorithm for computing the queues' costs. We give some rough intuition here. Thanks to the service priority rule, in the long run, the queues are tiered according to the rates of growth of their lengths: the faster growing ones have higher priority over the slower growing ones. Determining which queues grow the fastest is like a self-fulfilling prophecy: a group of queues grow the fastest even when they have the highest priority. The algorithm enumerates all possible "first tier" queues, and finds the one that fulfills the "prophecy." It then continues to find lower tiers, assuming that all higher tiers have priority. Nailing down this intuition involves intricate probabilistic arguments, and is a major technical accomplishment of [4].

In the step that picks Q_k, there is no ambiguity because the union of minimizers of $f(Q)$ can be shown to be another minimizer. The following results are

---

Algorithm 1: Computing the queues' costs given their strategies

---

Input: Queueing system $(S_1 \cup S_3, E, \boldsymbol{\lambda}, \boldsymbol{\mu})$, strategy profile $\mathbf{p}$

$I \leftarrow S_1, k \leftarrow 1$

while $I \neq \emptyset$ **do**

  Compute for each $Q \subseteq I$, $f(Q) = \frac{\sum_{j=1}^{m} \mu_j (1 - \prod_{i \in Q}(1 - p_{ij}))}{\sum_{i \in Q} \lambda_i}$.

  Let Q_k be the minimizer of $f(Q)$, breaking ties in favor of larger cardinality.

  if $f(Q_k) \geq 1$ **then**

    | For any $i \in I$, output $r_i(\mathbf{p}) = 0$, terminate.

  else

    | For each $i \in Q_k$, $r_i(\mathbf{p}) = 1 - f(Q_k)$;

    | Update $\mu_j \leftarrow \mu_j \prod_{i \in Q_k}(1 - p_{ij}), I \leftarrow I \setminus Q_k, k \leftarrow k + 1$.

  end

end

Output: a sequence of queueing groups $Q_1, \cdots, Q_K$ and $r_i(\mathbf{p})$ for each queue i.

---

straightforward generalizations of corresponding results in [4]. We state them without proofs. Theorem 6 and 7 correspond to Theorem 4.1 and 3.3 in [4], respectively. Theorem 8 is a generalization of Lemma 3.3 and Theorem 4.4 in [4].

Theorem 6. *For any strategy profile* $\mathbf{p}$, $c_i(\mathbf{p}) = \lim_{t \to \infty} \frac{T_t^i}{t} = r_i(\mathbf{p})$ *almost surely.*

Theorem 7. *If the cost function of each queue i is defined to be $r_i(\mathbf{p})$, then every system of the Patient Queueing Model admits a Nash Equilibrium.*

Theorem 8. *Given a queueing system and a strategy profile* $\mathbf{p}$, *let Q_1 be the first group of queues output by Algorithm 1. If $f(Q_1) > 1$, then the queueing system is stable under* $\mathbf{p}$.

5.3   Price of Anarchy in Patient Queueing Model

Our bicriteria result for general bipartite graphs is presented again in a form more comparable to the dual form of conditions for centralized stability (see Lemma 1).

Theorem 9. *Given a bipartite queueing system $(V, E, \boldsymbol{\lambda}, \boldsymbol{\mu})$, if for any $\boldsymbol{\alpha} = (\alpha_1, \cdots, \alpha_n) \in \{0, 1\}^n$, there is a matching matrix M, such that $\frac{1}{2}\boldsymbol{\alpha}^\top M \boldsymbol{\mu} > \boldsymbol{\alpha}^\top \boldsymbol{\lambda}$, then the system is stable in any Nash Equilibrium in the patient queue model.*

The proofs missing due to lack of space are deferred to the full version of this paper.

We remark that repeatedly playing a Nash Equilibrium strategy profile $\mathbf{p}$ may not be no-regret strategies (see an example in the full version), therefore Theorem 9 is not implied by our results in Sect. 3.

6   Conclusion

In this work generalize the decentralized queueing systems proposed by Gaitonde and Tardos [3,4]. In the full version of this paper, we also consider two more variants of the model with multiple layers, when packet arrivals are adversarial instead of probabilistic, and when packets themselves (rather than the queues they are from) determine which servers can process them. We show that the bicriteria results under no regret strategies are robust against these model modifications. Lastly, we provide a slightly tighter analysis for the Queue-CB model of [3] in the full version of the paper.

Since the bicriteria result in [3] for queues using no regret strategies is tight even for complete bipartite graphs, our results for the more general cases are also tight. On the other hand, we do not know if the factor 2 is tight in our result for patient queues in general bipartite graphs. It seems challenging to directly apply the deformation technique developed in [4] in this more general setting; we leave for future work to investigate the tight bound of this problem.

References

1. Borodin, A., Kleinberg, J.M., Raghavan, P., Sudan, M., Williamson, D.P.: Adversarial queuing theory. J. ACM **48**(1), 13–38 (2001)
2. Christodoulou, G., Kovács, A., Schapira, M.: Bayesian combinatorial auctions. J. ACM **63**(2), 11:1–11:19 (2016)
3. Gaitonde, J., Tardos, É.: Stability and learning in strategic queuing systems. In: Proceedings of the 21st ACM Conference on Economics and Computation, pp. 319–347 (2020)
4. Gaitonde, J., Tardos, E.: Virtues of patience in strategic queuing systems. In: Proceedings of the 22nd ACM Conference on Economics and Computation, pp. 520–540 (2021)
5. Johari, R., Tsitsiklis, J.N.: Efficiency loss in a network resource allocation game. Math. Oper. Res. **29**(3), 407–435 (2004)
6. Koutsoupias, E., Papadimitriou, C.: Worst-case equilibria. In: Meinel, C., Tison, S. (eds.) STACS 1999. LNCS, vol. 1563, pp. 404–413. Springer, Heidelberg (1999). https://doi.org/10.1007/3-540-49116-3_38
7. Pemantle, R., Rosenthal, J.S.: Moment conditions for a sequence with negative drift to be uniformly bounded in LR. Stoch. Process. Appl. **82**(1), 143–155 (1999)
8. Roughgarden, T., Tardos, É.: How bad is selfish routing? J. ACM **49**(2), 236–259 (2002)
9. Sentenac, F., Boursier, E., Perchet, V.: Decentralized learning in online queuing systems. Adv. Neural. Inf. Process. Syst. **34**, 18501–18512 (2021)

Optimal Prophet Inequality with Less than One Sample

Nick Gravin$^{(\boxtimes)}$ ⓘ, Hao Li, and Zhihao Gavin Tang ⓘ

ITCS, Shanghai University of Finance and Economics, 100 Wudong Road,
Shanghai 200433, China
{nikolai,tang.zhihao}@mail.shufe.edu.cn

Abstract. There is a growing interest in studying sample-based prophet inequality with the motivation stemming from the connection between the prophet inequalities and the sequential posted pricing mechanisms. Rubinstein, Wang, and Weinberg (ITCS 2021) established the optimal single-choice prophet inequality with only a single sample per each distribution. Our work considers the sample-based prophet inequality with *less than one sample* per distribution, i.e., scenarios with no prior information about some of the random variables. Specifically, we propose a p-sample model, where a sample from each distribution is revealed with probability $p \in [0,1]$ independently across all distributions. This model generalizes the single-sample setting of Rubinstein, Wang, and Weinberg (ITCS 2021), and the i.i.d. prophet inequality with a linear number of samples of Correa et al. (EC 2019). Our main result is the optimal $\frac{p}{1+p}$ prophet inequality for all $p \in [0,1]$.

Keywords: Prophet inequality · Online algorithms · Optimization with samples

1 Introduction

Prophet inequality is a fundamental problem in optimal stopping theory and online Bayesian optimization. Consider a sequence of n boxes arriving online, each box i associated with a random value X_i sampled from a priori known distribution $\mathcal{D}_i$. The actual value of X_i is observed upon the arrival of the box i and the algorithm decides immediately whether to accept it. If the box is accepted, the algorithm collects the observed value X_i and the game ends. Else, the algorithm proceeds to the next box. The goal is to maximize the value of the accepted box and to compete against the expected maximum value of all boxes,

This work is supported by Science and Technology Innovation 2030 - "New Generation of Artificial Intelligence" Major Project No. (2018AAA0100903), Innovation Program of Shanghai Municipal Education Commission, Program for Innovative Research Team of Shanghai University of Finance and Economics (IRTSHUFE) and the Fundamental Research Funds for the Central Universities. Zhihao Gavin Tang is supported by NSFC grant 61902233. Nick Gravin is supported by NSFC grant 62150610500.

K. A. Hansen et al. (Eds.): WINE 2022, LNCS 13778, pp. 115–131, 2022.
https://doi.org/10.1007/978-3-031-22832-2_7

i.e., $\mathbb{E}[\max_i X_i]$. The benchmark is also known as the prophet, since it can be interpreted as the expected value of an optimal algorithm that can look into the values of all boxes before making a choice. Krengel and Sucheston [23,24] first established an optimal $\frac{1}{2}$-competitive prophet inequality. Subsequently, Samuel-Cahn [27] provided a single-threshold algorithm with the same tight competitive ratio.

The classic single-choice prophet inequality is equivalent to the problem of designing revenue-maximizing sequential posted pricing mechanism [10,20]. That connection has inspired a number of generalizations in the field of algorithmic mechanism design to multi-choice settings such as matroids [4,16,22], matchings and combinatorial auctions [14,15,18], and general downward-closed constraints [25]. Furthermore, the sequential posted pricing motivates the study of prophet inequalities with limited information, as the complete knowledge of the prior distributions $(\mathcal{D}_i)_{i=1}^n$ is a rather strong and unrealistic assumption as was pointed out by Azar, Kleinberg, and Weinberg [1].

In the limited information setting, the algorithm may only access a limited number of samples per each distribution $\mathcal{D}_i$ instead of the complete description of $\mathcal{D}_i$ in the full-information case. This model is arguably more realistic than the full-information model, since samples are easy to collect, e.g., from historical data. Azar, Kleinberg, and Weinberg designed constant competitive algorithms with only a constant number of samples per distribution for various matroid and matching settings. Recently, Rubinstein, Wang, and Weinberg [26] proved that the optimal 1/2 competitive ratio for the single-choice prophet inequality can be achieved, with only a *single* sample per distribution. Furthermore, Caramanis et al. [3] explored the limit of single-sample prophet inequalities for matroids, matching, and combinatorial auctions. Correa et al. [5] also studied the prophet secretary problem with a single sample per distribution and obtained a 0.635-competitive algorithm.

1.1   Our Contributions

Model and Result. In the regime with little prior information, it is reasonable to assume that some distributions may be completely new, i.e., they have no samples whatsoever. We propose a new framework of p-sparse sample access parameterized by $p \in [0,1]$ and apply it to the single-choice prophet inequality. Specifically, we assume that independently for each box, the algorithm sees a sample from it with probability p. It generalizes the single-sample setting of Rubinstein, Wang, and Weinberg [26] when $p = 1$.

Our *less than one sample* regime also generalizes the model of Correa et al. [7] who studied the setting with linear βn and sublinear $o(n)$ number of samples for n *i.i.d. distributions*. They showed that no algorithm can achieve a competitive ratio better than $1/e$ when $\beta = o(1)$, and designed a $(1 - 1/e)$-competitive algorithm for $\beta = 1$. Subsequently, Correa et al. [8] achieved a tight competitive ratio of $\frac{1+\beta}{e}$ for all $\beta \leq \frac{1}{e-1}$, and improved the competitive ratio to 0.648 for $\beta = 1$.

To the best of our knowledge, we are the first to study the *less than one sample* setting for non identical distributions. Our main result is a *tight* $\frac{p}{1+p}$-competitive algorithm for the single-choice prophet inequality. Our algorithm is based on the simple Maximum-sample algorithm [26] that stops at the first value box i with a value X_i greater than the maximum sample. However, our version (see Definition 1 later in the paper) has a non trivial alternation.

Techniques. First, note that stochastic optimization with a constant number of samples makes problem so much harder than the full information case, e.g., in the closely related auction literature on revenue maximization with samples, designing mechanisms with 1 sample per distribution [12,17] is quite different from the full information setting. Moreover, it is a daunting task to get an improvement on the revenue guarantee from the setting with 1 sample to 2 samples per distribution (see, e.g., [2,11]).

At the technical level our analysis proceeds by reducing the original problem of maximizing the expected value to a simpler objective of stopping at any of the top k card values for each fixed $k \in \mathbb{N}$. We first studied the problem for $k = 1$ and identified a hard family of instances that already gives a tight upper bound of $\frac{p}{1+p}$ on the competitive ratio for any $p \in (0, 1]$.

Next, we found the right variation of the Max-Sample algorithm for the objective of stopping at the maximum ($k = 1$) with a matching lower bound of $\frac{p}{1+p}$ on the competitive ratio. Our proof proceeds by carefully constructing a set of disjoint events that would guarantee Max-Sample to win. A challenging part was to define/select the events in such a way that would keep the number of cases at a minimum. Lastly, we extended the analysis for $k = 1$ to arbitrary $k \in \mathbb{N}$ with a noticeably more elaborate set of winning events and larger case analysis.

1.2   Further Related Works

A closely related problem to prophet inequality is the celebrated secretary problem. In this setting, n elements arrive in a random arrival order. An online algorithm observes the value of each element and decide whether to take it immediately and irrevocably. Observe that in this setting, the algorithm has no prior information of the n values. Recently, Kaplan, Naori, and Raz [21] proposed a data-driven variant to the secretary problem. They assume that among the n values, a fraction p of the values are given as samples to the algorithm in advance; the remaining values either comes in an adversarial or in a random order. They designed an optimal algorithm for the adversarial arrivals and a near optimal algorithm for the random arrivals. Duetting et al. [13] generalized the latter setting to secretaries with advice and found an optimal algorithm for the random arrival variant. Correa et al. [6] proposed a slightly different model in which each value is sampled independently with probability p and designed optimal algorithms for all p. This model bridges the secretary problem (when $p = 0$) and the i.i.d. prophet inequality (when $p = 1$). These models have similar flavour to our problem, but are not directly comparable.

Another line of research in sample-based prophet inequality studies the sample complexity, i.e., how many samples are needed to almost (up to an ε error) match the competitive ratio in the full-information case. First, Correa et al. [7] proved that $O(n^2)$ samples are sufficient to get the competitive ratio of $0.745 - \varepsilon$ in the i.i.d. prophet inequality setting, where the optimal algorithm [9] with full information is 0.745-competitive. The sample complexity was later improved to $O(n/\varepsilon^6)$ by Rubinstein, Wang, and Weinberg [26]. Guo et al. [19] further improved the dependency on ε by establishing an upper bound of $O(n/\varepsilon^2)$.

2   Preliminaries

p-Sample Prophet Inequality. There are n boxes, whose values $\mathbf{v} = (X_1, \ldots, X_n)$ are drawn independently from $\mathcal{D}_1 \times \ldots \times \mathcal{D}_n$. In contrast to the classic prophet inequality, the algorithm does not have knowledge about the underlying distribution in advance. Instead, for each random variable X_i we observe a sample $\hat{X}_i$ independently for all $i \in [n]$ with probability p. For simplicity of notations, we assume $\hat{X}_i = 0$ when we do not see a sample. The goal is to maximize the expected value of the accepted box and to compete against the expected maximum $\mathbb{E}[\max_i X_i]$.

Our algorithm is defined as the following.

Definition 1 (Max-Sample algorithm). *Given as input samples $\hat{X}_1, \ldots, \hat{X}_n$, let $T = \max_{i \in [n]} \hat{X}_i$ and $i^* \overset{\text{def}}{=} \operatorname{argmax}_i \hat{X}_i$ ($i^* = 0$ if $T = 0$). Let X_j be the first observed random variable exceeding T.*

$$\text{If } j \neq i^*, \text{ take } X_j, \qquad \text{If } j = i^* \text{ then } \begin{cases} \text{take } X_j & \text{w.p. } \frac{2p}{1+p} \\ \text{skip } X_j, \text{ take next } X_\ell > T & \text{w.p. } \frac{1-p}{1+p} \end{cases}$$

Theorem 1. *Max-Sample is $\frac{p}{1+p}$-competitive for the p-sample prophet inequality problem. Moreover, the ratio is the best possible for any $p \in [0, 1]$.*

For the algorithmic part of the result, we shall focus on the following card model and analyze the performance of our algorithm. The card model is adapted from the work of Rubinstein, Wang, and Weinberg [26] and Correa et al. [5].

Card Model. Each box corresponds to a card $C_i = \{a_i, b_i\}$ with two unknown values written on either side. Each card is put on the table with one of its sides independently and uniformly at random facing down and the other side facing up. The card i corresponds to an ordered pair (v_i, s_i): a value at the bottom, and a value on top. I.e., $\mathbf{Pr}[v_i = a_i, s_i = b_i] = \mathbf{Pr}[v_i = b_i, s_i = a_i] = 0.5$. The online algorithm gets to see some of the top values in the initial stage before making any decisions. Each top value s_i is revealed (independently for all $i \in [n]$) to the algorithm with probability p, with remaining probability $1 - p$ the value is erased and is substituted with a blank. The online algorithm proceeds by flipping the cards one by one starting from C_1 and until the last card C_n. After flipping a card C_i, the algorithm observes the value v_i at the bottom and may either take it and stop, or discard the card C_i and continue. We denote the set of cards with revealed samples as $R \subseteq [n]$ and the distribution of the revealed samples $R \sim \mathcal{R}$. Also for each card $i \in [n]$ we use $r_i \in \{0, 1\}$ to indicate whether the sample on top of card i is revealed ($r_i = 1$) or not ($r_i = 0$). We denote by $\mathbf{r} \in \{0, 1\}^n$ the vector of revealed samples. We slightly abuse the notations and use $\mathcal{R}$ to denote the distribution of $\mathbf{r} \sim \mathcal{R}$. The algorithm sees revealed samples $\mathbf{s}(R)$ and aims to maximize the value of the accepted card and compete against the prophet in the card model: $\mathbb{E}_{\mathbf{v}, \mathbf{s}} [\max_{i \in [n]} v_i]$.

For analysis purpose, we sort the multi-set of values $V = \{a_i\}_{i=1}^n \cup \{b_i\}_{i=1}^n$ in decreasing order. We denote the elements in the sorted multi-set V as $w_1 \geq w_2 \geq \ldots \geq w_{2n}$. We use $\sigma : [2n] \to [n]$ to denote the indexes of the cards in V. Specifically, σ_1 is the index of the card with the largest values in V, σ_2 is the index of the card with the second largest value, etc.

It is straightforward to observe that a competitive algorithm for the card model preserves its competitive ratio in the p-sample prophet inequality setting, by setting the values $\{a_i, b_i\}$ to be independent samples of $\mathcal{D}_i$.

Roadmap. In Sect. 3, we consider the simpler task of stopping at the maximum value card. Built on it, we prove our main result in Sect. 4. Finally, in Sect. 5, we provide a matching hardness result.

3   Stopping at the Maximum

Consider the case when the largest value w_1 is much larger than the rest $w_i \in V$. In this case, the contribution of all other values to the expected reward of our algorithm and the prophet are negligibly small and the question is how often our algorithm stops at C_{σ_1} and gets $v_{\sigma_1} = w_1$.

Our objective then is to stop at the global maximum w_1 in V. The prophet gets w_1 with probability 0.5, whenever $v_{\sigma_1} = w_1$, i.e., when w_1 is at the bottom of the card C_{σ_1}. We show that Max-Sample stops at the maximum w_1 with probability at least $\frac{p}{2(1+p)}$, which gives us the desired guarantee in the special case when w_1 is much larger than all other $w_i \in V$. The analysis will be helpful for obtaining the result in general case.

Theorem 2. *Given that maximum w_1 is on the value side, i.e., $v_{\sigma_1} = w_1$, Max-Sample stops at the maximum with probability at least $\frac{p}{1+p}$.*

Remark 1. Before we proceed with the proof, we give an example demonstrating why the original algorithm of Rubinstein et al. [26] of accepting the first item above maximum sample has strictly worse performance than $\frac{p}{1+p}$. The instance has $n = 2$ boxes: the first box with distribution $F_1 = \mathtt{Uni}[1, 2]$, the second box with distribution $F_2 = \{v = 10000 \text{ w.p. } \frac{1}{100}, 0 \text{ w.p. } \frac{99}{100}\}$; let $p = \frac{1}{2}$. We may only consider the case when $X_2 = 10000$, $\hat{X}_2 = 0$ that contributes $10000 \cdot \frac{1}{100} \cdot \frac{199}{200} = 99.5$ to the expected value of the prophet, since the total contribution in all other cases is less than 3. In this case the algorithm of Rubinstein et al. gets X_1, if and only if $\hat{X}_1 > X_1$ which happens with probability $\frac{p}{2} < \frac{p}{1+p}$.

Proof. One difficulty in the analysis is that we know neither the order of cards, nor the pairings of $w_1, w_2, \ldots, w_{2n}$ (i.e., which pairs of them are on the same cards). Our approach in dealing with so many possibilities will be to describe a sequence of disjoint events that guarantee our algorithm to stop at w_1. We still need to consider a few cases, but only a small number.

We begin constructing these events by considering the cards with the largest values $w_2, \ldots, w_t$ until the next one $\sigma_{t+1} \in \{\sigma_1, \ldots, \sigma_t\}$, i.e., the first time when w_{t+1} is on the same card with one of the previous values $\{w_1, \ldots, w_t\}$. Let us first deal with the case when $\sigma_{t+1} \neq \sigma_1$, i.e., w_{t+1} is on the same card with one of the $\{w_2, \ldots, w_t\}$.

Case 1: $\sigma_{t+1} \neq \sigma_1$. We first consider the event $\mathcal{E}_1$ that w_2 is a visible sample $(s_{\sigma_2} = w_2, r_{\sigma_2} = 1)$, then Max-Sample sets the threshold $T = w_2$ and waits until w_1 (recall that w_1 must be at the bottom of its card C_{σ_1}) at which point the algorithm must stop and take w_1. We have $\mathbf{Pr}[\mathcal{E}_1] = \frac{p}{2}$.

Next, if $s_{\sigma_2} = w_2$ and the sample w_2 is *not revealed* $r_{\sigma_2} = 0$, then we can look at w_3. If w_3 is a revealed sample $(w_3 = s_{\sigma_3}, r_{\sigma_3} = 1)$ then Max-Sample must stop at w_1. This is our second event $\mathcal{E}_2$: $(s_{\sigma_2} = w_2, r_{\sigma_2} = 0)$, and $(w_3 = s_{\sigma_3}, r_{\sigma_3} = 1)$. Similarly, Max-Sample must stop at w_1 for each of the following events $\{\mathcal{E}_\ell\}_{\ell=1}^{t-1}$:

$$\mathcal{E}_\ell \stackrel{\text{def}}{=} \{\forall i \in [2, \ell] \ (s_{\sigma_i} = w_i, r_{\sigma_i} = 0), \text{ and } (w_{\ell+1} = s_{\sigma_{\ell+1}}, r_{\sigma_{\ell+1}} = 1)\}$$

$$\mathbf{Pr}[\mathcal{E}_\ell] = \left(\frac{1-p}{2}\right)^{\ell-1} \frac{p}{2}. \tag{1}$$

When we continue our sequence of events $\{\mathcal{E}_\ell\}_{\ell=1}^{t-1}$ to $\ell = t$, the value w_{t+1} appears on one of the previously fixed cards C_{σ_j} for $2 \leq j \leq t$. We note that Max-Sample algorithm skips the card with the maximum sample with probability $\frac{1-p}{1+p}$. Thus it may still stop at w_1 even when $v_{\sigma_j} = w_j$ (w_j is at the bottom of C_{σ_j} card). Finally, we define the last event $\mathcal{E}_t$ as follows:

$$\mathcal{E}_t \stackrel{\text{def}}{=} \left\{\forall i \in [t] \setminus \{1, j\} \ \begin{pmatrix} s_{\sigma_i} = w_i \\ r_{\sigma_i} = 0 \end{pmatrix} \text{ and } \begin{pmatrix} w_j = v_{\sigma_j}, & w_{t+1} = s_{\sigma_j} \\ r_{\sigma_j} = 1, & \text{alg. ignores } w_j \end{pmatrix} \right\}$$

$$\mathbf{Pr}[\mathcal{E}_t] = \left(\frac{1-p}{2}\right)^{t-2} \cdot \frac{p}{2} \cdot \frac{1-p}{1+p}. \tag{2}$$

As all events $\{\mathcal{E}_\ell\}_{\ell=1}^{t}$ are disjoint, we may combine (1) and (2) and get

$$\mathbf{Pr}[\text{alg. takes } w_1] \geq \sum_{\ell=1}^{t} \mathbf{Pr}[\mathcal{E}_\ell] = \frac{p}{2} \cdot \left[\left(\frac{1-p}{2}\right)^{t-2} \cdot \frac{1-p}{1+p} + \sum_{i=0}^{t-2} \left(\frac{1-p}{2}\right)^i\right]$$

$$= \frac{p}{1+p}\left(\frac{1-p}{2}\right)^{t-1} + \frac{p}{2} \cdot \frac{1 - \left(\frac{1-p}{2}\right)^{t-1}}{1 - \left(\frac{1-p}{2}\right)} = \frac{p}{1+p} \tag{3}$$

Case 2: $\sigma_{t+1} = \sigma_1$. We have the same events $\{\mathcal{E}_\ell\}_{\ell=1}^{t-1}$ as in (1). The $\mathcal{E}_t$ is now a little different, as we want Max-Sample algorithm to stop at card $C_{\sigma_{t+1}}$:

$$\mathcal{E}_t \stackrel{\text{def}}{=} \left\{ \forall i \in [t] \setminus \{1\} \ \begin{pmatrix} s_{\sigma_i} = w_i \\ r_{\sigma_i} = 0 \end{pmatrix} \text{ and } \begin{pmatrix} w_1 = v_{\sigma_1}, & w_{t+1} = s_{\sigma_1} \\ r_{\sigma_j} = 1, & \text{alg. takes } w_1 \end{pmatrix} \right\}$$

$$\mathbf{Pr}\left[\mathcal{E}_t\right] = \left(\frac{1-p}{2}\right)^{t-2} \cdot p \cdot \frac{2p}{1+p}. \tag{4}$$

We continue the sequence of events $\{\mathcal{E}_\ell\}_{\ell=1}^t$ after t by considering new cards $C_{\sigma_{t+2}}, C_{\sigma_{t+3}}, \ldots, C_{\sigma_k}$ until we get $\sigma_{k+1} \in \{\sigma_1, \ldots, \sigma_k\} \setminus \{\sigma_1, \sigma_{t+1}\}$, i.e., the first time w_{k+1} appears on the same card with one of the previous $\{w_1, \ldots, w_k\}$. Notice that w_1 and w_{t+1} are on the same card, and w_1 is always at the bottom of C_{σ_1}. We would like the sample $w_{t+1} = s_{\sigma_1}$ not to be revealed (i.e., $r_{\sigma_1} = 0$), which happens with probability $(1-p)$. We define $\{\mathcal{E}_\ell\}_{\ell=t+1}^{k-1}$ as follows

$$\mathcal{E}_\ell \stackrel{\text{def}}{=} \left\{ \begin{pmatrix} v_{\sigma_1} = w_1 \\ r_{\sigma_1} = 0 \end{pmatrix}, \forall i \in [\ell] \setminus \{1, t+1\} \begin{pmatrix} s_{\sigma_i} = w_i \\ r_{\sigma_i} = 0 \end{pmatrix}, \text{ and } \begin{pmatrix} w_\ell = s_{\sigma_\ell} \\ r_{\sigma_\ell} = 1 \end{pmatrix} \right\}$$

$$\mathbf{Pr}\left[\mathcal{E}_\ell\right] = (1-p) \cdot \left(\frac{1-p}{2}\right)^{\ell-2} \cdot \frac{p}{2}. \tag{5}$$

Finally, let j be the index such that w_{k+1} appears on one of the previously fixed cards C_{σ_j} for $2 \le j \le k$. We define the last event $\mathcal{E}_k$ similar to (2) as follows.

$$\mathcal{E}_k \stackrel{\text{def}}{=} \left\{ \forall i \in [k] \setminus \{j, t+1\} \ \begin{pmatrix} s_{\sigma_i} = w_i \\ r_{\sigma_i} = 0 \end{pmatrix}, \begin{pmatrix} w_j = v_{\sigma_j}, w_{k+1} = s_{\sigma_j} \\ r_{\sigma_j} = 1, \text{ alg. ignores } w_j \end{pmatrix} \right\}$$

$$\mathbf{Pr}\left[\mathcal{E}_k\right] = (1-p) \cdot \left(\frac{1-p}{2}\right)^{k-3} \cdot \frac{p}{2} \cdot \frac{1-p}{1+p}. \tag{6}$$

As all events $\{\mathcal{E}_\ell\}_{\ell=1}^k$ are disjoint, we combine (1), (4), (5), and (6) to get

$$\mathbf{Pr}[\text{alg. takes } w_1] \ge \sum_{\ell=1}^k \mathbf{Pr}[\mathcal{E}_\ell] = \frac{p}{2} \cdot \sum_{i=0}^{t-2} \left(\frac{1-p}{2}\right)^i + \left(\frac{1-p}{2}\right)^{t-2} \cdot \frac{2p^2}{1+p} +$$

$$p \cdot \sum_{\ell=t+1}^{k-1} \left(\frac{1-p}{2}\right)^{\ell-1} + \left(\frac{1-p}{2}\right)^{k-1} \cdot \frac{2p}{1+p} > \frac{p}{1+p}\left[1 - \left(\frac{1-p}{2}\right)^{t-1}\right]$$

$$+ \frac{p^2}{1+p}\left(\frac{1-p}{2}\right)^{t-1} + p \cdot \left(\frac{1-p}{2}\right)^t \frac{1-(\frac{1-p}{2})^{k-1-t}}{1-\frac{1-p}{2}} + \left(\frac{1-p}{2}\right)^{k-1} \cdot \frac{2p}{1+p}$$

$$= \frac{p}{1+p} - \frac{p}{1+p}\left(\frac{1-p}{2}\right)^{t-1}(1-p) + \frac{2p}{1+p}\left(\frac{1-p}{2}\right)^t \cdot \left[1 - \left(\frac{1-p}{2}\right)^{k-t-1}\right] +$$

$$\left(\frac{1-p}{2}\right)^{k-1} \frac{2p}{1+p} = \frac{p}{1+p}, \tag{7}$$

where to get the last inequality we simply decreased the term $\mathbf{Pr}[\mathcal{E}_t]$ in (4) to $\left(\frac{1-p}{2}\right)^{k-1} \frac{p^2}{1+p}$.

Theorem 2 follows from (3) in the case $\sigma_{t+1} \neq \sigma_1$ and from (7) in the case $\sigma_{t+1} = \sigma_1$.

4   Maximizing Expectation

In this section we prove our main result that Max-Sample algorithm achieves optimal competitive ratio of $\frac{p}{1+p}$ on arbitrary instances. To this end we consider a few special instances with top k values being almost equal to each other[1] and much larger than the remaining w_i for $i \in [2n] \setminus [k]$. We call it a top-k instance for any $k \leq 2n$. It turns out that restricting our attention only to the top-k instances is without loss of generality for the Max-Sample algorithm. We prove next that Max-Sample is a $\frac{p}{1+p}$ approximation to the prophet on any top-k instance for each $k \in [2n]$ using similar approach to what we did in Sect. 3 for the top-1 instances, but with a more elaborate case analysis.

Theorem 3. *Max-Sample algorithm is a $\frac{p}{1+p}$-approximation to the prophet for any value of $p \in (0, 1]$.*

Proof. We analyse Max-Sample in the card model and first show that restricting our attention only to the top-k instances is without loss of generality.

Claim. Suppose a *rank-based* ALG (i.e., ALG only uses comparisons ">, <" between variables and samples) is an $\alpha < 1$ approximation to the prophet on any top-k instance in the card model for each $k \geq 1$. Then ALG is an α-approximation to the prophet on every instance.

Proof. The fact that ALG is an α-approximation to the prophet on a top-k instance means that

$$\mathbf{Pr}_{\mathbf{v},\mathbf{r}}[\mathsf{ALG}(\mathbf{v},\mathbf{r}) \text{ gets } w_i \text{ for an } i \in [k]] \geq \alpha \cdot \mathbf{Pr}_{\mathbf{v}}[\exists i \in [k] \; v_{\sigma_i} = w_i] \quad (8)$$

for this instance. As ALG is an ordinal algorithm the same guarantee holds for *any instance* that is not necessarily a top-k. Expected performance of the ALG can be written as

$$\mathbb{E}[\mathsf{ALG}] = \sum_{k=1}^{2n} w_k \cdot \mathbf{Pr}[\mathsf{ALG} \text{ gets } w_k] = \sum_{k=1}^{2n} w_k \cdot \Big(\mathbf{Pr}[\mathsf{ALG} \text{ gets } w_i, i \in [k]]$$

$$- \mathbf{Pr}[\mathsf{ALG} \text{ gets } w_i, i \in [k-1]] \Big) = \sum_{k=1}^{2n} \mathbf{Pr}[\mathsf{ALG} \text{ gets } w_i, i \in [k]] \cdot (w_k - w_{k+1})$$

$$\geq \alpha \cdot \sum_{k=1}^{2n} \mathbf{Pr}[\exists i \in [k] \; v_{\sigma_i} = w_i] \cdot (w_k - w_{k+1}) = \alpha \cdot \mathbb{E}[\mathsf{Prophet}],$$

where $\mathbf{Pr}[\mathsf{ALG} \text{ gets } w_i, i \in [0]] = w_{2n+1} = 0$; we used (8) to get the inequality.

---

[1] E.g., $w_1 = w_2 + \varepsilon = \ldots = w_k + (k-1)\varepsilon$, for some negligibly small $\varepsilon > 0$.

To conclude the proof of Theorem 3, we only need to show that Max-Sample is a $\frac{p}{1+p}$-approximation to the prophet on a top-k instance for each $k \in [2n]$. Section 3 already gives the desired result for $k = 1$. For $k \geq 2$ we consider the sequence of cards $(\sigma_i)_{i=1}^{k}$ with the top k values. There are two cases: 1) there is a pair of w_i, w_j on the same card ($\sigma_i = \sigma_j$), or 2) all $(\sigma_i)_{i=1}^{k}$ are different.

Case 1. $\exists\ \sigma_j = \sigma_i$, $i, j \in [k]$. Let us consider the first time two top values appear on the same card, i.e., the smallest $i \leq k$ with $\sigma_i = \sigma_j$ for a $j < i$. Notice that the prophet can get one of the top-i values (either w_i or w_j), i.e., $\mathbf{Pr}[\exists j \in [i]\ v_{\sigma_j} = w_j] = 1$. On the other hand, for the Max-Sample it is only harder to stop at one of the top-i values. Hence, we can assume without loss of generality that $k = i$ and that $\sigma_j = \sigma_k$ is the only two top-k values on the same card. We distinguish three cases based on the index j.

Case 1.a. $k = 2$. Then $\sigma_1 = \sigma_2$, i.e., w_1 and w_2 are on the same card. Then consider the event $\mathcal{E}_0$ that $(v_{\sigma_1} = w_1, s_{\sigma_1} = w_2, r_{\sigma_1} = 1)$, then Max-Sample succeeds with probability $\frac{2p}{1+p}$. On the other hand, if $r_{\sigma_1} = 0$ (sample $s_{\sigma_1} \in \{w_1, w_2\}$ is not revealed) we can use the same events $(\mathcal{E}_i)_{i\geq 1}$ from Sect. 3 (count starts from w_3 instead of w_2) to guarantee that Max-Sample stops at w_1 or w_2 (whichever is at the bottom of C_{σ_1}). Overall, the events $(\mathcal{E}_i)_{i\geq 0}$ give us the desired guarantee

$$\mathbf{Pr}[\text{alg. wins}] \geq \frac{p}{2} \cdot \frac{2p}{1+p} + (1-p) \cdot \sum_{i \geq 1} \mathbf{Pr}[\mathcal{E}_i] \geq \frac{p^2}{1+p} + (1-p)\frac{p}{1+p} = \frac{p}{1+p}.$$

Case 1.b. $k > 2$ and $\sigma_j = \sigma_k \neq \sigma_1$. The Max-Sample wins in the event $\mathcal{E}_0$: $(v_{\sigma_1} = w_1, r_{\sigma_k} = 1)$. On the other hand, when $(v_{\sigma_1} = w_1, r_{\sigma_k} = 0)$, we can use the same events $(\mathcal{E}_i)_{i\geq 1}$ as in Sect. 3 with a small modification that w_j, w_k and their respective card C_{σ_j} are removed from the sequence $(w_i)_{i=1}^{2n}$ to guarantee the win of Max-Sample. Indeed, the card C_{σ_j} may only cause the algorithm to stop early at w_j or w_k, which is a win for the algorithm. We have

$$\mathbf{Pr}[\text{alg. wins}] \geq \mathbf{Pr}[\mathcal{E}_0] + \frac{1}{2}(1-p) \cdot \sum_{i \geq 1} \mathbf{Pr}[\mathcal{E}_i] \geq \frac{p}{2} + \frac{1-p}{2}\frac{p}{1+p} = \frac{p}{1+p}.$$

Case 1.c. $k > 2$ and $\sigma_k = \sigma_1$. We consider first what happens with the card C_{σ_1}. First, let us consider what happens when $(s_{\sigma_1} = w_k, r_{\sigma_1} = 0)$. The Max-Sample wins if at least one of the top values $w_i, i \in [2, k]$ is revealed as a sample $(s_{\sigma_i} = w_i, r_{\sigma_i} = 1)$. We define this event $\mathcal{E}_I$ as

$$\mathcal{E}_I \overset{\text{def}}{=} \{(s_{\sigma_1} = w_k, r_{\sigma_1} = 0),\ \exists\ 1 < i < k\ (s_{\sigma_i} = w_i, r_{\sigma_i} = 1)\}$$

$$\mathbf{Pr}[\mathcal{E}_I] = \frac{1-p}{2} \cdot \left(1 - \left(1 - \frac{p}{2}\right)^{k-2}\right) \tag{9}$$

Next, let us consider what happens when $(v_{\sigma_1} = w_1, s_{\sigma_1} = w_k, r_{\sigma_1} = 1)$. The algorithm is guaranteed to win when one of the w_i for $1 < i < k$ appears at the bottom, or as a revealed sample. The Max-Sample also wins when all of the w_i for $1 < i < k$ appear as a hidden samples and Max-Sample decides not to skip w_1 when it reaches $s_{\sigma_1} = w_k$. Formally, we define these two events $\mathcal{E}_{II}, \mathcal{E}_{III}$ as

$$\mathcal{E}_{II} \overset{\text{def}}{=} \{(s_{\sigma_1} = w_k, r_{\sigma_1} = 1), \; \exists\, 1 < i < k \; (v_{\sigma_i} = w_i \text{ or } s_{\sigma_i} = w_i, r_{\sigma_i} = 1)\}$$

$$\mathcal{E}_{III} \overset{\text{def}}{=} \{(s_{\sigma_1} = w_k, r_{\sigma_1} = 1), \; \forall\, 1 < i < k \; (s_{\sigma_i} = w_i, r_{\sigma_i} = 0), \mathsf{ALG} \text{ takes } w_1\}$$

$$\mathbf{Pr}\left[\mathcal{E}_{II} \sqcup \mathcal{E}_{III}\right] = \frac{p}{2} \cdot \left(1 - \left(\frac{1-p}{2}\right)^{k-2} + \left(\frac{1-p}{2}\right)^{k-2} \cdot \frac{2p}{1+p}\right) \qquad (10)$$

Finally, let us consider what happens when w_k is at the bottom of the card C_{σ_1} and w_1 is not revealed as a sample ($v_{\sigma_1} = w_k, r_{\sigma_1} = 0$). We would like to treat w_k as w_1 from Sect. 3 and construct events that guarantee Max-Sample to stop at w_k. The main problem is that if any of the w_i for $i \in [2, k-1]$ appears as a revealed sample, then the algorithm will never stop at w_k. To avoid this issue we will add the condition that no w_i for $i \in [2, k-1]$ is revealed as a sample (note that if w_i appears at the bottom of C_{σ_i}, it can only help Max-Sample to win). We use the events $\{\mathcal{E}_\ell\}_{\ell \geq 1}$ from Sect. 3 with a modification that w_k plays the role of w_1 and all $w_2, \ldots, w_{k-1}$ are ignored or equivalently treated as very small numbers (i.e., $(w_i)_{i \geq 2}$ in Sect. 3 correspond to $(w_i)_{i \geq k+1}$ in our instance, and w_1 in Sect. 3 corresponds to w_k here)[2]. Notice that if a card with $w_i, i \in [2, k-1]$ is used in an event $\mathcal{E}_\ell$, then the other value $w_j, j \geq k+1$ on the card C_{σ_i} must be a sample ($s_{\sigma_j} = w_j, v_{\sigma_j} = w_i$). I.e., we do not need to worry that w_i is revealed as a sample. Thus for each event $\mathcal{E}_\ell$ the algorithm wins in the event

$$\mathcal{E}'_\ell \overset{\text{def}}{=} \{(v_{\sigma_1} = w_k, r_{\sigma_1} = 0) \wedge \mathcal{E}_\ell \wedge \{\forall 1 < i < k \; v_{\sigma_i} = w_i \vee (s_{\sigma_i} = w_i, r_{\sigma_i} = 0)\}\}$$

$$\mathbf{Pr}\left[\mathcal{E}'_\ell\right] \geq \frac{1-p}{2} \cdot \left(1 - \frac{p}{2}\right)^{k-2} \cdot \mathbf{Pr}\left[\mathcal{E}_\ell\right]. \qquad (11)$$

When we combine the events defined by (9), (10), (11) we get

$$\mathbf{Pr}[\text{alg. wins}] \geq \mathbf{Pr}\left[\mathcal{E}_I \sqcup \mathcal{E}_{II} \sqcup \mathcal{E}_{III} \sqcup \bigsqcup_{\ell \geq 1} \mathcal{E}'_\ell\right] \geq \frac{1-p}{2} \cdot \left(1 - \left(1 - \frac{p}{2}\right)^{k-2}\right) +$$

$$\frac{p}{2} \cdot \left(1 - \left(\frac{1-p}{2}\right)^{k-2} + \left(\frac{1-p}{2}\right)^{k-2} \cdot \frac{2p}{1+p}\right) + \frac{1-p}{2}\left(1 - \frac{p}{2}\right)^{k-2} \cdot \frac{p}{1+p}$$

$$= \frac{1-p}{2} \cdot \left(1 - \frac{1}{1+p}\left(1 - \frac{p}{2}\right)^{k-2}\right) + \frac{p}{2} \cdot \left(1 - \frac{1-p}{1+p}\left(\frac{1-p}{2}\right)^{k-2}\right) \overset{k=3}{>}$$

$$\frac{1-p}{2} \cdot \left(1 - \frac{1}{1+p}\left(1 - \frac{p}{2}\right)\right) + \frac{p}{2} \cdot \left(1 - \frac{1-p}{1+p}\right) = \frac{p}{1+p},$$

---

[2] We can assume that once w_i is set to 0 for $1 < i < k$, it is small enough to not appear in the $\mathcal{E}_\ell$. Indeed, we can add a few dummy cards with both sides having negligibly small numbers in the beginning of the sequence that do not affect performance of Max-Sample, but still bigger than $w_i \leftarrow 0$.

where to get the last inequality we used that the previous expression is minimized for $k = 3$ (recall that $k > 2$) and also that $\left(\frac{1-p}{2}\right)^{k-2} < 1$. This concludes the proof for the case 1 as we have $\mathbf{Pr}[\text{alg. wins}] \geq \frac{p}{1+p}$ in each of the sub-cases 1.a, 1.b, and 1.c.

Case 2. $k \geq 2$ and $\forall\, 1 \leq i < j \leq k \ \sigma_i \neq \sigma_j$. The main difficulty in this case is that the value of the prophet $\mathbf{Pr}[\exists\, j \in [i]\ v_{\sigma_j} = w_j] = 1 - \frac{1}{2^k}$, which depends on k. On the positive side, there are no sub-cases here unlike case 1. We consider first what happens when at least one of the $w_i, i \in [k]$ is revealed as a sample. Let $j \in [k]$ be the first w_j with $(s_{\sigma_j} = w_j, r_{\sigma_j} = 1)$. For each $j \geq 2$, Max-Sample algorithm wins when at least one of the $w_i, i < j$ appears at the bottom of its card C_{σ_i}. Formally, we define these events $(\mathcal{E}_j^*)_{j \geq 2}^k$ as

$$\mathcal{E}_j^* \overset{\text{def}}{=} \left\{ (s_{\sigma_j} = w_j, r_{\sigma_j} = 1), \begin{array}{l} \forall\, i < j \ (v_{\sigma_i} = w_i \vee r_{\sigma_i} = 0) \\ \text{not } \forall\, i < j \ (s_{\sigma_i} = w_i \wedge r_{\sigma_i} = 0) \end{array} \right\}$$

$$\mathbf{Pr}\left[\mathcal{E}_j^*\right] = \frac{p}{2} \cdot \left(\left(1 - \frac{p}{2}\right)^{j-1} - \left(\frac{1-p}{2}\right)^{j-1} \right) \tag{12}$$

Next, consider the event $\mathcal{E}'$ that none of $w_i, i \in [k]$ is revealed as a sample and at least one of them is at the bottom of its card C_{σ_i}. In this case, we need to consider other $(w_i)_{i \geq k+1}$ to guarantee the win of Max-Sample. To this end, we would like to use the events $(\mathcal{E}_\ell)_{\ell \geq 1}$ from Sect. 3 with the following modification: all $w_1, \ldots, w_k$ are ignored, i.e., $(w_i)_{i \geq 2}$ from Sect. 3 correspond to $(w_i)_{i \geq k+1}$ in our instance. If event $\mathcal{E}_\ell$ does not specify position of any of the cards $C_{\sigma_1}, \ldots, C_{\sigma_k}$, then Max-Sample wins in the event $\mathcal{E}'_\ell$ defined as:

$$\mathcal{E}'_\ell \overset{\text{def}}{=} \left\{ \mathcal{E}_\ell , \begin{array}{l} \forall\, i \in [k] \ (v_{\sigma_i} = w_i \vee r_{\sigma_i} = 0) \\ \text{not } \forall\, i \in [k] \ (s_{\sigma_i} = w_i \wedge r_{\sigma_i} = 0) \end{array} \right\}$$

$$\mathbf{Pr}\left[\mathcal{E}'_\ell\right] = \left(\left(1 - \frac{p}{2}\right)^k - \left(\frac{1-p}{2}\right)^k \right) \cdot \mathbf{Pr}\left[\mathcal{E}_\ell\right] \tag{13}$$

Now, if $\mathcal{E}_\ell$ specifies the position of any of the cards $C_{\sigma_1}, \ldots, C_{\sigma_k}$, then let us consider the first time $j \geq k + 1$ when $\sigma_j = \sigma_i$, for an $i \in [k]$. Note that in this case w_i must be at the bottom of C_{σ_j} ($v_{\sigma_i} = w_i$). We can treat w_i as w_1 in the event $\mathcal{E}_\ell$ from Sect. 3 and ignore the remaining $w_t, t \in [k] \setminus \{i\}$. Then for every card $C_{\sigma_t}, t \in [k]$ that is specified in $\mathcal{E}_\ell$, we have $v_{\sigma_t} = w_t$. We immediately get $\neg \forall\, i \in [k] \ (s_{\sigma_i} = w_i \wedge r_{\sigma_i} = 0)$ and specifically for the card C_{σ_t} we get $(v_{\sigma_t} = w_t \vee r_{\sigma_t} = 0)$. We still need to check that $v_{\sigma_t} = w_t \vee r_{\sigma_t} = 0$ for the cards not specified in $\mathcal{E}_\ell$. Thus, for the event $\mathcal{E}'_\ell$ formally defined in (13) we get

$$\mathbf{Pr}[\mathcal{E}'_\ell] \geq \left(1 - \frac{p}{2}\right)^{k-1} \cdot \mathbf{Pr}[\mathcal{E}_\ell] \geq \left(\left(1 - \frac{p}{2}\right)^k - \left(\frac{1-p}{2}\right)^k \right) \cdot \mathbf{Pr}[\mathcal{E}_\ell] \tag{14}$$

Finally, we combine the events $\{\mathcal{E}_j^*\}_{j\geq 2}^k$ and $\{\mathcal{E}_\ell'\}_{\ell\geq 1}$ and use (12), (14) to get

$$
\mathbf{Pr}[\text{alg. wins}] \geq \mathbf{Pr}\left[\bigsqcup_{j=2}^{k} \mathcal{E}_j^* \sqcup \bigsqcup_{\ell\geq 1} \mathcal{E}_\ell'\right] \geq \frac{p}{2} \cdot \sum_{j=1}^{k-1}\left(\left(1-\frac{p}{2}\right)^j - \left(\frac{1-p}{2}\right)^j\right)
$$

$$
+ \left(\left(1-\frac{p}{2}\right)^k - \left(\frac{1-p}{2}\right)^k\right) \cdot \sum_{\ell\geq 1}\mathbf{Pr}[\mathcal{E}_\ell] \geq \frac{p}{2}\cdot\sum_{j=1}^{k-1}\left(\left(1-\frac{p}{2}\right)^j - \left(\frac{1-p}{2}\right)^j\right)
$$

$$
+ \left(\left(1-\frac{p}{2}\right)^k - \left(\frac{1-p}{2}\right)^k\right)\cdot\frac{p}{1+p} = \frac{p}{2}\left(1-\frac{p}{2}\right)\frac{1-\left(1-\frac{p}{2}\right)^{k-1}}{p/2}
$$

$$
- \frac{p}{2}\left(\frac{1-p}{2}\right)\frac{1-\left(\frac{1-p}{2}\right)^{k-1}}{1-\frac{1-p}{2}} + \left(\left(1-\frac{p}{2}\right)^k - \left(\frac{1-p}{2}\right)^k\right)\frac{p}{1+p}
$$

$$
= 1 - \frac{p}{2} - \frac{1}{1+p}\left(1-\frac{p}{2}\right)^k - \frac{p(1-p)}{2(1+p)} = \frac{1}{1+p}\left(1-\left(1-\frac{p}{2}\right)^k\right)
$$

$$(15)$$

We are left to verify that the right hand side of (15) is at least $\frac{p}{1+p}\cdot$ Prophet $=$ $\frac{p}{1+p}\cdot(1-\frac{1}{2^k})$. This is equivalent to showing that $f(p) \stackrel{\text{def}}{=} 1-(1-\frac{p}{2})^k \geq g(p) \stackrel{\text{def}}{=} p\cdot(1-\frac{1}{2^k})$. Now, observe that $f(0) = g(0)$, $f(1) = g(1)$, and $f'(p) - g'(p) = \frac{k}{2}(1-\frac{p}{2})^{k-1} - 1 + \frac{1}{2^k}$ is a decreasing function in p that is positive at $p = 0$ (recall that $k \geq 2$). These three conditions imply that $f(p) - g(p) \geq 0$ for any $p \in [0,1]$.

5   Matching Lower Bound

We give in this section a matching lower bound of $\frac{p}{1+p}$. Interestingly, our construction has the property that the maximum value among all n values together with all n samples (revealed or not) is almost surely much larger than the rest $2n - 1$ numbers, that is the upper bound of $\frac{p}{1+p}$ from Sect. 3 is tight.

Our construction is as follows for any fixed constant $p \in (0,1]$.

Example 1. Set $\varepsilon = o(p) > 0$. Let the number of variables $n = \Theta\left(\frac{1}{\varepsilon^2}\right)$. Define the distribution $F_0 = \{v = 0 \text{ w.p. } 1\}$ and distributions $F_i \stackrel{\text{def}}{=} \{v = \frac{1}{\varepsilon^i} \text{ w.p. } \varepsilon, v = 0 \text{ w.p. } 1 - \varepsilon\}$ for all $i \in [n]$. We construct the following mixture of n instances $\{I_i\}_{i=1}^n$.

$$
i\text{-th instance } I_i : \forall j \leq i \ \ \mathcal{D}_j = F_j, \forall j > i \ \ \mathcal{D}_j = F_0 \quad \mathbf{Pr}[I_i] = \frac{\varepsilon^{i-1}}{\sum_{j=0}^{n-1}\varepsilon^j}
$$

The next two claims describe an optimal online algorithm ALG for this instance. Let $i^* \stackrel{\text{def}}{=} \text{argmax}_i \hat{X}_i$ ($i^* = 1$ if all $\hat{X}_i = 0$).

Claim. There is an optimal ALG that does not take any X_i with $i < i^*$.

Proof. If ALG stops at any $i < i^*$, then its reward X_i is equal to or smaller than $\frac{1}{\varepsilon^{i^*-1}}$. On the other hand, ALG could wait until i^* and get a reward of at least $\frac{1}{\varepsilon^{i^*}}$ with probability ε, since the distribution $\mathcal{D}_{i^*} = F_i$. This gives at least as large expected reward of $\frac{1}{\varepsilon^{i^*}} \cdot \varepsilon$ as taking X_i.

Claim. The ALG that takes the first non zero X_i for $i \geq i^*$ is optimal.

Proof. First, we may assume that ALG does not stop before i^* by Claim 5. Also we can assume that ALG skips any $X_i = 0$. Note that the revealed samples up until i^* give no information about the variables after i^*. Thus ALG should only consider zero samples after i^*, which we denote by a vector $\mathbf{s}_R$. Then for each $j \geq i^*$

$$\mathbf{Pr}_{I_i}[\mathbf{s}_R = \mathbf{0} \mid I_i = I_j] = (1-\varepsilon)^{f(j)}, \text{ where } f(j) \overset{\text{def}}{=} |R \cap \{i^*, i^*+1, \ldots, j\}|.$$

Using Bayes rule and the law of total probability we can get

$$\mathbf{Pr}_{I_i}[I_i = I_j \mid i \geq i^*, \mathbf{s}_R = \mathbf{0}] = \frac{w_j}{\sum_{i \geq i^*} w_i}, \text{ where } w_j \overset{\text{def}}{=} \varepsilon^{j-i^*} \cdot (1-\varepsilon)^{f(j)} \quad (16)$$

We will prove that an optimal ALG should always take $X_t = \frac{1}{\varepsilon^t}$ for any $t \geq i^*$ and any $\mathbf{s}_R$ by backward induction on t. The base of induction for $t = n$ is trivial. We prove induction step for a $t < n$ assuming that the induction hypothesis holds for all $t' : t < t' \leq n$. Assume towards a contradiction that an optimal algorithm ALG$'$ does not take $X_t = \frac{1}{\varepsilon^t}$, then by the induction hypothesis ALG$'$ must wait until the next variable $X_{t'} > 0$ and stop. Next, we will show that the expected reward in this case is strictly smaller than $\frac{1}{\varepsilon^t}$ – the reward ALG would have by stopping at X_t.

$$\mathbb{E}[\text{ALG}'] = \sum_{j>t}^{n} \frac{1}{\varepsilon^j} \cdot \mathbf{Pr}[\forall t < i < j \; X_i = 0, X_j > 0] \cdot \mathbf{Pr}_{I_i}[i \geq j \mid i \geq t, \mathbf{s}_R = \mathbf{0}]$$

$$= \sum_{j>t}^{n} \frac{\varepsilon \cdot (1-\varepsilon)^{j-t-1}}{\varepsilon^j} \cdot \frac{\sum_{i \geq j}^{n} w_i}{\sum_{i \geq t}^{n} w_i} < \sum_{j>t}^{n} \frac{\varepsilon \cdot (1-\varepsilon)^{j-t-1}}{\varepsilon^j} \cdot \varepsilon^{j-t} < \frac{1}{\varepsilon^{t-1}} \sum_{i=0}^{\infty} (1-\varepsilon)^i$$

$$= \frac{1}{\varepsilon^t} = \mathbb{E}[\text{ALG}], \quad (17)$$

where to get the first inequality we observe that $\varepsilon^{j-t} w_t \geq w_j$, $\varepsilon^{j-t} w_{t+1} \geq w_{j+1}, \ldots, \varepsilon^{j-t} w_{n-j+t} \geq w_n$ by formula (16) and thus $\varepsilon^{j-t} \cdot \sum_{i \geq t}^{n} w_i > \sum_{i \geq j}^{n} w_i$; in the second inequality we simply extended the range of summation from $i = n-t-1$ to infinity. The strict inequality (17) shows that ALG$'$ cannot be optimal, and, therefore, an optimal ALG has to stop at X_t, which concludes the proof.

Now, we can compare the optimal online algorithm described by Claim 5 with the prophet.

Theorem 4. *The competitive ratio of any online algorithm with respect to the prophet is at least $\frac{p}{1+p}$ for the Example 1.*

Proof. First, we get the following lower bound on the expected reward of the prophet

$$\text{Prophet} \geq \sum_{\ell=1}^{n} \mathbf{Pr}_{I_i}[I_i = I_\ell] \cdot \mathbf{Pr}\left[X_\ell = \frac{1}{\varepsilon^\ell}\right] \cdot \frac{1}{\varepsilon^\ell} = \sum_{\ell=1}^{n} \frac{\varepsilon^{\ell-1} \cdot \varepsilon \cdot \frac{1}{\varepsilon^\ell}}{\sum_{i=0}^{n-1} \varepsilon^i} = n - o(n).$$

In what follows we get an upper bound on the expected reward ALG of the optimal online algorithm. Let us assume that the realized instance is $I_i = I_\ell$. We first observe that the total contribution from X_j with $1 \leq j < \ell$ is not more than

$$\sum_{j=1}^{\ell-1} \mathbf{Pr}[X_j > 0] \cdot \frac{1}{\varepsilon^j} = \varepsilon \cdot \sum_{j=1}^{\ell-1} \varepsilon^{-j} = O\left(\frac{1}{\varepsilon^{\ell-2}}\right). \tag{18}$$

As we will see later this turns out to be a negligibly small amount. Next we get an upper bound on the probability that ALG stops at X_ℓ when $I_i = I_\ell$ and $X_\ell > 0$.

$$\mathbf{Pr}[\text{ALG takes } X_\ell \mid I_i = I_\ell, X_\ell > 0] = \mathbf{Pr}[\forall 1 \leq i < \ell \; X_i = 0]$$

$$+ \sum_{j=1}^{\ell-1} \mathbf{Pr}[\forall j < i < \ell \; X_i = 0, X_j > 0] \cdot \mathbf{Pr}\left[\exists j < i \leq \ell \; (\hat{X}_i > 0, r_i = 1)\right]$$

$$= (1-\varepsilon)^{\ell-1} + \sum_{i=0}^{\ell-2} \varepsilon \cdot (1-\varepsilon)^i \cdot (1 - (1-p\varepsilon)^{i+1}) \tag{19}$$

We further estimate the term $A_\ell \stackrel{\text{def}}{=} \sum_{i=0}^{\ell-2} \varepsilon \cdot (1-\varepsilon)^i \cdot (1 - (1-p\varepsilon)^{i+1})$ in (19). We give an upper bound on A_ℓ by analysing a simple Markov chain M that corresponds to this summation. Markov chain M has 4 states $\{S, I, \text{End}, \text{Win}\}$; the random walk starts in the S state, and from there we can go either to Win with probability $p\varepsilon$, or to I with the remaining probability $1 - p\varepsilon$; from state I we can either go back to S with probability $1 - \varepsilon$, or go to End with remaining probability ε; finally, both states Win and End are terminal states, i.e., once the random walk gets in one of them, it stays there forever. The Win state represents that ALG successfully reaches X_ℓ and End state represents that ALG stops at an earlier random variable. Observe that

$$\mathbf{Pr}[\text{reach Win}] = 1 - \mathbf{Pr}[\text{reach End}] = 1 - \sum_{\ell=0}^{\infty} \varepsilon(1-\varepsilon)^\ell(1-p\varepsilon)^{\ell+1}$$

$$= \sum_{\ell=0}^{\infty} \varepsilon(1-\varepsilon)^\ell - \sum_{\ell=0}^{\infty} \varepsilon(1-\varepsilon)^\ell(1-p\varepsilon)^{\ell+1} = \sum_{\ell=0}^{\infty} \varepsilon(1-\varepsilon)^\ell(1-(1-p\varepsilon)^{\ell+1}) \geq A_\ell$$

On the other hand, we have a simple recurrent equation for $\mathbf{Pr}[\text{reach Win}] = p\varepsilon + (1-p\varepsilon) \cdot (1-\varepsilon) \cdot \mathbf{Pr}[\text{reach Win}]$, which gives us

$$\frac{p + o(p)}{1+p} = \frac{p}{1+p-\varepsilon p} = \mathbf{Pr}[\text{reach Win}] \geq A_\ell. \tag{20}$$

For the other term $B_\ell \overset{\text{def}}{=} (1-\varepsilon)^{\ell-1}$ in (19) we will use that $n = \Omega(\frac{1}{\varepsilon^2})$ is rather large and thus for almost all ℓ the term B_ℓ is negligibly small. Now we can combine the bounds (18), (19), and (20) together to get the lower bound on expected reward ALG of the optimal online algorithm.

$$\mathsf{ALG} \le \sum_{\ell=1}^{n} \mathbf{Pr}_{I_i}[I_i = I_\ell] \cdot \left[\frac{\mathbf{Pr}[X_\ell > 0]}{\varepsilon^\ell} \cdot \mathbf{Pr}[\text{ take } X_\ell \mid I_i = I_\ell, X_\ell > 0] \right.$$

$$\left. + O\left(\frac{1}{\varepsilon^{\ell-2}}\right) \right] = \sum_{\ell=1}^{n} \frac{\varepsilon^{\ell-1}}{\sum_{i=0}^{n-1} \varepsilon^i} \cdot \left[\frac{\varepsilon}{\varepsilon^\ell} \cdot \left(A_\ell + (1-\varepsilon)^{\ell-1} \right) + O\left(\frac{1}{\varepsilon^{\ell-2}}\right) \right] \le$$

$$\sum_{\ell=1}^{n} \varepsilon^{\ell-1} \left[\frac{A_\ell}{\varepsilon^{\ell-1}} + \frac{(1-\varepsilon)^{\ell-1}}{\varepsilon^{\ell-1}} + O\left(\frac{1}{\varepsilon^{\ell-2}}\right) \right] < n \cdot A_n + O(n\varepsilon) + \sum_{\ell=1}^{\infty} (1-\varepsilon)^{\ell-1}$$

$$\le \frac{np + o(np)}{1+p} + O(n\varepsilon) + \frac{1}{\varepsilon} = n \cdot \left(\frac{p}{1+p} + o(1) \right).$$

Combining this upper bound on ALG with a lower bound on the prophet we get the desired bound $\mathsf{ALG} \le (1+o(1))\frac{p}{1+p}\mathsf{Prophet}$.

6   Conclusions and Open Directions

In this paper we analysed the impact of sparse statistical information on the stochastic optimization in the classic prophet inequality (PI) setting. The optimization from samples framework allows one to nicely capture the sparsity of the information with a single parameter $p \in [0,1]$. At least in the PI setting, it appears that finding good algorithms is more challenging than in the one-sample-per-distribution regime, but our regime also seems to be more tractable than the few-samples-per-distribution regime.

There are a many interesting directions for future research in the sparse sample regime. For example

1. An immediate question is to analyse other PI settings in the sparse samples regime: e.g., a natural candidate is the k-unit prophet inequality.
2. It is interesting to see the interplay between sparse statistical information and another commonly used secretary assumption (i.e., random order arrival models). Specifically, it is unclear if one can improve the performance in secretary models with an additional statistical information. E.g., would it be possible to strictly improve the online algorithm in classic secretary problem for any value of $p > 0$?
3. Many auction design problems have been analysed in the optimization from samples framework and there have been a few results in one-sample-per-distribution regime. These settings are natural candidates to be studied in the sparse samples regime as well.

References

1. Azar, P.D., Kleinberg, R., Weinberg, S.M.: Prophet inequalities with limited information. In: SODA, pp. 1358–1377. SIAM (2014)
2. Babaioff, M., Gonczarowski, Y.A., Mansour, Y., Moran, S.: Are two (samples) really better than one? In: EC, p. 175. ACM (2018)
3. Caramanis, C., et al.: Single-sample prophet inequalities via greedy-ordered selection. In: SODA, pp. 1298–1325. SIAM (2022)
4. Chawla, S., Hartline, J.D., Malec, D.L., Sivan, B.: Multi-parameter mechanism design and sequential posted pricing. In: STOC, pp. 311–320. ACM (2010)
5. Correa, J.R., Cristi, A., Epstein, B., Soto, J.A.: The two-sided game of googol and sample-based prophet inequalities. In: SODA, pp. 2066–2081. SIAM (2020)
6. Correa, J.R., Cristi, A., Feuilloley, L., Oosterwijk, T., Tsigonias-Dimitriadis, A.: The secretary problem with independent sampling. In: SODA, pp. 2047–2058. SIAM (2021)
7. Correa, J.R., Dütting, P., Fischer, F.A., Schewior, K.: Prophet inequalities for I.I.D. random variables from an unknown distribution. In: EC, pp. 3–17. ACM (2019)
8. Correa, J.R., Dütting, P., Fischer, F.A., Schewior, K., Ziliotto, B.: Unknown I.I.D. prophets: better bounds, streaming algorithms, and a new impossibility (extended abstract). In: ITCS. LIPIcs, vol. 185, pp. 86:1–86:1. Schloss Dagstuhl - Leibniz-Zentrum für Informatik (2021)
9. Correa, J.R., Foncea, P., Hoeksma, R., Oosterwijk, T., Vredeveld, T.: Posted price mechanisms and optimal threshold strategies for random arrivals. Math. Oper. Res. **46**(4), 1452–1478 (2021)
10. Correa, J.R., Foncea, P., Pizarro, D., Verdugo, V.: From pricing to prophets, and back! Oper. Res. Lett. **47**(1), 25–29 (2019)
11. Daskalakis, C., Zampetakis, M.: More revenue from two samples via factor revealing SDPS. In: EC, pp. 257–272. ACM (2020)
12. Dhangwatnotai, P., Roughgarden, T., Yan, Q.: Revenue maximization with a single sample. Games Econ. Behav. **91**, 318–333 (2015)
13. Dütting, P., Lattanzi, S., Leme, R.P., Vassilvitskii, S.: Secretaries with advice. In: EC, pp. 409–429. ACM (2021)
14. Ezra, T., Feldman, M., Gravin, N., Tang, Z.G.: Online stochastic max-weight matching: prophet inequality for vertex and edge arrival models. In: EC, pp. 769–787. ACM (2020)
15. Feldman, M., Gravin, N., Lucier, B.: Combinatorial auctions via posted prices. In: SODA, pp. 123–135. SIAM (2015)
16. Feldman, M., Svensson, O., Zenklusen, R.: Online contention resolution schemes. In: SODA, pp. 1014–1033. SIAM (2016)
17. Fu, H., Immorlica, N., Lucier, B., Strack, P.: Randomization beats second price as a prior-independent auction. In: EC, p. 323. ACM (2015)
18. Gravin, N., Wang, H.: Prophet inequality for bipartite matching: merits of being simple and non adaptive. In: EC, pp. 93–109. ACM (2019)
19. Guo, C., Huang, Z., Tang, Z.G., Zhang, X.: Generalizing complex hypotheses on product distributions: auctions, prophet inequalities, and pandora's problem. In: COLT. Proceedings of Machine Learning Research, vol. 134, pp. 2248–2288. PMLR (2021)
20. Hajiaghayi, M.T., Kleinberg, R.D., Sandholm, T.: Automated online mechanism design and prophet inequalities. In: AAAI, pp. 58–65. AAAI Press (2007)

21. Kaplan, H., Naori, D., Raz, D.: Competitive analysis with a sample and the secretary problem. In: SODA, pp. 2082–2095. SIAM (2020)
22. Kleinberg, R., Weinberg, S.M.: Matroid prophet inequalities. In: STOC, pp. 123–136. ACM (2012)
23. Krengel, U., Sucheston, L.: Semiamarts and finite values. Bull. Am. Math. Soc. **83**(4), 745–747 (1977)
24. Krengel, U., Sucheston, L.: On semiamarts, amarts, and processes with finite value. In: Probability on Banach Spaces, vol. 4, pp. 197–266 (1978)
25. Rubinstein, A.: Beyond matroids: secretary problem and prophet inequality with general constraints. In: STOC, pp. 324–332. ACM (2016)
26. Rubinstein, A., Wang, J.Z., Weinberg, S.M.: Optimal single-choice prophet inequalities from samples. In: ITCS. LIPIcs, vol. 151, pp. 60:1–60:10. Schloss Dagstuhl - Leibniz-Zentrum für Informatik (2020)
27. Samuel-Cahn, E.: Comparison of threshold stop rules and maximum for independent nonnegative random variables. Ann. Probab. 1213–1216 (1984)

Exploiting Extensive-Form Structure
in Empirical Game-Theoretic Analysis

Christine Konicki[✉], Mithun Chakraborty, and Michael P. Wellman

University of Michigan, Ann Arbor, MI 48109, USA
`{ckonicki,dcsmc,wellman}@umich.edu`

Abstract. Empirical game-theoretic analysis (EGTA) is a general framework for reasoning about complex games using agent-based simulation. Data from simulating select strategy profiles is employed to estimate a cogent and tractable game model approximating the underlying game. To date, EGTA methodology has focused on game models in normal form; though the simulations play out in sequential observations and decisions over time, the game model abstracts away this temporal structure. Richer models of *extensive-form games* (EFGs) provide a means to capture temporal patterns in action and information, using tree representations. We propose *tree-exploiting EGTA* (TE-EGTA), an approach to incorporate EFG models into EGTA. TE-EGTA constructs game models that express observations and temporal organization of activity, albeit at a coarser grain than the underlying agent-based simulation model. The idea is to exploit key structure while maintaining tractability. We establish theoretically and experimentally that exploiting even a little temporal structure can vastly reduce estimation error in strategy-profile payoffs compared to the normal-form model. Further, we explore the implications of EFG models for iterative approaches to EGTA, where strategy spaces are extended incrementally. Our experiments on several game instances demonstrate that TE-EGTA can also improve performance in the iterative setting, as measured by the quality of equilibrium approximation as the strategy spaces are expanded.

1 Introduction

Empirical game-theoretic analysis (EGTA) (Wellman 2016) employs agent-based simulation to induce a game model over a restricted set of strategies. The methodology is salient for games that are too complex for analytic description and reasoning. Complexity in dynamics and information can be expressed in a simulator, but abstracted from the game model. In typical EGTA practice, simulation data is used to estimate a *normal-form game* (NFG) model, associating a payoff vector with each combination of strategies available to the agents. But game theory offers richer model forms that capture sequentiality in agent play and conditional information. Specifically, *extensive-form game* (EFG) models represent the game as a tree, where nodes or sets of nodes represent states,

---

Supplementary Information The online version contains supplementary material available at https://doi.org/10.1007/978-3-031-22832-2_8.

K. A. Hansen et al. (Eds.): WINE 2022, LNCS 13778, pp. 132–149, 2022.
https://doi.org/10.1007/978-3-031-22832-2_8

and edges represent player moves and chance events. Whereas NFGs treat agent strategies as atomic objects, EFGs afford a finer-grained expression of the observations and actions that define these strategies, capturing structure that may be shared among many strategies. The goal of this work is to take advantage of extensive-form structure, at flexible granularity, for complex game environments described by agent-based simulation. Our approach, *Tree-Exploiting EGTA* (TE-EGTA), follows the basic framework of EGTA, but employs a parameterized EFG model to leverage part of the game's tree structure.

Taking advantage of extensive form necessitates two key modifications to the EGTA process. First, we require methods to estimate the more complex model form: an abstracted game tree parameterized by player utilities at terminal nodes and probability distributions over successors for stochastic events represented by *chance nodes* in the tree. These stochastic events, together with *information-set* structure, model the *imperfect information* available to the players. We introduce straightforward techniques to estimate these game-tree parameters, and describe how the structure intuitively affords more effective use of available simulation data. Second, we require methods for extending extensive-form models as the strategy space is expanded, across iterations of the EGTA process. We introduce techniques for iterative augmentation of empirical game-tree models with new (best-response) strategies, within a standard approach that incorporates deep RL within EGTA (Lanctot et al. 2017).

To establish the benefits of tree-exploitation for EGTA, we show that an extensive-form empirical game model provides (with high probability) a more accurate approximation of the true game than a normal-form model constructed from the same simulation data. As it is generally intractable to construct a game tree expressing the full fidelity of the game simulated, our approach is designed to operate on highly abstracted models capturing only selected tree structure. To ground the meaning of such abstractions, we provide an algorithm that produces a coarsened model given the full game and a description of what to abstract away. We demonstrate the efficacy of TE-EGTA through experiments on three stylized games, and over varying levels of abstraction. We compare TE-EGTA to normal-form EGTA on two key performance measures. The first is the average error incurred from estimating the true player payoffs for all strategy combinations in the empirical game. The second is the *regret* of empirical-game solutions with respect to the full multiagent scenario, computed over successive empirical game models in an iterative EGTA process.

Outline. Section 2 provides technical preliminaries, including a formal exposition of the EFG representation and precise elaboration of the EGTA framework and process. Sect. 3 delineates our algorithmic contribution, TE-EGTA, starting with the structure of an extensive-form empirical game model and how to estimate its parameters from simulation data (Sect. 3.1). We then give a theoretical procedure for generating a (usually) coarsened extensive-form model from the underlying game (Sect. 3.2), and explain how to iteratively refine the model via simulation-aided strategy exploration (Sect. 3.3). In Sect. 4, we present theoretical results on the advantage of TE-EGTA over normal-form EGTA in approximating true payoffs given a set of strategy profiles. All proofs are available in the full version. In Sect. 5, we report experiments that demonstrate the improvement in strategy-profile payoff estimation (Sect. 5.1) and in model refinement using the PSRO approach (Lanctot et al. 2017) (Sect. 5.2) produced via tree exploitation. Sect. 6 concludes.

2   Preliminaries

2.1   Extensive-Form Games (EFGs)

An *extensive-form game* (EFG) is a standard model for strategic multi-agent scenarios where agents act *sequentially* with potentially varying degrees of *imperfect information* about the history of game play. Early algorithmic work on EFGs showed how to generalize the Lemke-Howson method for computing Nash equilibria (NE) for two-player games with perfect recall (Koller et al. 1996). Well-known game-theoretic methods such as replicator dynamics (Gatti et al. 2013) and fictitious self-play (Heinrich et al. 2015) have also been adapted for EFGs. The task of successful abstraction with exploitability guarantees has also been investigated: Kroer and Sandholm (2018) gave a framework for analyzing abstractions of large-scale EFGs, and Zhang and Sandholm (2020) introduced the notion of small certificates carrying proofs of approximate NE. Other works have developed algorithms that search for optimal strategies or approximate equilibria that minimize exploitability (Johanson et al. 2012; Lockhart et al. 2019). In this paper, we will only consider games with perfect recall, so no player can forget what it observed or knew earlier.

Tree Structure. Formally, a finite, imperfect-information EFG is a tuple $G :=$ $\langle N, H, V, \{\mathcal{I}_j\}_{j=0}^n, \{\Pi_j\}_{j=1}^n, X, P, u \rangle$. The components of G are defined as follows (see Fig. 1 for an illustrative example):

- $N = \{0, \ldots, n\}$ is the set of *players*. Player 0 represents *Nature*, a non-strategic agent responsible for stochastic events that impact the course of play; the remaining players are strategic rational agents.
- H is the finite game tree, rooted at a node h_0, that captures the dynamic nature of interactions. Each node $h \in H$ represents a *state* of the game, also identified with a history of actions (see below) beginning at the *initial state* h_0 which corresponds to the null history $\emptyset$. The leaves or *terminal nodes* $T \subset H$ represent possible end-states of the game. We refer to the non-terminal nodes of H as *decision nodes*, represented by the set $D = H \setminus T$.
- $V : D \to N$ assigns a player to each decision node h.
- For each player $j \in N$, $\mathcal{I}_j$ is a partition of $V^{-1}(j)$ where each $I \in \mathcal{I}_j$ is an *information set (infoset)* of j. All nodes $h \in I$ are indistinguishable from the viewpoint of player j.
- At each information set $I \in \mathcal{I}_j$, player j has a set of available actions $\Pi_j(I)$.
- A node h where $V(h) = 0$ is called a *chance node*. $X(h)$ is the set of actions available to Nature (i.e., possible outcomes of the stochastic event) at h, and $P(\cdot \mid h)$ is the probability distribution over $X(h)$.
- The *utility function* $u : T \to \mathbb{R}^n$ maps each terminal node to a real-valued vector of players' utilities $\{u_j(t)\}_{j=1}^n$.

The directed edge connecting any $h \in I$ to its child child[h] represents a state transition resulting from $V(h)$'s move, and is labeled with an action $\pi \in \Pi_{V(h)}(I)$ if $V(h) \neq 0$, or an outcome $x \in X(h)$ otherwise. We denote by $\varphi(h, j)$ the history of actions belonging to player j up to node h.

Strategies and Payoffs. A *pure strategy* for player $j \in N \setminus \{0\}$ specifies the action $\pi_j \in \Pi_j(I)$ that j selects at information set $I \in \mathcal{I}_j$. More generally, a *mixed strategy* or simply *strategy* $\sigma_j(\cdot \mid I)$ defines a probability distribution over $\Pi_j(I)$ at each information set of agent j; that is, action π_j is selected with probability $\sigma_j(\pi_j \mid I)$. The vector $\boldsymbol{\sigma} = (\sigma_1, \ldots, \sigma_n)$ is called a *strategy profile*, and $\boldsymbol{\sigma}_{-j}$ represents the combination of strategies for players other than j. We denote the set of all strategies available to player j by Σ_j and the space of joint strategy profiles by $\Sigma = \times_{j=1}^{n} \Sigma_j$. Let $r_j(t, \sigma_j)$ denote the probability that node t is reached if player j adopts strategy σ_j and all other players (including Nature) always choose actions that lead to h when possible; the probability that t is reached under strategy profile $\boldsymbol{\sigma}$ is given by its *reach probability*, $r(t, \boldsymbol{\sigma}) = \prod_{j \in N} r_j(t, \sigma_j)$. Likewise, the contribution of Nature to the reach probability of t is $r_0(t) = \prod_{h \in H, e \in X(h) \cap \varphi(t, 0)} P(e \mid h)$. We define the *payoff* from joint strategy profile $\boldsymbol{\sigma}$ to player j as its expected utility over all end-states: $U_j(\boldsymbol{\sigma}) := \sum_{t \in T} u_j(t) r(t, \boldsymbol{\sigma})$.

Best Response Formulation and Regret. A *best response (BR)* of player $j \in N \setminus \{0\}$ to $\boldsymbol{\sigma}_{-j}$ is a strategy $\sigma_j \in \arg\max_{\sigma'_j \in \Sigma_j} U_j(\sigma'_j, \boldsymbol{\sigma}_{-j})$ that maximizes the payoff for j given $\boldsymbol{\sigma}_{-j}$. The *regret* of player j from playing $\boldsymbol{\sigma}$ is given by $\text{Reg}_j(\boldsymbol{\sigma}) = \max_{\sigma_j \in \Sigma_j} U_j(\sigma_j, \boldsymbol{\sigma}_{-j}) - U_j(\boldsymbol{\sigma})$. The total regret of the strategy profile $\boldsymbol{\sigma}$ is the sum: $\text{Reg}(\boldsymbol{\sigma}) = \sum_{j=1}^{n} \text{Reg}_j(\boldsymbol{\sigma})$. For $\varepsilon > 0$, an *ε-Nash equilibrium* is a strategy profile $\boldsymbol{\sigma}$ such that $\text{Reg}_j(\boldsymbol{\sigma}) \leq \varepsilon$ for every player $j \in N \setminus \{0\}$; a strategy profile $\boldsymbol{\sigma}$ with $\text{Reg}(\boldsymbol{\sigma}) = 0$ is a *Nash equilibrium*.

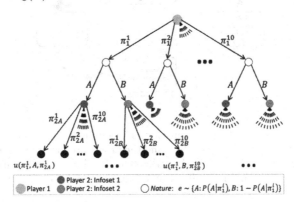

Fig. 1. EFG representation of GAME1, our running example also used in our experiments. Dashed lines indicate outgoing edges to nodes omitted from this illustration.

Running Example. Consider the two-agent strategic scenario depicted in Fig. 1, which we call GAME1. First, Player 1 chooses an action from $\Pi_1 =$

$\{\pi_1^i\}_{i=1}^{10}$; then, a single stochastic event $X(\pi_1^i) \in \{A, B\}$ occurs, outcome A having probability $P(A \mid \pi_1^i)$ dependent on Player 1's choice π_1^i. Player 2 observes the outcome $e \in \{A, B\}$ but not Player 1's chosen action, which induces two information sets for Player 2. Player 2 also has ten actions to choose from in each information set, $\Pi_{2A} = \{\pi_{2A}^i\}_{i=1}^{10}$ and $\Pi_{2B} = \{\pi_{2B}^i\}_{i=1}^{10}$. Each leaf with history $(\pi_1^i, e, \pi_{2e}^{i'})$ is labeled with the 2-dimensional vector of Player 1 and 2's realized utilities. Neither the conditional probabilities $P(A \mid \pi_1^i)$ nor the leaf utilities $u(\pi_1^i, e, \pi_{2e}^{i'})$ are known *a priori* to the game analyst.

2.2 Empirical Game-Theoretic Analysis (EGTA)

The framework of EGTA was developed for the application of game-theoretic reasoning to scenarios too complex for analytic description, accessible only in the form of a procedural simulation (Wellman 2016). Over the years, EGTA has been applied to multifarious problem domains including recreational strategy games (Tuyls et al. 2020), security games (Wang et al. 2019), social dilemmas (Leibo et al. 2017), and auctions (Wellman 2020). There is also substantial work on methodological questions such as how to decide which strategy profiles to simulate (Fearnley et al. 2015; Jordan et al. 2008), and how to reason statistically about estimated game models (Areyan Viqueira et al. 2020; Tuyls et al. 2020; Vorobeychik 2010). Recently, EGTA has received newfound attention, as the simulation-based approach meshes well with powerful new strategy generation methods from deep reinforcement learning (RL) (Lanctot et al. 2017).

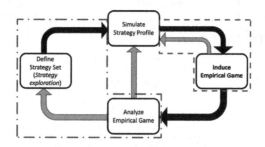

Fig. 2. Schematic illustration of EGTA. TE-EGTA modifies two subprocesses to incorporate the tree structure of EFGs: accumulation of simulation data into the game model (enclosed in blue, described in Sect. 3.1); and the procedure for augmenting $\hat{\Sigma}$ with new strategies (enclosed in red, described in Sect. 3.3). Black (resp. grey) arrows represent the sequence of operations (resp. direction of possible information flow). (Color figure online)

The main feature of EGTA is its construction of an *empirical game model* $\hat{G}$ of a much larger game of interest, called the *true game* G, from simulation data. A typical EGTA process (see Fig. 2) iteratively refines and extends $\hat{G}$ by cumulative simulation over an incrementally growing strategy space. $\hat{G}$ is a *simplification* of the underlying G since: (1) it is defined on restricted subsets $\hat{\Sigma}_j \subset \Sigma_j$ of the players' true-game strategy spaces, and the *restricted strategy profile space*, given by $\hat{\Sigma} = \times_{j=1}^n \hat{\Sigma}_j$, is typically a vast reduction of Σ; (2) some information revelation and conditioning structure may be abstracted away. Moreover, we assume that G is accessible only through a high-fidelity but expensive *simulator* that executes a given strategy profile in G and outputs limited observation histories and noisy utility samples. $\hat{G}$ is thus also an *approximation* of G since its parameters must be estimated from this simulation data.

Almost all EGTA literature to date expresses game models in *normal form*, given by a (multi-dimensional) matrix of payoff estimates for combinations of agents' strategies from the restricted set. The multi-agent scenarios themselves are typically dynamic in nature, as represented by an agent-based simulator; agent strategies are generally conditional on partial observations. For example, a normal-form game model for GAME$_1$ in Sect. 2.1 would treat each pure strategy $\pi_2^{i'}$ of player 2 as atomic, abstracting away the nuanced conditioning on whether A or B happened, and record estimated utility vectors for strategy combinations of the form $(\pi_1^i, \pi_2^{i'})$ from restricted set.

As our objective is to extend EGTA to extensive-form modeling, we will call this normal-form baseline *NF-EGTA*. In NF-EGTA, the sole simulator output of concern is the noisy sample of players' payoffs, from which we compute estimates $\{\hat{U}_j^{NF}(\boldsymbol{\sigma})\}_{j=1}^n$ of the true utilities $\{U_j(\boldsymbol{\sigma})\}_{j=1}^n$ to obtain the empirical game model $\hat{G}$. We then analyze or solve this tractable, multi-dimensional game matrix by standard techniques to obtain a result for the next iteration. Termination may be decided by a criterion such as the *true-game regret* of a solution (i.e., the maximum payoff increase achievable by any player j by deviating to a strategy in Σ_j rather than $\hat{\Sigma}_j$) falling below a specified threshold. If termination criteria are not met, we expand the restricted strategy sets through a process called *strategy exploration* (Balduzzi et al. 2019; Jordan et al. 2010), and update $\hat{G}$ through further simulation and model induction.

Game Model Estimation. Consider the process of estimating a normal-form model for an underlying extensive-form game implicitly represented by traces from the simulator. Suppose we simulate each strategy profile in $\hat{\Sigma}$ m times. Each simulated play traces a path through the game tree ending at some undisclosed terminal node $t \in T$ and returns a vector of noisy payoffs for all players sampled from a distribution with expectation $u(t)$. Let $\{\bar{u}_j^i\}_{j=1}^n$ denote the realized payoff sample at the end of the ith simulation for $i = 1, \ldots, m$; Typically, NF-EGTA's payoff estimate $\hat{U}_j^{NF}(\boldsymbol{\sigma})$ is the simple average of these samples. $\hat{U}_j^{NF}(\boldsymbol{\sigma})$ is an unbiased estimator of the true payoff, as shown in Proposition 1. In practice, the number of samples m that can be acquired is limited by the computational cost of simulation. This begs the question: can incorporating tree structure into $\hat{G}$ improve the accuracy of estimated payoffs, relative to *NF-EGTA*, for a fixed simulation budget m? We address this question in Sect. 3.1.

Proposition 1. *For every player* $j \in N \setminus \{0\}$ *and strategy profile* $\boldsymbol{\sigma} \in \hat{\Sigma}$, $\mathbb{E}\left[\hat{U}_j^{NF}(\boldsymbol{\sigma})\right] = U_j(\boldsymbol{\sigma})$.

Policy-Space Response Oracles (PSRO). A fully automated implementation of the iterative EGTA framework of Fig. 2 requires the ability to automatically generate new strategies based on analysis of the empirical game model at a given point. Phelps et al. (2006) first introduced automated strategy generation to EGTA via genetic search, and Schvartzman and Wellman (2009) first employed RL for this purpose. The advent of deep RL methods brought significant new power to this approach, which is now the predominant means of accumulating a set of restricted strategies in EGTA algorithms.

Lanctot et al. (2017) developed a general framework for interleaving empirical game modeling with deep RL techniques, which they termed *policy-space response oracles*. A key idea of PSRO is that of a *meta-strategy solver* (MSS), an abstract operation that implements the "Analyze Empirical Game" block of Fig. 2. The output of an MSS is a strategy profile, which provides the other-agent context for a BR calculation performed by deep RL. The policy generated by RL as a BR to the MSS result is then added as a new strategy to expand the current restricted strategy space, leading to another round of simulation and induction for the next EGTA iteration. The MSS concept provides a useful abstraction for expressing a variety of approaches to strategy exploration (Wang et al. 2022). For example, using Nash equilibrium as an MSS yields the double oracle (DO) algorithm (McMahan et al. 2003). If the MSS simply returns the uniform distribution over the restricted strategy sets, the algorithm reduces to fictitious play.

Prior work has extended the DO algorithm to exploit game-tree structure. Bošanský et al. (2014) developed a *sequence-form double-oracle* algorithm for zero-sum EFGs that maintains a restricted game model based on partial action sequences. The XDO algorithm of McAleer et al. (2021) for two-player zero-sum games computes a mixed BR at each information set, as compared to normal-form DO which mixes policies only at the root level. It modifies PSRO for EFGs while still using a normal-form empirical model. The benefits over normal-form demonstrated by these works suggest EGTA can be similarly extended to exploit game-tree structure beyond the strategy exploration block.

3   Tree-Exploiting EGTA

We call our approach for augmenting empirical game models to incorporate extensive-form game elements *tree-exploiting EGTA* (TE-EGTA). In the typical normal-form treatment of EGTA, the underlying game is parameterized by entries in a payoff matrix $\{U_j(\boldsymbol{\sigma})\}_{j \in N \setminus \{0\}, \boldsymbol{\sigma} \in \Sigma}$.[1] TE-EGTA instead parameterizes the underlying game to capture the EFG tree structure through a set of *leaf utilities* $\{u(t)\}_{t \in T}$, and *conditional probability distributions* that are dependent on possibly unobserved previous choices made in the game play and estimated from observations of stochastic events.

---

[1] More general approaches based on regression have been proposed (Sokota et al. 2019; Vorobeychik et al. 2007), which also amount to parameterized representations of a payoff function.

We assume that the structure of decisions and stochastic events in the empirical EFG model is given (typically a high-level abstraction of the game tree implicitly represented by the simulator, as discussed in Sect. 3.2). This ensures that the order of player choices and stochastic events in the empirical game tree matches the order in the true game, from root to leaf. In particular, the true game's information sets must be a refinement of the empirical game's information sets. Given this structure, we treat observations of Nature's actions as conditioned on past game play. The empirical game tree therefore must associate with each chance node a conditional probability distribution over the relevant set of outgoing edges. Leaves of the tree are associated with payoff estimates, which depend on the entire path from the root.

Each simulation of a strategy profile yields sample payoffs, as well as a trace of publicly or privately *observable actions* from both the players and Nature that are made over the course of the game. This is a key point of contrast with the normal-form model, for which only payoffs are relevant. The trace of actions tells us which leaf node in the abstract model is reached and what stochastic event outcomes were realized along the way.

To explain our tree-exploiting estimation approach, we first restate the expression for $U_j(\boldsymbol{\sigma})$ in a way that explicitly factors in probabilities of specific *observations* of stochastic events. We assume that a game theorist working with the black-box simulator's partial observations in order to formulate an empirical model is aware of the game's rules, and so can surmise where in the game the observation has occurred. We also assume that the observation labels used by the simulator allow the game theorist to distinguish the observations from each other and associate them with the appropriate chance nodes. A stochastic observation during gameplay is captured in the tree by an edge $e \in \varphi(t,0)$ from a chance node h such that $V(h) = 0$ to a node with history he. The reach probability of he from the perspective of Nature is $r_0(he) = P(e \mid h)$, and recall $r_0(t)$ is the joint probability of Nature's choices along the path from the root to t. Hence,

$$U_j(\boldsymbol{\sigma}) = \sum_{t \in T} u_j(t) \prod_{k=1}^{n} r_k(t, \sigma_k) r_0(t). \tag{1}$$

3.1 TE-EGTA Game Model Estimation

The probabilities $r_k(t, \sigma_k)$, for all terminal nodes t, are directly determined by the strategy profile $\boldsymbol{\sigma}$. Hence, to estimate $U_j(\boldsymbol{\sigma})$ based on Eq. (1), we need estimates for $u(t)$ and $\{r_0(t)\}_{t \in T}$. These are, in fact, the game parameters for TE-EGTA (leaf utilities and conditional probabilities respectively) that we introduced above. We denote the respective estimates by $\{\hat{u}_j(t)\}_{j=1}^{n}$ and $\{\hat{r}_0(t)\}_{t \in T}$.

A key feature of TE-EGTA is that, in modeling the payoff of strategy profile $\boldsymbol{\sigma}$, we estimate the parameters using *all* relevant simulation data, not just the data from simulating $\boldsymbol{\sigma}$. Different strategy profiles may lead to overlapping or identical paths being taken through the game tree, with some probability. We compute $\hat{u}_j(t)$ as the sample average of player j's payoffs across simulation runs that terminate at node t. Similarly, we estimate chance node probabilities using all simulations. Suppose a chance node h is reached m_h times across all simulation data, and the node with history he (reflecting Nature's choice e) is reached $m_{he} < m_h$ times. The empirical probability of observing the stochastic outcome represented by e in the game tree is $\frac{m_{he}}{m_h}$. Note that m_h can never be zero because the algorithm for constructing the empirical game model includes only nodes that are reached in simulation. Finally, we give player j's estimated payoff for strategy profile $\boldsymbol{\sigma}$:

$$\hat{U}_j^{TE}(\boldsymbol{\sigma}) = \sum_{t \in T} \hat{u}_j(t) \prod_{k=1}^{n} r_k(t, \sigma_k) \left(\prod_{e \in \varphi(t,0)} \frac{m_{he}}{m_h} \right).$$

Recall that each strategy profile $\boldsymbol{\sigma}$ in $\hat{\Sigma}$ is simulated m times, resulting in m game play sequences for each. Some strategies that end at different terminal nodes t_1 and t_2 may still include the same node h in their respective paths and result in the same observation $e \in \hat{X}(h)$. The observation occurs with the same probability for both strategies since their histories diverge only at node he. This feature is what allows the empirical game model to take into account the role of different decision points in the formulation of player strategies in a way that the normal-form model does not.

To illustrate the difference in model estimation between NF- and TE-EGTA, consider the following example from GAME1. Suppose we simulate the strategy profile (π_1^1, π_2^1) 10 times, and obtain the following payoff samples for Player 1: $99, 95, 100, 96, 95, 100, 92, 95, 93, 94$; we also observe outcome A of the stochastic event in the first 6 of these 10 simulations. NF-EGTA would simply average the 10 payoff samples and record $\hat{U}_1^{NF}(\pi_1^1, \pi_2^1) = 95.9$. In contrast, TE-EGTA distinguishes the 6 samples corresponding to the leaf (π_1^1, A, π_{2A}^1) from the 4 samples corresponding to the leaf (π_1^1, B, π_{2B}^1), and separately averages them to get the estimates $\hat{u}_1(\pi_1^1, A, \pi_{2A}^1) = 97.5$ and $\hat{u}_1(\pi_1^1, B, \pi_{2B}^1) = 93.5$. Now, suppose we also have data from 10 simulations of another strategy profile (π_1^1, π_2^2), $\pi_2^2 \neq \pi_2^1$, A being realized in 5 of these simulations. From this experience, our overall estimated probability of A conditioned on π_1^1 is $\frac{6+5}{10+10} = 0.55$. Thus, using all relevant sample data, $\hat{U}_1^{TE}(\pi_1^1, \pi_2^1) = 0.55 \times 97.5 + (1 - 0.55) \times 93.5 = 95.7$.

The following proposition shows that, like NF-EGTA, TE-EGTA produces unbiased estimates of strategy-profile payoffs. However, our theoretical results in Sect. 4 suggest that TE-EGTA offers more accurate payoff estimates with a high probability.

Proposition 2. *For every player* $j \in N \setminus \{0\}$ *and strategy profile* $\boldsymbol{\sigma} \in \hat{\Sigma}$, $\mathbb{E}_{t \sim r(T, \boldsymbol{\sigma})} \left[\hat{U}_j^{TE}(\boldsymbol{\sigma}) \right] = U_j(\boldsymbol{\sigma}).$

3.2   The Game Model as an Abstraction

Abstraction methods have extended the state of the art in solving imperfect-information games over the years (Sandholm 2010), particularly poker. An abstraction algorithm takes as input a complete game description and produces a simpler version of the tree. TE-EGTA incorporates some of the tree structure from the true game into the empirical game model; in order to ground this game model as a coarse abstraction of the underlying game, we describe **Coarsen**, an algorithm that coarsens a game tree by abstracting away chance nodes.

We express *coarseness* as the fraction of chance nodes from the true game that are included in the empirical game model. An empirical game that matches the true game's structure would include all of them; conversely, an empirical game in normal-form would include none of them. We are primarily concerned with games represented by agent-based simulation where the representation of the true game as an EFG is intractable, and thus we would not expect to obtain a coarsened model by actually applying **Coarsen**. Our intent is to contextualize a coarsened game as one that could in principle be produced by abstracting away chance nodes.

Coarsen Algorithm for coarsening an input game G

Require: Input game G, partition $C' \subseteq C$ and map $\rho : C' \to X$
    Copy $H' = H$, with each node h represented by its history
    for $c \in C'$, beginning at the chance node furthest from the root **do**
        Let $I_j(c)$ be the set of infosets induced by each event $e \in X(c)$ for player j.
        Compute power set Z^* of intersections $Z = \bigcap_{I \in I_j(c)} \{h \mid he \in I\}$ of all the histories h across $I_j(c)$.
        for $Z \in Z^*$ **do**
            $\{I'_j, \Pi'_j(I'_j)\}, H' = $ **CoarsenInfosets** $(I_j(c), Z, \rho, G)$
            Assign $X'(c) = X(c) \setminus \rho(c)$
            $\mathcal{I}'_j, \Pi'_j = $ **CondenseBranching** $(\{I'_j, \Pi'_j(I'_j)\}, \mathcal{I}'_j)$
        end for
    end for
    Assign $X'(c) = X(c)$ for all $c \notin C'$.
    Assign all player j's infosets not conditioned on chance events from any $c \in C'$ to $\mathcal{I}'_j$
    For all nodes h that preceded or did not follow any nodes in C', assign $V'(h) = V(h)$

    return $G' = (N, H', V', \{\mathcal{I}'\}_{j=1}^n, \{\Pi'_j\}_{j=1}^n, X')$

The algorithm is given a partition of both G's chance nodes $C = \{h \in H \mid V(h) = 0\}$ and the set of outcomes $X(h)$ for each chance node, denoting what to exclude from the coarsened tree. One important restriction on G is that the child nodes of a given chance node in C' must all belong to the same player so that they can be collapsed into one node. We denote the abstracted game by $G' = \langle N, H', V', \{\mathcal{I}'_j\}_{j=1}^n, \{\Pi'_j\}_{j=1}^n, X' \rangle$ whose components are defined as in Sect. 2. The nodes identified, information sets, and action spaces will necessarily differ from those of G, depending on what information is coarsened and where. Without loss of generality, **Coarsen** treats both G and G' as binary trees in order to limit the branching factor of G'. **CoarsenInfosets** transforms the intersecting information sets of the children of each $c \in C'$ into a new information set for G'

whose action space is the Cartesian product of the old infosets' action spaces. To keep the branching factor equal to 2, **CondenseBranching** transforms these action spaces (comprised of tuples) into binary (sub-)trees where each edge is part of an action tuple.

3.3 Tree-Exploiting PSRO

Recall the PSRO framework for iterative EGTA with deep RL, introduced in Sect. 2.2. Like EGTA more generally, past work within the PSRO framework has relied on normal-form representations of the empirical game, even though the games of interest are inherently sequential. We call PSRO that uses a normal-form (resp. tree-exploiting) empirical game NF-PSRO (TE-PSRO). In addition to exploiting extensive structure for estimation (Sect. 3.1), TE-PSRO also takes advantage of the tree representation for managing the restricted strategy space. A single pure strategy profile can result in multiple different paths depending on Nature's choices. If a new best response for a given infoset is part of the profile, new paths with their own new utilities and stochastic distributions at Nature's decision points are discovered and added to the empirical game tree. If one of those paths includes moves from other players that are already part of the game tree, then additional samples from this new combination can be included in the (tighter) estimation of the old parameters pertinent to that path.

Consider the empirical game in Fig. 3a with restricted strategy sets $\hat{\Pi}_1, \hat{\Pi}_{2A}$, and $\hat{\Pi}_{2B}$ for each information set as shown; the true game here is GAME₁. Let $BR_1(\sigma_{2A}, \sigma_{2B})$ and $(BR_{2A}(\sigma_1), BR_{2B}(\sigma_1))$ denote the respective best responses from GAME₁ (the true game) to the strategy profile $(\sigma_1, (\sigma_{2A}, \sigma_{2B}))$. Suppose, in an iteration, $BR_1(\sigma_{2A}, \sigma_{2B}) = \pi_1^2$, $BR_{2A}(\sigma_1) = \pi_{2A}^2$, and $BR_{2B}(\sigma_1) = \pi_{2B}^1$.

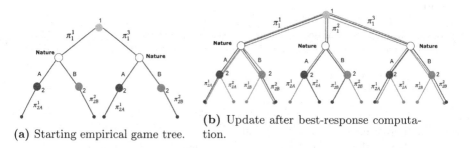

(a) Starting empirical game tree.

(b) Update after best-response computation.

Fig. 3. Two successive steps of possible TE-PSRO instantiation on GAME₁.

In the next round, the new best-response elements are considered in conjunction with the pre-existing strategy combinations from the restricted set, as well as other players' new best responses. The resulting trajectories are shown in Fig. 3b: (1) $BR_1 \times \Pi_{2A} \times \Pi_{2B}$ highlighted in yellow; (2) $\Pi_1 \times BR_{2A} \times \Pi_{2B}$ highlighted in blue; (3) $\Pi_1 \times \Pi_{2A} \times BR_{2B}$ highlighted in orange; and (4) (BR_1, BR_{2A}, BR_{2B}) highlighted in purple. See the full paper for more detail. This expansion of the empirical game tree captures finer-grained structural information about the true game than simply adding a matrix entry for each new best-response combination.

4   Payoff Estimation Improvement: Theoretical Results

To develop a formal framework for comparing the efficacy of payoff estimation (Sect. 3.1) by TE-EGTA and NF-EGTA, we apply the concept of *uniform approximation of a game* (Areyan Viqueira et al. 2020) to our setting. Consider a true EFG G and an empirical game $\hat{G}$ with the same set of players and with restricted set $\hat{\Sigma}$ constructed from accumulated simulation data upon termination of EGTA. Let $\hat{U}_j(\boldsymbol{\sigma})$ be the estimate in $\hat{G}$ of an arbitrary player j's true payoff under strategy profile $\boldsymbol{\sigma}$.

Definition 1. *The ℓ_∞-norm between games G and $\hat{G}$ is given by*

$$\| G - \hat{G} \|_\infty = \max_{j \in N \setminus \{0\}, \boldsymbol{\sigma} \in \hat{\Sigma}} |U_j(\boldsymbol{\sigma}) - \hat{U}_j(\boldsymbol{\sigma})|.$$

If $\| G - \hat{G} \|_\infty \le \varepsilon$, then $\hat{G}$ is said to be a unifor ε-approximation of G.

Note that in this definition, the maximization is only over the restricted set $\hat{\Sigma} \subseteq \Sigma$. An important consequence of $\hat{G}$ being a uniform approximation of G upon EGTA's termination is that a strategy profile that is an approximate Nash equilibrium in $\hat{G}$ is an approximate Nash equilibrium in G as well:

Proposition 3. *If $\hat{G}$ is a uniform ε-approximation of G and $\boldsymbol{\sigma}$ is a γ-Nash equilibrium of $\hat{G}$ for some $\gamma \ge 0$, then $\mathsf{Reg}_j(\boldsymbol{\sigma}) \le 2\varepsilon + \gamma$ for each player $j \in N \setminus \{0\}$ upon the termination of EGTA.*

The main result of this section is that for a given EFG, under reasonable assumptions, TE-EGTA induces an empirical game model that is a tighter uniform approximation of the EFG than that induced by NF-EGTA, with a high probability. Given an arbitrary true game G, let $\hat{G}_{NF}$ and $\hat{G}_{TE}$ denote respectively the empirical game models induced by the application of NF-EGTA and TE-EGTA to G over the same restricted set $\hat{\Sigma}$.[2] We further assume an upper and a lower bound for all agent payoff samples returned by the simulator. Let c be the number of strategy profiles from the restricted set that, after each profile is sampled m times, result in a path taken through the tree that includes the first edge of $\varphi(t)$. c can be as small as 1 and as large as $O(|\Sigma_j|)$ for some $j \in N$ depending on the game structure and when the selected EGTA method terminates. Combined with Proposition 3, we have the following result, which also

---

[2] In the iterative application of EGTA, the NF- and TE- variants may produce different choices of strategies to add; hence, strategy sets covered at a given iteration number tend to diverge. However, for comparing model estimation accuracy, however, it makes sense to start with a common baseline of strategy space. Our experiments (Sect. 5) provide empirical corroboration that the benefits accrue as well when we examine the trajectory of models produced within the iterative PSRO framework.

implies a tighter upper bound for player regret in G under approximate equilibria in the empirical game model computed using payoffs estimated through TE-EGTA.

Theorem 1. *For any $\delta \in (0,1)$ and the same number m of game simulation repetitions in each iteration of either type of EGTA, there exist positive constants ε_{NF} and ε_{TE} such that $\frac{\varepsilon_{TE}}{\varepsilon_{NF}} = \frac{1}{\sqrt{c}}$, and with probability at least $1 - \delta$ w.r.t. the randomness in the simulator payoff output, $\hat{G}_{NF}$ (respectively, $\hat{G}_{TE}$) is a uniform ε_{NF}-approximation (respectively, ε_{TE}-approximation) of G.*

5   Experiments

We conducted two sets of experiments comparing TE-EGTA with varying levels of tree structure exploitation to NF-EGTA. Each set used three different EFGs, chosen so that the corresponding empirical game models induced by our flexible tree-exploiting framework would vary in size and complexity. We implemented a simulator for each game that produced observations in accordance with the corresponding stochastic events, and end-state payoff samples that were normally distributed about the true utilities at the respective terminal nodes with a noise variance $\epsilon = 0.1$. The first game was GAME_1 (Sect. 2.1). In our experiments, for each instance of GAME_1, we randomly assigned $P(A \mid \pi_1^i)$ from $U[0,1]$ for each $\pi_1^i \in \Pi_1$ and $u(t)$ from $\{0, 0.25, \ldots, 4.75, 5\}$ for each leaf utility. During each game play sample, the simulator returned the realized outcome A or B of the single stochastic event and a noisy payoff vector.

The second game was GAME_2, an extension of GAME_1 having a second stochastic event $e_2 \in \{C, D\}$ after Player 2's turn and a second turn for Player 1 afterward. Player 1 only observes its first action and the second event e_2. Thus Player 2 has 2 information sets whereas Player 1 has $1 + 2 \cdot 10 = 21$. For its second turn, Player 1 has ten options depending on which outcome of e_2 it observed: $\Pi_{1C} = \{\pi_{1C}^i\}_{i=1}^{10}$ and $\Pi_{1D} = \{\pi_{1D}^i\}_{i=1}^{10}$. See the full version of this paper for an illustration. For each instance of GAME_2 and each π_{2A}^i (respectively, π_{2B}^i), we sampled $P(C \mid A, \pi_{2A}^i)$ (respectively, $P(C \mid B, \pi_{2B}^i)$) from $U[0,1]$. Each leaf utility was chosen uniformly at random from the set $\{0, 0.1, \ldots, 9.9, 10\}$. We experimented with two game model forms: one for when the simulator returned a noisy payoff vector and e_1 only, and one for when it returned the vector and outcomes of both events.

The final game was GAME_3, which begins with a stochastic event $e_1 \in \{A, B, C, D\}$. Player 1 observes the event and then takes a turn, choosing one of four possible actions. Next, Player 2 observes the event (but not Player 1's action) and also chooses from four possible actions. This 3-round sequence is repeated twice, but in each subsequent sequence, the only outcomes available to Nature and the agents are the remaining ones that have not yet been chosen. For instance, if $e_1 = A$, then Nature can only output $e_2 \in \{B, C, D\}$ during its second turn and $e_3 \in \{B, C, D\} \setminus \{e_2\}$ during its third. Likewise, the players are restricted to the actions that they have not yet played in the previous 3-round sequence(s). Since the players are only unable to observe the other player's actions during the *current* 3-round sequence, each player has $4 + 4^3 \cdot 3 + 4^3 \cdot 3^3 \cdot 2 = 3652$ information sets. To compare the effects of varying degrees of tree exploitation, we examined three different game model forms: (1) simulator reports observation e_1 only; (2) simulator reports e_1 and e_2 only; and (3) simulator reports all three events. We believe that a model that includes only the first stochastic event would generally yield only a negligible difference in accuracy from a model that includes only the second (or third) stochastic event.

Each iteration of EGTA had a fixed budget of 500 total samples available for all strategy combinations to be fed into the simulator for GAME₁ and GAME₂. Due to the larger size, we allotted 5000 total samples for GAME₃. We ran the experiments for GAME₁ on a standard laptop (Quad-Core Intel Core i7 Processor, 2.7 GHz, 16 GB RAM). Each repetition of both TE-PSRO and NF-PSRO for GAME₁ finished in less than 1 min. We ran the experiments for GAME₂ and GAME₃ on a single core of the Great Lakes Slurm cluster at the University of Michigan, with 786MB of memory. NF-PSRO on GAME₂ consistently finished within 6 min, and took 4–90 min for GAME₃. TE-PSRO required between 3 min and 5 h for GAME₂ (depending on the MSS used, see Sect. 5.2), and at most 1 h for GAME₃. All figures include the metrics' initial values at time-step 0.

5.1 TE-EGTA Payoff Estimation

The aim of the first set of experiments was to assess the improvement in strategy profile payoff estimation produced by incorporating the EFG tree structure into the empirical game model. We ran NF-EGTA and TE-EGTA on each true game with the same number $m = 500$ of simulations for each strategy-profile payoff vector estimation. To update the game model for either variant of EGTA, we implemented the PSRO framework using an oracle that returns the best response to the other player's strategy for GAME₁ and GAME₂. However, the size of GAME₃ made a best response oracle infeasible, so we instead used Q-learning to compute an approximate best response from the true game. For newly selected strategy profiles that were simulated in each iteration, we computed estimated payoffs $\hat{U}_j^{NF}(\boldsymbol{\sigma})$ (resp. $\hat{U}_j^{TE}(\boldsymbol{\sigma})$) for NF-EGTA (resp. TE-EGTA) from accumulated simulation data using the approach described in Sect. 2.2 (resp. Sect. 3.1). We evaluated the *estimation error* for that iteration of either variant as the average absolute difference between true and estimated payoffs for all players over all strategy combinations in the current empirical game. We repeated this operation for 25 initial restricted sets, each consisting of a single randomly chosen policy, and reported the estimation error averaged over all 25 repetitions for each iteration of PSRO in Fig. 4.

As the plots show, TE-EGTA achieves significantly lower payoff estimation error compared to NF-EGTA across all games. It is also clear that while the vast number of infosets in GAME₃ led NF-EGTA to perform worse as more strategy combinations were added despite an unchanging sample budget m; such was not the case for TE-EGTA, which converged very quickly. We attribute this to the relatively small number of actions (2, 3, or 4) available at each information set, as

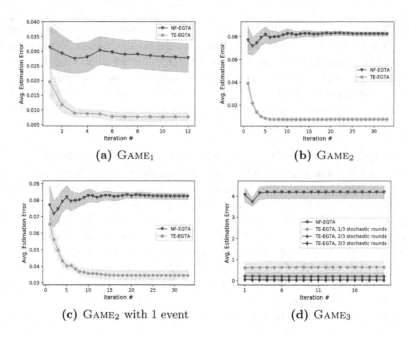

Fig. 4. Average estimation error of strategy payoffs over the course of EGTA's runtime. Shaded areas represent the standard error of the mean. The estimation errors at iteration 0 are identical since the restricted sets for both models contain the same randomly chosen policy; hence, they are omitted.

well as the large number of infosets relative to the total number of game paths. Q-learning returned a best response for every infoset that could be reached, given σ, so the empirical game ceased growing after only a few iterations. Finally, we note that the more stochastic events included in $\hat{G}$, the more tree structure is exploited by TE-EGTA, and the lower the resulting payoff error. In fact, the inclusion of even a single stochastic event or round in the model dramatically decreased the payoff error in comparison to NF-EGTA.

5.2   Iterative Model Refinement in PSRO

Our second set of experiments compared the power of NF-PSRO and TE-PSRO to iteratively explore the EFG's strategy space and fine-tune their respective empirical game models. PSRO terminates once no new best responses can be added to $\hat{\Sigma}$. To evaluate the efficacy of this iterative fine-tuning, we computed the regret $\mathsf{Reg}(\sigma)$ (as defined in Sect. 2.1) in the true game G of the solution σ returned by the MSS in every iteration.

For NF-PSRO, we used the Python-Gambit interface to represent the empirical game and used Gambit's `lcp` solver as the MSS. The solver takes as input an NFG or EFG, converts it into a linear complementarity program, and solves for all NE. We also used the `lcp` as the TE-PSRO solver for GAME$_1$ and GAME$_2$. It

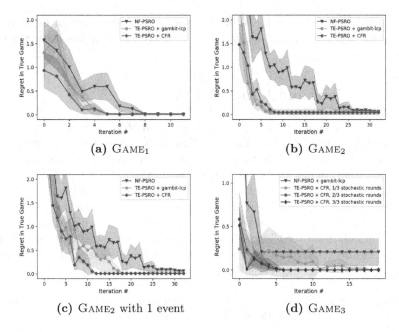

Fig. 5. Average regret of solution profiles over the course of PSRO's runtime. Shaded areas represent standard error of the mean.

is important to note that Gambit's solvers can become intractable for medium or large game trees. However, when possible, we intentionally chose an MSS that finds exact solutions to the empirical game in order to minimize any error/variability in the solutions resulting from the iterative process of adding strategies and fine-tuning the empirical game models. For medium-to-large game trees like GAME3, we used counterfactual regret minimization (CFR) (Zinkevich et al. 2007) to find an approximate NE and Q-learning to learn an approximate best response from the true game. We used CFR as the MSS for GAME2 as well for comparison to the exact lcp solver. As in Sect. 5.1, we repeated PSRO for 25 different restricted sets, each consisting of a single, randomly chosen strategy profile. We report the regret curves, averaged over 25 repetitions, in Fig. 5.

TE-PSRO converged on average to a regret at least as tight as NF-PSRO using the same simulation budget and regardless of which pure σ the initial restricted set contained. It also converged in fewer iterations, particularly in GAME2 and GAME3 as more tree structure was included in $\hat{G}$. Additional plots in the full version of this paper demonstrate the same result for different numbers of samples. However, the standard error shadings for GAME1 overlap mainly due to the high volatility in NF-PSRO regret in earlier iterations. Since, in each iteration, we add new, pertinent best responses to $\hat{\Sigma}$, we hypothesize that their absence from the previous strategy space caused the regret to increase. A one-sided two-sample t-test on each of the iterations of GAME1's regret curves established that TE-PSRO's regret improvement was statistically significant.

These results suggest that including even some tree structure in $\hat{G}$ results in PSRO converging at least as quickly and to a solution that has lower regret in the true game.

6 Conclusions and Future Work

This study represents a first step towards the goal of leveraging extensive-form structure within the EGTA framework. Our work complements prior research that showed benefits of exploiting tree structure in game reasoning and learning, for example studies that demonstrated advantages of extensive form in techniques based on the double oracle algorithm (Bošanský et al. 2014; McAleer et al. 2021). In future work, we hope to draw on further insights from this line of work, combining the best features of techniques from game reasoning, machine learning, and simulation-based game modeling. One particularly fruitful direction may be consideration of strategy exploration methods that explicitly consider extensive structure in the currently defined strategy space.

Acknowledgments. This work was supported in part by a grant from the Effective Altruism Foundation, and by the US National Science Foundation under CRII Award 2153184.

References

Areyan Viqueira, E., Cousins, C., Greenwald, A.: Improved algorithms for learning equilibria in simulation-based games. In: 19th International Conference on Autonomous Agents and Multi-Agent Systems (2020)

Balduzzi, D., et al.: Open-ended learning in symmetric zero-sum games. In: 36th International Conference on Machine Learning (2019)

Bošanský, B., Kiekintveld, C., Lisý, V., Pěchouček, M.: An exact double-oracle algorithm for zero-sum extensive-form games with imperfect information. J. Artif. Intell. Res. **51**, 829–866 (2014)

Fearnley, J., Gairing, M., Goldberg, P., Savani, R.: Learning equilibria of games via payoff queries. J. Mach. Learn. Res. **16**, 1305–1344 (2015)

Gatti, N., Panozzo, F., Restelli, M.: Efficient evolutionary dynamics with extensive-form games. In: 27th AAAI Conference on Artificial Intelligence (2013)

Heinrich, J., Lanctot, M., Silver, D.: Fictitious self-play in extensive-form games. In: 32nd International Conference on Machine Learning (2015)

Johanson, M., Bard, N., Burch, N., Bowling, M.: Finding optimal abstract strategies in extensive-form games. In: 26th AAAI Conference on Artificial Intelligence (2012)

Jordan, P.R., Vorobeychik, Y., Wellman, M.P.: Searching for approximate equilibria in empirical games. In: 7th International Conference on Autonomous Agents and Multi-Agent Systems (2008)

Jordan, P.R., Schvartzman, L.J., Wellman, M.P.: Strategy exploration in empirical games. In: 9th International Conference on Autonomous Agents and Multi-Agent Systems, pp. 1131–1138 (2010)

Koller, D., Megiddo, N., von Stengel, B.: Efficient computation of equilibria for extensive two-person games. Games Econom. Behav. **14**, 247–259 (1996)

Kroer, C., Sandholm, T.: A unified framework for extensive-form game abstraction with bounds. In: 32nd Conference on Neural Information Processing Systems (2018)

Lanctot, M., et al.: A unified game-theoretic approach to multiagent reinforcement learning. In: 31st Annual Conference on Neural Information Processing Systems (2017)

Leibo, J.Z., Zambaldi, V., Lanctot, M., Marecki, J., Graepel, T.: Multi-agent reinforcement learning in sequential social dilemmas. In: 16th International Conference on Autonomous Agents and Multi-Agent Systems (2017)

Lockhart, E., et al.: Computing approximate equilibria in sequential adversarial games by exploitability descent. In: 28th International Joint Conference on Artificial Intelligence (2019)

McAleer, S., Lanier, J., Wang, K.A., Baldi, P., Fox, R.: XDO: a double oracle algorithm for extensive-form games. In: 35th Annual Conference on Neural Information Processing Systems (2021)

McMahan, H.B., Gordon, G.J., Blum, A.: Planning in the presence of cost functions controlled by an adversary. In: 20th International Conference on Machine Learning, pp. 536–543 (2003)

Phelps, S., Marcinkiewicz, M., Parsons, S., McBurney, P.: A novel method for automatic strategy acquisition in n-player non-zero-sum games. In: 5th International Joint Conference on Autonomous Agents and Multi-Agent Systems, pp. 705–712 (2006)

Sandholm, T.: The state of solving large incomplete-information games, and application to poker. AI Mag. **31**(4), 13–32 (2010)

Schvartzman, L.J., Wellman, M.P.: Stronger CDA strategies through empirical game-theoretic analysis and reinforcement learning. In: 8th International Conference on Autonomous Agents and Multi-Agent Systems, pp. 249–256 (2009)

Sokota, S., Ho, C., Wiedenbeck, B.: Learning deviation payoffs in simulation-based games. In: 33rd AAAI Conference on Artificial Intelligence (2019)

Tuyls, K.: Bounds and dynamics for empirical game-theoretic analysis. Auton. Agents Multi-Agent Syst. **34**(7) (2020)

Vorobeychik, Y.: Probabilistic analysis of simulation-based games. ACM Trans. Modeling Comput. Simul. **20**(3), 16:1–16:25 (2010)

Vorobeychik, Y., Wellman, M.P., Singh, S.: Learning payoff functions in infinite games. Mach. Learn. **67**, 145–168 (2007)

Wang, Y.: Deep reinforcement learning for green security games with real-time information. In: 33rd AAAI Conference on Artificial Intelligence (2019)

Wang, Y., Ma, Q., Wellman, M.P.: Evaluating strategy exploration in empirical game-theoretic analysis. In: 21st International Conference on Autonomous Agents and Multi-Agent Systems (2022)

Wellman, M.P.: Putting the agent in agent-based modeling. Auton. Agent. Multi-Agent Syst. **30**(6), 1175–1189 (2016). https://doi.org/10.1007/s10458-016-9336-6

Wellman, M.P.: Economic reasoning from simulation-based game models. Œconomia **10**, 257–278 (2020)

Zhang, B.H., Sandholm, T.: Small Nash equilibrium certificates in very large games. In: 34th Annual Conference on Neural Information Processing Systems (2020)

Zinkevich, M., Johanson, M., Bowling, M.H., Piccione, C.: Regret minimization in games with incomplete information. In: 21st Conference on Neural Information Processing Systems (2007)

Constructing Demand Curves
from a Single Observation of Bundle Sales

Will Ma[1]($\boxtimes$) and David Simchi-Levi[2]

[1] Graduate School of Business, Columbia University, New York, NY 10027, USA
wm2428@gsb.columbia.edu
[2] Institute for Data, Systems, and Society, Department of Civil and Environmental
Engineering, and Operations Research Center, Massachusetts Institute of Technology,
Cambridge, MA 02139, USA
dslevi@mit.edu

Abstract. Firms typically require multiple sales observations under varying prices to understand how the demand for their items respond to price. In this paper, our partner online retailer is faced with the problem of reconstructing demand curves when only a *single* point on each curve has been historically observed. We show how a second point on each curve can be extracted from the sales of their discounted bundles, after which we help them estimate linear demand curves for their items. We perform this extraction by fitting a multi-item valuation model from the bundle pricing literature, introducing a new iterative procedure for solving this fitting problem. Our extraction process reveals a new insight on the relationship between an item's relative frequency of bundle sales vs. the steepness of its demand curve around its current price. We validate this insight on the data provided by our partner firm.

Keywords: Learning valuation distributions · Bundling

1 Introduction

Demand prediction is critical to a firm's operations, allowing for the advance planning of production, distribution, and fulfillment. This paper is concerned with modeling how the demand of an individual item responds to price, when only a *single* point on this curve has been historically observed. As a motivating example taken from the data of our partner online retailer, a recently-released pressure cooker has been consistently priced at $207 and its weekly demand has been 222. The firm wants to know—how would the weekly demand change if the price is reduced to $190, as planned for the upcoming promotional week surrounding Black Friday?

The reader may pause for a second here and convince themselves that it is difficult to say anything without further information. Would the demand that week skyrocket to 500, or would there be little boost in demand from the price reduction? To answer these questions, some contextual knowledge about pressure cookers and how similar items fared during previous promotions would be

K. A. Hansen et al. (Eds.): WINE 2022, LNCS 13778, pp. 150–166, 2022.
https://doi.org/10.1007/978-3-031-22832-2_9

required. Or more commonly, if one had access to additional data points on the demand curve, then it is possible to perform curve-fitting—e.g., if there had already been a week where the price dropped by $7 and the demand increased by 70, then fitting a linear demand curve would allow one to extrapolate something like *"the weekly demand increases by 10 for every $1 dropped in price"*.

However, our partner online retailer does not have detailed contextual information for most of its items, nor is it known to vary its prices over the course of its regular yearly operations. Instead, it is famous for "discounts" that come in the form of yearlong-available *bundles*, where different items are packaged together and sold cheaply. In this example, the pressure cooker is offered together with a recipe book for $212, with the recipe book individually costing $14. In effect, this bundle offers a discount of $207 + $14 − $212 = $9 for purchasing both items, and our partner has shared with us the weekly sales of both individual items as well as the bundle. Using only this information, as depicted in Table 1, can we say anything about the items' demands after a price drop?

Table 1. The limited data from which we attempt to construct demand curves.

	Price	Average weekly sales
Pressure cooker	$207	222
Recipe book	$14	30
Bundle	$212 ($9 discount)	29

We answer this question with a surprising "Yes", with the intuition being that the bundle sales contains information about the demand of the pressure cooker after a $9 discount, providing a second point on the demand curve. We show how to extract this data point from the bundle sales by fitting a multi-item valuation model, after which we construct a linear demand curve out of the two data points. This allows us to make a statement of the form *"the demand increases by x for every $1 dropped in price"*. While the numerical details for our fitting are deferred to Sect. 2, here we explain how the main insight from our multi-item valuation fitting applies to Table 1.

> **Main Insight**: If item i is bundled with item j and j is purchased relatively frequently (resp. infrequently) without i, then the demand curve of i is flat (resp. steep) around its current price.

Applying the insight to Table 1, we see that the pressure cooker was purchased relatively frequently without the recipe book (222 times, compared to 29 times with the recipe book), and hence the book's demand curve is flat around its current price. On the other hand, we see that the recipe book was purchased 30 times without the pressure cooker (relative to 29 times with), and hence the cooker's demand curve is steeper around its current price.

These conclusions from our insight can be explained as follows. Every week, 222 people purchase the pressure cooker and decline to "add on" the recipe book at a $9 discount in the form of the bundle. This suggests that an individual item discount is also unlikely to increase the sales of the recipe book, corresponding to a flat demand curve. By contrast, only 30 people purchase the recipe book and decline the pressure cooker at a \$9 discount. Consequently, an individual item discount on the pressure cooker is more likely to boost sales, corresponding to a steeper demand curve.

At this point the reader may have various reservations about our insight, which we discuss. First, it is possible to formulate alternate explanations for the data in Table 1—e.g., the low individual sales of the recipe book could be a consequence of it being a complementary product to the pressure cooker, and have nothing to do with the demand curve of either item. Second, even if the pressure cooker did have a steep demand curve around its current price, this may not generalize to the largely discounted Black Friday price which is far away from the current price. Finally, some customers who purchased the pressure cooker may not have even been aware of the recipe book or carefully evaluated the bundle deal, invalidating the explanation behind our main insight.

While we acknowledge all of these reservations, in Sect. 3 we validate our insight to generally hold on the data provided by our partner online retailer. To be specific, for a large number of items and bundles containing them, we are given their fixed prices and average weekly sales before Black Friday. We calculate the ratios suggested by our insight and generate a ranking of the items by our estimated steepness of their demand curves. To evaluate this ranking, we sort the items by their observed price elasticities after their discounts and sales on Black Friday. We find our ranking to be correct on a statistically-significant fraction of pairwise comparisons, and that in fact it provides an accurate first-order segmentation of items as either elastic and inelastic.

We provide some explanations for this, in spite of the aforementioned reservations. First, we note that the online retailer also offers many bundles whose constituents are not complements, such as a pressure cooker and a rice cooker, alleviating the first concern. Second, since we only validated through sorting by elasticity (as opposed to exact demand estimation), some of the error caused by the non-linearity of demand curves would have been hidden, sidestepping the second concern. Finally, the online retailer's website automatically displays and recommends bundles containing any viewed items, supporting our assumption that customers are aware of the bundle deals.

We conclude by emphasizing that we are not suggesting bundle discounts as a go-to method for learning demand curves, since as just discussed, they merely provide an educated guess instead of an exact prediction. If a firm's main goal is learning, then they should experiment with price, extract product features, conduct surveys, etc.—tried-and-true methods for accurate demand learning which we discuss in Sect. 1.1. However, for firms like our partner online retailer which have not invested in these methods, we showed them how to still glean valuable information from their bundle sales. As a general takeaway, our paper establishes that bundle discounts essentially provide a form of *price variation*, which makes it *possible* to construct linear demand curves, even when the data does not contain any price changes or useful covariates for comparing items.

1.1 Literature Review

Our work is related to several streams of literature on demand learning.

Learning from Limited Information. Our paper is related to the problem of understanding customer demand for new products with very little historical data. Previous works have suggested comparing the features of new products to existing ones [3,20], or efficient methods for eliciting additional information [6,12]. By contrast, our paper assumes that sales have already been observed at a single price, and extracts a second point on each demand curve based on discount purchases, with no further covariate or survey information.

We note that the situation of having access to only one point on a demand curve is practically well-motivated [8] beyond the example of our online retailer. For this situation, without further information, robust pricing solutions have been proposed [2]. Our paper suggests that records of discounted sales could be a subtle source for further information.

Learning from Bundle Sales. The technical part of our paper is related to the challenging estimation problem where items can be sold individually or in bundles. Existing work on this topic has focused on discrete choice models, where a bundle S is explicitly modeled as an "alternative", as surveyed in [27]. The valuation of a customer segment t for a bundle S could then depend on only the items in S [14], the pairs of items in S so that pairwise complementarity is modeled [23], or covariates based on both t and S in accordance to the *balance* model [19] which allows for customer heterogeneity and complex correlations [16]. The random utility derived by segment t from alternative S is then the valuation minus β_S times the price, where β_S is a bundle-dependent price elasticity parameter, plus a Gumbel noise term. Once the model is specified, the parameters which best fit the data can be computed using a Hierarchical Bayes framework [11]. Although these models considered are richer than ours, their parameters are identifiable only if there is sufficient variation in the prices and covariates—something we do not have.

We note that progress [13] has been made on this problem since our paper. Their work focuses more on the preference estimation problem than the insight that bundle sales can contain richer information about consumers.

Modeling Customer Choice for Bundles. We fit the standard multi-item valuation model from the bundle pricing literature [1,5], making the common modeling assumptions of valuations being *additive* and *independent* [4,22], and also *uniform* [9,18] in some of our results. Fitting this parsimonious model leads to our main insight, which is empirically validated to be relevant even beyond the assumptions. To the best of our knowledge, we are the first to fit this as a structural model, instead of using it for price optimization.

We should mention that there are other, more explicit ways of modeling customer valuations for bundles in specific industries [15,17,21] or in the product versioning literature [10,28]. However, these models also require rich data and industry-specific knowledge to fit.

Demand Learning Through Moment Matching. The way in which we fit our multi-item valuation model can be described as a Method of Moments, which (in its Generalized form) has been classically been used for estimating price elasticities as part of the BLP procedure [7,26]. Similar as before, BLP cannot be used in our setting because it requires differentiated markets with price variation. Finally, we should mention that although we just assume demand curves to be linear after recovering their two quantiles, the general problem of fitting distributions from quantiles is discussed in [25].

2   Fitting a Multi-item Valuation Model

We introduce our multi-item valuation model and fitting problem. Recall that the goal is to identify a second point on the demand curve of every item, given only a single set of bundle sales numbers. Consequently, we must restrict the multi-item valuation model to have few enough parameters such that it is fully identifiable from the bundle sales. We summarize all of our restrictions/assumptions and provide a thorough discussion of them in Sect. 2.5. We also explain how to draw our *main insight* in Sect. 2.5, and empirically validate it in Sect. 3.

Input Data. There are n different items, denoted by $[n] = \{1, \ldots, n\}$. For every $S \subseteq [n]$, we observe the number of customers N_S who purchased exactly the subset S from the items in $[n]$. It is assumed that we observe $N_\emptyset$, the number of customers who visited the store without making a purchase; therefore, we know the total number of customers $N = \sum_{S \subseteq [n]} N_S$. It is also assumed that the same customer never purchases multiple copies of the same item.

For every $S \subseteq [n]$, we are also told the fixed price $P(S)$ a customer has to pay to obtain the set of items S. We assume that $P(S)$ takes the form

$$P(S) = \begin{cases} \sum_{i \in S} P_i, & S \neq [n] \\ \sum_{i=1}^{n} P_i - d, & S = [n] \end{cases}$$

where $P_i \geq 0$ denotes the individual price of item i and $d \in (0, \sum_{i=1}^{n} P_i]$ denotes the *bundle discount* for purchasing all of $[n]$. This set of items $[n]$ should be interpreted as a small group of items being sold together and it is assumed that there are no partial discounts for purchasing part of $[n]$.

Multi-item Valuation Model. A customer has a *valuation* $v(S) \geq 0$ for each subset S of items, with $v(\emptyset) = 0$. A customer purchases the S maximizing her *surplus* $v(S) - P(S)$, where we allow ties to be broken arbitrarily. It is assumed that $v(S)$ is *additive*, taking the form

$$v(S) = \sum_{i \in S} x_i \qquad \forall S \subseteq [n]$$

where $x_i \geq 0$ denotes the customer's atomic valuation for item i.

For each of the N customers visiting the store, her valuation vector $(x_1, \ldots, x_n)$ is drawn IID from a multi-dimensional distribution D, which captures the heterogeneity of valuations across the population. It is also assumed that $D = D_1 \times \ldots \times D_n$ is a *product distribution*, with atomic valuation x_i being drawn *independently* from marginal distribution D_i for all items $i \in [n]$.

Constructing a Linear Demand Curve for Each Item. Our goal is to compute estimates of

$$\left\{ \Pr[x_i < P_i - d], \qquad \Pr[P_i - d \le x_i < P_i], \qquad \Pr[P_i \le x_i] : i \in [n] \right\} \quad (1)$$

which best fit the observed sales $\{N_S : S \subseteq [n]\}$. A linear demand curve can then be constructed for each item $i \in [n]$ as follows. We know that the demand at price P_i is $N \cdot \Pr[x_i \ge P_i]$, and that the demand at the "discounted" price $P_i - d$ is $N \cdot (\Pr[P_i - d \le x_i < P_i] + \Pr[P_i \le x_i])$, where $N = \sum_{S \subseteq [n]} N_S$ is the known total customer population. From these two points we can linearly extrapolate the demand $D(P)$ at a price P to be

$$D(P) = N \cdot \frac{\Pr[P_i - d \le x_i < P_i]}{d} (P_i - P) + N \cdot \Pr[x_i \ge P_i].$$

2.1 Formulation of Fitting Problem

Having explained how computing (1) would allow us to fit linear demand curves, we now focus our attention on computing (1), based on the observed bundle sales $\{N_S : S \subseteq [n]\}$. Generally speaking, our approach can be described as a *Method of Moments* where the aim is to find distributions $\{D_i : i \in [n]\}$ such that the expected fraction of the population to choose each subset $S \subseteq [n]$ matches the empirically observed fraction N_S/N. In this subsection, we derive formulas which allow us to express these expected fractions solely in terms of the probabilities in (1).

Definition 1 (Underlying Parameters). *For all items $i \in [n]$, define*

$$q_i^* = \Pr[x_i \ge P_i] \qquad and \qquad a_i^* = \Pr[x_i \ge P_i - d | x_i < P_i].$$

These represent the underlying model parameters and note that all of the probabilities in (1) can be computed from them. q_i^* represents the probability of a customer being willing to buy item i at its individual price P_i, while a_i^* represents the probability of a customer being willing to buy item i at price $P_i - d$, conditioned on her not being willing to buy item i at price P_i.

Definition 2 (Observed Values). *For all $S \subseteq [n]$, define $\hat{p}_S = N_S/N$, which denotes the fraction of customers observed to choose subset S. These represent the empirical values from which we will try to recover the parameters q_i^* and a_i^*. For all $S \subseteq [n]$, let p_S^* denote the expected value of $\hat{p}_S$.*

The following sequence of lemmas leads to a system of equations that expresses each p_S^* in terms of $\{q_i^*, a_i^* : i \in [n]\}$. The proofs of these lemmas are deferred to the full version of this paper [24].

Lemma 1. *For a customer with valuations $x_1, \ldots, x_n$, the surplus-maximizing subset is the full, discounted set of items $[n]$ if and only if*

$$\sum_{i=1}^{n} \max\{P_i - x_i, 0\} \leq d. \tag{2}$$

We can interpret $\max\{P_i - x_i, 0\}$ as the *deficit* incurred by the customer had she been forced to buy item i at price P_i, which equals 0 if $x_i \geq P_i$. This allows us to characterize the customer's purchase decision. Indeed, she first checks inequality (2), i.e. whether her total deficit from buying all the items is covered by the bundle discount. If so, then she buys all the items; otherwise, she selects the items i for which her valuation is at least the individual price P_i and buys this subset.

Lemma 2. *For all subsets $S \neq \emptyset$, define*

$$F_S^* = \Pr\left[\sum_{i \notin S}(P_i - x_i) \leq d \,\middle|\, P_i - d \leq x_i < P_i \; \forall i \in S\right]. \tag{3}$$

Then, for all subsets $S \neq [n]$, the probability p_S^* of a customer selecting subset S satisfies

$$p_S^* = \left(\prod_{i \in S} q_i^*\right)\left(\prod_{i \notin S}(1 - q_i^*)\right)\left(1 - F_{[n]\setminus S}^* \prod_{i \notin S} a_i^*\right). \tag{4}$$

Note that Eq. (4) in Lemma 2 only holds for subsets $S \neq [n]$, but $p_{[n]}^*$ can be found by computing $1 - \sum_{S \subsetneq [n]} p_S^*$. Intuitively, the three parentheses in expression (4) can be interpreted as follows. For a subset $S \neq [n]$ to be selected, the customer has to: (i) value the items $i \in S$ above their individual prices; (ii) value the items $i \notin S$ below their individual prices; and (iii) not prefer the full set of items over S, i.e. incur greater than d deficit from buying the items $i \notin S$.

Following the Method of Moments, our goal is now to replace p_S^* and $F_{[n]\setminus S}^*$ in (4) with observed values, in order to solve for the underlying parameters q_i^* and a_i^*. It is natural to replace each p_S^* with the empirically observed fraction $\hat{p}_S$. However, without any distributional assumptions, estimating $F_{[n]\setminus S}^*$ is difficult, because it depends on the exact conditional distributions of x_i when $x_i \in [P_i - d, P_i)$ for $i \notin S$. As a saving grace, we observe that $F_{[n]\setminus S}^*$ can be exactly computed when these distributions are *uniform*. We note that having uniform valuations is equivalent to having linear demand curves, an assumption made up-front in this paper due to limited data.

Lemma 3. *For a subset $S \neq \emptyset$, suppose that the conditional distribution of x_i on $[P_i - d, P_i)$ is uniform, for all $i \in S$. Then $F_S^* = \frac{1}{|S|!}$.*

Using the uniform distribution as a benchmark, we replace $F^*_{[n]\setminus S}$ with $\frac{1}{(n-|S|)!}$ in (4) and attempt to solve the system. Even though the true distributions may not be uniform, we analyze the error in our solution as a function of a *non-uniformity* parameter (Sect. 2.3). We also show that our method numerically yields a good solution for common non-uniform distributions (Sect. 2.4).

Definition 3 (Fitting Problem). *For all $S \neq \emptyset$, let F_S be shorthand for $\frac{1}{|S|!}$. Then our fitting problem is to solve for $q_1, \ldots, q_n$ and $a_1, \ldots, a_n$ from the following system of equations:*

$$\Big(\prod_{i \in S} q_i\Big)\Big(\prod_{i \notin S}(1 - q_i)\Big)\Big(1 - F_{[n]\setminus S} \prod_{i \notin S} a_i\Big) = \hat{p}_S \qquad \forall S \neq [n]. \qquad (5)$$

2.2 Iterative Fitting Algorithm

The system which we need to solve, (5), consists of intractable high-dimensional polynomial equations. Moreover, since there are $2^n - 1$ equations and only $2n$ variables, the system will generally be overdetermined, with no solution unless the given values $\hat{p}_S$ and $F_{[n]\setminus S}$ exactly match the true values p^*_S and $F^*_{[n]\setminus S}$ (in which case setting $q_i = q^*_i$ and $a_i = a^*_i$ will be a solution).

To cope, we propose an iterative algorithm that hopes to quickly arrive at a "reasonable" solution, which we will justify both theoretically (Sect. 2.3) and numerically (Sect. 2.4). In the full version [24], we compare with the established *Generalized Method of Moments (GMM)* and *Maximum Likelihood (ML)* methods for fitting an overdetermined system of equations, with the main advantage of our method being that it runs much faster. Indeed, to our knowledge, the GMM and ML estimators for our problem can only be computed via brute force, and in the full version [24] we numerically demonstrate why this quickly becomes impractical.

Our algorithm is based on the following observation: given a set of candidate values for $(q_i)_{i \in [n]}$, it is possible to derive from (5) a closed-form expression for each a_i; and vice-versa, i.e. given $(a_i)_{i \in [n]}$ we can express each q_i in terms of $(a_i)_{i \in [n]}$. These expressions are given in the description of our algorithm, in Fig. 1. Our algorithm is parameterized by $\{\mathcal{S}_i : i \in [n]\}$, where each $\mathcal{S}_i$ is a collection of *strict* subsets of $[n] \setminus \{i\}$. Equations in (5) where $S = S'$ or $S = S' \cup \{i\}$, with $S' \in \mathcal{S}_i$, are what the algorithm uses to update its current solution $(q_i^{(k)})_{i \in [n]}$ during each iteration $k \geq 0$. We provide a full intuitive explanation of our algorithm's update rules, (6)–(8), in the full version [24].

2.3 Theoretical Correctness of Algorithm

We now show that our iterative algorithm correctly solves system (5) when it is possible, i.e. when

1. The observed fractions $\hat{p}_S$ equal the true probabilities p^*_S; and

For $i = 1, \ldots, n$ initialize

$$q_i^{(0)} \leftarrow \Big(1 + \Big(\prod_{S \in \mathcal{S}_i} \frac{\hat{p}_{S \cup \{i\}}}{\hat{p}_S}\Big)^{-1/|\mathcal{S}_i|}\Big)^{-1}. \tag{6}$$

for $k = 0, 1, \ldots$ **do**
    For $i = 1, \ldots, n$ set

$$a_i^{(k)} \leftarrow \max\Big\{1 - \frac{\hat{p}_{[n]\backslash\{i\}}}{\big(\prod_{j \neq i} q_j^{(k)}\big)\big(1 - q_i^{(k)}\big)}, 0\Big\}. \tag{7}$$

    For $i = 1, \ldots, n$ set

$$q_i^{(k+1)} \leftarrow \Big(1 + \Big(\prod_{S \in \mathcal{S}_i} \Big(\frac{\hat{p}_{S \cup \{i\}}}{\hat{p}_S} \cdot \frac{1 - F_{[n]\backslash S} \prod_{j \notin S} a_j^{(k)}}{1 - F_{[n]\backslash S\backslash\{i\}} \prod_{j \notin S, j \neq i} a_j^{(k)}}\Big)\Big)^{-1/|\mathcal{S}_i|}\Big)^{-1}. \tag{8}$$

end for

Fig. 1. Iterative algorithm, on input $\{\hat{p}_S : S \neq [n]\}$ and $\{\mathcal{S}_i : i \in [n]\}$

2. The values $F_S = \frac{1}{|S|!}$ based on the uniform distribution equal the true values F_S^*.

In this case, our algorithm converges upon the true parameters q_i^* and a_i^* for all items i. In fact, our theoretical analysis is only for the algorithm using the parameters $\mathcal{S}_1 = \cdots = \mathcal{S}_n = \{\emptyset\}$, in which case the conditions above only need to hold for a subfamily of relevant subsets S.

We explain why we *cannot expect convergence* to the true parameters without the two conditions above. For the first, it is easy to see that if $\hat{p}_S$ is allowed to differ from $\hat{p}_S$, then it is possible for $\{\hat{p}_S : S \subseteq [n]\}$ to be perturbed in a way such that the system (5) being fitted has a feasible solution which is different from $\{q_i^*, a_i^* : i \in [n]\}$. There would be no way to know that this solution was incorrect. To justify the second condition, note that if the valuations were non-uniform, then the probability F_S^* would depend on the exact positioning of prices P_i, d relative to the distribution (instead of having the simple formula $F_S^* = \frac{1}{|S|!}$ in the uniform case). Therefore, F_S in our system (5) would become an unknown, even if the functional form of the valuation (e.g. Normal) was known, leading to a system with more unknowns than equations.

Therefore, we can only expect exact convergence when the two conditions above do hold for the subsets S used by the algorithm, and we now formally state this result.

Theorem 1 (Exact Convergence Theorem). *Suppose $n \geq 3$, and that $\hat{p}_S = p_S^*$ for all S with $|S| \in \{0, 1, n-1\}$, and that $F_S = F_S^*$ for all S with $|S| \geq n-1$ (a sufficient condition for $F_S = F_S^*$ is the distributions being uniform). Furthermore suppose that $q_i^*, a_i^* \in (0, 1)$ for all i. Then the estimates returned by our algorithm* in Fig. 1, using parameters $\mathcal{S}_i = \{\emptyset\}$ for all i, satisfies

$$\lim_{k \to \infty} q_i^{(k)} = q_i^* \qquad and \qquad \lim_{k \to \infty} a_i^{(k)} = a_i^* \qquad \forall i \in [n].$$

We note that the condition $n \geq 3$ in Theorem 1 is also necessary, because if $n \leq 2$, then our fitting problem (5) has more unknowns than equations.

More generally, in the full version [24], we relax the two conditions above and analyze the multiplicative errors in our algorithm's estimates $q_i^{(k)}$ and $a_i^{(k)}$ after some number of iterations k, where the bound depends on the error in the values of $\hat{p}_S$ and F_S used by algorithm. While all of the analysis is deferred to the full version [24], we provide a brief summary of our technique. We first derive an expression for the multiplicative error in the algorithm's initial estimates $q_i^{(0)}$. We carefully bound the error propagation over iterations, using the tightest possible inequalities on how errors can multiply. This allows us to bound the error in our algorithm's estimates after many iterations, as a function of the errors in $\hat{p}_S$ and in our estimates $F_S = \frac{1}{|S|!}$ of F_S^*. Importantly, we show that if both of these input errors are 0, then the error in our algorithm's output converges to 0, thereby establishing Theorem 1.

Finally, we remark that our convergence result resembles a *contraction mapping* argument. However, in our case we know that the fixed point will be $\left(q_i^*\right)_{i \in [n]}$, and hence we directly establish convergence to that point. We believe this to be easier than trying to show that our algorithm's mapping from $\left(q_i^{(k)}\right)_{i \in [n]}$ to $\left(q_i^{(k+1)}\right)_{i \in [n]}$ is generally contractive over the feasible space $[0, 1]^n$.

2.4 Numerically Testing Algorithm on Synthetic Instances

In our numerical experiments, we relax two assumptions (no noise in $\hat{p}_S$, uniform valuations) necessary for the exact convergence result in Theorem 1, and test how our algorithm performs without them. We generate synthetic instances from the test bed of [15], on which we can compare our algorithm's output to the ground truth. Although quantitative results are deferred to the full version [24], we summarize our findings as follows:

- By iterating our algorithm with a simple stopping criterion based on comparing $a^{(k)}$ to $a^{(k-1)}$, approximately-correct parameters can be reached in less than 10 iterations on average;
- The error due to noise in $\hat{p}_S$ can be mitigated by increasing the cardinality of $\mathcal{S}_i$;
- The increase in error, if our algorithm does not observe no-purchases, is negligible for $n \geq 4$;
- The error due to distributions being non-uniform is less than 2%, as long as the distributions are *unimodal*—we provide an explanation for this surprising accuracy in the full version [24];
- GMM and ML are generally inapplicable because their optimization problems are intractable to solve by brute force, even if $n = 3$;

– "Rigging" the instances so that the true parameters lie on the search grids of GMM and ML, we find that our algorithm's performance is worse under noisy $\hat{p}_S$, but quickly becomes indistinguishable as this noise is reduced.

2.5   Discussion of Assumptions and Derivation of Main Insight

In Table 2, we list the eight assumptions made to the end of Sect. 2.3, summarizing that most of them are *necessary*, in order for the model to be identifiable from limited data (Assumptions 1–4) and for exact convergence to be theoretically possible (Assumptions 7–8). In conjunction with Sect. 2.4, which shows that Assumptions 6–8 are not needed for good numerical performance, we believe this justifies our choice of model and proposed fitting algorithm.

Of course, we also have to discuss whether Assumptions 1–8 hold in practice, which we do so in the second column of Table 2. We emphasize that even when they do not hold, a *main insight* from our model and algorithm does empirically hold independent of these assumptions, through real-world validation in the full version [24]. We now explain the main insight from our model, first repeating its statement from the Introduction.

Main Insight: If item i is bundled with item j and j is purchased relatively frequently (resp. infrequently) without i, then the demand curve of i is flat (resp. steep) around its current price.

Explanation: Fixing an item i, the items j it is bundled with being *purchased relatively frequently without i* corresponds to $\hat{p}_{[n]\setminus\{i\}}$ being *high*. In instruction (7) of our algorithm, this causes the recovered value $a_i^{(k)}$ to be *low* on every iteration k. Assuming that $a_i^{(k)}$ is close to the true value a_i^* after some number of iterations k, we see that by definition of a_i^*, there must be a *low* probability of the customer's valuation for i being higher than $P_i - d$, conditioned on it being lower than P_i. Therefore, discounting item i by d yields a *low* increase in purchase probability, corresponding to a *flat* demand curve.

3   Empirical Validation of Insight

In this section we test how the main insight from our model applies to a real-world data set.

We consider sales data provided by a large Latin American online retailer, on their home and kitchen items, many of which are sold in discounted bundles. For each item and bundle, we are given its price and sales for 26 weeks starting June 1st, 2016. We emphasize that this online retailer does not generally vary its prices over the regular year, so the price of each item and bundle is identical across the first 25 weeks. On the other hand, the last of these 26 weeks is a major promotional week containing Black Friday during which many items are heavily marked down.

We use an item's sales boost after a Black Friday markdown to measure its price elasticity, and test whether this agrees with the steepness of the item's

Table 2. Categorizing all of our assumptions, discussing their theoretical necessity and practical justifiability.

Assumption	Theoretical necessity	True in practice?
1. Customer purchases at most one of each item **2. Valuations are Additive** i.e. $v(S) = \sum_{i \in S} x_i$ for some $x_1, \ldots, x_n$ **3. Valuations are Independent** i.e. $D = D_1 \times \ldots \times D_n$	Completely necessary. Assumption 1 is necessary to even have the concept of a valuation $v(S)$ for a subset S. Assumptions 2 and 3 are necessary to prevent model from having more parameters than equations	Common modeling simplifications which are difficult to validate exactly. Customer valuations are believed to be super-additive for bundles of complements (e.g. pressure cooker, recipe book) and sub-additive for bundles of substitutes (e.g. pressure cooker, rice cooker)
4. $n \geq 3$	If $n = 2$, then 4 parameters but only 3 equations. We show an example of solving the system by brute force when one parameter is given, in the full version [24]	Some bundles in practice have $n = 2$ items
5. Single Bundle Discount i.e. $P(S) = \begin{cases} \sum_{i \in S} P_i, & S \neq [n] \\ \sum_{i=1}^{n} P_i - d, & S = [n] \end{cases}$ for some $P_1, \ldots, P_n, d$	Not necessary, but made for mathematical tractability. We also solve the case where $P(\cdot)$ describes a two-part tariff, in the full version [24]	Usually true, since discounts for sub-bundles $S \subsetneq [n]$ make pricing too complex. However, bundles in practice can overlap
6. No-purchases are Observed	Not necessary, but made for mathematical tractability. Not needed for good numerical performance unless $n = 3$. We show how to analytically deduce $N_\emptyset$ if it is not given, in the full version [24]	Can approximate by counting the number of receipts containing no items from the bundle $[n]$ in question (brick-and-mortar retailer), or the number of page views (online retailer)
7. No Noise in Observations i.e. $\hat{p}_S = \Pr_{v \sim D}[S = \operatorname{argmax}_{S'} v(S')]$ for all S **8. Valuations are Uniform** i.e. $D_i = \operatorname{Unif}[l_i, u_i]$ for all i	Only used in our exact convergence result (Theorem 1), and necessary for that result. Not used in our general, and also not needed for good numerical performance.	Not always the case, but we note that a uniform valuation distribution (equivalently, a linear demand curve) is a commonly-used model of demand

demand curve as indicated by its bundle sales prior to Black Friday. We revisit the example from the Introduction, where we now also display the Black Friday markdown amounts, in Table 3. Our goal is to use the items' bundle sales to speculate on their sales boosts after the Black Friday markdowns.

Table 3. An example from the data. Can we speculate on the "?" values using only the information in the table? (Answers in text.)

	First 25 weeks		Black Friday week	
	Price in R\$	Average weekly sales	Price in R\$	Sales
Pressure cooker	\$207	222	\$190 (\$17 markdown)	?
Recipe book	\$14	30	\$3 (\$11 markdown)	?
Bundle	\$212 (\$9 discount)	29	\$212 (dominated by markdowns)	

A naive observation of the data might suggest that neither item is particularly elastic, because the discounted bundle did not sell many copies (only 29) relative to the weekly sales of the items (222 and 30, respectively). However, the analysis in our paper would suggest that the recipe book is *much less elastic* than the pressure cooker. This is because a large number of people (222) paid \$207 for the pressure cooker alone, declining to add the recipe book at the "highly discounted" price of \$5 (normally they would have to pay \$14 for the recipe book, but the bundle allows them to pay \$5). By contrast, much fewer (30) people purchased the recipe book alone.

In the end, the recipe book sold only 176 copies during Black Friday week, representing a dismal price elasticity given the massive 79% Black Friday markdown. Meanwhile, the pressure cooker sold 1153 copies during Black Friday week, achieving a similar percentage increase from a much smaller 8% markdown, and hence representing a much better price elasticity. This is exactly consistent with our predictions from the bundle sales before Black Friday.

3.1 Relationship Between Bundle Sales and Price Elasticity

We test whether the relationships from the example in Table 3 generalize to the other items in the data set. Most of the items have been offered in multiple bundles, so for each item, we consider aggregate statistics over the bundles containing it.

Definition 4. *Define the following for each item i, based on the 25 weeks before Black Friday.*

- PctBund(i): *out of all copies of i sold, the fraction of which came from a bundle purchase.*
- PctBundPartners(i): *the weighted average value of* PctBund *over the partner items of i, where an item j is considered a partner of i if there is some bundle containing both i and j.*
- AvgBundDisc(i): *the % discount provided by the bundles containing i, taking a weighted average.*

For full details on how we processed these statistics from the data, we defer to the full version [24].

To illustrate these statistics, we calculate them for the example in Table 3. In the full data, the (pressure) cooker was only part of the bundle with the (recipe) book, while the book was part of many bundles with different kitchen appliances. An additional 235 copies of it were sold weekly in these other bundles. Therefore,

$$\text{PctBund(Cooker)} = \frac{29}{29 + 222} \approx 12\% \quad \text{and} \quad \text{PctBund(Book)} = \frac{29 + 235}{29 + 30 + 235} \approx 90\%.$$

The cooker's sole partner is the book, so PctBundPartners(Cooker) = PctBund(Book) = 90%. Meanwhile, PctBundPartners(Book) = 24% after averaging over all of its partners. Finally, AvgBundDisc considers the % discounts provided by the bundles, which for this one is $1 - \frac{212}{207+14} \approx 4\%$. Hence AvgBundDisc(Cooker) = 4%. AvgBundDisc(Book) takes an average over all of the bundles containing the book, ending up at 8%.

For each item i, we let PriceElas(i) denote its price elasticity, measured as its % increase in sales during Black Friday week, divided by its % decrease in individual price. We are interested in the relationship of PriceElas(i) with both

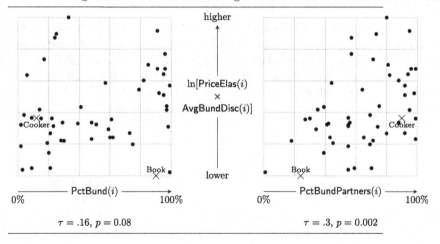

Fig. 2. Relationship between bundle sales and price elasticity. The example items from Table 3 are labeled with an "x". The exact functional form chosen (with the vertical axis on a log-scale) is purely for visual clarity and does not affect the Kendall's τ measure of statistical significance.

$\frac{\mathsf{PctBund}(i)}{\mathsf{AvgBundDisc}(i)}$ and $\frac{\mathsf{PctBundPartners}(i)}{\mathsf{AvgBundDisc}(i)}$, where we have divided by the bundle discount to penalize bundle sales percentages that came from highly discounted bundles. We will only investigate the *ordering* of the items according to these statistics, i.e., do the items with the highest values of $\frac{\mathsf{PctBund}(i)}{\mathsf{AvgBundDisc}(i)}$ or $\frac{\mathsf{PctBundPartners}(i)}{\mathsf{AvgBundDisc}(i)}$ tend to have the highest values of $\mathsf{PriceElas}(i)$ after Black Friday? Consequently, we can equivalently consider the relationship between $\ln(\mathsf{PriceElas}(i) \times \mathsf{AvgBundDisc}(i))$ vs. both $\mathsf{PctBund}(i)$ and $\mathsf{PctBundPartners}(i)$, which we plot in Fig. 2, for the 51 items that were marked down during Black Friday and had prior bundle sales. We emphasize that the functional form plotted in Fig. 2 was chosen purely for visual clarity, and does not affect the Kendall's τ measure of correlation used, which only considers the *rankings* of the items according to $\mathsf{PriceElas}(i)$, $\frac{\mathsf{PctBund}(i)}{\mathsf{AvgBundDisc}(i)}$, and $\frac{\mathsf{PctBundPartners}(i)}{\mathsf{AvgBundDisc}(i)}$.

Looking at Fig. 2, $\mathsf{PctBund}(i)$ shows a positive correlation with price elasticity, with $\tau = .16$ (i.e. 58% of the pairs were concordant), which for 51 data points results in a two-sided p-value of .08. However, the correlation is weak with many outliers, as exemplified by the recipe book. That is, it is possible for an "add-on" item like the recipe book to be inelastic to a Black Friday discount, even though most of its previous sales occurred in discounted bundles.

Meanwhile, $\mathsf{PctBundPartners}(i)$ demonstrates a highly significant correlation with price elasticity, with $\tau = .3$ and $p = .002$. This is exactly as explained by our model: if many people bought the *partners* of item i without adding i at a bundle discount, then they also wouldn't buy item i after an individual discount; on the other hand, if nobody bought the partners of item i individually, then a logical hypothesis is that everybody was willing to add i to the bundle.

It is also interesting to note that in Fig. 2, the uncertainty for a given value of PctBundPartners(i) appears to be one-sided. That is, a high PctBundPartners(i) could still imply a low price elasticity, likely caused by bundle sales that were driven by highly complementary items. However, a low PctBundPartners(i) never implied a high price elasticity, likely because substitutes being bundled together (which would cause this effect) occurs less frequently in practice.

4   Conclusion

In this paper we are faced with our online retailer's problem of constructing demand curves for items which have only been sold at a single price. We intuit that a second point on each item's demand curve can be extracted by looking at its selling frequency in discounted bundles. To formalize this, we propose a fitting problem for a parsimonious model (with many assumptions, but necessarily so) from the bundle pricing literature and derive an iterative algorithm for matching the moments. This algorithm unveils a "main insight": a directional relationship between the frequency of an item's bundle sales and the slope of its demand curve. We test this insight on the data from our online retailer and verify that (even though the assumptions may not hold) the directional relationship holds.

Of course, our approach merely shows how to make an "educated guess" when faced with limited data (no price variation, no covariates), so we mostly see it being used in one of two ways. First, it can be used as a starting point to identify the items more likely to be elastic, which should be further investigated and considered for a price drop. Indeed, as we saw in Fig. 2, a markdown was much more likely to be effective on items with a high value of PctBundPartners, as opposed to a low value of PctBundPartners. Second, we see the simplicity and interpretability of our PctBundPartners statistic as a benefit to managers. For example, suppose after further analysis (possibly further data collection) that an item's elasticity is much lower than that indicated by its PctBundPartners statistic. This identifies items that are frequently sold in bundles with complementary products, and these items should be considered for inclusion in more bundles, instead of considered for a price drop.

Next, we discuss the higher-order managerial question of why our online retailer did not collect better data in the first place. General reasons for refraining from changing prices include operational overhead, potentially negative customer response, and encouraging customers to strategically wait [29]. Our partner online retailer tends to focus their discounts for Black Friday week, where it is expected that prices will drop and then rise back up. We would also like to point out that in our industry experience, even for firms that do change prices, often the system tracks these changes poorly and the only data available is of the form (average price, sales).

Finally, we mention two other generally challenging issues, which are beyond the scope of our analysis. First, there could be endogeneity in the price elasticities constructed from the data—the goods that were discounted on Black Friday and the discount amounts were intentionally chosen by a revenue-maximizing firm. Second, the demand curve for Black Friday could be fundamentally different from the demand curves seen during the rest of the year. However, we do not have access to data of more regular discounts that might also display less endogeneity—our partner firm generally does not offer these. Our validation using the Black Friday data made use of what was available to us.

Acknowledgement. The anonymous reviewers from WINE22 are thanked for their insightful comments which improved the final version of the paper.

References

1. Adams, W.J., Yellen, J.L.: Commodity bundling and the burden of monopoly. Q. J. Econ. 475–498 (1976)
2. Allouah, A., Bahamou, A., Besbes, O.: Optimal pricing with a single point. In: Proceedings of the 22nd ACM Conference on Economics and Computation, pp. 50–50 (2021)
3. Baardman, L., Levin, I., Perakis, G., Singhvi, D.: Leveraging comparables for new product sales forecasting. SSRN 3086237 (2017)
4. Babaioff, M., Immorlica, N., Lucier, B., Weinberg, S.M.: A simple and approximately optimal mechanism for an additive buyer. J. ACM (JACM) **67**(4), 1–40 (2020)
5. Bakos, Y., Brynjolfsson, E.: Bundling information goods: pricing, profits, and efficiency. Manag. Sci. **45**(12), 1613–1630 (1999)
6. Ben-Akiva, M.E., McFadden, D., Train, K., et al.: Foundations of stated preference elicitation: consumer behavior and choice-based conjoint analysis. Now (2019)
7. Berry, S., Levinsohn, J., Pakes, A.: Automobile prices in market equilibrium. Econometrica: J. Econometric Soc. 841–890 (1995)
8. Besbes, O., Elmachtoub, A.N., Sun, Y.: Pricing analytics for rotable spare parts. INFORMS J. Appl. Anal. **50**(5), 313–324 (2020)
9. Bhargava, H.K.: Mixed bundling of two independently valued goods. Manag. Sci. **59**(9), 2170–2185 (2013)
10. Bhargava, H.K., Choudhary, V.: Research note-when is versioning optimal for information goods? Manag. Sci. **54**(5), 1029–1035 (2008)
11. Bradlow, E.T., Rao, V.R.: A hierarchical Bayes model for assortment choice. J. Mark. Res. **37**(2), 259–268 (2000)
12. Cao, X., Zhang, J.: Preference learning and demand forecast. Market. Sci. (2020)
13. Chen, N., Farajollahzadeh, S., Wang, G.: Learning consumer preferences from bundle sales data. arXiv preprint arXiv:2209.04942 (2022)
14. Chiu, I.H.S., Russell, G.J., Gruca, T.S.: Predicting bundle preference using configuration data. In: Robert Mittelstaedt Doctoral Symposium Proceedings, p. 53 (2016)
15. Chu, C.S., Leslie, P., Sorensen, A.: Bundle-size pricing as an approximation to mixed bundling. Am. Econ. Rev. 263–303 (2011)
16. Chung, J., Rao, V.R.: A general choice model for bundles with multiple-category products: application to market segmentation and optimal pricing for bundles. J. Mark. Res. **40**(2), 115–130 (2003)
17. Crawford, G.S.: The discriminatory incentives to bundle in the cable television industry. Quant. Mark. Econ. **6**(1), 41–78 (2008)

18. Eckalbar, J.C.: Closed-form solutions to bundling problems. J. Econ. Manag. Strateg. **19**(2), 513–544 (2010)
19. Farquhar, P.H., Rao, V.R.: A balance model for evaluating subsets of multiattributed items. Manag. Sci. **22**(5), 528–539 (1976)
20. Ferreira, K.J., Lee, B.H.A., Simchi-Levi, D.: Analytics for an online retailer: demand forecasting and price optimization. Manuf. Serv. Oper. Manag. **18**(1), 69–88 (2016)
21. Foubert, B., Gijsbrechts, E.: Shopper response to bundle promotions for packaged goods. J. Mark. Res. **44**(4), 647–662 (2007)
22. Hart, S., Nisan, N.: Approximate revenue maximization with multiple items. J. Econ. Theory **172**, 313–347 (2017)
23. Kamakura, W.A., Kwak, K.: Menu-choice modeling. Available at SSRN 2162019 (2012)
24. Ma, W., Simchi-Levi, D.: Constructing demand curves from a single observation of bundle sales. Available at SSRN 2754279 (2016)
25. Morgenthaler, S., Tukey, J.W.: Fitting quantiles: doubling, HR, HQ, and HHH distributions. J. Comput. Graph. Stat. **9**(1), 180–195 (2000)
26. Rasmusen, E.: The BLP method of demand curve estimation in industrial organization (2007)
27. Seetharaman, P., et al.: Models of multi-category choice behavior. Mark. Lett. **16**(3–4), 239–254 (2005)
28. Shugan, S.M., Moon, J., Shi, Q., Kumar, N.S.: Product line bundling: why airlines bundle high-end while hotels bundle low-end. Mark. Sci. **36**(1), 124–139 (2017)
29. Zbaracki, M.J., Ritson, M., Levy, D., Dutta, S., Bergen, M.: Managerial and customer costs of price adjustment: direct evidence from industrial markets. Rev. Econ. Stat. **86**(2), 514–533 (2004)

Mechanism Design

Fair and Efficient Multi-resource Allocation for Cloud Computing

Xiaohui Bei[1], Zihao Li[1], and Junjie Luo[2(✉)]

[1] Nanyang Technological University, Singapore, Singapore
xhbei@ntu.edu.sg, zihao004@e.ntu.edu.sg
[2] Beijing Jiaotong University, Beijing, China
jjluo1@bjtu.edu.cn

Abstract. We study the problem of allocating multiple types of resources to agents with Leontief preferences. The classic Dominant Resource Fairness (DRF) mechanism satisfies several desired fairness and incentive properties, but is known to have poor performance in terms of social welfare approximation ratio. In this work, we propose a new approximation ratio measure, called *fair-ratio*, which is defined as the worst-case ratio between the optimal social welfare (resp. utilization) among all *fair* allocations and that by the mechanism, allowing us to break the lower bound barrier under the classic approximation ratio. We then generalize DRF and present several new mechanisms with two and multiple types of resources that satisfy the same set of properties as DRF but with better social welfare and utilization guarantees under the new benchmark. We also demonstrate the effectiveness of these mechanisms through experiments on both synthetic and real-world datasets.

Keywords: Fair division · Mechanism design · Cloud computing

1 Introduction

In order to offer flexible resources and economies of scale, in cloud computing systems, a fundamental problem is to efficiently allocate heterogeneous computing resources, such as CPU time and memory, to agents with different demands. This resource allocation problem presents several significant challenges from a technical perspective. For example, how to balance the efficiency of the system and fairness among users? How to incentivize agents to participate and truthfully reveal their private information? These are all delicate issues that need to be carefully considered when designing a resource allocation algorithm.

One of the most widely used mechanisms for multi-type resource allocation is the *Dominant Resource Fairness (DRF)* mechanism proposed by [7]. This work assumes that agents in the system have *Leontief* preferences, which means they demand to receive resources of each type in fixed proportions. Under such preferences, the proposed DRF mechanism generalizes the max-min allocation by equalizing the share of the most demanded resource, called *dominant share*, for

© The Author(s), under exclusive license to Springer Nature Switzerland AG 2022
K. A. Hansen et al. (Eds.): WINE 2022, LNCS 13778, pp. 169–186, 2022.
https://doi.org/10.1007/978-3-031-22832-2_10

all agents. [7] show that DRF satisfies a set of desirable properties. These include fairness properties: (i) *share incentive (SI)*, all agents should be at least as happy as if each resource is equally allocated to all agents, and (ii) *envy-freeness (EF)*, no agent should prefer the allocation of another agent; efficiency properties: (iii) *Pareto optimality (PO)*, it is impossible to increase the allocation of one agent without decreasing the allocation of another agent; as well as incentive properties: (iv) *strategy-proofness (SP)*, no agent can benefit from reporting a false demand. Consequently, DRF has received significant attention with many variants proposed to tackle different restrictions occurred in practice.

Despite the above attractive properties, however, DRF is known to have poor performance in terms of utilitarian social welfare, which is defined as the sum of utilities of all agents. Many alternative mechanisms have then been proposed to tackle this issue and balance the trade-off between fairness and efficiency [2, 3, 8, 10, 11, 13, 22, 23]. Most of these mechanisms still satisfy SI, EF, and PO. However, none of them satisfy SP. Recently, [10] propose the so called 2-dominant resource fairness (2-DF) to balance fairness and efficiency. Different from other mechanisms, 2-DF satisfies SP and PO, but does not satisfy SI and EF generally. On the other hand, [19] justify this worst-case performance of DRF by showing that any mechanism satisfying any of the three properties SI, EF, and SP cannot guarantee more than $\frac{1}{m}$ of the optimal social welfare, which is also what DRF can achieve. Here m denotes the number of resource types. This characterization seems to suggest that from a worst-case viewpoint, DRF has the best possible social welfare guarantee among all fair or truthful mechanisms.

In this work, we aim to design new mechanisms that satisfy the same set of properties with DRF but with better efficiency guarantees. In order to get around the theoretical barrier set by [19], we first propose and justify a new benchmark to measure the social welfare guarantee of a mechanism. Note that [19] and many other works use the *approximation ratio*, which is defined as the worst-case ratio between the *optimal social welfare among all allocations* and the mechanism's social welfare, as the performance measure of a mechanism. However, since SI and EF are both fairness properties that place significant constraints on feasible allocations, it is not surprising that any allocation satisfying SI or EF would incur a large approximation ratio of m. On the other hand, one can show that any mechanism satisfying SI has approximation ratio at most m. This means all mechanisms satisfying SI and EF will have the same worst-case approximation ratio, which renders the approximation ratio notion meaningless in systems where these fairness conditions are hard constraints that must be satisfied. Since fairness is a hard constraint in many practical applications, we argue that it is more reasonable to compare the mechanism's social welfare to the optimal social welfare among *all allocations that satisfy SI and EF*. To this end, we modify the approximation ratio definition and propose this according variant. The new definition allows us to get pass the lower bound barrier from [19] and design mechanisms with better social welfare approximation ratio guarantees.

1.1   Our Results

We design new resource allocation mechanisms that satisfy properties such as SI, EF, PO, and SP, and at the same time achieve high efficiency. The efficiency is measured by two objectives: *social welfare*, defined as the sum of utilities of all agents, and *utilization*, defined as the minimum utilization rate among all resources. Social welfare is an indicator commonly used to measure efficiency, while improving utilization rate is also an important goal for cloud providers for cost-saving (see, e.g., Amazon[1], IBM[2]). In academia, utilization has been studied by [12,13,16]. For the performance measure, we define *fair-ratio* for social welfare (resp. utilization) of a mechanism as the worst-case ratio between the social welfare (resp. utilization) achieved by the optimal mechanism *satisfying SI and EF* and that by the mechanism. See formal definitions in Sect. 2.

We first focus on the setting where all agents' dominant resources fall into two types. This is the most basic and arguably also the most important setting in cloud computing and other application domains such as high performance computing. For example, most existing commercial cloud computing services, such as Azure, Amazon EC2, and Google Cloud, work with only two (dominant) resources: CPU and memory. Two-resource setting can also be used to model the coupled CPU-GPU architectures where CPU and GPU are integrated into a single chip to achieve high performance computing [22]. In this setting, we present three new mechanisms UNB, BAL, and BAL*, all with better fair-ratio guarantees than DRF. Different from DRF which equalizes the dominant share of all agents, the idea behind our new mechanisms is to partition all agents into two groups according to their dominant resources and carefully increase the share of agents with the smallest fraction of their non-dominant resource in each group. Mechanism UNB satisfies all four properties (SI, EF, PO, and SP) and has a fair-ratio of $\frac{3}{2}$ for social welfare and 2 for utilization. Mechanism BAL further improves the fair-ratio for social welfare to $\frac{4}{3}$. However, BAL satisfies SI, EF, and PO, but not SP. Finally, we generalize BAL to a new mechanism BAL* which satisfies all the four properties and has the same asymptotic fair-ratio as BAL when the number of agents n goes to infinity. We further provide a more fine-grained analysis of the fair-ratio parameterized by a *minority population ratio* parameter $\alpha \in (0, \frac{1}{2}]$, which is defined as the fraction of agents in the smaller group classified by their dominant resources. Table 1 lists a summary of the fair-ratios of different mechanisms in the worst case and in terms of α. We also compare our mechanisms with DRF by conducting experiments on both synthetic and real-world data. Our results match well with the theoretical bounds of fair-ratios and show that both UNB and BAL* achieve better social welfare and utilization than DRF.

Next we move to the general situation with $m \geq 2$ resources. We first give a family $\mathcal{F}$ of mechanisms, containing DRF as a special case, that satisfy all the four properties. This answers the question posed by [7] that *"whether DRF*

---

[1]   https://aws.amazon.com/blogs/aws/cloud-computing-server-utilization-the-environment/.

[2]   https://www.ibm.com/cloud/learn/cloud-computing.

Table 1. Fair-ratio results for $m = 2$ resources overview.

	Social welfare	Utilization
DRF (Lemma 1)	$2\,(2-\alpha)$	$\infty\,\left(\frac{1}{\alpha}\right)$
UNB (Theorem 1)	$\frac{3}{2}\,(1+\alpha)$	$2\,\left(\frac{1}{1-\alpha}\right)$
BAL (Theorem 2)	$\frac{4}{3}\,\left(\frac{4-2\alpha}{3-\alpha}\right)$	$2\,\left(\frac{2}{1+\alpha}\right)$
BAL* (Theorem 3)	$\left[\frac{4-2\alpha}{3-\alpha}, \frac{4-2\alpha}{3-\alpha-\frac{1}{n}}\right]$	$\left[\frac{2}{1+\alpha}, \frac{2}{1+\alpha-\frac{1}{n}}\right]$

Table 2. Price of SP results overview.

	Social welfare	Utilization
$m = 2$ (Theorem 6)	$[1, 3-\sqrt{3}+\frac{1}{2n}]$	$[1, \frac{3}{2-\frac{1}{n}}]$
$m = 3$ (Theorem 7)	$[2, 3]$	∞
$m \geq 4$ (Theorem 7)	m	∞

is the only possible strategy-proof policy for multi-resource fairness, given other desirable properties such as Pareto efficiency". Unfortunately, as we will see in the next part, for general m all mechanisms that satisfy the four properties will have the same fair-ratio as DRF. Nevertheless, we show that a generalization of UNB still satisfies the four properties and its fair-ratio is always weakly better than DRF.

Finally, we investigate the efficiency loss caused by incentive constraints. We define the *price of strategyproofness* (Price of SP) for social welfare (resp. utilization) as the best fair-ratio for social welfare (resp. utilization) among all mechanisms which satisfy SI, EF, PO, and SP. Our results are summarized in Table 2. For the case with $m = 2$ resources, we show that the price of SP is at most $3 - \sqrt{3} + \frac{1}{2n}$ for social welfare, and at most $\frac{3}{2-\frac{1}{n}}$ for utilization. When $m = 3$, the price of SP is ∞ for utilization and between 2 and 3 for social welfare. Finally, when $m \geq 4$, the price of SP is ∞ for utilization and m for social welfare, which implies that in the general setting all mechanisms that satisfy the four properties have the same fair-ratio as DRF.

Due to the lack of space, we refer the reader to the long version of the paper [1] for proof details.

1.2   Related Work

Since its introduction by [7], DRF has been extended in multiple directions, including the setting with weighted agents or indivisible tasks [19], the setting when resources are distributed over multiple servers with placement constraints [21,25] or without placement constraints [5,24], a dynamic setting when agents arrive at different times [14] and the case when agents' demands are limited [16,17]. In contrast to these works, we consider the original setting and aim to design mechanisms with better efficiency guarantees than DRF. Notably, [16] generalize DRF to the limited demand setting, and study the approximation ratio of the generalized mechanism by comparing it with the optimal allocation satisfying PO, SI and EF. Essentially, their results implies that for two resources, the fair-ratio of DRF is 2 for social welfare and ∞ for utilization, which can be seen as a special case of our more fine-grained result in Lemma 1 parameterized by α. [4] advocate a different fairness notion called *Bottleneck Based Fairness (BBF)* for multi-resource allocation with Leontief preferences and show that a BBF allocation always exists. [9] extend DRF and BBF for a larger family of utilities and give a polynomial time algorithm to compute a BBF solution. Characterization of mechanisms satisfying a set of desirable properties under Leontief preferences has been studied in economics literature [6,15,18]. However, they consider different properties than what we consider.

2 Preliminaries

2.1 Multi-resource Allocation

We start by introducing the formal model of multi-resource allocation. The notations are mainly adopted from [19]. Given a set of agents $N = \{1, 2, \ldots, n\}$ and a set of resources R with $|R| = m$, each agent i has a *resource demand vector* $\mathbf{D}_i = \{D_{i1}, D_{i2}, \ldots, D_{im}\}$, where D_{ir} is the ratio between the demand of agent i for resource r to complete one task and the total amount of that resource. The *dominant resource* of an agent i is the resource r_i^* such that $r_i^* \in \arg\max_{r \in R} D_{ir}$. For simplicity, we assume that all agents have positive demands, i.e., $D_{ir} > 0, \forall i \in N, \forall r \in R$. For each agent i and each resource r, define $d_{ir} = \frac{D_{ir}}{D_{ir_i^*}} \in (0, 1]$ as the *normalized demand* and denote the normalized demand vector of agent i by $\mathbf{d}_i = \{d_{i1}, d_{i2}, \ldots, d_{im}\}$. An *instance* of the multi-resource allocation problem with n agents and m resource is a matrix $\mathbf{I}$ of size $n \times m$ with each row representing a normalized demand vector.

To help better understand these notions, consider a cloud computing scenario where two agents share a system with 9 CPUs and 18 GB RAM. Each task agent 1 runs require $\langle 1 \text{ CPUs}, 4 \text{ GB}\rangle$, and each task agent 2 runs require $\langle 3 \text{ CPUs}, 1 \text{ GB}\rangle$. Since each task of agent 1 demands $\frac{1}{9}$ of the total CPU and $\frac{2}{9}$ of the total RAM, the demand vector for agent 1 is $\mathbf{D}_1 = \{\frac{1}{9}, \frac{2}{9}\}$, with RAM being its dominant resource, and the corresponding normalized demand vector is $\mathbf{d}_1 = \{\frac{1}{2}, 1\}$. Similarly, for agent 2 we have $\mathbf{D}_2 = \{\frac{1}{3}, \frac{1}{18}\}$, $\mathbf{d}_2 = \{1, \frac{1}{6}\}$, and its dominant resource is CPU.

Given problem instance $\mathbf{I}$, an *allocation* $\mathbf{A}$ is a matrix of size $n \times m$ which allocates a fraction A_{ir} of resource r to agent i. We assume all resources are divisible. An allocation $\mathbf{A}$ is *feasible* if no resource is required more than available, i.e., $\sum_{i \in N} A_{ir} \leq 1, \forall r \in R$. We assume agents have *Leontief preferences* and the *utility* of an agent with its allocation vector $\mathbf{A}_i$ is defined as

$$u_i(\mathbf{A}_i) = \max\{y \in \mathbb{R}_+ : \forall r \in R, A_{ir} \geq y \cdot d_{ir}\}.$$

We say an allocation is *non-wasteful* if for each agent $i \in N$ there exists $y \in \mathbb{R}_+$ such that $A_{ir} = y \cdot d_{ir}, \forall r \in R$. In words, for each agent, the amount of allocated resources are proportional to its normalized demand vector. The *dominant share* of an agent i under a non-wasteful allocation $\mathbf{A}$ is $A_{ir_i^*}$, where r_i^* is i's dominant resource.

Denote the set of all instances by $\mathcal{I}$, and the set of all feasible allocations by $\mathcal{A}$. A *mechanism* is a function $f : \mathcal{I} \to \mathcal{A}$ that maps every instance to a feasible allocation. We use $f_i(\mathbf{I})$ to denote the allocation vector to agent i under instance $\mathbf{I}$. A mechanism is non-wasteful if the allocation of the mechanism on any instance is non-wasteful. We only consider non-wasteful mechanisms.

2.2   Dominant Resource Fairness (DRF)

The *DRF* mechanism [7] works by maximizing and equalizing the dominant shares of all agents, subject to the feasible constraint. Let x be the dominant share of each agent, DRF solves the following linear program:

$$\text{maximize} \quad x$$
$$\text{subject to} \quad \sum_{i \in N} x \cdot d_{ir} \leq 1, \quad \forall r \in R$$

This linear program can be rewritten as $x^* = \frac{1}{\max_{r \in R} \sum_{i \in N} d_{ir}}$. Then, for agent i the allocation $\mathbf{A}_i = x^* \cdot \mathbf{d}_i$.

2.3   Properties of Mechanisms

In this work we are interested in the following properties of a resource allocation mechanism.

Definition 1 (Share Incentive (SI)). *An allocation $\mathbf{A}$ is SI if $u_i(\mathbf{A}_i) \geq \frac{1}{n}, \forall i \in N$. A mechanism f is SI if for any instance $\mathbf{I} \in \mathcal{I}$ the allocation $f(\mathbf{I})$ is SI.*

Definition 2 (Envy Freeness (EF)). *An allocation $\mathbf{A}$ is EF if $u_i(\mathbf{A}_i) \geq u_i(\mathbf{A}_j), \forall i, j \in N$. A mechanism f is EF if for any instance $\mathbf{I} \in \mathcal{I}$ the allocation $f(\mathbf{I})$ is EF.*

Definition 3 (Pareto Optimality (PO)). *An allocation $\mathbf{A}$ is PO if it is not dominated by another allocation $\mathbf{A}'$, i.e., there is no $\mathbf{A}'$ such that $\exists i_0 \in N : u_{i_0}(\mathbf{A}'_{i_0}) > u_{i_0}(f_{i_0}(\mathbf{I}))$ and $\forall i \in N : u_i(\mathbf{A}'_{i_0}) \geq u_i(f_i(\mathbf{I}))$. A mechanism f is PO if for any instance $\mathbf{I} \in \mathcal{I}$ the allocation $f(\mathbf{I})$ is PO.*

Definition 4 (Strategyproofness (SP)). *A mechanism f is SP if no agent can benefit by reporting a false demand vector, i.e., $\forall \mathbf{I} \in \mathcal{I}, \forall i \in N, \forall \mathbf{d}'_i, u_i(f_i(\mathbf{I})) \geq u_i(f_i(\mathbf{I}'))$, where $\mathbf{I}'$ is the resulting instance by replacing agent i's demand vector by $\mathbf{d}'_i$.*

Notice that SI, EF, and PO are defined for both allocations and mechanisms, while SP is only defined for mechanisms. It is easy to verify that a non-wasteful mechanism satisfies PO if and only if at least one resource is used up in the allocation returned by the mechanism.

[7] shows that DRF satisfies all of these desirable properties.

2.4 Approximation Ratio

We define *social welfare (SW)* of an allocation $\mathbf{A}$ as the sum of the utilities of all agents,

$$SW(\mathbf{A}) = \sum_{i \in N} u_i(\mathbf{A}_i).$$

As in [16], we define *utilization* of an allocation $\mathbf{A}$ as the minimum utilization rate of m resources,

$$U(\mathbf{A}) = \min_{r \in R} \sum_{i \in N} A_{ir}.$$

As discussed in the introduction, we use a revised notion of approximation ratio to measure the efficiency performance of a mechanism, where we use the optimal *fair* allocation as the benchmark instead of the original benchmark which is based on the optimal allocation.

Definition 5. *The* fair-ratio *for social welfare (resp. utilization) of a mechanism f is defined as, among all instances $\mathbf{I} \in \mathcal{I}$, the maximum ratio of the optimal social welfare (resp. utilization) among all allocations that satisfy SI and EF over the social welfare (resp. utilization) of $f(\mathbf{I})$, i.e.,*

$$FR_{SW} = \max_{\mathbf{I} \in \mathcal{I}} \frac{\max\limits_{\mathbf{A} \text{ is } SI, EF} SW(\mathbf{A})}{SW(f(\mathbf{I}))} \quad and \quad FR_{Util} = \max_{\mathbf{I} \in \mathcal{I}} \frac{\max\limits_{\mathbf{A} \text{ is } SI, EF} U(\mathbf{A})}{U(f(\mathbf{I}))}.$$

3 Two Types of Resources

In this section we focus on the case where there are only two competing resources. More specifically, we assume that among the m types of resources, there exists $r_1, r_2 \in R$, such that for any agent i and any other resource $r \neq r_1, r_2$, we have $d_{ir_1} \geq d_{ir}$ and $d_{ir_2} \geq d_{ir}$. This means in any allocation, other resources will not run out before r_1 or r_2 runs out. Thus it is equivalent to assume that R contains only two resources r_1 and r_2.

We partition all agents into two groups G_1 and G_2, where $G_i(i = 1, 2)$ consists of all agents whose dominant resource is r_i. Agents with demand vector $(1, 1)$ are considered to be in G_1. Denote $n_1 = |G_1|$ and $n_2 = |G_2|$. Without loss of generality, we assume that $n_1 \geq \frac{n}{2}$ (otherwise we can rename the two resources).

We now let

$$\alpha := \frac{n_2}{n} \in (0, \frac{1}{2}]$$

Algorithm 1: $\text{UNB}(\mathbf{d_1}, \mathbf{d_2}, \ldots, \mathbf{d_n})$

---

1 $C \leftarrow (c_1, c_2) = (1, 1)$ // remaining resources
2 $G_1 \leftarrow \{i \mid d_{i,1} = 1\}; G_2 \leftarrow \{i \mid d_{i,1} < 1\}$
3 **foreach** $i \in N$ **do**
4 $\quad$ $\mathbf{A}_i \leftarrow \frac{1}{n}\mathbf{d}_i$ // every agent receives $\frac{1}{n}$ dominant share
5 $\quad$ $C \leftarrow C - \mathbf{A}_i$

6 **while** $c_1 > 0$ *and* $c_2 > 0$ **do**
7 $\quad$ $P \leftarrow \arg\min_{i \in G_2} A_{i,1}$ // agents with the smallest fraction of resource r_1
8 $\quad$ $\delta_0 \leftarrow \min_{i \in N \backslash P} A_{i,1} - \min_{i \in P} A_{i,1}$ // increasing step when 2nd smallest fraction of
$\quad$ resource r_1 is reached
9 $\quad$ $\delta_1 \leftarrow \frac{c_1}{|P|}, \delta_2 \leftarrow \frac{c_2}{\sum_{i \in P} \frac{1}{d_{i,1}}}$ // increasing step when resource r_1 (or r_2) is
$\quad$ used up
10 $\quad$ $\delta^* \leftarrow \min\{\delta_0, \delta_1, \delta_2\}$
11 $\quad$ **foreach** $i \in P$ **do**
12 $\quad\quad$ $\mathbf{A}_i \leftarrow \mathbf{A}_i + (\delta^*, \frac{\delta^*}{d_{i,1}})$ // increase resource r_1 by the same δ^*
13 $\quad\quad$ $C \leftarrow C - (\delta^*, \frac{\delta^*}{d_{i,1}})$

14 **return A**

---

be the fraction of agents in the smaller group and we call α the *minority population ratio*. We assume that $\alpha > 0$, because when $\alpha = 0$ the only allocation satisfying SI is to give every agent $\frac{1}{n}$ of the first resource (and the corresponding amount of the second resource). As we will see in the following, α is crucial in analyzing the fair-ratio of a mechanism.

We start by analyzing the fair-ratio of DRF.

Lemma 1. *With 2 resources, for instances with minority population ratio α, we have*

$$\text{FR}_{\text{SW}}(\text{DRF}) = 2 - \alpha \quad and \quad \text{FR}_{\text{Util}}(\text{DRF}) = \frac{1}{\alpha}.$$

When α approaches 0, we have $\text{FR}_{\text{SW}}(\text{DRF}) \to 2$ and $\text{FR}_{\text{Util}}(\text{DRF}) \to \infty$. Notice that with 2 resources $\text{FR}_{\text{SW}}(f)$ for any mechanism f satisfying SI is at most 2 as the mechanism can always achieve at least 1 in SW.

In the following, we present two new mechanisms with the same set of properties as DRF but with better fair-ratios.

3.1 Mechanism UNB

The more detailed analysis of Lemma 1 shows that when the population of two groups are unbalanced, i.e., when α is close to 0, it is better to allocate more resources to agents in the minor group G_2 with smaller $d_{i,1}$. This idea leads to mechanism UNB, described in Algorithm 1. The mechanism has two steps. In step 1, the mechanism allocates every agent $\frac{1}{n}\mathbf{d}_i$ of resources such that each

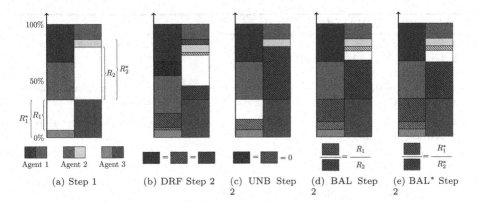

Fig. 1. Allocations under DRF, UNB, BAL and BAL* in Example 1. The shaded area represents the added parts in respective Step 2.

agent has a dominant share of $\frac{1}{n}$, which ensures SI. In step 2, the mechanism repeats the following process till one resource is used up: Select a set of agents from G_2 who have the smallest fraction t_1 of resource r_1, denoted by P, and increase their fractions of resource r_1 at the same speed (δ^*) till the fraction reaches the second smallest fraction t_2 in G_2 ($\delta^* = \delta_0$) or one resource is used up ($\delta^* = \delta_1$ for resource r_1 and $\delta^* = \delta_2$ for resource r_2).

Example 1. Consider an instance with 3 agents who have demand vectors $\mathbf{d}_1 = (1, \frac{2}{5})$, $\mathbf{d}_2 = (1, \frac{1}{5})$ and $\mathbf{d}_3 = (\frac{1}{5}, 1)$. We compare the allocation under UNB and DRF. Notice that DRF can also be viewed as a two-step mechanism, where in step 1 every agent gets $\frac{1}{n}$ dominant share (the same as UNB) and in step 2 we increase the dominant share of every agent at the same speed till one resource is used up. For the above instance, in step 1 all 3 agents get $\frac{1}{3}$ dominant share, and the remaining resource is $C = (\frac{4}{15}, \frac{7}{15})$, corresponding to Fig. 1a. In step 2, under DRF, all agents have the same dominant share $x^* = \frac{1}{\max\{\frac{11}{5}, \frac{8}{5}\}} = \frac{5}{11}$ and the final allocation vectors are $A_1 = (\frac{5}{11}, \frac{2}{11})$, $A_2 = (\frac{5}{11}, \frac{1}{11})$ and $A_3 = (\frac{1}{11}, \frac{5}{11})$, corresponding to Fig. 1b. Under UNB, we increase the allocation of agent 3, who currently has the smallest fraction $\frac{1}{15}$ of resource r_1, till the second resource r_2 is used up and we have $A_3 = (\frac{4}{25}, \frac{4}{5})$, corresponding to Fig. 1c. The SW under DRF is $\frac{5}{11} \times 3 \approx 1.36$, while the SW under UNB is $\frac{1}{3} + \frac{1}{3} + \frac{4}{5} \approx 1.47$.

We show that UNB satisfies all four properties and has a better fair-ratio than DRF.

Theorem 1. *With* 2 *resources, mechanism* UNB *can be implemented in polynomial time, satisfies SI, EF, PO, and SP, and has*

$$\mathrm{FR}_{\mathrm{SW}}(\mathrm{UNB}) = 1 + \alpha \quad and \quad \mathrm{FR}_{\mathrm{Util}}(\mathrm{UNB}) = \frac{1}{1 - \alpha}.$$

Because $\alpha \in (0, 1/2]$, we have $\mathrm{FR}_{\mathrm{SW}}(\mathrm{UNB}) \le 3/2$ and $\mathrm{FR}_{\mathrm{Util}}(\mathrm{UNB}) \le 2$, both of which are significantly better than DRF.

Algorithm 2: BAL($\mathbf{d}_1, \mathbf{d}_2, \ldots, \mathbf{d}_n$)

---
1 $C \leftarrow (c_1, c_2) = (1,1)$ // remaining resources
2 $G_1 \leftarrow \{i \mid d_{i,1} = 1\}; G_2 \leftarrow \{i \mid d_{i,1} < 1\}$
3 foreach $i \in N$ **do**
4 $\mathbf{A}_i \leftarrow \frac{1}{n}\mathbf{d}_i$ // every agent receives $\frac{1}{n}$ dominant share
5 $C \leftarrow C - \mathbf{A}_i$

6 $(R_1, R_2) \leftarrow C$ // remaining resources after step 1
7 while $c_1 > 0$ *and* $c_2 > 0$ **do**
8 $P_1 \leftarrow \arg\min_{i \in G_1} A_{i,2}$
9 $P_2 \leftarrow \arg\min_{i \in G_2} A_{i,1}$
10 $(\delta_1^*, \delta_2^*) \leftarrow$ CalcStep ()   // calculate increasing steps
11   **foreach** $k = 1, 2$ **do**
12   **foreach** $i \in P_k$ **do**
13 $\mathbf{A}_i \leftarrow \mathbf{A}_i + \frac{\delta_k^*}{d_{i,3-k}}\mathbf{d}_i$ // increase the non-dominant resource by the same δ_k^*
14 $C \leftarrow C - \frac{\delta_k^*}{d_{i,3-k}}\mathbf{d}_i$

15 return A

The intuition behind $\mathrm{FR_{SW}(UNB)} \leq 1 + \alpha$ is that under UNB agents in G_1 get at most α less utility than the optimal allocation and agents in G_2 get no less utility than the optimal allocation. For the lower bound $\mathrm{FR_{SW}(UNB)} \geq 1 + \alpha$, consider instances where after step 1 the remaining resource is $C = (\alpha - \epsilon, \epsilon)$ with $\epsilon \to 0$. In step 2 UNB can only increase the allocations of agents in G_2 and get SW at most $1 + \epsilon$, while the optimal allocation can increase the allocation of agents in G_1 with the smallest $d_{i,2}$ and get SW of $1 + \alpha - \epsilon$.

3.2 Mechanism BAL and BAL*

According to Theorem 1, UNB has the worst performance when the population of two groups are balanced, i.e., when α is close to $\frac{1}{2}$, because in step 2 it only increases allocations of agents in one group (G_2). In this case, a better strategy in step 2 is to increase allocations of agents from both groups.

Following this intuition, we propose mechanism BAL, described in Algorithm 2. Mechanism BAL also has two steps. Step 1 is the same as UNB, where every agent gets $\frac{1}{n}$ dominant share. In step 2, the mechanism increases allocations of agents from both groups, and within each group the method is the same as in UNB, that is, within each group, only agents who have the smallest amount of the non-dominant resource will be allocated more resources, and they will be allocated the same fraction (δ_1^* or δ_2^*) of the non-dominant resource. In addition, BAL controls the relative allocation rates (δ_1^*, δ_2^*) of two groups such that the ratio between the increased dominant shares of two groups is proportional to the ratio between the remaining amounts of two resources after step 1. Formally, let ΔS_1 and ΔS_2 be the sum of increased dominant share of agents in G_1 and G_2 in

---

Algorithm 3: CalcStep ()

---

1 **foreach** $k = 1, 2$ **do**

2     $\delta_k \leftarrow \min\limits_{i \in N \setminus P_k} A_{i,3-k} - \min\limits_{i \in P_k} A_{i,3-k}$ // increasing step when 2nd smallest

     fraction of resource r_{3-k} is reached

3     $D_k \leftarrow \sum_{i \in P_k} \frac{1}{d_{i,3-k}}$

4     $\overline{\delta_k} \leftarrow \dfrac{c_{3-k}}{|P_k| + D_k \frac{R_{3-k}}{R_k}}$ // increasing step when resource r_{3-k} is used up

5     $\delta_k^* \leftarrow \min\{\delta_k, \overline{\delta_k}\}$

6 **if** $\frac{\delta_1^* D_1}{\delta_2^* D_2} \leq \frac{R_1}{R_2}$ **then**

7     $\delta_2^* \leftarrow \delta_1^* \cdot \frac{D_1}{D_2} \cdot \frac{R_2}{R_1}$ // decrease δ_2^* according to δ_1^*

8 **else**

9     $\delta_1^* \leftarrow \delta_2^* \cdot \frac{D_2}{D_1} \cdot \frac{R_1}{R_2}$ // decrease δ_1^* according to δ_2^*

10 **return** (δ_1^*, δ_2^*)

---

step 2 respectively. Let $R_1 = 1 - \frac{n_1}{n} - \frac{1}{n} \sum_{i \in G_2} d_{i,1}$ and $R_2 = 1 - \frac{n_2}{n} - \frac{1}{n} \sum_{i \in G_1} d_{i,2}$ be the amount of remaining resources after step 1, then BAL ensures that

$$\frac{\Delta S_1}{\Delta S_2} = \frac{R_1}{R_2}. \tag{1}$$

This condition is crucial to guarantee the good performance of BAL.

To compute the increasing steps (δ_1^*, δ_2^*) (CalcStep () in line 10 of Algorithm 2), we calculate the largest increasing steps (δ_1^*, δ_2^*) such that condition (1) is satisfied and one of the following conditions is satisfied: (a) one resource has been used up; (b) one agent has to be added into P_1 or P_2. The concrete algorithm is given in Algorithm 3.

We now show that BAL satisfies SI, EF, and PO, and its fair-ratio for SW is at most $\frac{4}{3}$.

Theorem 2. *With 2 resources, mechanism BAL can be implemented in polynomial time, satisfies SI, EF, and PO, and has*

$$\mathrm{FR_{SW}(BAL)} = \frac{4 - 2\alpha}{3 - \alpha} \quad and \quad \mathrm{FR_{Util}(BAL)} = \frac{2}{1 + \alpha}.$$

The intuition behind $\mathrm{FR_{SW}(BAL)} \leq \frac{4-2\alpha}{3-\alpha}$ is that compared with the optimal allocation, where in step 2 the sum of increased dominant share of agents in G_1 and G_2 are ΔS_1^* and ΔS_2^* respectively, we can show that either $\Delta S_1 \geq \Delta S_1^*$ or $\Delta S_2 \geq \Delta S_2^*$, and $\Delta S_i \geq \frac{1}{2} \Delta S_i^*$ for any $i \in \{1, 2\}$. Combining them with the fact that $\max\{\Delta S_1^*, \Delta S_2^*\} \leq 1 - \alpha$, we get

$$\mathrm{FR_{SW}(BAL)} = \frac{1 + \Delta S_1^* + \Delta S_2^*}{1 + \Delta S_1 + \Delta S_2} \leq \max_{i \in \{1,2\}} \frac{1 + \Delta S_i^*}{1 + \frac{1}{2} \Delta S_i^*} \leq \frac{2 - \alpha}{1 + \frac{1-\alpha}{2}} = \frac{4 - 2\alpha}{3 - \alpha}.$$

For the lower bound $\mathrm{FR_{SW}(BAL)} \geq \frac{4-2\alpha}{3-\alpha}$, consider instances where after step 1 the remaining resource is $C = (\epsilon, 1 - \alpha - \epsilon)$ with $\epsilon \to 0$. In step 2 the optimal allocation can allocate all the remaining resource to agent i^* in G_2 who has demand vector $(\frac{\epsilon}{1-\alpha-\epsilon}, 1)$ and get SW of $2 - \alpha - \epsilon$, while for BAL, because of the condition (1), we can only give about half of the remaining resource r_1 to i^* and the other half to agents in G_1 such that $\frac{\Delta S_1}{R_1} = \frac{\Delta S_2}{R_2} \approx \frac{1}{2}$, where the SW is about $1 + \frac{1-\alpha}{2} = \frac{1}{2}(3-\alpha)$.

However, BAL does not satisfy SP as the agent with the minimum $d_{i,1}$ (or minimum $d_{i,2}$) could influence the ratio $\frac{R_1}{R_2}$ by modifying its demand vector to get more resources in step 2, as shown in the following example.

Example 2. Consider an instance with two agents who have demand vectors $\mathbf{d}_1 = (1, \frac{1}{2})$ and $\mathbf{d}_2 = (\frac{1}{4}, 1)$. According to BAL, in step 1 agent 1 gets $(\frac{1}{2}, \frac{1}{4})$ and agent 2 gets $(\frac{1}{8}, \frac{1}{2})$. Then the remaining resources is $(\frac{3}{8}, \frac{1}{4})$ and the increasing speed ratio is $\frac{3}{2}$. In step 2, agent 1 gets $(\frac{3}{14}, \frac{3}{28})$ and agent 2 gets $(\frac{1}{28}, \frac{1}{7})$, and resource 2 is used up. Overall agent 2 gets $(\frac{9}{56}, \frac{9}{14})$. However, if agent 2 reports another demand vector $\mathbf{d}'_2 = (\frac{1}{2}, 1)$, then both agents will get the same dominant share $\frac{2}{3}$ under BAL. In particular, agent 2 will get $(\frac{1}{3}, \frac{2}{3})$, which is strictly better than $(\frac{9}{56}, \frac{9}{14})$. Therefore, BAL is not SP.

Fortunately, we can make BAL satisfy SP with a small modification. In the following we propose a slightly different mechanism BAL* that replaces the condition (1) by the following condition:

$$\frac{\Delta S_1}{\Delta S_2} = \frac{R_1^*}{R_2^*} = \frac{R_1 + \frac{1}{n}d_{i^*,1}}{R_2 + \frac{1}{n}d_{j^*,2}}, \tag{2}$$

where i^* is an agent in G_2 with the minimum $d_{i,1}$ and j^* is an agent in G_1 with the minimum $d_{i,2}$. That is, the ratio between ΔS_1 and ΔS_2 is proportional to the ratio between the remaining amounts of two resources when all agents except i^* and j^* get $\frac{1}{n}$ dominant share. Intuitively, for agent i^*, this modification prevents it from increasing $d_{i^*,1}$ to influence $\frac{R_1^*}{R_2^*}$, unless $d_{i^*,1}$ becomes larger than the second smallest $d_{i,1}$, for which case we can show that i^* cannot benefit.

We show that BAL* satisfies all four properties including SP, and its fair-ratio is very close to that of BAL.

Theorem 3. *With 2 resources, mechanism BAL* can be implemented in polynomial time, satisfies SI, EF, PO, and SP, and has*

$$\mathrm{FR}_{\mathrm{SW}}(\mathrm{BAL}^*) \in \left[\frac{4-2\alpha}{3-\alpha}, \frac{4-2\alpha}{3-\alpha-\frac{1}{n}}\right]; \quad \mathrm{FR}_{\mathrm{Util}}(\mathrm{BAL}^*) \in \left[\frac{2}{1+\alpha}, \frac{2}{1+\alpha-\frac{1}{n}}\right].$$

Example 1 (Continued). We compare BAL and BAL* for the instance in Example 1. Step 1 is the same as before and we have $\frac{R_1}{R_2} = \frac{\frac{4}{15}}{\frac{7}{15}} = \frac{4}{7}$ and $\frac{R_1^*}{R_2^*} = \frac{\frac{4}{15}+\frac{1}{15}}{\frac{7}{15}+\frac{1}{15}} = \frac{5}{8}$. In step 2, under BAL, we increase the allocation of agent 2 by $(\frac{16}{81}, \frac{16}{405})$, and that of agent 3 by $(\frac{28}{405}, \frac{28}{81})$ such that resource r_1 is used up. Notice that $\Delta S_1 = \frac{16}{81}$ and $\Delta S_2 = \frac{28}{81}$ satisfy $\frac{\Delta S_1}{\Delta S_1} = \frac{R_1}{R_2}$. Under BAL*, we increase the

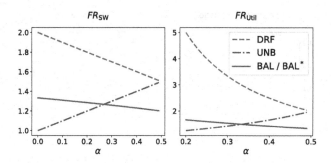

Fig. 2. Fair-ratio of mechanisms as a function of α. As $\mathrm{FR}_{\mathrm{Util}}(DRF) \to \infty$ when $\alpha \to 0$, for better visualization, we only show $\mathrm{FR}_{\mathrm{Util}}$ for $\alpha \in [0.2, 0.5]$.

allocation of agent 2 by $(\frac{20}{99}, \frac{4}{99})$, and that of agent 3 by $(\frac{32}{495}, \frac{32}{99})$. Notice that $\Delta S_1 = \frac{20}{99}$ and $\Delta S_2 = \frac{32}{99}$ satisfy $\frac{\Delta S_1}{\Delta S_1} = \frac{R_1^*}{R_2^*}$. These two corresponds to Fig. 1d and 1e. The SW under BAL and BAL* is ≈ 1.54 and ≈ 1.53 respectively, which is larger than 1.36 under DRF and 1.47 under UNB.

Figure 2 shows fair-ratios of DRF, UNB, BAL, and BAL* (when $n \to \infty$) as a function of α. Notice that all three new mechanisms have better fair-ratio than DRF for any $\alpha \in (0, \frac{1}{2})$. Among new mechanisms, UNB has better fair-ratio than BAL (BAL*) when α is close to 0 while BAL (BAL*) has better fair-ratio than UNB when α is close to 0.5. Note that we can combine these two mechanisms to achieve a better fair-ratio, which will be further discussed in Sect. 5.

3.3   Experimental Evaluation

The above analysis of fair-ratio shows that our mechanism UNB and BAL* have better performance than DRF from the worst-case perspective. In this section, we compare the performance of DRF, UNB and BAL* when $m = 2$ using both synthetic instances and real-world instances based on Google cluster-usage traces [20]. Our results are shown in Fig. 3, where we plot the ratio between the optimal allocation (satisfying SI and EF) and the allocation under compared mechanisms. Our results match well with the above fair-ratios and show that both UNB and BAL* achieve better social welfare and utilization than DRF.

Random Instances with Different α. First we compare mechanisms on random instances with fixed $n = 100$ and different $\alpha \in \{0.05, 0.10, \ldots, 0.50\}$. For each α, we average over 1000 instances to get the data point. To control the value of α, we choose $n(1 - \alpha)$ agents and set $d_{i,1} = 1$ for them, and for the remaining agents we set $d_{i,2} = 1$. The other entries of the demand vectors are sampled uniformly from $\{0.01, 0.02, \ldots, 1.00\}$.

The result is shown in the first row of Fig. 3. For SW, BAL* is very close to the optimal solution (the ratio is close to 1) and BAL* is always better than DRF for different values of α. UNB also outperforms DRF for most values of α except when $\alpha \in [0.45, 0.5]$. Comparing UNB and BAL*, similarly to the crossing point of their theoretical fair-ratios in Fig. 2, their performance on random instances

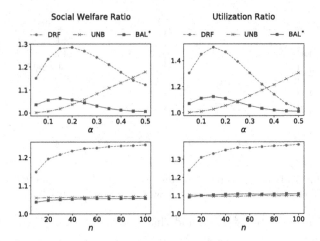

Fig. 3. Performance ratio between the optimal allocation and allocations under DRF, UNB, BAL* on synthetic instances with different α (1st row) and real-world instances with different n (2nd row).

also cross when $\alpha \approx 0.25$ in Fig. 3, confirming that when $\alpha \to 0$, UNB is better than BAL*, and when $\alpha \to 0.5$, BAL* is better than UNB. When $\alpha \geq 0.2$, the performance trend of three mechanisms matches well with the fair-ratio. More precisely, when α increases, BAL* and DRF perform better while UNB performs worse. The comparison of three mechanisms in utilization is almost the same as in SW.

Instances Generated from Google Trace. Next we test mechanisms on instances that are generated according to the real demands of tasks from the Google traces. The Google traces record the demands for CPU and memory of each submitted task. We normalize these demands to get a pool of normalized demand vectors. Then we generate instances by randomly sampling demand vectors from this pool. We compare mechanisms on instances with different number $n \in \{10, 20, \ldots, 100\}$ of agents. For each n, we average over 1000 instances to get the data point.

The result is shown in the second row of Fig. 3. For both SW and utilization, UNB and BAL* outperform DRF and the improvements are more than 10%. The performance of UNB and BAL* are very close, because in the demand vector pool more agents (about 67%) have CPU as the dominant resource and hence the generated instances have α close to 0.33. Notice that the fair-ratios of UNB and BAL* are indeed very close when $\alpha = 0.33$ (see Fig. 2).

4   Multiple Types of Resources

We move to the general case with $m \geq 2$ types of resources.

4.1   A Family of Mechanisms

We start by presenting a large family of mechanisms that satisfy the four desired properties SI, EF, PO, and SP, which includes DRF as a special case. This is in response to the question asked in [7] that *"whether DRF is the only possible strategy-proof policy for multi-resource fairness, given other desirable properties such as Pareto efficiency"*. Although many mechanisms based on DRF have been proposed for different settings and there are characterizations of mechanisms satisfying desirable properties under Leontief preferences [6,15,18], to the best of our knowledge, there is no work that directly answers this question.

We call a function g that maps vectors $\mathbf{v} \in [0,1]^m$ to $\mathbb{R}$ *monotone*, if it satisfies that for any two vectors $\mathbf{v}_1, \mathbf{v}_2$ with $\mathbf{v}_1 > \mathbf{v}_2$, $g(\mathbf{v}_1) > g(\mathbf{v}_2)$. Here $\mathbf{v}_1 > \mathbf{v}_2$ means $\mathbf{v}_1$ is element-wise strictly larger than $\mathbf{v}_2$. Denote $\mathcal{G}$ the set of all monotone functions. Now we define a family of mechanisms $\mathcal{F}$ based on monotone functions. For each monotone function $g \in \mathcal{G}$, we define a mechanism $f_g \in \mathcal{F}$ as follows. The mechanism contains two steps that have the same flavor as UNB. In step 1, every agent receives $\frac{1}{n}$ dominant share. In step 2, we increase the allocation for agents that have the minimum value of $g(\mathbf{A}_i)$ till some resource is used up. We show that all mechanisms in $\mathcal{F}$ satisfy the four desired properties.

Theorem 4. *For any m, every mechanism $F_g \in \mathcal{F}$ satisfies SI, EF, PO, and SP.*

With the large family of mechanisms at hand, the next question is to check if there exists any mechanism from $\mathcal{F}$ that can achieve better efficiency than DRF. Unfortunately, as we will see in the next part, all mechanisms from $\mathcal{F}$ will have the same approximation guarantee for general m. This means from a worst-case analysis point of view, no mechanism has a provable better SW or utilization than DRF. Thus a more fine-grained analysis is needed to find better mechanisms. In the next section, we analyze a special mechanism from $\mathcal{F}$, which can be seen as a generalization of UNB, by considering two parameters.

4.2   Generalization of UNB

Similar to the case with 2 resources, we first partition all agents into m groups $G_i(i \in [m])$ according to their dominant resources and choose an arbitrary group (say G_1) as a special group. Then, we let $\alpha := 1 - \frac{|G_1|}{n}$ be the fraction of agents not in G_1, and let $\beta := \sum_{i \in N \setminus G_1} \frac{d_{i,1}}{n\alpha}$ be the average demand of agents not in G_1 for resource r_1.

UNB can be generalized as follows. In step 1, each agent gets $\frac{1}{n}$ dominant share. In step 2, we increase the allocation of agents who have the smallest fraction of resource r_1 in the same speed for resource r_1, till some resource is used up. With slight abuse of notation, we still call this generalized mechanism UNB. Note that this mechanism is equivalent to the mechanism from the family $\mathcal{F}$ with monotone function $g(\mathbf{v}) = \mathbf{v}_1$. We prove the fair-ratio of UNB and DRF parameterized by α and β in the following theorem.

Theorem 5. *With $m \geq 3$ resources, mechanism UNB can be implemented in polynomial time, satisfies SI, EF, PO, and SP, has $\mathrm{FR}_{\mathrm{Util}}(\mathrm{UNB}) = \mathrm{FR}_{\mathrm{Util}}(\mathrm{DRF}) = \infty$, and*

$$\mathrm{FR}_{\mathrm{SW}}(\mathrm{UNB}) = \max\left\{m - \alpha\beta - (1-\alpha), \frac{m - \alpha\beta}{1 + \frac{1-\beta}{\beta}\alpha}\right\},$$

compared to

$$\mathrm{FR}_{\mathrm{SW}}(\mathrm{DRF}) = \max\left\{m - \alpha\beta - (1-\alpha), (m - \alpha\beta)(1 - \alpha(1-\beta))\right\}.$$

In particular, one can show that for any $\alpha, \beta \in (0,1)$, $1 - \alpha(1-\beta) > \frac{1}{1 + \frac{1-\beta}{\beta}\alpha}$. This means $\mathrm{FR}_{\mathrm{SW}}(\mathrm{UNB})$ is always weakly better than $\mathrm{FR}_{\mathrm{SW}}(\mathrm{DRF})$. We also conduct experiment to compare UNB and DRF on random instances with $m = 3, 4, 5$. Our results show that UNB outperforms DRF for a wide range of α and β, especially when α and β are small. See the long version [1] for more details.

5   Price of Strategyproofness

At last, we investigate the efficiency loss caused by incentive constraints. In [19], it is shown that any mechanism that satisfies one of SI, EF and SP can only guarantee at most $1/m$ of the social welfare. However, the benchmark used in [19] is the optimal social welfare among all allocations. Recall that the fair-ratio defined in this paper is benchmarked against the best social welfare (resp. utilization) among all allocations that satisfy SI and EF. In other words, we are working entirely in the domain of fair allocations. Moreover, note that any optimal allocation satisfying SI and EF must also satisfy PO. Therefore, the lower bound of the fair-ratio characterizes the efficiency loss caused by strategyproofness. Accordingly, we can define the price of strategyproofness as follows.

Definition 6. *The* Price of Strategyproofness (Price of SP) *for social welfare (resp. utilization) is defined as the best fair-ratio for social welfare (resp. utilization) among all mechanisms which satisfy SI, EF, PO and SP.*

We now study the price of SP and start with the case with two resources. Recall that $\mathrm{FR}_{\mathrm{SW}}(\mathrm{UNB})$ is increasing with α while $\mathrm{FR}_{\mathrm{SW}}(\mathrm{BAL}^*)$ is decreasing with α (see Fig. 2). By choosing the better one from UNB and BAL^* for each value of α, we get a new mechanism with a better fair-ratio than both UNB and BAL^* and show that the price of SP is at most $3 - \sqrt{3}$ for SW.

Theorem 6. *With 2 resources and $n \geq 2$ agents, the price of SP is at most $3 - \sqrt{3} + \frac{1}{2n} \xrightarrow{n \to \infty} 3 - \sqrt{3}$ for SW and at most $\frac{3}{2 - \frac{1}{n}} \xrightarrow{n \to \infty} \frac{3}{2}$ for utilization.*

We then show that for general m except one special case, all mechanisms satisfying SI, EF, PO and SP have the same fair-ratio.

Theorem 7. *For social welfare, the price of SP is m when $m \geq 4$ and between 2 and 3 when $m = 3$. For utilization, the price of SP is ∞ for any $m \geq 3$.*

One of the main results of [19] is that any mechanism that satisfies SP can only guarantee at most $1/m$ of the social welfare. Theorem 7 strengthens this by showing that the result still holds for $m \geq 4$ even if we use fair-ratio as our benchmark. Our proof follows a similar framework as [19, Theorem 4.1], but requires a different construction to incorporate EF and SI. For the case when $m = 3$, we show that the price of SP is still ∞ for utilization, while for SW we can only get a lower bound of 2. We leave the gap as an open question.

6   Conclusion

In this paper, we investigate the multi-type resource allocation problem. Generalizing the classic DRF mechanism, we propose several new mechanisms in the two-resource setting and in the general m-resource setting. The new mechanisms satisfy the same set of desirable properties as DRF but with better efficiency guarantees. For future works, we hope to extend these mechanisms to handle more realistic assumptions, such that when agents have limited demands or indivisible tasks, and when agents arrive at different times.

Acknowledgements. This work was supported by the Ministry of Education, Singapore, under its Academic Research Fund Tier 1 (RG23/20).

References

1. Bei, X., Li, Z., Luo, J.: Fair and efficient multi-resource allocation for cloud computing. CoRR abs/2210.05237 (2022). https://arxiv.org/abs/2210.05237
2. Bonald, T., Roberts, J.: Enhanced cluster computing performance through proportional fairness. Perform. Eval. **79**, 134–145 (2014)
3. Bonald, T., Roberts, J.: Multi-resource fairness: objectives, algorithms and performance. In: Proceedings of the 2015 ACM SIGMETRICS International Conference on Measurement and Modeling of Computer Systems, pp. 31–42 (2015)
4. Dolev, D., Feitelson, D.G., Halpern, J.Y., Kupferman, R., Linial, N.: No justified complaints: on fair sharing of multiple resources. In: Innovations in Theoretical Computer Science 2012, pp. 68–75 (2012)
5. Friedman, E., Ghodsi, A., Psomas, C.A.: Strategyproof allocation of discrete jobs on multiple machines. In: Proceedings of the 15th ACM Conference on Economics and Computation (EC), pp. 529–546 (2014)
6. Friedman, E.J., Ghodsi, A., Shenker, S., Stoica, I.: Strategyproofness, leontief economies and the Kalai-Smorodinsky solution (2011)
7. Ghodsi, A., Zaharia, M., Hindman, B., Konwinski, A., Shenker, S., Stoica, I.: Dominant resource fairness: fair allocation of multiple resource types. In: Proceedings of the 8th USENIX Conference on Networked Systems Design and Implementation (NSDI), pp. 323–336 (2011)
8. Grandl, R., Ananthanarayanan, G., Kandula, S., Rao, S., Akella, A.: Multi-resource packing for cluster schedulers. ACM SIGCOMM Comput. Commun. Rev. **44**(4), 455–466 (2014)

9. Gutman, A., Nisarr, N.: Fair allocation without trade. In: Proceedings of the 11th International Conference on Autonomous Agents and Multiagent Systems (AAMAS), pp. 816–823 (2012)

10. Jiang, S., Wu, J.: Multi-resource allocation in cloud data centers: a trade-off on fairness and efficiency. Concurr. Comput. Pract. Exp. **33**(6), e6061 (2021)

11. Jin, Y., Hayashi, M.: Efficiency comparison between proportional fairness and dominant resource fairness with two different type resources. In: 2016 Annual Conference on Information Science and Systems (CISS), pp. 643–648 (2016)

12. Jin, Y., Hayashi, M.: Trade-off between fairness and efficiency in dominant alpha-fairness family. In: INFOCOM 2018 - IEEE Conference on Computer Communications Workshops, pp. 391–396 (2018)

13. Joe-Wong, C., Sen, S., Lan, T., Chiang, M.: Multiresource allocation: fairness-efficiency tradeoffs in a unifying framework. IEEE/ACM Trans. Network. **21**(6), 1785–1798 (2013)

14. Kash, I., Procaccia, A.D., Shah, N.: No agent left behind: dynamic fair division of multiple resources. J. Artif. Intell. Res. **51**, 579–603 (2014)

15. Li, J., Xue, J.: Egalitarian division under leontief preferences. Econ. Theor. **54**(3), 597–622 (2013)

16. Li, W., Liu, X., Zhang, X., Zhang, X.: Multi-resource fair allocation with bounded number of tasks in cloud computing systems. In: National Conference of Theoretical Computer Science (NCTCS), pp. 3–17 (2017)

17. Narayana, S., Kash, I.A.: Fair and efficient allocations with limited demands. In: Proceedings of the 35th AAAI Conference on Artificial Intelligence (AAAI), vol. 35, pp. 5620–5627 (2021)

18. Nicoló, A.: Efficiency and truthfulness with leontief preferences. A note on two-agent, two-good economies. Rev. Econ. Des. **8**(4), 373–382 (2004)

19. Parkes, D.C., Procaccia, A.D., Shah, N.: Beyond dominant resource fairness: extensions, limitations, and indivisibilities. ACM Trans. Econ. Comput. (TEAC) **3**(1), 1–22 (2015)

20. Reiss, C., Wilkes, J., Hellerstein, J.L.: Google cluster-usage traces: format + schema. In: White Paper, pp. 1–14. Google Inc. (2011)

21. Tahir, Y., Yang, S., Koliousis, A., McCann, J.: UDRF: multi-resource fairness for complex jobs with placement constraints. In: 2015 IEEE Global Communications Conference (GLOBECOM), pp. 1–7 (2015)

22. Tang, S., He, B., Zhang, S., Niu, Z.: Elastic multi-resource fairness: balancing fairness and efficiency in coupled CPU-GPU architectures. In: SC 2016: Proceedings of the International Conference for High Performance Computing, Networking, Storage and Analysis, pp. 875–886 (2016)

23. Tang, S., Yu, C., Li, Y.: Fairness-efficiency scheduling for cloud computing with soft fairness guarantees. IEEE Trans. Cloud Comput. 1–1 (2020)

24. Wang, W., Li, B., Liang, B.: Dominant resource fairness in cloud computing systems with heterogeneous servers. In: IEEE INFOCOM 2014-IEEE Conference on Computer Communications, pp. 583–591 (2014)

25. Wang, W., Li, B., Liang, B., Li, J.: Multi-resource fair sharing for datacenter jobs with placement constraints. In: SC 2016: Proceedings of the International Conference for High Performance Computing, Networking, Storage and Analysis, pp. 1003–1014 (2016)

Optimal Impartial Correspondences

Javier Cembrano[1(✉)], Felix Fischer[2], and Max Klimm[1]

[1] Institut für Mathematik, Technische Universität Berlin, Berlin, Germany
cembrano@math.tu-berlin.de, klimm@math.tu-berlin.de
[2] School of Mathematical Sciences, Queen Mary University of London, London, UK
felix.fischer@qmul.ac.uk

Abstract. We study mechanisms that select a subset of the vertex set of a directed graph in order to maximize the minimum indegree of any selected vertex, subject to an impartiality constraint that the selection of a particular vertex is independent of the outgoing edges of that vertex. For graphs with maximum outdegree d, we give a mechanism that selects at most $d + 1$ vertices and only selects vertices whose indegree is at least the maximum indegree in the graph minus one. We then show that this is best possible in the sense that no impartial mechanism can only select vertices with maximum degree, even without any restriction on the number of selected vertices. We finally obtain the following trade-off between the maximum number of vertices selected and the minimum indegree of any selected vertex: when selecting at most k vertices out of n, it is possible to only select vertices whose indegree is at least the maximum indegree minus $\lfloor (n - 2)/(k - 1) \rfloor + 1$.

Keywords: Voting · Impartial selection · Mechanism design

1 Introduction

Impartial selection is the problem of selecting vertices with large indegree in a directed graph, in such a way that the selection of a particular vertex is independent of the outgoing edges of that vertex. The problem models a situation where agents nominate one another for selection and are willing to offer their true opinion on other agents as long as this does not affect their own chance of being selected.

The selection of a single vertex is governed by strong impossibility results. For graphs with maximum outdegree one, corresponding to situations where each agent submits a single nomination, every impartial selection rule violates one of two basic axioms [11] and as a consequence must fail to provide a non-trivial multiplicative approximation to the maximum indegree. For graphs with arbitrary outdegrees, corresponding to situations where each agent can submit multiple nominations, impartial rules violate an even weaker axiom and cannot provide a non-trivial approximation in a multiplicative or additive sense [1,8]. These impossibilities largely remain in place if rather than a single vertex we

© The Author(s), under exclusive license to Springer Nature Switzerland AG 2022
K. A. Hansen et al. (Eds.): WINE 2022, LNCS 13778, pp. 187–203, 2022.
https://doi.org/10.1007/978-3-031-22832-2_11

want to select any fixed number of vertices, but positive results can be obtained if we relax the requirement that the same number of vertices must be selected in every graph [4,18].

From a practical point of view, the need for such a relaxation should not necessarily be a cause for concern. Indeed, situations in the real world to which impartial selection is relevant often allow for a certain degree of flexibility in the number of selected agents. The exact number of papers accepted to an academic conference is usually not fixed in advance but depends on the number and quality of submissions. Best paper awards at conferences are often given in overlapping categories, and some awards may only be given if this is warranted by the field of candidates. The Fields medal is awarded every four years to two, three, or four mathematicians under the age of 40. Examples at the more extreme end of the spectrum of flexibility include the award of job titles such as vice president or deputy vice-principal. Such titles can often be given to a large number of individuals at a negligible cost per individual, but should only be given to qualified individuals so as not to devalue the title.

Tamura and Ohseto [18] specifically studied what they call nomination correspondences, i.e., rules that may select an arbitrary set of vertices in any graph. For graphs with maximum outdegree one a particular such rule, plurality with runners-up, satisfies impartiality and appropriate versions of the two axioms of Holzman and Moulin [11]. The rule selects any vertex with maximum indegree; if there is a unique such vertex, any vertex whose indegree is smaller by one and whose outgoing edge goes to the vertex with maximum indegree is selected as well. An appropriate measure for the quality of rules that select varying numbers of vertices is the difference in the worst case between the best vertex and the worst selected vertex, and we can call a rule α-min-additive if the maximum difference, taken over all graphs, between these two quantities is at most α. In this terminology, plurality with runners-up is 1-min-additive.

As Tamura and Ohseto [18] point out, it may be desirable in practice to ensure that the maximum number of vertices selected is not too large, a property that plurality with runners-up clearly fails. It is therefore interesting to ask whether there exist rules that are α-min-additive and never select more than k vertices, for some fixed α and k. For graphs with outdegree one, Tamura and Ohseto [18] answer this question in the affirmative: a variant of plurality with runners-up that breaks ties according to a fixed ordering of the vertices remains 1-min-additive but never selects more than two vertices.

Our Contribution. Our first result provides a generalization of the result of Tamura and Ohseto [18] to graphs with larger outdegrees: for graphs with maximum outdegree d, it is possible to achieve 1-min-additivity while selecting at most $d + 1$ vertices. For the particular case of graphs with unbounded outdegrees we obtain a slight improvement, by guaranteeing 1-min-additivity without ever selecting all vertices. Our second result establishes that 1-min-additivity is best possible, thus ruling out the existence of impartial mechanisms that only select vertices with maximum indegree. This holds even when no restrictions are imposed on the number of selected vertices, and is shown alongside analogous impossibility results concerning the maximization of the median or mean indegree of the selected vertices instead of their minimum indegree. Our third result provides a trade-off between the maximum number of vertices selected, where smaller is better, and the minimum indegree of any selected vertex, where larger is better: if we are allowed to select at most k vertices out of n, we can guarantee α-min-additivity for $\alpha = \lfloor (n-2)/(k-1) \rfloor + 1$. This is achieved by removing a subset of the edges from the graph before plurality with runners-up is applied, in order to guarantee impartiality while selecting fewer vertices. We do not know whether this last result is tight and leave open the interesting question for the optimal trade-off between the number and quality of selected vertices.

Related Work. Impartiality as a property of an economic mechanism was introduced by de Clippel et al. [9], and first applied to the selection of vertices in a directed graph by Alon et al. [1] and Holzman and Moulin [11]. Whereas Holzman and Moulin [11] gave axiomatic characterizations for mechanisms selecting a single vertex when all outdegrees are equal to one, Alon et al. [1] studied the ability of impartial mechanisms to approximate the maximum indegree for any fixed number of vertices when there are no limitations on outdegrees.

Both sets of authors obtained strong impossibility results, which a significant amount of follow-up work has since sought to overcome. Randomized mechanisms providing non-trivial multiplicative guarantees had already been proposed by Alon et al. [1], and Fischer and Klimm [10] subsequently achieved the best possible such guarantee for the selection of one vertex. Starting from the observation that worst-case instances for randomized mechanisms have small indegrees, Bousquet et al. [5] developed a mechanism that is asymptotically optimal as the maximum indegree grows, and Caragiannis et al. [6,7] initiated the study of mechanisms providing additive rather than multiplicative guarantees. Cembrano et al. [8] subsequently identified a deterministic mechanism that provides non-trivial additive guarantees whenever the maximum outdegree is bounded and established that no such guarantees can be obtained with unbounded outdegrees. Randomized mechanisms have been also studied from an axiomatic point of view by Mackenzie [14,15].

Bjelde et al. [4] gave randomized mechanisms with improved multiplicative guarantees for the selection of more than one vertex and observed that when selecting at most k vertices rather than exactly k, deterministic mechanisms can in fact achieve non-trivial guarantees. An axiomatic study of Tamura and Ohseto [18] for the outdegree-one case came to the same conclusion: when allowing for the selection of a varying number of vertices, the impossibility result of Holzman and Moulin [11] no longer holds. Tamura [17] subsequently characterized a mechanism proposed by Tamura and Ohseto [18], which in some cases selects all vertices, as the unique minimal mechanism satisfying impartiality, anonymity, symmetry, and monotonicity.

Impartial mechanisms have finally been proposed for various problems other than selection, including peer review [2,13,16,20], rank aggregation [12], progeny maximization [3,21], and network centralities [19].

2   Preliminaries

For $n \in \mathbb{N}$, let $[n] = \{1, 2, \ldots, n\}$, and let

$$\mathcal{G}_n = \left\{ (V, E) : V = [n], E \subseteq (V \times V) \setminus \bigcup_{v \in V} \{(v, v)\} \right\}$$

be the set of directed graphs with n vertices and no loops. Let $\mathcal{G} = \bigcup_{n \in \mathbb{N}} \mathcal{G}_n$. For $G = (V, E) \in \mathcal{G}$ and $v \in V$, let $N^+(v, G) = \{u \in V : (v, u) \in E\}$ be the out-neighborhood and $N^-(v, G) = \{u \in V : (u, v) \in E\}$ the in-neighborhood of v in G. Let $\delta^+(v, G) = |N^+(v, G)|$ and $\delta^-(v, G) = |N^-(v, G)|$ denote the outdegree and indegree of v in G, and $\Delta(G) = \max_{v \in V} \delta^-(v, G)$ the maximum indegree of any vertex in G. When the graph is clear from the context, we will sometimes drop G from the notation and write $N^+(v)$, $N^-(v)$, $\delta^+(v)$, $\delta^-(v)$, and Δ. Let $\mathrm{top}(G) = \max\{v \in V : \delta^-(v) = \Delta(G)\}$ denote the vertex of G with the largest index among those with maximum indegree. For $n \in \mathbb{N}$ and $d \in [n-1]$, let $\mathcal{G}_n(d) = \{(V, E) \in \mathcal{G}_n : \delta^+(v) \leq d$ for every $v \in V\}$ be the set of graphs in $\mathcal{G}_n$ with maximum outdegree at most d, and $\mathcal{G}(d) = \bigcup_{n \in \mathbb{N}} \mathcal{G}_n(d)$.

A k-*selection mechanism* is then given by a family of functions $f : \mathcal{G}_n \to 2^{[n]}$, one for each $n \in \mathbb{N}$, mapping each graph to a subset of its vertices, where we require that $|f(G)| \leq k$ for all $G \in \mathcal{G}$. In a slight abuse of notation, we will use f to refer to both the mechanism and to individual functions of the family. Given $G = (V, E) \in \mathcal{G}$ and $v \in V$, let $\mathcal{N}_v(G) = \{(V, E') \in \mathcal{G} : E \setminus (\{v\} \times V) = E' \setminus (\{v\} \times V)\}$ be the set neighboring graphs of G with respect to v, in the sense that they can be obtained from G by changing the outgoing edges of v. Mechanism f is *impartial* on $\mathcal{G}' \subseteq \mathcal{G}$ if on this set of graphs the outgoing edges of a vertex have no influence on its selection, i.e., if for every graph $G = (V, E) \in \mathcal{G}'$, $v \in V$, and $G' \in \mathcal{N}_v(G)$, it holds that $f(G) \cap \{v\} = f(G') \cap \{v\}$. Given a k-selection mechanism f and an aggregator function $\sigma : 2^{\mathbb{R}} \to \mathbb{R}$ such that $\sigma(\emptyset) = 0$ and, for every $S \subseteq \mathbb{R}$ with $|S| \geq 1$, $\min\{x \in S\} \leq \sigma(S) \leq \max\{x \in S\}$, we say that f is α-σ-*additive* on $\mathcal{G}' \subseteq \mathcal{G}$, for $\alpha \geq 0$, if for every graph in $\mathcal{G}'$ the function σ evaluated on the choice of f differs from the maximum indegree by at most α, i.e., if

$$\sup_{G \in \mathcal{G}'} \left\{ \Delta(G) - \sigma \left(\{\delta^-(v, G)\}_{v \in f(G)} \right) \right\} \leq \alpha.$$

We will specifically be interested in the cases where σ is the minimum, the median, and the mean, and respectively call a mechanism α-*min-additive*, α-*median-additive*, and α-*mean-additive* in these cases.

3   Plurality with Runners-Up

Focusing on the case with maximum outdegree one, Tamura and Ohseto [18] proposed a mechanism they called *plurality with runners-up*. The mechanism, which we describe formally in Algorithm 1, selects all vertices with maximum indegree; if there is a unique such vertex, then any vertex with an outgoing edge

---

Algorithm 1: Plurality with runners-up

Input: Digraph $G = (V, E) \in \mathcal{G}_n(1)$
Output: Set $S \subseteq V$ of selected vertices
Let $S = \{v \in V : \delta^-(v) = \Delta(G)\}$;
if $S = \{v\}$ *for some* $v \in V$ **then**
    $S \leftarrow S \cup \{u \in V : \delta^-(u) = \Delta(G) - 1 \text{ and } (u, v) \in E\}$
end
Return S

---

to that vertex whose indegree is smaller by one is selected as well. The idea behind this mechanism is that vertices in the latter category would be among those with maximum degree if their outgoing edge was deleted, and thus any impartial mechanism seeking to select the vertices with maximum degree would also have to select those vertices. Plurality with runners-up is impartial on $\mathcal{G}(1)$, and in any graph with n vertices selects between 1 and n vertices whose degree is equal to the maximum degree or the maximum degree minus one. It is thus an impartial and 1-min-additive n-selection mechanism on $\mathcal{G}_n(1)$ for every $n \in \mathbb{N}$. It is natural to ask whether a similar additive guarantee can be obtained for more general settings. In this section, we answer this question in the affirmative, and in particular study for which values of n, k, and d there exists an impartial and 1-min-additive k-selection mechanism on $\mathcal{G}_n(d)$. We will see later, in Sect. 4, that 1-min-additivity is in fact best possible for all cases covered by our result, with the exception of the boundary case where $n = 2$.

While Tamura and Ohseto [18] do not limit the maximum number of selected vertices, they discuss briefly a modification of their mechanism that retains impartiality and 1-min-additivity but selects at most 2 vertices. Instead of all vertices with maximum indegree, the modified mechanism breaks ties in favor of a single maximum-degree vertex using a fixed ordering of the vertices. In order to guarantee impartiality, the modified mechanism then also selects any vertex that would be selected in the graph obtained by deleting the outgoing edge of that vertex. The assumption that every vertex has at most one outgoing edge means that at most one additional vertex is selected. There thus exists a 1-min-additive k-selection mechanism on $G(1)$ for every $k \geq 2$.

Our first result generalizes this mechanism to settings with arbitrary outdegrees, as long as the maximum number of selected vertices is large enough. To this end we show that when the maximum outdegree is d, to achieve impartiality, at most d vertices have to be selected in addition to the one with maximum indegree and highest priority.[1] We formally describe the resulting mechanism in Algorithm 2, and will refer to it as *asymmetric plurality with runners-up* and

---

[1] In this mechanism and wherever ties are broken in the rest of the paper, we break ties in favor of greater index, so top(G) is the vertex with maximum indegree and highest priority in graph G. Naturally, any other deterministic tie-breaking rule could be used instead.

Algorithm 2: Asymmetric plurality with runners-up $\mathcal{P}(G)$

Input: Digraph $G = (V, E) \in \mathcal{G}_n$
Output: Set $S \subseteq V$ of selected vertices
Let $S = \emptyset$;
for $v \in V$ **do**
    Let $G_v = (V, E \setminus (\{v\} \times V))$;
    if $\mathrm{top}(G_v) = v$ **then**
        $S \leftarrow S \cup \{v\}$
    end
end
Return S

---

denote its output on graph G by $\mathcal{P}(G)$. We obtain the following theorem, which generalizes the known result for the outdegree-one case.

Theorem 1. *For every $n \in \mathbb{N}$, $d \in [n-1]$, and $k \in \{d+1, \ldots, n\}$, there exists an impartial and 1-min-additive k-selection mechanism on $\mathcal{G}_n(d)$.*

We will be interested in the following in comparing vertices both according to their indegree and to their index, and we will use regular inequality symbols, as well as the operators max and min, to denote the lexicographic order among pairs of the form $(\delta^-(v), v)$. The following lemma characterizes the structure of the set of vertices selected by Algorithm 2, and provides the main technical ingredient to the proof of Theorem 1.

Lemma 1. *Let $G = (V, E) \in \mathcal{G}$ and $v \in V$. Then, $v \in \mathcal{P}(G)$ if and only if*

(a) for every $w \in V$ with $(\delta^-(w), w) > (\delta^-(v), v)$ it holds $(v, w) \in E$; and
(b) one of the following holds:
    (i) $\delta^-(v) = \Delta(G)$; or
    (ii) $\delta^-(v) = \Delta(G) - 1$ and $v > w$ for every $w \in V$ with $\delta^-(w) = \Delta(G)$.

Proof. We first show that, if $v \in \mathcal{P}(G)$ for a given graph G, then (a) and (b) follow. Let $G = (V, E) \in \mathcal{G}$, and let $v \in \mathcal{P}(G)$. To see (a), suppose there is $w \in V$ with $(\delta^-(w, G), w) > (\delta^-(v, G), v)$. Since $v \in \mathcal{P}(G)$, we have $v = \mathrm{top}(G_v)$ with $G_v = (V, E \setminus (\{v\} \times V))$. This implies $(\delta^-(v, G_v), v) > (\delta^-(w, G_v), w)$ and therefore $\delta^-(w, G) > \delta^-(w, G_v)$, because $\delta^-(v, G) = \delta^-(v, G_v)$. Since G and G_v only differ in the outgoing edges of v, we conclude that $(v, w) \in E$. To prove (b), we note that for every $w \in V$ we have

$$(\delta^-(v, G), v) = (\delta^-(v, G_v), v) > (\delta^-(w, G_v), w) \geq (\delta^-(w, G) - 1, w), \quad (1)$$

where the last inequality comes from the fact that each vertex has at most one incoming edge from v. If there is no $w \in V \setminus \{v\}$ with $\delta^-(w) = \Delta(G)$, the maximum indegree must be that of v, so $\delta^-(v) = \Delta(G)$ and (i) follows. Otherwise, for each $w \in V \setminus \{v\}$ with $\delta^-(w) = \Delta(G)$, (1) yields $(\delta^-(v, G), v) > (\Delta(G) - 1, w)$. We conclude that either $\delta^-(v, G) > \Delta(G) - 1$, in which case (i) holds, or both $\delta^-(v) = \Delta(G) - 1$ and $v > w$, which implies (ii).

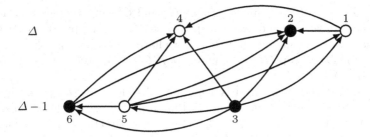

Fig. 1. Example of a set of vertices selected by Algorithm 2. In this illustration and throughout the paper, vertices are arranged vertically according to indegree and horizontally according to index, so that vertices on the left are favored in case of ties. The vertices selected by the mechanism are drawn in white, those not selected in black. Vertices with indegree below $\Delta - 1$, as well as edges incident to such vertices, are not shown. Denoting the graph as $G = (V, E)$, and letting $G_v = (V, E \setminus (\{v\} \times V))$ for each vertex v, the selected vertices v are those for which $\mathrm{top}(G_v) = v$. Specifically, vertices 2, 3, and 6 are not selected because $\mathrm{top}(G_2) = 4$, $\mathrm{top}(G_3) = 4$, and $\mathrm{top}(G_6) = 1$.

We now prove the other direction. Let $G = (V, E) \in \mathcal{G}$ and $v \in V$ such that both (a) and (b) hold. Let $G_v = (V, E \setminus (\{v\} \times V))$. We have to show that $\mathrm{top}(G_v) = v$, i.e., that for every $w \in V \setminus \{v\}$, $(\delta^-(v, G_v), v) > (\delta^-(w, G_v), w)$. Let w be a vertex in $V \setminus \{v\}$. If $(\delta^-(v, G), v) > (\delta^-(w, G), w)$, we can conclude immediately since $\delta^-(v, G_v) = \delta^-(v, G)$ and $\delta^-(w, G_v) \leq \delta^-(w, G)$. Otherwise, we know from (a) that $(v, w) \in E$ and thus $\delta^-(w, G_v) = \delta^-(w, G) - 1$. If v satisfies (i), this yields

$$\delta^-(v, G_v) = \delta^-(v, G) = \Delta(G) \geq \delta^-(w, G) = \delta^-(w, G_v) + 1,$$

so $(\delta^-(v, G_v), v) > (\delta^-(w, G_v), w)$. On the other hand, if v satisfies (ii), then

$$\delta^-(v, G_v) = \delta^-(v, G) = \Delta(G) - 1 \geq \delta^-(w, G) - 1 = \delta^-(w, G_v),$$

and $v > w$ implies $(\delta^-(v, G_v), v) > (\delta^-(w, G_v), w)$ as well.   □

Observe that Lemma 1 implies in particular that $\mathrm{top}(G) \in \mathcal{P}(G)$ for every graph G. Figure 1 provides an example of the characterization given by Lemma 1, in terms of indegrees, tie-breaking order, and edges among selected vertices.

We are now ready to prove Theorem 1.

Proof of Theorem 1. We show that for every $n \in \mathbb{N}$ and $d \in [n-1]$, asymmetric plurality with runners-up is impartial and 1-min-additive on $\mathcal{G}_n(d)$, and that for every $G = (V, E) \in \mathcal{G}_n(d)$, it selects at most $d + 1$ vertices. If this is the case, then for every $k \in \{d + 1, \ldots, n\}$ the mechanism would satisfy the statement of the theorem. Therefore, let n and d be as mentioned.

Impartiality follows from the definition of the mechanism, because the outgoing edges of a vertex are not taken into account when deciding whether the vertex is taking part on the selected set or not. If we let $G = (V, E)$, $v \in V$, and

$G' = (V, E') \in \mathcal{N}_v(G)$, then the graphs G_v and G'_v constructed when running the mechanism with each of these graphs G and G' as an input, respectively, are the same because by definition of $\mathcal{N}_v(G)$ we have $E \setminus (\{v\} \times V) = E' \setminus (\{v\} \times V)$. Since $v \in \mathcal{P}(G) \Leftrightarrow \text{top}(G_v) = v$, and $v \in \mathcal{P}(G') \Leftrightarrow \text{top}(G'_v) = v$, we conclude $v \in \mathcal{P}(G) \Leftrightarrow v \in \mathcal{P}(G')$.

To see that the mechanism is 1-min-additive, let $G \in \mathcal{G}_n(d)$ and first note that $\mathcal{P}(G) \neq \emptyset$ since Lemma 1 implies that $\text{top}(G) \in \mathcal{P}(G)$. From this lemma we also know that for every $v \in \mathcal{P}(G)$, $\delta^-(v) \geq \Delta(G) - 1$. We conclude that $\min\{\{\delta^-(v)\}_{v \in \mathcal{P}(G)}\} \geq \Delta(G) - 1$, and since this holds for every $G \in \mathcal{G}_n(d)$, the mechanism is 1-min-additive.

Finally, let $G = (V, E) \in \mathcal{G}_n(d)$, and suppose that $|\mathcal{P}(G)| > d+1$. If we denote $v_L = \operatorname{argmin}_{v \in \mathcal{P}(G)}\{(\delta^-(v), v)\}$, from Lemma 1 we know that $(v_L, w) \in E$ for every $w \in V$ with $(\delta^-(w), w) > (\delta^-(v_L), v_L)$, thus $\delta^+(v_L) \geq |\mathcal{P}(G)| - 1 > d$, a contradiction. We conclude that $|\mathcal{P}(G)| \leq d + 1$.

The following result, concerning mechanisms that may select an arbitrary number of vertices, follows immediately from Theorem 1.

Corollary 1. *For every $n \in \mathbb{N}$, there exists an impartial and 1-min-additive n-selection mechanism on $\mathcal{G}_n$.*

On $\mathcal{G}_n$, i.e., in the case of unbounded outdegrees, this result can in fact be improved slightly to guarantee 1-min-additivity while selecting only at most $n-1$ vertices. The improvement is achieved by a more intricate version of asymmetric plurality with runners-up, which we call *asymmetric plurality with runners-up and pivotal vertices*. We formally describe this mechanism in Algorithm 3 and denote its output for graph G by $\mathcal{P}^P(G)$. Given a graph $G = (V, E)$, call a vertex $u \in \mathcal{P}(G)$ *pivotal* for $v \in \mathcal{P}(G)$ if there exists a graph $G_{uv} \in \mathcal{N}_u(G)$ such that $v \notin \mathcal{P}(G_{uv})$, i.e., if the outgoing edges of u can be changed in such a way that v is no longer selected by asymmetric plurality with runners-up. Asymmetric plurality with runners-up and pivotal vertices then selects every vertex in $\mathcal{P}(G)$ that is pivotal for every other vertex in $\mathcal{P}(G)$. The mechanism turns out to inherit impartiality and 1-min-additivity, and to never select all vertices. The proof is omitted due to space constraints.

Theorem 2. *For every $n \in \mathbb{N}$ and $k \in \{n - 1, n\}$, there exists an impartial and 1-min-additive k-selection mechanism on $\mathcal{G}_n$.*

4   An Impossibility Result

When we established the existence of an impartial and 1-min-additive k-selection mechanism on $\mathcal{G}(d)$ whenever $k \geq d+1$, we claimed this result to be best possible in the sense that the additive guarantee cannot be improved. We will prove this claim, that impartiality is incompatible with the requirement to only select vertices with maximum indegree, as a corollary of a more general result.

While selecting only vertices with maximum indegree is a natural goal for mechanisms that select varying numbers of vertices, other natural objectives

Algorithm 3: Asymmetric plurality with runners-up and pivotal vertices $\mathcal{P}^{\mathrm{P}}(\mathsf{G})$

Input: Digraph $G = (V, E) \in \mathcal{G}_n$
Output: Set $S \subseteq V$ of selected vertices with $|S| \leq n - 1$
Let $S \leftarrow \emptyset$;
for $u \in \mathcal{P}(G)$ **do**
    if *for every* $v \in \mathcal{P}(G) \setminus \{u\}$ *there exists* $G_{uv} \in \mathcal{N}_u(G)$ *such that* $v \notin \mathcal{P}(G_{uv})$
    then
        $S \leftarrow S \cup \{u\}$
    end
end
Return S

---

exist for such mechanisms such as maximizing the median or mean indegree of the selected vertices. For both of these objectives, the mechanisms discussed in the previous section immediately provide upper bounds: if a k-selection mechanism always selects one vertex with maximum indegree and is α-min-additive then it is clearly α-median-additive and $\left(\frac{k-1}{k}\alpha\right)$-mean-additive; Theorem 1 thus implies the existence of a 1-median-additive and $\frac{k-1}{k}$-mean-additive k-selection mechanism on $\mathcal{G}(d)$, whenever $k \geq d + 1$. To improve on 1-median-additivity, it would be acceptable to select vertices with low indegree as long as a greater number of vertices with maximum indegree is selected at the same time. To improve on $\frac{k-1}{k}$-mean-additivity, it would suffice to select more than one vertex with maximum indegree whenever this is possible, and to otherwise select only a sublinear number in k of vertices with indegree equal to the maximum indegree minus one. The following result shows that no such improvements are possible.

Theorem 3. *Let* $n \in \mathbb{N}$, $n \geq 3$, $k \in [n]$, *and* $d \in [n-1]$. *Let* f *be an impartial* k-selection mechanism. If f is α_1-median-additive on $\mathcal{G}_n(d)$, then $\alpha_1 \geq 1/2(1 + \mathbb{1}(d \geq 3))$. If f is α_2-mean-additive on $\mathcal{G}_n(d)$, then $\alpha_2 \geq \lfloor\frac{d+1}{2}\rfloor / (\lfloor\frac{d+1}{2}\rfloor + 1)$.

Proof. Let n, k, and d be as in the statement of the theorem. In the following we suppose that there is an impartial k-selection mechanism f which is either α_1-median-additive on $\mathcal{G}_n(d)$ with $\alpha_1 < 1/2(1 + \mathbb{1}(d \geq 3))$, or α_2-mean-additive on $\mathcal{G}_n(d)$ with $\alpha_2 < \lfloor\frac{d+1}{2}\rfloor / (\lfloor\frac{d+1}{2}\rfloor + 1)$.

We first prove the result for the case $d = 1$. We consider the graph $G = (V, E) \in \mathcal{G}_n(1)$ with $E = \{(1,2), (2,3), (3,1)\}$, consisting of a 3-cycle and $n - 3$ isolated vertices. We consider as well, for $v \in \{1, 2, 3\}$, the graph $G_v = (V, E_v)$ where v deviates from the 3-cycle by changing its outgoing edge to the previous vertex in the cycle, i.e.,

$$E_1 = \{(1,3), (2,3), (3,1)\}, E_2 = \{(1,2), (2,1), (3,1)\}, E_3 = \{(1,2), (2,3), (3,2)\}.$$

Since f is α_1-median-additive with $\alpha_1 < 1/2$ or α_2-mean-additive with $\alpha_2 < 1/2$, we have that $f(G_1) = \{3\}$, $f(G_2) = \{1\}$, and $f(G_3) = \{2\}$. In particular, for $v \in \{1, 2, 3\}$, $v \notin f(G_v)$. Since for each $v \in \{1, 2, 3\}$ it holds $E_v \setminus (\{v\} \times$

$V) = E \setminus (\{v\} \times V)$, we conclude by impartiality that $v \notin f(G)$, and thus $f(G) \cap \{1, 2, 3\} = \emptyset$. This implies that both the median and the mean indegree of the vertices in $f(G)$ are 0, which contradicts the additive guarantee of this mechanism because $\Delta(G) = 1$.

In the following, we assume $d \geq 2$. We denote $D = [d+1]$ and consider in what follows two families of graphs with n vertices, K_v for each $v \in D$ and K_{uv} for each $u, v \in D, u \neq v$. They are constructed from a complete subgraph on D but deleting the outgoing edges of v, in the case of K_v, and the outgoing edges of u and v, in the case of K_{uv}. All the other vertices remain isolated. Formally, taking $V = [n]$ we define

$$K_v = (V, (D \setminus \{v\}) \times D) \text{ for every } v \in D,$$
$$K_{uv} = (V, (D \setminus \{u, v\}) \times D) \text{ for every } u, v \in D \text{ with } u \neq v.$$

If there is $v \in D$ such that $v \notin f(K_v)$, then

$$\text{median}\left(\{\delta^-(w, K_v)\}_{w \in f(K_v)}\right) \leq d - 1 = \Delta(K_v) - 1,$$
$$\text{mean}\left(\{\delta^-(w, K_v)\}_{w \in f(K_v)}\right) \leq d - 1 = \Delta(K_v) - 1,$$

which is a contradiction, so the result follows immediately. Therefore, in the following we assume that for every $v \in D$ we have $v \in f(K_v)$. We claim that for every $v \in D$,

$$|\{u \in D \setminus \{v\} : u \in f(K_v)\}| \geq \left\lfloor \frac{d+1}{2} \right\rfloor.$$

Let us see why the result follows if the claim holds. If this is the case, f selects one vertex with maximum indegree d in K_v and at least $\left\lfloor \frac{d+1}{2} \right\rfloor$ vertices with indegree $d - 1$. This yields both

$$\text{median}\left(\{\delta^-(w, K_v)\}_{w \in f(K_v)}\right) \leq \begin{cases} d - \frac{1}{2} \text{ if } d = 2 \\ d - 1 \text{ otherwise,} \end{cases}$$

and

$$\text{mean}\left(\{\delta^-(w, K_v)\}_{w \in f(K_v)}\right) \leq \frac{d + (d-1)\left\lfloor \frac{d+1}{2} \right\rfloor}{\left\lfloor \frac{d+1}{2} \right\rfloor + 1} = d - \frac{\left\lfloor \frac{d+1}{2} \right\rfloor}{\left\lfloor \frac{d+1}{2} \right\rfloor + 1},$$

which is a contradiction since $\Delta(K_v) = d$.

Now we prove the claim. Suppose that for every $v \in D$ we have $v \in f(K_v)$ and

$$|\{u \in D \setminus \{v\} : u \in f(K_v)\}| < \left\lfloor \frac{d+1}{2} \right\rfloor. \tag{2}$$

Let $v \in D$ and $u \in D \setminus \{v\}$ such that $u \notin f(K_v)$. Observing that

$$((D \setminus \{v\}) \times D) \setminus (\{u\} \times V) = ((D \setminus \{u, v\}) \times D) \setminus (\{u\} \times V),$$

we obtain from impartiality that $u \notin f(K_{uv})$. From the bounds on α_1 or α_2 that f satisfies by assumption, this mechanism has to select a vertex with maximum indegree in this graph; otherwise, both the median and the mean of the selected set would be at most $\Delta(K_{uv}) - 1$. Since $\delta^-(w) < \Delta(K_{uv})$ for every $w \notin \{u,v\}$, it holds $v \in f(K_{uv})$. Using impartiality once again, we conclude $v \in f(K_u)$. We have shown the following property:

$$\text{For every } u, v \in D : u \notin f(K_v) \Longrightarrow v \in f(K_u). \tag{3}$$

Consider now the graph $H = (D, F)$, where for each $u, v \in D$ with $u \neq v$, $(u, v) \in F$ if and only if $u \notin f(K_v)$. Property (2) implies that

$$\delta^-(v, H) > d - \left\lfloor \frac{d+1}{2} \right\rfloor \iff \delta^-(v, H) \geq d + 1 - \left\lfloor \frac{d+1}{2} \right\rfloor$$

for each $v \in D$. In particular, there has to be a vertex $v^* \in D$ such that $\delta^+(v^*, H) \geq d + 1 - \lfloor (d+1)/(2) \rfloor$ as well. For this vertex we have

$$\delta^+(v^*, H) + \delta^-(v^*, H) \geq 2 \left(d + 1 - \left\lfloor \frac{d+1}{2} \right\rfloor \right) \geq d + 1.$$

Since H has $d + 1$ vertices, this implies the existence of $w^* \in D$ for which $\{(v^*, w^*), (w^*, v^*)\} \subset F$, i.e., both $v^* \notin f(K_{w^*})$ and $w^* \notin f(K_{v^*})$. This contradicts (3), so we conclude the proof of the claim and the proof of the theorem.

□

Figure 2 provides an illustration of Theorem 3 for the case where $n = 3$, Fig. 3 for the case where $n = 4$.

The median of any set of numbers is an upper bound on their minimum. Therefore, if no impartial mechanism exists that is α-median-additive on $\mathcal{G}' \subseteq \mathcal{G}$ for $\alpha < \bar{\alpha}$, then no impartial mechanism can exist that is α-min-additive on $\mathcal{G}'$ for $\alpha < \lceil \bar{\alpha} \rceil$. We thus obtain the following impossibility result, which we have claimed previously.

Corollary 2. Let $n \in \mathbb{N}$, $n \geq 3$, and $k \in [n]$. Let f be an α-min-additive impartial k-selection mechanism on $\mathcal{G}_n$. Then $\alpha \geq 1$.

The impossibility results imply that for $k \geq d + 1$, the mechanisms of Sect. 3 are best possible for the minimum and median objectives except in a few boundary cases. When $n = 2$, selecting each of the two vertices if and only if it has an incoming edge is impartial and achieves 0-min-additivity and 0-median-additivity. When $n = 3$, it is possible to select in an impartial way at least one vertex with maximum indegree and at most one vertex with indegree equal to the maximum indegree minus one, thus guaranteeing 1/2-median-additivity. For the mean objective, the mechanisms of Sect. 3 are best possible asymptotically under the additional assumption that $k = O(d)$.

It is worth pointing out that the proof of the impossibility result uses graphs in which some vertices, in particular those with maximum indegree, do not have any outgoing edges. However, the impossibility extends naturally to the case where this cannot happen, corresponding to the practically relevant case in which abstentions are not allowed, as long as $n \geq 4$ and $d \geq 3$. For this it is enough to define $D = [d]$, add a new vertex with outgoing edges to every vertex in D and incoming edges from the vertices in D which do not have any outgoing edge, and construct a cycle containing the vertices in $V \setminus D$.

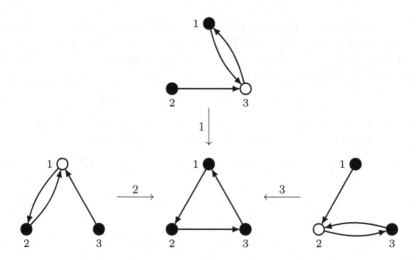

Fig. 2. Counterexample to the existence of an impartial 3-selection mechanism that is α-median-additive or α-mean-additive on $\mathcal{G}_3$ for $\alpha < 1/2$. Vertices drawn in white have to be selected, vertices in black cannot be selected. For the graphs at the top, on the left, and on the right, this follows from α-median-additivity or α-mean-additivity for $\alpha < 1/2$. An arrow with label v from one graph to another indicates that one can be obtained from the other by changing the outgoing edges of vertex v; by impartiality, the vertex thus has to be selected in both graphs or not selected in both graphs. It follows that no vertices are selected in the graph at the center, a contradiction to the claimed additive guarantee.

5 Trading Off Quantity and Quality

We have so far given impartial selection mechanisms for settings where the maximum outdegree d is smaller than the maximum number k of vertices that can be selected, and have shown that the mechanisms provide best possible additive guarantees in such settings. We will now consider settings where $d \geq k$, such that asymmetric plurality with runners-up selects too many vertices and therefore cannot be used directly. We obtain the following result.

Theorem 4. *For every $n \in \mathbb{N}$ and $k \in \{2, \ldots, n\}$, there exists an impartial and $(\lfloor (n-2)/(k-1) \rfloor + 1)$-min-additive k-selection mechanism on $\mathcal{G}_n$.*

The result is obtained by a variant of asymmetric plurality with runners-up in which some edges are deleted before the mechanism is run. In principle, deleting a certain number of edges can affect the additive guarantee by the same amount, if all of the deleted edges happen to be directed at the same vertex. By studying the structure of the set of vertices selected by the mechanism, we will instead be able to delete edges to distinct vertices and thus keep the negative impact on the additive guarantee under control.

The modified mechanism, which we call *asymmetric plurality with runners-up and edge deletion*, is formally described in Algorithm 4. It deletes any edges from

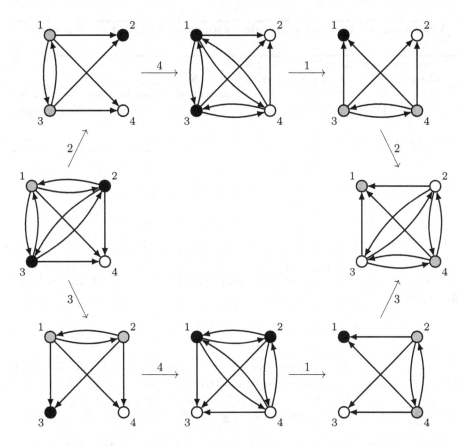

Fig. 3. Counterexample to the existence of an impartial 4-selection mechanism that is α_1-median-additive on $\mathcal{G}_4(3)$ for $\alpha_1 < 1$ or α_2-mean-additive on $\mathcal{G}_4$ for $\alpha_2 < 2/3$. Vertices drawn in white have to be selected, vertices in black cannot be selected, and vertices in gray may or may not be selected. For the graph on the left, this follows from α_1-median-additivity for $\alpha_1 < 1$ or α_2-mean-additivity for $\alpha_2 < 2/3$: under these assumptions at most one of the vertices with indegree 2 can be selected, which without loss of generality we can assume to be vertex 1. For the other graphs, it then follows by impartiality, and for the graph on the right yields a contradiction to the claimed additive guarantees. (Color figure online)

a vertex to the $\lfloor (n-2)/(k-1) \rfloor$ vertices preceding that vertex in the tie-breaking order, and applies asymmetric plurality with runners-up to the resulting graph. The following lemma shows that without such edges, the maximum number of vertices selected is reduced to k.

Lemma 2. *Let* $n \in \mathbb{N}$, $k \in \{2, \ldots, n\}$, *and* $r \in \mathbb{N}$ *with* $r \geq \lfloor (n-2)/(k-1) \rfloor$. *Let* $G = (V, E) \in \mathcal{G}_n$ *be such that for every* $u \in \{1, \ldots, n-1\}$ *and every* $v \in \{u+1, \ldots, \min\{u+r, n\}\}$, $(u, v) \notin E$. *Then,* $|\mathcal{P}(G)| \leq k$.

Algorithm 4: Asymmetric plurality with runners-up and edge deletion $\mathcal{P}^D(G)$

Input: Digraph $G = (V, E) \in \mathcal{G}_n$, $k \in \{2, \ldots, n\}$
Output: Set $S \subseteq V$ of selected vertices with $|S| \le k$
Let $r = \lfloor (n-2)/(k-1) \rfloor$;   // number of outgoing edges to remove
Let $R = \bigcup_{u=1}^{n-1} \bigcup_{v=u+1}^{\min\{u+r,n\}} \{(u,v)\}$;   // edges to be removed
Let $\bar{G} = (V, E \setminus R)$;
Return $\mathcal{P}(\bar{G})$

Proof. We let $S^i = \{v \in \mathcal{P}(G) : \delta^-(v) = \Delta(G) - i\}$ and $n^i = |S^i|$ for $i \in \{0, 1\}$, and we denote its elements in increasing order by v_j^i for $j \in [n^i]$, i.e.,

$$S^i = \{v_j^i\}_{j=1}^{n^i} \text{ with } v_1^i < v_2^i \cdots < v_{n^i}^i \text{ for each } i \in \{0, 1\}.$$

From Lemma 1, we know that $\mathcal{P}(G) = S^0 \cup S^1$, that for $i \in \{0, 1\}$ we have $(v_j^i, v_k^i) \in E$ for every j, k with $j < k$, and that $v_1^1 > v_{n^0}^0$. This allows to define, for $i \in \{0, 1\}$,

$$\bar{S}^i = \{v \in V \setminus S_i : v_1^i < v < v_{n^i}^i\}, \quad \bar{n}^i = |\bar{S}^i|,$$

such that $\bar{S}^0 \cap \bar{S}^1 = \emptyset$.

Fix $i \in \{0, 1\}$ and suppose that $n^i \ge 2$. Combining both the fact that $(v_j^i, v_k^i) \in E$ for every j, k with $j < k$, and that for every $u \in \{1, \ldots, n-1\}$ and $v \in \{u+1, \ldots, \min\{u+r, n\}\}$, $(u, v) \notin E$, we have that for every $j \in [n^i - 1]$ it holds $v_{j+1}^i - v_j^i \ge r+1$. Summing over j yields $v_{n^i}^i - v_1^i \ge (n^i - 1)(r+1)$, hence

$$\bar{n}^i = v_{n^i}^i - v_1^i + 1 - n^i \ge (n^i - 1)(r+1) + 1 - n^i = (n^i - 1)r,$$

where the first equality comes from the definition of the set $\bar{S}^i$. This implies $n^i \le 1 + \bar{n}^i/r$. We can now lift the assumption $n^i \ge 2$, since when $n^i = 1$ we have $\bar{n}^i = 0$ and the inequality holds as well, and write the following chain of inequalities:

$$|\mathcal{P}(G)| = n^0 + n^1 \le 2 + \frac{\bar{n}^0 + \bar{n}^1}{r} \le 2 + \frac{n - |\mathcal{P}(G)|}{r},$$

where the last inequality comes from the fact that all the sets S^0, S^1, $\bar{S}^0$, $\bar{S}^1$ are disjoint and therefore their cardinalities sum up to at most n. This bounds the number of selected vertices as $|\mathcal{P}(G)| \le (2r + n)/(r + 1)$.

Suppose now that $|\mathcal{P}(G)| \ge k + 1$. Using the previous bound, this yields

$$2r + n \ge (k+1)(r+1) \iff r \le \frac{n - k - 1}{k - 1} = \frac{n - 2}{k - 1} - 1,$$

which contradicts the lower bound on r in the statement of the lemma.   □

Figure 4 illustrates the argument and notation of Lemma 2. We are now ready to prove Theorem 4.

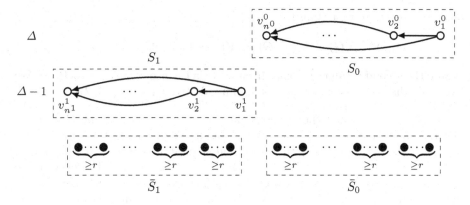

Fig. 4. Illustration of Lemma 2. There are no edges from a vertex to any of the r vertices to its left, which means that for each vertex in S_0 or S_1, except for the leftmost vertex, there exist are at least r vertices outside these sets. Such vertices are not arranged according to their indegrees, and edges from vertices in S_1 to every vertex in S_0 have been omitted for clarity.

Proof of Theorem 4. We show that Algorithm 4 satisfies the conditions of the theorem. Let $n \in \mathbb{N}$ and $k \in \{2, \ldots, n\}$. Impartiality follows from the fact that Algorithm 2 is impartial, thus the potential deletion of outgoing edges of a given vertex cannot affect the fact of selecting this vertex or not. Formally, if $G = (V, E)$, $v \in V$ and $G' = (V, E') \in \mathcal{N}_v(G)$, then defining $\bar{G} = (V, \bar{E})$ and $\bar{G}' = (V, \bar{E}')$ as the graphs constructed when running Algorithm 4 with G and G' as input graphs, respectively, we have

$$\bar{E} \setminus (\{v\} \times V) = (E \setminus (\{v\} \times V)) \setminus \left(\bigcup_{u=1}^{n-1} \bigcup_{w=u+1}^{\min\{u+r,n\}} \{(u,w)\} \right)$$

$$= (E' \setminus (\{v\} \times V)) \setminus \left(\bigcup_{u=1}^{n-1} \bigcup_{w=u+1}^{\min\{u+r,n\}} \{(u,w)\} \right)$$

$$= \bar{E}' \setminus (\{v\} \times V),$$

where we use that $G' \in \mathcal{N}_v(G)$. Impartiality then follows directly from impartiality of plurality with runners-up. For the following, let $G = (V, E) \in \mathcal{G}_n$ and define r and $\bar{G}$ as in the mechanism. Since the first step of the mechanism ensures that for every $u \in \{1, \ldots, n-1\}$ and every $v \in \{u+1, \ldots, \min\{u+r, n\}\}$, $(u, v) \notin E$, Lemma 2 implies that $|\mathcal{P}^D(G)| = |\mathcal{P}(\bar{G})| \leq k$. Finally, in order to show the additive guarantee we first note that, for every $v \in V$, $\delta^-(v, G) \leq \delta^-(v, \bar{G}) + r$, since at most $|\{v-r, \ldots, v-1\} \cap V| \leq r$ incoming edges of v are deleted when defining $\bar{G}$ from G. In particular, $\Delta(G) \leq \Delta(\bar{G}) + r$. Using this observation and denoting $v^* \in \operatorname{argmin}_{v \in \mathcal{P}^D(G)} \{\delta^-(v, G)\}$ an arbitrary element with minimum indegree among those selected by asymmetric plurality with runners-up and edge deletion, we obtain that

$$\delta^-(v^*, G) \geq \delta^-(v^*, \bar{G}) \geq \Delta(\bar{G}) - 1 \geq \Delta(G) - r - 1,$$

where the second inequality comes from Lemma 1, since v^* belongs to $\mathcal{P}(\bar{G})$. We conclude that the mechanism is $(r+1)$-min-additive for $r = \lfloor (n-2)/(k-1) \rfloor$. $\square$

It is easy to see that the previous analysis is tight from a graph $G = (V, E)$ where exactly $r = \lfloor (n-2)/(k-1) \rfloor$ incoming edges of the top-voted vertex are deleted, and a vertex with the second highest indegree u such that $u >$ top(G), $(u, \text{top}(G)) \in E$, and $\delta^-(u) = \Delta(G) - r - 1$ is selected. However, we do not know whether the tradeoff provided by Theorem 4 is best possible for any impartial mechanism, and the question for the optimum tradeoff is an interesting one. Currently, when $d \geq k$ a gap remains between the upper bound of $\lfloor (n-2)/(k-1) \rfloor + 1$ and a lower bound of 1, which is relatively large when the number k of vertices that can be selected is small. We may, alternatively, also ask for the number of vertices that have to be selected in order to guarantee 1-min-additivity. Currently, the best upper bound on this number is $n - 1$.

In addition to the question about the performance of the mechanism introduced in this section, the sole fact that sometimes it does not select vertices with indegree strictly higher than the one of other selected vertices may seem unfair. Unfortunately, this is unavoidable whenever $d \geq k$ and α-min-additivity is imposed for some $\alpha < d$, as one can see from a graph consisting of a complete subgraph on $d + 1$ vertices and $n - (d+1)$ isolated vertices. For any k-selection mechanism, a vertex in the complete subgraph is not selected, and impartiality forces us to not select it either when its outgoing edges are deleted and it is the unique top-voted vertex.

Acknowledgments. The authors have benefitted from discussions with David Hannon. Research was supported by the Deutsche Forschungsgemeinschaft under project number 431465007 and by the Engineering and Physical Sciences Research Council under grant EP/T015187/1.

References

1. Alon, N., Fischer, F., Procaccia, A., Tennenholtz, M.: Sum of us: strategyproof selection from the selectors. In: Proceedings of the 13th Conference on Theoretical Aspects of Rationality and Knowledge, pp. 101–110 (2011)
2. Aziz, H., Lev, O., Mattei, N., Rosenschein, J.S., Walsh, T.: Strategyproof peer selection using randomization, partitioning, and apportionment. Artif. Intell. **275**, 295–309 (2019)
3. Babichenko, Y., Dean, O., Tennenholtz, M.: Incentive-compatible selection mechanisms for forests. In: Proceedings of the 21st ACM Conference on Economics and Computation, pp. 111–131 (2020)
4. Bjelde, A., Fischer, F., Klimm, M.: Impartial selection and the power of up to two choices. ACM Trans. Econ. Comput. **5**(4), 1–20 (2017)

5. Bousquet, N., Norin, S., Vetta, A.: A near-optimal mechanism for impartial selection. In: Liu, T.-Y., Qi, Q., Ye, Y. (eds.) WINE 2014. LNCS, vol. 8877, pp. 133–146. Springer, Cham (2014). https://doi.org/10.1007/978-3-319-13129-0_10

6. Caragiannis, I., Christodoulou, G., Protopapas, N.: Impartial selection with additive approximation guarantees. In: Fotakis, D., Markakis, E. (eds.) SAGT 2019. LNCS, vol. 11801, pp. 269–283. Springer, Cham (2019). https://doi.org/10.1007/978-3-030-30473-7_18

7. Caragiannis, I., Christodoulou, G., Protopapas, N.: Impartial selection with prior information. arXiv preprint arXiv:2102.09002 (2021)

8. Cembrano, J., Fischer, F., Hannon, D., Klimm, M.: Impartial selection with additive guarantees via iterated deletion. arXiv preprint arXiv:2205.08979 (2022)

9. De Clippel, G., Moulin, H., Tideman, N.: Impartial division of a dollar. J. Econ. Theory **139**(1), 176–191 (2008)

10. Fischer, F., Klimm, M.: Optimal impartial selection. SIAM J. Comput. **44**(5), 1263–1285 (2015)

11. Holzman, R., Moulin, H.: Impartial nominations for a prize. Econometrica **81**(1), 173–196 (2013)

12. Kahng, A., Kotturi, Y., Kulkarni, C., Kurokawa, D., Procaccia, A.D.: Ranking wily people who rank each other. In: Proceedings of the 32nd AAAI Conference on Artificial Intelligence (2018)

13. Kurokawa, D., Lev, O., Morgenstern, J., Procaccia, A.D.: Impartial peer review. In: Proceedings of the 24th International Joint Conference on Artificial Intelligence (2015)

14. Mackenzie, A.: Symmetry and impartial lotteries. Games Econom. Behav. **94**, 15–28 (2015)

15. Mackenzie, A.: An axiomatic analysis of the papal conclave. Econ. Theor. **69**, 713–743 (2020)

16. Mattei, N., Turrini, P., Zhydkov, S.: PeerNomination: relaxing exactness for increased accuracy in peer selection. arXiv preprint arXiv:2004.14939 (2020)

17. Tamura, S.: Characterizing minimal impartial rules for awarding prizes. Games Econom. Behav. **95**, 41–46 (2016)

18. Tamura, S., Ohseto, S.: Impartial nomination correspondences. Soc. Choice Welfare **43**(1), 47–54 (2014)

19. Wąs, T., Rahwan, T., Skibski, O.: Random walk decay centrality. In: Proceedings of the AAAI Conference on Artificial Intelligence, vol. 33, pp. 2197–2204 (2019)

20. Xu, Y., Zhao, H., Shi, X., Zhang, J., Shah, N.B.: On strategyproof conference peer review. arXiv preprint arXiv:1806.06266 (2018)

21. Zhang, X., Zhang, Y., Zhao, D.: Incentive compatible mechanism for influential agent selection. In: Caragiannis, I., Hansen, K.A. (eds.) SAGT 2021. LNCS, vol. 12885, pp. 79–93. Springer, Cham (2021). https://doi.org/10.1007/978-3-030-85947-3_6

Improved Approximation to First-Best Gains-from-Trade

Yumou Fei[✉]

Peking University, Beijing, China
2000010769@stu.pku.edu.cn

Abstract. We study the two-agent single-item bilateral trade. Ideally, the trade should happen whenever the buyer's value for the item exceeds the seller's cost. However, the classical result of Myerson and Satterthwaite showed that no mechanism can achieve this without violating one of the Bayesian incentive compatibility, individual rationality and weakly balanced budget conditions. This motivates the study of approximating the trade-whenever-socially-beneficial mechanism, in terms of the expected gains-from-trade. Recently, Deng, Mao, Sivan, and Wang showed that the random-offerer mechanism achieves at least a $1/8.23$ approximation. We improve this lower bound to $1/3.15$ in this paper. We also determine the exact worst-case approximation ratio of the seller-pricing mechanism assuming the distribution of the buyer's value satisfies the monotone hazard rate property.

Keywords: Bilateral trade · Mechanism design · Approximation algorithm

1 Introduction

Two-sided markets, with strategic players on both the sell-side and the buy-side, has been an important research topic in economics. This paper considers the simplest model of such market, the two-agent single-item bilateral trade.

1.1 Model

Suppose there is a single seller and a single buyer on the market. There is an item held by the seller and to be sold to the buyer. The buyer's private value for the item is a random variable v with cumulative distribution function (CDF) F, and the seller's private cost of selling it out is a random variable c with CDF G. We assume that v and c are independent, and that F and G have finite first moments. We need to design a mechanism which, on input v and c, decides whether the transaction should happen. Let $x(\cdot, \cdot)$ be a (Borel-measurable) function from $\mathbb{R}^2$ to $[0, 1]$, which denotes the trading probability decided by the mechanism, or the "allocation rule". The *gains-from-trade* (GFT), or the expected social utility gain from trading, is defined as

$$\mathsf{GFT} := \mathop{\mathbb{E}}_{\substack{v \sim F \\ c \sim G}} \left[(v - c) \cdot x(v, c) \right].$$

Thinking of F and G as given, what choice of the function x maximizes GFT? It is clear that whatever F and G are, the quantity GFT is always maximized at $x^*(v, c) := \mathbb{1}\{v \geq c\}$. However, in reality, we have to take into consideration the strategic behavior of both sides of the market. As in many mechanism design problems, in order to make an allocation rule Bayesian incentive compatible (BIC) and individually rational (IR), we need to include a "money transfer" rule. But unlike in the case of one-sided auctions, the "payment" from the sell-side can be negative in two-sided markets. Thus, in addition to the commonly studied BIC and IR conditions, the "balance of budget" is another important consideration in the design of bilateral trade mechanisms. In particular, we are concerned with the following requirement:

Definition 1. *If the payment from the buyer is always at least the revenue of the seller, we say the mechanism is weakly budget balanced (WBB).*

Myerson and Satterthwaite [12] show that for very general F and G, the ideal allocation rule $x^*(v, c) = \mathbb{1}\{v \geq c\}$ cannot be made into a BIC, IR and WBB mechanism. But on the other hand, this idealism does offer a benchmark which we can try to approximate using mechanisms with these properties. We call the gains-from-trade achieved by $x^*(v, c)$ the *first-best* GFT. Namely, define

$$\mathsf{FB} := \mathop{\mathbb{E}}_{\substack{v \sim F \\ c \sim G}} \left[(v - c) \cdot \mathbb{1}\{v \geq c\} \right].$$

The maximum GFT achievable by Bayesian incentive compatible, individually rational and weakly budget balanced mechanisms is denoted SB, or the *second-best*. Myerson and Satterthwaite [12] also describe a BIC, IR and WBB mechanism that actually achieves the second-best gains-from-trade. However, this second-best mechanism is complicated and difficult to implement in practice. As a consequence, the understanding of the second-best mechanism and the search for simple and practical alternatives have become major research problems. The following are some of the most studied simple mechanisms:

- The fixed-price mechanism. A fixed price p is set and the transaction happens iff $v \geq p \geq c$. Formally, the allocation rule is $x(v, c) = \mathbb{1}\{v \geq p \geq c\}$ and the buyer pays the seller the price p. The resulting gains-from-trade is

$$\mathsf{FixedP} = \sup_{p \in \mathbb{R}} \int_{-\infty}^{p} \int_{p}^{+\infty} (v - c) \, \mathrm{d}F(v) \, \mathrm{d}G(c).$$

The fixed price mechanism has the additional advantage that it's dominant strategy incentive compatible (DSIC).

- The seller-pricing mechanism. The seller gets the right to set the price and the buyer can only decide whether or not to buy the item and pay this price. The buyer chooses to buy it iff $v \geq p$. Knowing his private cost c, the seller sets the price p that maximizes $(p - c)(1 - F(p^-))$, where $F(p^-) := \lim_{x \to p^-} F(x)$. The resulting gains-from-trade is

$$\mathsf{SellerP} = \int_{-\infty}^{+\infty} \int_{p_c}^{\infty} (v - c) \, dF(v) \, dG(c), \quad \text{where } p_c \in \arg\max_p (p - c)(1 - F(p^-)).$$

- The buyer-pricing mechanism. The buyer gets the right to set the price and the seller can only decide whether or not to sell the item and take this price. The seller chooses to sell it iff $c \leq p$. Knowing his private value v, the buyer sets the price p that maximizes $(v - p)G(p)$. The resulting gains-from-trade is

$$\mathsf{BuyerP} = \int_{-\infty}^{+\infty} \int_{-\infty}^{p_v} (v - c) \, dG(c) \, dF(v), \quad \text{where } p_v \in \arg\max_p (v - p)G(p).$$

- The random-offerer mechanism. Flip a fair coin to decide who gets to set the price. This is a 50:50 mixture of the seller pricing mechanism and the buyer price mechanism, which extracts a gains-from-trade of

$$\mathsf{RandOff} = \frac{1}{2}\mathsf{SellerP} + \frac{1}{2}\mathsf{BuyerP}.$$

It's well-known that the quantities BuyerP and SellerP are symmetric to each other. To make it clear, we make the following formal statement:

Proposition 1. *Let F and G be the CDFs of two distributions on $\mathbb{R}$, each with finite first moment. If we let $F^{\text{neg}}(x) = 1 - F((-x)^-)$ be the CDF for the negated distribution and similarly let $G^{\text{neg}} = 1 - G((-x)^-)$, we have*

$$\mathsf{SellerP}(G, F) = \mathsf{BuyerP}(F^{\text{neg}}, G^{\text{neg}})$$

and of course

$$\mathsf{FB}(G, F) = \mathsf{FB}(F^{\text{neg}}, G^{\text{neg}}).$$

Remark 1. Notice that in our description of the model, the distribution of the buyer's or the seller's value is not assumed to be nonnegative, so negating them will not be a problem.

Proof. Since

$$\arg\max_p (v - p)G^{\text{neg}}(p)$$

$$= \arg\max_p (v - p)\left(1 - G((-p)^-)\right)$$

$$= \arg\max_p \left((-p) - (-v)\right)\left(1 - G((-p)^-)\right)$$

$$= -\arg\max_p \left(p - (-v)\right)\left(1 - G(p^-)\right),$$

for each possible function $v \mapsto p_v$ in the definition of $\mathsf{BuyerP}(F^{\mathrm{neg}}, G^{\mathrm{neg}})$, the function $c \mapsto -p_{-c}$ is a valid function in the definition of $\mathsf{SellerP}(G, F)$. So

$$
\begin{aligned}
\mathsf{SellerP}(G, F) &= \int_{-\infty}^{+\infty} \int_{-p_{-c}}^{\infty} (v - c)\, \mathrm{d}F(v)\, \mathrm{d}G(c) \\
&= \int_{-\infty}^{+\infty} \int_{-\infty}^{p_{-c}} (-c - v)\, \mathrm{d}F^{\mathrm{neg}}(v)\, \mathrm{d}G(c) \\
&= \int_{-\infty}^{+\infty} \int_{-\infty}^{p_c} (c - v)\, \mathrm{d}F^{\mathrm{neg}}(v)\, \mathrm{d}G^{\mathrm{neg}}(c) \\
&= \mathsf{BuyerP}(F^{\mathrm{neg}}, G^{\mathrm{neg}}).
\end{aligned}
$$

Remark 2. The preceding proposition is obvious from the intuitive understanding that in terms of gains-from-trade, the role of the buyer and the seller is symmetric. It's worth noting that if another well-studied quantity–the welfare (see Sect. 1.4) is concerned, then the symmetry no longer holds.

1.2 Our Results

In this paper, we focus on the following problems:

Problem 1. Among all problems surrounding the bilateral trade setting, one basic open problem is to determine the worst-case approximation ratio of SB to FB, i.e.

$$
\inf_{F,G} \frac{\mathsf{SB}}{\mathsf{FB}}.
$$

Problem 2. Seeing that the second best mechanism is usually too complex to be practical, it's natural to ask the same thing for simple mechanisms: Are there some simple BIC, IR and WBB mechanisms that achieve good approximations to the first best GFT?

It was not until the recent groundbreaking work of Deng, Mao, Sivan, and Wang [6] that the second-best gains-from-trade was shown to provide a constant approximation to the first-best, i.e. $\inf_{F,G}(\mathsf{SB}/\mathsf{FB}) > 0$. In fact, they attack Problem 1 by attacking Problem 2 instead: they show that $\mathsf{FB} \leq 8.23 \cdot \mathsf{RandOff}$. Since $\mathsf{SB} \geq \mathsf{RandOff}$, this implies that $\mathsf{FB} \leq 8.23 \cdot \mathsf{SB}$. In other words, we can write

$$
\inf_{F,G} \frac{\mathsf{SB}}{\mathsf{FB}} \geq \inf_{F,G} \frac{\mathsf{RandOff}}{\mathsf{FB}} \geq \frac{1}{8.23} \approx 0.121.
$$

In the hardness direction, Leininger, Linhart, and Radner [9] and Blumrosen and Mizrahi [3] exhibit distributions for which the ratio SB/FB is $2/e \approx 0.736$. Babaioff, Dobzinski, and Kupfer [1] show that $\inf_{F,G}(\mathsf{RandOff}/\mathsf{FB}) \leq 0.495$. Cai, Goldner, Ma, and Zhao [5] (in their arxiv version) independently give a proof that $\inf_{F,G}(\mathsf{RandOff}/\mathsf{FB}) \leq \frac{1}{2} - \varepsilon$ for some constant $\varepsilon > 0$.

This leaves a relatively large gap between the lower bound of 0.121 and the upper bound of 0.736 for Problem 1. For the approximation ratio of RandOff, the gap is left at $[0.121, 0.495]$. In Sect. 2, we improve the lower bounds of both problems from $1/8.23 \approx 0.121$ to $1/3.15 \approx 0.317$, as another step towards the ultimate goal of closing the gaps.

Theorem 1. *For any pair of distributions F and G, the inequality* FB $\leq 3.15 \cdot$ RandOff *holds.*

In contrast to the constant approximation provided by the random-offerer mechanism, most of the other simple mechanisms, including the buyer-pricing mechanism, the seller-pricing mechanism and the fixed-price mechanism, cannot approximate FB in general. For the seller-pricing mechanism, it's well known that taking G to be a single point mass at 0 and F an equal-revenue-distribution makes the ratio SellerP/FB tend to 0. Since BuyerP is symmetric to SellerP, the buyer-pricing mechanism also has an approximation ratio of 0. The case of the fixed-price mechanism follows from taking F and G to be exponential distributions, as is shown by Blumrosen and Dobzinski [2]. Partly compensating for this hardness of approximation, there is another type of results concerned with the setting where F and G are subject to some restrictions. For example, by McAfee [10] and Kang and Vondrák [7],

$$\inf_{F=G} \frac{\mathsf{FixedP}}{\mathsf{FB}} = \frac{1}{2}.$$

As another example, Blumrosen and Mizrahi [3] consider a "monotone hazard rate" condition (see Definition 2) on F, and prove that

$$0.368 \approx \frac{1}{e} \leq \inf_{\substack{F \in \mathcal{MHR} \\ G}} \frac{\mathsf{SellerP}}{\mathsf{FB}}.$$

In Sect. 3, we show that the approximation ratio of SellerP to FB assuming monotone hazard rate of the buyer's distribution can be determined exactly (to be $1/(e-1) \approx 0.582$):

Theorem 2. *If $F \in \mathcal{MHR}$, then* FB $\leq (e-1) \cdot$ SellerP, *and the constant $(e-1)$ is optimal.*

1.3   Main New Ideas

The main new idea involved in our improvements is that the power of Fubini-Tonelli theorem, or equivalently integration by parts, can be better utilized.

It is fair to say that Fubini-Tonelli theorem is visually "obvious", but certainly it's not mathematically trivial. In a visual image of a double integral, applying the theorem often represents a change of viewpoint, for example from the buyer's point of view to the seller's (see Lemma 1, which already appears in [6]).

However, somewhat neglected is the power of successive applications of the theorem. If, after a Fibini-type transformation, the same quantity can be represented in a different visual image, then another application of the theorem might combine with the previous transformation into a deeper nontrivial effect, and lead to unexpected new equality results. This is what happens in our Lemma 2.

Our proof in Sect. 2 is based on the same framework as Deng, Mao, Sivan, and Wang's work [6], and our Lemma 1 is directly taken from their work. The difference between this paper and theirs lies in our Lemma 2 and Lemma 3, which replace the "Lemma 3.1" of their paper, which is the technically major part of the analysis.

The proof in Sect. 3 is relatively easier, as there is essentially only one visual image where changes of integration order take place. To get a sense of the image, see Fig. 1. Although the result is a sharpening of Blumrosen and Mizrahi's work [3], our proof in Sect. 3 is basically independent of their work, and most of the involved equations and lemmas are new, except for one place, as explained in Remark 5.

1.4 More Related Work

One might also ask about how much gains-from-trade would be lost if we use simple mechanisms instead of the known second-best mechanism. That is, the worst-case approximation ratios of simple mechanisms to SB are also of interest. Brustle, Cai, Wu, and Zhao [4] show that $\inf_{F,G}(\mathsf{RandOff}/\mathsf{SB}) = \frac{1}{2}$.

There is also a line of research concerning the approximation of welfare, instead of gains-from-trade. Since we have

$$\mathsf{Welfare} = \mathop{\mathbb{E}}_{\substack{v\sim F \\ c\sim G}} \left[v \cdot x(v,c) + c \cdot \left(1 - x(v,c)\right)\right] = \mathop{\mathbb{E}}_{c\sim G}[c] + \mathsf{GFT},$$

maximizing welfare is equivalent to maximizing gains from trade. In addition, providing a constant approximation to the first-best GFT also provides a constant approximation to the first-best welfare (but not vice versa). Blumrosen and Dobzinski [2] show that the fixed-price mechanism provides a $\left(1 - \frac{1}{e}\right)$-approximation to the first-best welfare (improved to $1 - \frac{1}{e} + 0.0001$ by Kang, Pernice and Vondrák [8]). Blumrosen and Mizrahi [3] show that the first-best welfare is inapproximable to above a fraction of 0.934 by BIC, IR and WBB mechanisms.

2 Bounding the Approximation Ratio

In this section, we will prove the main result $\mathsf{FB} \leq 3.15 \cdot \mathsf{RandOff}$. As pointed out by Deng, Mao, Sivan, and Wang [6], as far as SellerP, BuyerP and FB are concerned, it is without loss of generality to assume that the distribution of v and c are supported on $[0,1]$ and have continuous and positive densities.

2.1 Notational Preparations

1. The major advantage of assuming the existence of positive densities is that we can define the following "quantile function", a powerful tool for analysis introduced by Deng, Mao, Sivan, and Wang [6]: for each $x \in [0,1]$, define $\mu(x)$ to be the $(1 - \lambda)$-quantile of $F|_{\geq x}$, i.e.

$$\mu(x) = F^{-1}\left(1 - \lambda + \lambda F(x)\right),$$

where $\lambda \in (0,1)$ is a parameter to be chosen. Since F is strictly increasing and continuous on $[0,1]$, so are F^{-1} and μ. Then, since μ is strictly increasing and continuous on $[0,1]$, we can also define its inverse μ^{-1} on $[\mu(0), \mu(1)] = [\mu(0), 1]$.

2. Let SProfit denote the maximum profit that the seller can gain in a seller-pricing mechanism, and let BProfit denote the maximum utility that the buyer can gain in a buyer-pricing mechanism. Obviously, SProfit $\leq$ SellerP and BProfit $\leq$ BuyerP. Since the quantities SProfit and BProfit are much easier to handle than SellerP and BuyerP, we will prove the result FB $\leq 3.15 \cdot \left(\frac{1}{2}\text{SellerP} + \frac{1}{2}\text{BuyerP}\right)$ by showing FB $\leq 3.15 \cdot \left(\frac{1}{2}\text{SProfit} + \frac{1}{2}\text{BProfit}\right)$ instead.

2.2 Proof of Main Theorem

The main theme in the proof is to put everything as an integration over the seller's cost c. Note that the definition of SProfit is already of this form, and FB can also easily be expressed in this form. Deng, Mao, Sivan, and Wang [6] notice that BProfit can also be put into this form (at the expense of possibly shrinking it a little). We state it as follows:

Lemma 1. *Let μ be the quantile function defined by any $\lambda \in (0,1)$. Then*

$$\int_0^1 \int_{\mu(c)}^1 \left(s - \mu^{-1}(s)\right) dF(s)\, dG(c) \leq \text{BProfit}.$$

Proof. By Fubini-Tonelli theorem, we can change the order of integration as follows:

$$\int_0^1 \int_{\mu(c)}^1 \left(s - \mu^{-1}(s)\right) dF(s)\, dG(c)$$

$$= \int_{\mu(0)}^1 \int_0^{\mu^{-1}(v)} \left(v - \mu^{-1}(v)\right) dG(c)\, dF(v)$$

$$= \int_{\mu(0)}^1 \left(v - \mu^{-1}(v)\right) G(\mu^{-1}(v))\, dF(v)$$

$$\leq \int_{\mu(0)}^1 \max_p (v - p) G(p)\, dF(v)$$

$$\leq \text{BProfit}.$$

In light of Lemma 1, we can now put all three quantities FB, SProfit and BProfit into integrations over $dG(c)$. The integrands are:

$$\text{FB}(c) := \int_c^1 (v - c)\, dF(v),$$

$$\text{SProfit}(c) := \max_p (p - c)(1 - F(p)),$$

$$\text{BProfit}(c) := \int_{\mu(c)}^1 \left(s - \mu^{-1}(s)\right) dF(s).$$

We have (the first two directly follow from the definition of FB and SProfit, while the third follows from Lemma 1)

$$\mathsf{FB} = \int_0^1 \mathsf{FB}(c)\,\mathrm{d}G(c),$$

$$\mathsf{SProfit} = \int_0^1 \mathsf{SProfit}(c)\,\mathrm{d}G(c),$$

$$\mathsf{BProfit} \geq \int_0^1 \mathsf{BProfit}(c)\,\mathrm{d}G(c).$$

However, the expression of $\mathsf{BProfit}(c)$ looks nothing like $\mathsf{FB}(c)$ or $\mathsf{SProfit}(c)$. Our next step is to transform it into a more familiar form:

Lemma 2. *Let μ be the quantile function defined by any $\lambda \in (0,1)$. Then for any $c \in [0,1]$,*

$$\mathsf{BProfit}(c) = (1-\lambda) \cdot \mathsf{FB}(c) - \int_c^{\mu(c)} (v-c)\,\mathrm{d}F(v).$$

Proof. This follows from successive uses of Fubini-Tonelli theorem:

$\mathsf{BProfit}(c)$

$$= \int_{\mu(c)}^1 (s - \mu^{-1}(s))\,\mathrm{d}F(s)$$

$$= \int\int_{\substack{\mu(c)\leq s\leq 1 \\ \mu^{-1}(s)\leq t\leq s}} 1\,\mathrm{d}t\,\mathrm{d}F(s)$$

$$= \int\int_{\substack{c\leq t\leq \mu(c) \\ \mu(c)\leq s\leq \mu(t)}} 1\,\mathrm{d}t\,\mathrm{d}F(s) + \int\int_{\substack{\mu(c)<t<1 \\ t\leq s\leq \mu(t)}} 1\,\mathrm{d}t\,\mathrm{d}F(s)$$

$$= \int_c^{\mu(c)} \Big(F(\mu(t)) - F(\mu(c))\Big)\,\mathrm{d}t + \int_{\mu(c)}^1 \Big(F(\mu(t)) - F(t)\Big)\,\mathrm{d}t$$

$$= \int_c^{\mu(c)} \Big(F(\mu(t)) - F(t)\Big)\,\mathrm{d}t - \int_c^{\mu(c)} \Big(F(\mu(c)) - F(t)\Big)\,\mathrm{d}t + \int_{\mu(c)}^1 \Big(F(\mu(t)) - F(t)\Big)\,\mathrm{d}t$$

$$= \int_c^1 \Big(F(\mu(t)) - F(t)\Big)\,\mathrm{d}t - \int_c^{\mu(c)} \Big(F(\mu(c)) - F(t)\Big)\,\mathrm{d}t$$

$$= \int_c^1 (1-\lambda)\Big(1 - F(t)\Big)\,\mathrm{d}t - \int_c^{\mu(c)} \Big(F(\mu(c)) - F(t)\Big)\,\mathrm{d}t$$

$$= (1-\lambda)\int_c^1 \int_t^1 1\,\mathrm{d}F(v)\,\mathrm{d}t - \int_c^{\mu(c)} \int_t^{\mu(c)} 1\,\mathrm{d}F(v)\,\mathrm{d}t$$

$$= (1-\lambda)\int_c^1 \int_c^v 1\,\mathrm{d}t\,\mathrm{d}F(v) - \int_c^{\mu(c)} \int_c^v 1\,\mathrm{d}t\,\mathrm{d}F(v)$$

$$= (1-\lambda)\int_c^1 (v-c)\,\mathrm{d}F(v) - \int_c^{\mu(c)} (v-c)\,\mathrm{d}F(v)$$

$$= (1-\lambda) \cdot \mathsf{FB}(c) - \int_c^{\mu(c)} (v-c)\,\mathrm{d}F(v).$$

Remark 3. The work by Deng, Mao, Sivan, and Wang [6] also contains a major part that transforms $\mathsf{BProfit}(c)$ into a nicer form. A key step in their transformation is to use summations to bound integrations, which incurs a loss in the constant. In comparison, our Lemma 2 only involves integrations and is completely lossless.

Note that the minuend on the right hand side of the preceding lemma is already a fraction of $\mathsf{FB}(c)$. The next lemma shows that the diminution caused by the subtrahend is under control:

Lemma 3. *Let μ be the quantile function defined by any $\lambda \in (0,1)$. Then for any $c \in [0,1]$,*

$$\int_c^{\mu(c)} (v-c)\, \mathrm{d}F(v) \le \ln\frac{1}{\lambda} \cdot \mathsf{SProfit}(c).$$

Proof. We resort to the definition of $\mathsf{SProfit}(c)$:

$$\int_c^{\mu(c)} (v-c)\, \mathrm{d}F(v) \le \int_c^{\mu(c)} \frac{1}{1-F(v)} \cdot \max_p (p-c)(1-F(p))\, \mathrm{d}F(v)$$

$$= \mathsf{SProfit}(c) \cdot \int_c^{\mu(c)} \frac{\mathrm{d}F(v)}{1-F(v)}$$

$$= \mathsf{SProfit}(c) \cdot \ln\left(\frac{1-F(c)}{1-F(\mu(c))}\right)$$

$$= \ln\frac{1}{\lambda} \cdot \mathsf{SProfit}(c)$$

Now we can prove the theorem by combining the preceding three lemmas.

Theorem 3. $\mathsf{FB} \le 3.15 \cdot \mathsf{RandOff}$.

Proof. Let μ be the quantile function defined by any $\lambda \in (0,1)$. We have

$$\mathsf{FB}(c) = \frac{1}{1-\lambda}\left((1-\lambda)\cdot\mathsf{FB}(c) - \int_c^{\mu(c)}(v-c)\,\mathrm{d}F(v)\right) + \frac{1}{1-\lambda}\int_c^{\mu(c)}(v-c)\,\mathrm{d}F(v)$$

$$\le \frac{1}{1-\lambda}\mathsf{BProfit}(c) + \frac{1}{1-\lambda}\ln\frac{1}{\lambda}\cdot\mathsf{SProfit}(c) \quad \text{(By Lemma 2 and Lemma 3)}.$$

Therefore,

$$\mathsf{FB} = \int_0^1 \mathsf{FB}(c)\,\mathrm{d}G(c)$$

$$\le \frac{1}{1-\lambda}\int_0^1 \mathsf{BProfit}(c)\,\mathrm{d}G(c) + \frac{1}{1-\lambda}\ln\frac{1}{\lambda}\int_0^1 \mathsf{SProfit}(c)$$

$$\le \frac{1}{1-\lambda}\mathsf{BProfit} + \frac{1}{1-\lambda}\ln\frac{1}{\lambda}\cdot\mathsf{SProfit}$$

$$\le \frac{1}{1-\lambda}\mathsf{BuyerP} + \frac{1}{1-\lambda}\ln\frac{1}{\lambda}\cdot\mathsf{SellerP}.$$

By the symmetry described in Remark 1, we also have

$$\mathsf{FB} \le \frac{1}{1-\lambda}\mathsf{SellerP} + \frac{1}{1-\lambda}\ln\frac{1}{\lambda} \cdot \mathsf{BuyerP}.$$

Adding up the preceding two inequalities, we get

$$\mathsf{FB} \le \left(\frac{1}{1-\lambda} + \frac{1}{1-\lambda}\ln\frac{1}{\lambda}\right)\left(\frac{1}{2}\mathsf{SellerP} + \frac{1}{2}\mathsf{BuyerP}\right)$$

$$= \left(\frac{1}{1-\lambda} + \frac{1}{1-\lambda}\ln\frac{1}{\lambda}\right)\mathsf{RandOff}.$$

Note that the above holds for all $\lambda \in (0,1)$. Thus the conclusion follows by calculating

$$\min_{0<\lambda<1}\left(\frac{1}{1-\lambda} + \frac{1}{1-\lambda}\ln\frac{1}{\lambda}\right) \approx 3.1462.$$

3   Approximation Ratio Under the MHR Condition

In Sect. 2, we use the quantities SProfit and BProfit to lower-bound SellerP and BuyerP, a (painful) compromise we make in the face of the difficulty in analyzing SellerP and BuyerP themselves. In this section, we will see that by imposing a restriction on the distribution of the buyer's value, one can significantly reduce the difficulty of analysis.

3.1   Preliminaries

We state the definition of the *hazard rate* and the *virtual value function*, which are commonly studied in auction theory since the work of Myerson [11].

Definition 2. *A distribution on $[0,1]$ with CDF F and continuous and positive density function f is said to have the monotone hazard rate (MHR) property if the hazard rate*

$$h(x) = \frac{f(x)}{1-F(x)}$$

is a monotone non-decreasing function of x.

Definition 3. *Define the virtual value function of the buyer to be*

$$\varphi(x) = x - \frac{1-F(x)}{f(x)}.$$

It can be seen from this definition that the MHR property of F implies that φ is strictly increasing, and hence its inverse function φ^{-1} exists on $[\varphi(0),1] \supset [0,1]$. What greatly reduces the difficulty of analysis is the following fact:

Proposition 2. *If the distribution of the buyer's value satisfies the MHR property, then the price p_c that the seller would set given his cost c is exactly $\varphi^{-1}(c)$.*

Proof. This is immediate from Myerson's auction theory [11]. Here we give a short explanation, for the sake of completeness. Given $c \in [0,1]$, we have for each $p \in [c,1]$

$$\frac{d}{dp}(p-c)(1-F(p)) = (1-F(p)) - (p-c)f(p) = f(p)(c - \varphi(p)).$$

Since φ is strictly increasing, the function $p \mapsto (p-c)(1-F(p))$ has a unique maximum $p^* = \varphi^{-1}(c)$.

Let $\mathcal{MHR}$ be the collection of all MHR distributions on $[0,1]$. Our goal in this section is to show that

$$\inf_{\substack{F \in \mathcal{MHR} \\ G}} \frac{\mathsf{SellerP}}{\mathsf{FB}} = \frac{1}{e-1}.$$

In Sect. 3.2, we will prove that when $F \in \mathcal{MHR}$, the inequality $\mathsf{FB} \leq (e-1) \cdot \mathsf{SellerP}$ holds. Then, in Sect. 3.3 we will prove that the constant $(e-1)$ is optimal in the above inequality.

Remark 4. We can assume that the distribution defined by G is supported on $[0,1]$ as well. Indeed, if we "truncate" G into $G_{\text{truncate}}(x) = \begin{cases} 0 & \text{if } x < 0 \\ G(x) & \text{if } 0 \leq x < 1 \\ 1 & \text{if } x \geq 1 \end{cases}$, the ratio $\mathsf{SellerP}/\mathsf{FB}$ will not increase. Since we are concerned with the infimum of this ratio, this truncation is without loss of generality.

3.2   Proof of Lower Bound

In light of Proposition 2, we can define

$$\mathsf{SellerP}(c) = \int_{\varphi^{-1}(c)}^{1} (v-c)\,\mathrm{d}F(v),$$

and from the definition of $\mathsf{SellerP}$ we have

$$\mathsf{SellerP} = \int_{0}^{1} \mathsf{SellerP}(c)\,\mathrm{d}G(c).$$

Since we also have

$$\mathsf{FB} = \int_{0}^{1} \mathsf{FB}(c)\,\mathrm{d}G(c),$$

it suffices to show that $\mathsf{FB}(c) \leq (e-1) \cdot \mathsf{SellerP}(c)$ for each $c \in [0,1]$. Integrating by parts, we have

$$\mathsf{FB}(c) = \int_{c}^{1} (v-c)\,\mathrm{d}F(v) = (1-c) - \int_{c}^{1} F(v)\,\mathrm{d}(v-c) = \int_{c}^{1} (1-F(v))\,\mathrm{d}v$$

$$= \int_{c}^{\varphi^{-1}(c)} (1-F(v))\,\mathrm{d}v + \int_{\varphi^{-1}(c)}^{1} (1-F(v))\,\mathrm{d}v$$

$$\tag{1}$$

and

$$\text{SellerP}(c) = \int_{\varphi^{-1}(c)}^{1} (v - c) \, dF(v)$$

$$= (1 - c) - \left(\varphi^{-1}(c) - c\right) F\left(\varphi^{-1}(c)\right) - \int_{\varphi^{-1}(c)}^{1} F(v) \, d(v - c)$$

$$= \int_{c}^{1} 1 \, dv - \int_{c}^{\varphi^{-1}(c)} F\left(\varphi^{-1}(c)\right) \, dv - \int_{\varphi^{-1}(c)}^{1} F(v) \, dv \tag{2}$$

$$= \int_{c}^{\varphi^{-1}(c)} \left(1 - F\left(\varphi^{-1}(c)\right)\right) \, dv + \int_{\varphi^{-1}(c)}^{1} (1 - F(v)) \, dv.$$

Note that the right hand sides of Eq. 2 and Eq. 1 already have the second term in common. The next lemma relates their first terms also to each other.

Lemma 4. *When* $0 \leq c \leq v \leq \varphi^{-1}(c)$, *we have*

$$\frac{1 - F(v)}{1 - F\left(\varphi^{-1}(c)\right)} \leq \exp\left(\frac{\varphi^{-1}(c) - v}{\varphi^{-1}(c) - c}\right).$$

Proof. By definition of the function φ,

$$c = \varphi\left(\varphi^{-1}(c)\right) = \varphi^{-1}(c) - \frac{1}{h\left(\varphi^{-1}(c)\right)}.$$

Hence we get a nice expression for the hazard rate at $\varphi^{-1}(c)$:

$$h\left(\varphi^{-1}(c)\right) = \frac{1}{\varphi^{-1}(c) - c}.$$

Define a function $H(x) = -\ln(1 - F(x))$ on $[0, 1)$ (known as the *cumulative hazard function*), we have for each $x \in [c, \varphi^{-1}(c)]$

$$H'(x) = \frac{f(x)}{1 - F(x)} = h(x) \leq h\left(\varphi^{-1}(c)\right) = \frac{1}{\varphi^{-1}(c) - c}.$$

Integrating with respect to x both sides of the above inequality from v to $\varphi^{-1}(c)$, we get

$$H\left(\varphi^{-1}(c)\right) - H(v) \leq \frac{\varphi^{-1}(c) - v}{\varphi^{-1}(c) - c},$$

or

$$\frac{1 - F(v)}{1 - F\left(\varphi^{-1}(c)\right)} \leq \exp\left(\frac{\varphi^{-1}(c) - v}{\varphi^{-1}(c) - c}\right).$$

Remark 5. The proof in Blumrosen and Mizrahi's paper [3] also contains a step that is equivalent to the special case $v = c$ of the preceding lemma, but they apply it in a different way to a different framework of analysis.

Theorem 4. *When $F \in \mathcal{MHR}$, the inequality FB $\leq (e-1) \cdot$ SellerP holds.*

Proof. By Lemma 4, for each $0 \leq c < 1$ and $c \leq v \leq \varphi^{-1}(c)$,

$$(1 - F(v)) \leq \exp\left(\frac{\varphi^{-1}(c) - v}{\varphi^{-1}(c) - c}\right) \cdot \left(1 - F\left(\varphi^{-1}(c)\right)\right).$$

Integrating with respect to v the above inequality from c to $\varphi^{-1}(c)$, we get

$$\int_c^{\varphi^{-1}(c)} (1 - F(v)) \, dv$$

$$\leq \left(1 - F\left(\varphi^{-1}(c)\right)\right) \cdot \int_c^{\varphi^{-1}(c)} \exp\left(\frac{\varphi^{-1}(c) - v}{\varphi^{-1}(c) - c}\right) dv$$

$$= \left(1 - F\left(\varphi^{-1}(c)\right)\right) \cdot \left(\varphi^{-1}(c) - c\right) \cdot \int_0^1 \exp(x) \, dx \quad \left(\text{substituting } x = \frac{\varphi^{-1}(c) - v}{\varphi^{-1}(c) - c}\right)$$

$$= (e-1) \cdot \int_c^{\varphi^{-1}(c)} \left(1 - F\left(\varphi^{-1}(c)\right)\right) dv.$$

Therefore, combining the above with Eq. 1 and Eq. 2, we have for each $c \in [0,1]$

$$\text{FB}(c) = \int_c^{\varphi^{-1}(c)} (1 - F(v)) \, dv + \int_{\varphi^{-1}(c)}^1 (1 - F(v)) \, dv$$

$$\leq (e-1) \int_c^{\varphi^{-1}(c)} \left(1 - F\left(\varphi^{-1}(c)\right)\right) dv + \int_{\varphi^{-1}(c)}^1 (1 - F(v)) \, dv$$

$$\leq (e-1) \int_c^{\varphi^{-1}(c)} \left(1 - F\left(\varphi^{-1}(c)\right)\right) dv + (e-1) \int_{\varphi^{-1}(c)}^1 (1 - F(v)) \, dv$$

$$= (e-1) \cdot \text{SellerP}(c).$$

It follows that FB $\leq (e-1) \cdot$ SellerP.

3.3   Proof of Upper Bound

Theorem 5. *The constant $(e-1)$ in Theorem 4 is optimal.*

Proof. We need only to construct examples of F and G such that the ratio FB/SellerP can be infinitely close to $(e-1)$. Let G be a single point mass at 0, and F be the piecewise function:

$$F(x) = \begin{cases} 1 - e^{-x} & \text{if } x \leq 1 - \delta \\ 1 - e^{-1+\delta}(1-x)\left(\frac{1-\delta}{\delta^2} x - \frac{1-3\delta+\delta^2}{\delta^2}\right) & \text{if } 1 - \delta \leq x \leq 1 \end{cases},$$

where $\delta > 0$ is very close to 0. The function F is defined to be the CDF of an exponential distribution on $[0, 1 - \delta]$ and a quadratic function on $[1 - \delta, 1]$ that smoothly connects the exponential part and the point $F(1) = 1$:

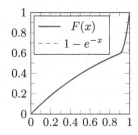

Fig. 1. The CDF of the distribution F

For $x \leq 1 - \delta$, the CDF F has a constant hazard rate $f(x)/(1 - F(x)) = e^{-x}/e^{-x} = 1$, while on $[1 - \delta, 1]$, the function $F' = f$ is monotone increasing and hence the hazard rate of F is monotone increasing on $[1 - \delta, 1]$. Thus F, as is defined above, satisfies the MHR property. For $x \leq 1 - \delta$,

$$\varphi(x) = x - \frac{1 - F(x)}{f(x)} = x - 1 < 0,$$

and hence we must have $\varphi^{-1}(0) > 1 - \delta$. This implies that

$$\mathsf{SellerP} = \mathsf{SellerP}(0) = \int_{\varphi^{-1}(0)}^{1} (v - 0) \, \mathrm{d}F(v) \leq \int_{1-\delta}^{1} v \, \mathrm{d}F(v) \leq \int_{1-\delta}^{1} 1 \, \mathrm{d}F(v) = F(1) - F(1 - \delta) = e^{-1+\delta},$$

which tends to e^{-1} as $\delta \to 0$. But we also have according to Equation 1

$$\mathsf{FB} = \mathsf{FB}(0) = \int_{0}^{1} (1 - F(v)) \, \mathrm{d}v,$$

which clearly tends to $\int_{0}^{1} e^{-x} \, \mathrm{d}x = 1 - e^{-1}$ when $\delta \to 0$. This shows that

$$\inf_{\substack{F \in \mathcal{MHR} \\ G}} \frac{\mathsf{SellerP}}{\mathsf{FB}} \leq \frac{e^{-1}}{1 - e^{-1}} = \frac{1}{e - 1}.$$

Remark 6. Note that the hard case given above is actually when the seller has no cost. In this case, the gains-from-trade is equal to the welfare. Since any lower bound of the gains-from-trade approximation ratio always applies to the welfare approximation ratio as well, we can combine this observation with the lower bound proved in Sect. 3.2 to conclude that, assuming MHR of the buyer's distribution, the welfare approximation ratio of the seller-pricing mechanism to the first-best mechanism is also equal to $(e - 1)$.

Remark 7. If we take the buyer's CDF F to be the one in the preceding proof, and the seller's CDF to be $G(x) = 1 - F(1 - x)$, it's easy to find that $\mathsf{RandOff}/\mathsf{FB} \to 1 - 1/e$ when the parameter $\delta \to 0$. Combining this with the lower bound in Sect. 3.2, we have

$$0.582 \approx \frac{1}{e - 1} \leq \inf_{F, G^{R} \in \mathcal{MHR}} \frac{\mathsf{RandOff}}{\mathsf{FB}} \leq 1 - \frac{1}{e} \approx 0.632,$$

where $G^{\mathrm{R}} := 1-G(1-x)$. This shows that $\inf_{F,G^{\mathrm{R}}\in\mathcal{MHR}}(\mathsf{RandOff}/\mathsf{FB})$ is strictly larger than $\inf_{F,G}(\mathsf{RandOff}/\mathsf{FB})$, which lies in $[0, 317, 0.495]$ (see Sect. 1.2).

Acknowledgement. The author would like to thank Kangning Wang and Zhaohua Chen for reading an earlier draft, discussion about the content, and their helpful suggestions on the presentation of the paper.

References

1. Babaioff, M., Dobzinski, S., and Kupfer, R.: A note on the gains from trade of the random-offerer mechanism, CoRR abs/2111.07790 (2021). arXiv:2111.07790
2. Blumrosen, L., Dobzinski, S.: (Almost) efficient mechanisms for bilateral trading. Games and Economic Behavior, vol. 130, pp. 369–383. Elsevier (2021). https://doi.org/10.1016/j.geb.2021.08.011
3. Blumrosen, L., Mizrahi, Y.: Approximating gains-from-trade in bilateral trading. In: WINE, pp. 400–413 (2016). https://doi.org/10.1007/978-3-662-54110-4_28
4. Brustle, J., Cai, Y., Wu, F., Zhao, M.: Approximating gains from trade in two-sided markets via simple mechanisms. In: EC, pp. 589–590 (2017). https://doi.org/10.1145/3033274.3085148
5. Cai, Y., Goldner, K., Ma, S., Zhao, M.: On multi-dimensional gains from trade maximization. In: SODA, pp. 1079–1098 (2021). https://doi.org/10.1137/1.9781611976465.67
6. Deng, Y., Mao, J., Sivan, B., Wang, K.: Approximately efficient bilateral trade. In STOC, pp. 718–721 (2022). https://doi.org/10.1145/3519935.3520054
7. Kang, Z., Vondrák, J.: Fixed-price approximations to optimal efficiency in bilateral trade. Available at SSRN 3460336 (2019)
8. Kang, Z., Pernice, F., Vondrák, J.: Fixed-price approximations in bilateral trade. In: SODA, pp. 2964–2985 (2022). https://doi.org/10.1137/1.9781611977073.115
9. Leininger, W., Linhart, P., Radner, R.: Equilibria of the sealed-bid mechanism for bargaining with incomplete information. Journal of Economic Theory, vol. 48, pp. 63–106. Elsevier (1989). https://doi.org/10.1016/0022-0531(89)90120--8
10. McAfee, R.: The gains from trade under fixed price mechanisms. Appl. Econ. Res. Bull. **1**, 1–10 (2008)
11. Myerson, R: Optimal auction design. Mathematics of Operations Research, vol. 6, pp. 58–73. INFORMS (1981). https://doi.org/10.1287/moor.6.1.58
12. Myerson, R., Satterthwaite, M.: Efficient mechanisms for bilateral trading. J. Econ. Theory **29**, 265–281. Elsevier (1983). https://doi.org/10.1016/0022-0531(83)90048--0

Better Approximation for Interdependent SOS Valuations

Pinyan Lu[1], Enze Sun[2], and Chenghan Zhou[3(✉)]

[1] ITCS, Shanghai University of Finance and Economics, Shanghai, China
lu.piyan@mail.shufe.edu.cn
[2] Department of Computer Science, The University of Hong Kong, Hong Kong, China
sunenze@connect.hku.hk
[3] Department of Computer Science, Princeton University, Princeton, New Jersey, USA
chenghanzh@princeton.edu

Abstract. Submodular over signal (SOS) defines a family of interesting functions for which there exist truthful mechanisms with constant approximation to the social welfare for agents with interdependent valuations. The best-known truthful auction is of 4-approximation and a lower bound of 2 was proved. We propose a new and simple truthful mechanism to achieve an approximation ratio of 3.315. In particular, we first generalize the random sampling auction in [9]. Then, we proposed a brand new auction that is simple, efficient to implement and easy to verify the truthfulness. We call our mechanism the contribution-based mechanism. Our proposed mechanism with better approximation runs a convex combination of the above two mechanisms. Since the random sampling mechanism performs well when the second largest value is comparable to the largest one while contribution-based mechanism performs well when the largest one is much larger than all other values, their combination achieves a good balance for all instances. The approximation of our final mechanism is 3.315. This improves the previous 4-approximation mechanism for the first time. Besides the new auction, we also investigate the relation with SOS and strong-SOS, a stronger notion of SOS which was also introduced in [9]. We build a reduction and prove that strong-SOS is as difficult as SOS in terms of approximation ratio for single item setting. This means that it is fine to design mechanisms for strong-SOS valuation only if it is easier since the mechanism can be transformed to a mechanism for general SOS valuations with almost same approximation ratio. The full version of our paper can be found here: https://arxiv.org/abs/2210.06507.

Keywords: Mechanism design · Combinatorial auction · Approximation algorithms

1 Introduction

In most study of auction theory, it is assumed that valuations are agents' private information and they know their own values when they submit their bids to the auctioneer. However, this is not usually the case in real life. For example, when one buys a house or an art work by auction, his valuation largely depends on other's valuations since they

K. A. Hansen et al. (Eds.): WINE 2022, LNCS 13778, pp. 219–234, 2022.
https://doi.org/10.1007/978-3-031-22832-2_13

will impact the item's resale value later on. When an advertiser bids an impression or click in internet, the value largely depends on that particular customer for which other bidders may have more information. To describe the valuation interdependence between different bidders, a model is proposed by Milgrom and Weber [19]. Each bidder i holds some private information about the item, denoted by a signal $s_i \in \mathbb{R}^+$. Agent i's valuation when receiving the item is a public-known function $v_i(\mathbf{s})$ that depends on the signals of all bidders. This has become the standard model for interdependent value settings (IDV) and has been studied in the economics literature for a few decades [4, 7–10, 12, 14, 20].

For this model without any restriction on the valuation functions, it is impossible to design a truthful auction[1] with good social welfare guarantee [15]. This is in strong contrasts to the private valuation model, for which the VCG mechanism can achieve truthfulness and optimal welfare simultaneously [6, 13, 18, 21]. A natural extension of VCG mechanism only works when the valuation functions satisfy a technical condition called single-crossing condition [2, 3, 5, 16, 22]. However, there are many relevant settings where the single-crossing condition does not hold [7, 8, 17].

A different and beautiful perspective is proposed by Eden et al. [9]. They introduced a new condition of the valuation functions called submodular over signal (SOS) property. Submodular captures a natural diminishing returns property, which is very common in economics settings. They designed a simple random sampling auction to achieve an approximation of 4 and proved that no truthful auction can do better than 2-approximation.

This 4-approximation remains the state-of-art for general SOS settings. The only improvement was made for the very special case of binary signal, where the signal for each agent only has two possible values. For this special binary signal setting, it is proved that there exists a tight 2-approximation auction [1]. It is an existential proof rather than an efficient design. To construct such a 2-approximation auction may need exponential time.

Our Contributions

Firstly, we generalize the random sampling auction in [9]. In [9], they evenly divided the agents into two sets and their analysis paired a set of agents with its complement to prove their approximation ratio. We generalize this to sampling with arbitrary probability p and get a similar approximation ratio in terms of p. Although the result is as expected, the proof is completely new since the original pairing trick does not work for biased sampling. We also get a more careful analysis of the approximation with a term involving the ratio of the largest and second largest values. This part is easy but crucial to the final improvement of the mechanism. The observation is quite intuitive: the random sampling mechanism performs much better when the second largest value is comparable to the largest one.

---

[1] The truthfulness notion here is ex-post IC & IR rather than DSIC since DSIC is not possible for interdependent valuation.

Secondly, we proposed a brand new auction. Our mechanism is very simple, it allocates the item to agent i with the probability

$$\frac{1}{2}\left(1 - \frac{\max_{j\in[n],j\neq i} v_j(\mathbf{s}_{-i},0)}{\max_{j\in[n]} v_j(\mathbf{s})}\right).$$

We call our mechanism the contribution-based mechanism. The intuition and the meaning of the name will be discussed in Sect. 3. This mechanism is simple, efficient to implement and easy to verify the truthfulness since the allocation rule is monotone. However, the tricky part is to verify that it is indeed a well-defined mechanism, namely the overall allocation probability cannot exceed one. This proof crucially uses the property of SOS. In particular, we obtain a lemma from the SOS property which is the key of the proof. This lemma may be of independent interests. For example, the lemma is used in the analysis of the random sampling auction. This contribution-based mechanism's approximation ratio is at least $\frac{1}{2}\left(1 - \frac{v_{(2)}(\mathbf{s})}{v_{(1)}(\mathbf{s})}\right)$, where $v_{(1)}(\mathbf{s})$ and $v_{(2)}(\mathbf{s})$ are the largest and the second largest values respectively given the signal profile s. From this expression we can see that it achieves good ratio when the largest value is much larger than others. This is in the opposite direction with the random sampling mechanism.

Finally, we run a convex combination of the above two mechanisms. Since the random sampling mechanism performs well when the second largest value is comparable to the largest one while contribution-based mechanism performs well when the largest one is much larger than all other values, their combination achieves a good balance for all instances. The approximation of our final mechanism is 3.315. This improves the previous 4-approximation mechanism for the first time.

Besides the new auction, we also investigate the relation with SOS and strong-SOS, a stronger notion of SOS which was also introduced in [9]. We build a reduction and prove that strong-SOS is as difficult as SOS in terms of approximation ratio for single item setting. This means that it is fine to design mechanisms for strong-SOS valuation only if it is easier since the mechanism can be transformed to a mechanism for general SOS valuations with almost same approximation ratio. In [9], a better approximation ratio was given for strong-SOS when the size of signal space is restricted. This does not contradict to our result since our reduction will enlarge the signal space greatly.

Related Works

In this paper, we only focus on the canonical single item setting. The original paper [9] studied SOS valuations in a much broader combinatorial auction setting. Their 4-approximation works for any single-parameter downward-closed setting with single-dimensional signals and SOS valuations. They also studied multi-dimensional signal with separable SOS valuations and gave a 4-approximation. We defined an extend version of SOS called d-SOS. We are not going to define all these extended versions but focus on single item and SOS valuation for simplicity. Interested readers can find these extensions in paper [9]. The above mentioned 2-approximation auction [1] can extend to systems with matroid constraints. A recent paper also studied single item setting but with private SOS valuations [11]. In our setting, the signals are private while the valuation functions are public.

2 Preliminaries

We consider a single-item auction with n bidders. In the interdependent setting, each bidder i holds some private information about the item, denoted by a signal $s_i \in \mathbb{R}^+$. The signals of all bidders participating in the auction can be collected as a vector $\mathbf{s} = (s_1, s_2, \cdots, s_n)$. We sometimes use $(\mathbf{s}_{-i}, s_i)$ to emphasize the signal of agent i. The signal space is denoted by $\mathcal{S}$. By convention, we assume that $s_i = 0$ is the minimum signal in each agent i's signal space.

Agent i's valuation when receiving the item is a public-known function $v_i(\mathbf{s})$ that depends on the signals of all bidders. By convention, we assume that $v_i(\mathbf{s})$ is non-negative, weakly increasing in all signals and strongly increasing in bidder i's signal. $v_{(k)}(\mathbf{s})$ denotes the kth largest valuation of a single agent when the signals of all bidders is $\mathbf{s}$. We focus on the case where the valuation function for each agent is *submodular over signals (SOS)*:

Definition 1 (Submodularity over signals). *A valuation function $v(\mathbf{s})$ is submodular over signals if for all bidders i, $\mathbf{s}'_{-i} \succeq \mathbf{s}_{-i}$ and $s'_i \geq s_i$,*

$$v(\mathbf{s}'_{-i}, s'_i) - v(\mathbf{s}'_{-i}, s_i) \leq v(\mathbf{s}_{-i}, s'_i) - v(\mathbf{s}_{-i}, s_i).$$

A mechanism $M = (\mathbf{x}, \mathbf{p})$ decides the allocation rule $\mathbf{x}$ and payment $\mathbf{p}$. Without loss of generality, we consider direct mechanisms where bidders report their private signals $\tilde{\mathbf{s}}$ as bids. The mechanism then allocates the item to bidder i with probability $x_i(\tilde{\mathbf{s}})$ and asks for payment $p_i(\tilde{\mathbf{s}})$. $\mathbf{x}$ satisfies feasibility constraint $\sum_{i \in [n]} x_i(\mathbf{s}) \leq 1$ for all signal profiles $\mathbf{s}$.

Throughout our analysis, we adopt the solution concepts of *ex-post IC & IR* mechanisms, defined as follows:

Definition 2. *An ex-post incentive compatible (IC) mechanism means that each bidder does not regret reporting his private signal s_i truthfully after knowing all the other bidders' reported signals $\mathbf{s}_{-i}$. Formally, let the signal profile be $\mathbf{s} = (s_i, \mathbf{s}_{-i})$. For all signals s'_i,*

$$x_i(\mathbf{s}_{-i}, s_i) v_i(\mathbf{s}) - p_i(\mathbf{s}_{-i}, s_i) \geq x_i(\mathbf{s}_{-i}, s'_i) v_i(\mathbf{s}) - p_i(\mathbf{s}_{-i}, s'_i)$$

An ex-post individual rational (IR) mechanism satisfies

$$x_i(\mathbf{s}_{-i}, s_i) v_i(\mathbf{s}) - p_i(\mathbf{s}_{-i}, s_i) \geq 0$$

In this paper, we mainly focus on the allocation rule since our goal is to maximize social welfare. The following characterization allows us to design allocation rule alone with an additional monotone constraint. The payment can be deviated from the allocation rule by the standard method and will be omitted in this paper [20].

Lemma 1. *For an allocation rule $\mathbf{x}$, there exists a payment rule $\mathbf{p}$ to make the mechanism $M = (\mathbf{x}, \mathbf{p})$ ex-post IC & IR iff it satisfies the following monotonicity: for any bidder i, $\mathbf{s}_{-i}$ and $s'_i > s_i$, we have $x_i(\mathbf{s}_{-i}, s'_i) \geq x_i(\mathbf{s}_{-i}, s_i)$.*

3   Contribution-Based Mechanism

Let agent i^* be the agent with the maximum valuation at signal s. The optimal social welfare is $v_{i^*}(s)$. However, we do not view that this social welfare is contributed by the agent i^* alone since it also depends on other agents' signals. We view the contribution of agent $i \in [n]$ (including i^*) as $v_{i^*}(s) - \max_{j \in [n], j \neq i} v_j(\mathbf{s}_{-i}, 0)$, where $\max_{j \in [N], j \neq i} v_j(\mathbf{s}_{-i}, 0)$ is the optimal social welfare when agent i is not in the game (his signal is "zeroed out" and his valuation is excluded). This difference of social welfare is the contribution brought by agent i to the game. We want to allocate the item to the agents proportional to their contributions. That is why we call our mechanism Contribution-Based Mechanism. The ideal allocation probability for agent i should be $\frac{v_{i^*}(s) - \max_{j \in [n], j \neq i} v_j(\mathbf{s}_{-i}, 0)}{v_{i^*}(s)}$. Unfortunately, this is not a valid mechanism since the total probability may exceed 1. However, we are able to prove that the total probability will never exceed 2 due to the SOS property of the functions. Therefore, we can half the probability and get a valid mechanism.

Contribution-Based Mechanism: Let agent i^* be the agent with maximum valuation at signal s. For every agent i, we allocate the item to him with the probability

$$x_i(\mathbf{s}) = \frac{v_{i^*}(s) - \max_{j \in [n], j \neq i} v_j(\mathbf{s}_{-i}, 0)}{2v_{i^*}(s)}$$

Before we prove that it is indeed a valid mechanism and analyse its approximation, we first prove an important property of the SOS functions. (This is also discovered in [11].)

Lemma 2. *Let $T \subseteq [n]$ be a subset of bidders. Signals* s *and* s' *satisfy* $\forall t \in T, s'_t \leq s_t$ *and* $\mathbf{s}_{-T} = \mathbf{s}'_{-T}$. *For each bidder* $i \in [n]$,

$$\sum_{t \in T} (v_i(\mathbf{s}) - v_i(\mathbf{s}_{-t}, s'_t)) \leq v_i(\mathbf{s})$$

Proof. Denote $T = \{t_1, t_2, \cdots, t_{|T|}\}$.

$$\sum_{j=1}^{|T|} \left(v_i(\mathbf{s}) - v_i(\mathbf{s}_{-t_j}, s'_{t_j}) \right)$$

$$\leq \sum_{j=1}^{|T|} v_i \left(\mathbf{s}_{-\{t_1, \cdots, t_{j-1}\}}, s'_{t_1}, \cdots, s'_{t_{j-1}} \right) - v_i \left(\mathbf{s}_{-\{t_1, \cdots, t_j\}}, s'_{t_1}, \cdots, s'_{t_j} \right)$$

$$= v_i(\mathbf{s}) - v_i(\mathbf{s}')$$

$$\leq v_i(\mathbf{s})$$

The first inequality comes from the SOS property.

Theorem 1. *Contribution-Based Mechanism is an ex-post IC & IR mechanism with* $\frac{1}{2} \left(1 - \frac{v_{(2)}(\mathbf{s})}{v_{(1)}(\mathbf{s})} \right)$-*approximation, where* $v_{(1)}(\mathbf{s})$ *and* $v_{(2)}(\mathbf{s})$ *are the largest and the second largest values respectively given the signal profile* s.

Proof. **Correctness:** We prove that $\sum_{i\in[n]} x_i(\mathbf{s}) \leq 1$. It is clear that $x_{i^*}(\mathbf{s}) \leq 1/2$. So we only need to show that $\sum_{i\in[n],i\neq i^*} x_i(\mathbf{s}) \leq 1/2$.

$$
\begin{aligned}
\sum_{i\in[n],i\neq i^*} x_i(\mathbf{s}) &= \sum_{i\in[n],i\neq i^*} \frac{v_{i^*}(\mathbf{s}) - \max_{j\in[n],j\neq i} v_j(\mathbf{s}_{-i},0)}{2v_{i^*}(\mathbf{s})} \\
&\leq \sum_{i\in[n],i\neq i^*} \frac{v_{i^*}(\mathbf{s}) - v_{i^*}(\mathbf{s}_{-i},0)}{2v_{i^*}(\mathbf{s})} \\
&= \frac{\sum_{i\in[n],i\neq i^*}(v_{i^*}(\mathbf{s}) - v_{i^*}(\mathbf{s}_{-i},0))}{2v_{i^*}(\mathbf{s})} \\
&\leq \frac{v_{i^*}(\mathbf{s})}{2v_{i^*}(\mathbf{s})} \\
&= \frac{1}{2}
\end{aligned}
$$

The second inequality comes from Lemma 2.

Monotonicity: For any agent i, any $\mathbf{s}_{-i}$ and two signals $s'_i \geq s_i$, by the weak monotonicity of valuation functions, we have

$$
\max_{j\in[n]} v_j(\mathbf{s}_{-i}, s'_i) \geq \max_{j\in[n]} v_j(\mathbf{s}_{-i}, s_i).
$$

As a result

$$
\begin{aligned}
x_i(\mathbf{s}_{-i}, s'_i) &= \frac{\max_{j\in[n]} v_j(\mathbf{s}_{-i}, s'_i) - \max_{j\in[n],j\neq i} v_j(\mathbf{s}_{-i},0)}{2\max_{j\in[n]} v_j(\mathbf{s}_{-i}, s'_i)} \\
&\geq \frac{\max_{j\in[n]} v_j(\mathbf{s}_{-i}, s_i) - \max_{j\in[n],j\neq i} v_j(\mathbf{s}_{-i},0)}{2\max_{j\in[n]} v_j(\mathbf{s}_{-i}, s_i)} \\
&= x_i(\mathbf{s}_{-i}, s_i)
\end{aligned}
$$

Therefore, our mechanism always assigns a higher probability to agent i when its signal is stronger. There exists a payment rule to make it ex-post IC and IR.

Approximation: We simply verify that the agent i^* get the item with the probability at least $\frac{1}{2}\left(1 - \frac{v_{(2)}(\mathbf{s})}{v_{(1)}(\mathbf{s})}\right)$ since $v_{i^*}(\mathbf{s})$ is the targeted optimal social welfare. By definition, we have $v_{(1)}(\mathbf{s}) = v_{i^*}(\mathbf{s})$ and

$$
v_{(2)}(\mathbf{s}) = \max_{j\in[n],j\neq i^*} v_j(\mathbf{s}) \geq \max_{j\in[n],j\neq i^*} v_j(\mathbf{s}_{-i^*},0).
$$

Therefore,

$$
x_{i^*}(\mathbf{s}) = \frac{1}{2}\left(1 - \frac{\max_{i\in[n],i\neq i^*} v_i(\mathbf{s}_{-i^*},0)}{v_{i^*}(\mathbf{s})}\right) \geq \frac{1}{2}\left(1 - \frac{v_{(2)}(\mathbf{s})}{v_{(1)}(\mathbf{s})}\right).
$$

4   Random Sampling Auction

Our random sampling auction is a generalization of the auction in [9]. Their auction is a special case of ours by choosing $p = 0.5$.
Random Sampling Mechanism

- Each agent i is allocated to set A with probability p and set B with probability $1 - p$.
- For $i \in B$, let $w_i = v_i(\mathbf{s}_A, s_i, \mathbf{0}_{B \setminus \{i\}})$.
- Allocate to bidder $\operatorname{argmax}_{i \in B} w_i$

It is clear that this is an ex-post IC & IR mechanism since the allocation rule is monotone. In [9], they proved that the approximation ratio of the Random Sampling Mechanism is $1/4$ when $p = 1/2$. We shall prove that the ratio is $p(1 - p)$ for general p and further refine the ratio in terms of $\frac{v_{(2)}(\mathbf{s})}{v_{(1)}(\mathbf{s})}$.

Theorem 2. *For every signal* $\mathbf{s}$, *Random Sampling Mechanism is an ex-post IC & IR mechanism with the approximation ratio of*

$$p(1 - p)\left(1 + p \cdot \frac{v_{(2)}(\mathbf{s})}{v_{(1)}(\mathbf{s})}\right).$$

We start with the following lemma.

Lemma 3. *For any* $a \in [n - 2]$ *and* $i \in [n]$,

$$\frac{1}{a \cdot \binom{n-1}{a}} \sum_{A:|A|=a} v_i(\mathbf{s}_A, \mathbf{0}_B, s_i) \geq \frac{1}{(a + 1) \cdot \binom{n-1}{a+1}} \sum_{A:|A|=a+1} v_i(\mathbf{s}_A, \mathbf{0}_B, s_i).$$

Proof. Before the formal proof, let us give some intuition. $\frac{1}{\binom{n-1}{a}} \sum_{A:|A|=a} v_i(\mathbf{s}_A, \mathbf{0}_B, s_i)$ is the expected value of $v_i(\mathbf{s}_A, \mathbf{0}_B, s_i)$ when A is an uniform random set of size a. So, $\frac{1}{a \cdot \binom{n-1}{a}} \sum_{A:|A|=a} v_i(\mathbf{s}_A, \mathbf{0}_B, s_i)$ can be viewed as an amortized expected value. This lemma says that this amortized expected value decreases in terms of the set size a. This is not surprising given the SOS property of the valuation function.

Now we prove it formally. After canceling common factors in the binomial coefficients of both sides, the inequality is equivalent to the following one

$$\sum_{A:|A|=a} v_i(\mathbf{s}_A, \mathbf{0}_B, s_i) \geq \frac{a}{n - 1 - a} \sum_{A:|A|=a+1} v_i(\mathbf{s}_A, \mathbf{0}_B, s_i).$$

We shall prove this inequality in the remaining of the proof. First we have

$$\sum_{A:|A|=a} v_i(\mathbf{s}_A, \mathbf{0}_B, s_i) = \frac{1}{n - 1 - a} \sum_{A:|A|=a+1} \sum_{j \in A} v_i(\mathbf{s}_{A \setminus \{j\}}, \mathbf{0}_{B \cup \{j\}}, s_i). \quad (1)$$

This identity holds since each term in the LHS is counted $n - 1 - a$ times in RHS. Every set of size a can be extended to $n - 1 - a$ different sets of size $a + 1$.

By applying Lemma 2, we get

$$\sum_{j\in A}(v_i(\mathbf{s}_A,\mathbf{0}_B,s_i) - v_i(\mathbf{s}_{A\setminus\{j\}},\mathbf{0}_{B\cup\{j\}},s_i)) \le v_i(\mathbf{s}_A,\mathbf{0}_B,s_i).$$

After rearranging the terms we get

$$\sum_{j\in A}v_i(\mathbf{s}_{A\setminus\{j\}},\mathbf{0}_{B\cup\{j\}},s_i) \ge (|A|-1)v_i(\mathbf{s}_A,\mathbf{0}_B,s_i). \qquad (2)$$

Connecting (1) and (2), we get

$$\sum_{A:|A|=a}v_i(\mathbf{s}_A,\mathbf{0}_B,s_i) \ge \frac{a}{n-1-a}\sum_{A:|A|=a+1}v_i(\mathbf{s}_A,\mathbf{0}_B,s_i)$$

This concludes the proof.

We can keep applying this monotonicity lemma and bound all these summations by the value of $v_i(\mathbf{s})$.

Corollary 1. *For any* $0 \le a \le n-1$

$$\sum_{A:|A|=a}v_i(\mathbf{s}_A,\mathbf{0}_B,s_i) \ge \frac{a\cdot\binom{n-1}{a}}{n-1}v_i(\mathbf{s})$$

Proof. This inequality is trivial when $a = n-1$. For $a \in [n-2]$, we can keep using the above lemma to get the proof.

$$\frac{1}{a\cdot\binom{n-1}{a}}\sum_{A:|A|=a}v_i(\mathbf{s}_A,\mathbf{0}_B,s_i)$$

$$\ge \frac{1}{(a+1)\cdot\binom{n-1}{a+1}}\sum_{A:|A|=a+1}v_i(\mathbf{s}_A,\mathbf{0}_B,s_i)$$

$$\ge \cdots$$

$$\ge \frac{1}{(n-1)\cdot\binom{n-1}{n-1}}\sum_{A:|A|=n-1}v_i(\mathbf{s}_A,\mathbf{0}_B,s_i)$$

$$= \frac{1}{n-1}v_i(\mathbf{s}).$$

Lemma 4. *Let A be a random subset of $[n]\setminus\{i\}$, where each bidder in A is chosen with probability p, and let $B := ([n]\setminus\{i\})\setminus A$. For any $\mathbf{s}$,*

$$\mathbb{E}_A[v_i(\mathbf{s}_A,\mathbf{0}_B,s_i)] \ge p\cdot v_i(\mathbf{s})$$

Proof.

$$\mathbb{E}_A[v_i(\mathbf{s}_A, \mathbf{0}_B, s_i)]$$

$$= \sum_A p^{|A|}(1-p)^{|B|} \cdot v_i(\mathbf{s}_A, \mathbf{0}_B, s_i)$$

$$= \sum_{a=1}^{n-1} p^a(1-p)^{n-1-a} \cdot \sum_{A:|A|=a} v_i(\mathbf{s}_A, \mathbf{0}_B, s_i)$$

$$\geq \sum_{a=1}^{n-1} p^a(1-p)^{n-1-a} \cdot \frac{a}{n-1} \binom{n-1}{a} v_i(\mathbf{s})$$

$$= p \cdot v_i(\mathbf{s}) \cdot \sum_{a=1}^{n-1} p^{a-1}(1-p)^{n-1-a} \binom{n-2}{a-1}$$

$$= p \cdot v_i(\mathbf{s}) \cdot (p+(1-p))^{n-2}$$

$$= p \cdot v_i(\mathbf{s})$$

Proof of Theorem 2. Without loss of generality, we assume that the agent 1 and 2 achieve the largest and the second largest values respectively given the signal profile s. We calculate the social welfare of the auction from two disjoint events $1 \in B$ and $2 \in B \wedge 1 \in A$.

$$\mathbb{E}\left[\max_{i \in B} w_i\right]$$

$$\geq \mathbb{E}\left[\max_{i \in B} w_i \cdot 1_{1 \in B}\right] + \mathbb{E}\left[\max_{i \in B} w_i \cdot 1_{2 \in B \wedge 1 \in A}\right]$$

$$= \mathbb{E}\left[w_1 \cdot 1_{1 \in B}\right] + \mathbb{E}\left[w_2 \cdot 1_{2 \in B \wedge 1 \in A}\right]$$

$$= \mathbb{E}\left[v_1(s_1, \mathbf{s}_A, \mathbf{0}_{B_{-1}}) \mid 1 \in B\right] \cdot \Pr(1 \in B) + \mathbb{E}\left[v_2(s_2, \mathbf{s}_A, \mathbf{0}_{B_{-2}}) \mid 2 \in B \wedge 1 \in A\right] \cdot \Pr(2 \in B \wedge 1 \in A)$$

$$\geq p \cdot v_1(\mathbf{s}) \cdot (1-p) + p \cdot v_2(\mathbf{s}) \cdot p(1-p),$$

where the last inequality uses Lemma 4.

Since the optimal social welfare is $v_{(1)}(\mathbf{s})$, the approximation ratio of the random sampling mechanism is at least

$$p(1-p)\left(1 + p \cdot \frac{v_{(2)}(\mathbf{s})}{v_{(1)}(\mathbf{s})}\right).$$

$\square$

5   Mechanism

Theorem 3. *For agents with SOS valuations, there is a polynomial time, ex-post IC & IR mechanism that gives 3.31543-approximation to the optimal welfare.*

Proof. The final mechanism is a convex combination of the above two mechanisms: run the contribution-base mechanism with probability q and the random sampling mechanism with probability $1 - q$. We note that the sampling probability p within the random sampling mechanism and this combination probability q are two parameters of the mechanism to be fixed later.

It is obvious that this mechanism is polynomial time and ex-post IC & IR since both contribution base mechanism and random sampling mechanism are.

The approximation ratio is just the convex combination of the two mechanisms.

$$\frac{1}{2}\left(1 - \frac{v_{(2)}(\mathbf{s})}{v_{(1)}(\mathbf{s})}\right)q + p(1-p)\left(1 + p \cdot \frac{v_{(2)}(\mathbf{s})}{v_{(1)}(\mathbf{s})}\right)(1-q)$$
$$= \frac{q}{2} + p(1-p)(1-q) + \left(p^2(1-p)(1-q) - \frac{q}{2}\right)\frac{v_{(2)}(\mathbf{s})}{v_{(1)}(\mathbf{s})}.$$

By choosing $q = \frac{2p^2(1-p)}{1+2p^2(1-p)}$, The coefficient of $\frac{v_{(2)}(\mathbf{s})}{v_{(1)}(\mathbf{s})}$ vanishes since $p^2(1-p)(1-q) - \frac{q}{2} = 0$. The approximation ratio is then

$$\frac{q}{2} + p(1-p)(1-q) = \frac{p(1-p^2)}{1+2p^2(1-p)}.$$

Let p be the non-negative real solution of $2x^4 - 4x^3 + 5x^2 - 1 = 0$ ($p \approx 0.54056$ and thus $q \approx 0.21167$), we get the final approximation ratio $0.30162 = \frac{1}{3.31543}$. This concludes the proof of our main result.

6 Strong-SOS

In [9], a stronger notion called Strong-SOS was also proposed.

Definition 3 (Strong-SOS). *A valuation function* $v(\mathbf{s})$ *is strong submodular over signals if for all bidders* i, $\mathbf{s}' \succeq \mathbf{s}$ *and* $\delta \geq 0$,

$$v(\mathbf{s}'_{-i}, s'_i + \delta) - v(\mathbf{s}'_{-i}, s'_i) \leq v(\mathbf{s}_{-i}, s_i + \delta) - v(\mathbf{s}_{-i}, s_i).$$

Although these two definitions look similar and seem to only differ in a small technical condition, we shall argue that the concept of SOS is much more natural and robust than strong-SOS. Signal is an abstract of some private information of the agents and the number represents the strength of the signal. In many cases, only the relative order of different signals rather than their concrete numbers matter since different representations of the signal may have complete different numbers. In particular, if there is a monotone mapping $\phi_i : S_i \to S'_i$ to change one representation of the signals to another, this should not change the problem at all since a mechanism for one representation can be directly transformed to the other with exactly the same performance and behavior. The property of SOS is also invariant for different representations: the valuation is SOS before the monotone mapping iff it is SOS after the mapping. This is desirable and shows the robustness of the definition. However, this invariant does not hold for the definition of strong-SOS. As a result, strong-SOS is not a property for the valuation function alone but a property for valuation function combined with a particular representation of the signal space.

Another advantage for the definition of SOS is that it does not require any additional structure or property in the signal space other than the ordering structure. For strong-SOS, it requires an additional metric structure so that we can define addition. Furthermore, it requires the space to be continuous such as an interval of real numbers or integer numbers, otherwise it may trivialize the definition. For example, if the space contains four numbers $S_i = \{0, 1, 3, 7\}$, there does not exists any $s_i' \neq s_i \in S_i$ and $\delta \neq 0$ such that both $s_i' + \delta$ and $s_i + \delta$ are in S_i. When $\delta = 0$, the condition in the strong-SOS property is trivial; when $s_i' = s_i$, the property degenerates to SOS.

The above observation says that the property of strong-SOS crucially depends on the property of the signal space. So, it may not be that robust and widely applicable. In the following, we argue that it is not that special either. The informal statement is that for any SOS valuation, there exists a monotone mapping of the signal space such that it becomes strong-SOS after the mapping. The take away here is that one may abandon the concept of strong-SOS and focus mainly on SOS. On the other hand, we can also interpret it positively: one can make use of the strong-SOS property freely when designing mechanisms if it is helpful. Then the mechanism can be transformed to general SOS functions.

Theorem 4. *If there exists a mechanism with α-approximation for strong-SOS valuations, then there exists one with $(1 - O(\epsilon))\alpha$-approximation for general SOS valuations with finite discrete signal spaces.*

We only prove the result for finite discrete signal spaces for simplicity. We believe that it also holds for continuous space (maybe under some smoothness condition for the valuation functions such as Lipschitz condition). The detailed formal proof below is not that informative since most of the technical effort is to deal with the oddness for the definition of strong-SOS. The high level idea is simple: just find a mapping. As long as this mapping grows very fast (so we choose exponential functions here), it becomes strong-SOS. However, the space is not continuous after the mapping. We need to fill the holes, we use convex combination to fill the holes.

Proof. Assume that in the auction, each bidder's valuation function v_i is SOS. First, we convert the SOS valuation functions $\{v_i\}_{i \in [n]}$ into a new set of strong-SOS valuation functions $\{\bar{v}_i\}_{i \in [n]}$. Based on the results of $\bar{x}$ on $\{\bar{v}_i\}_{i \in [n]}$, we construct a new mechanism $\mathcal{M}$ that achieves $\alpha(1 - O(\epsilon))$-approximation for the original SOS valuation functions $\{v_i(s)\}_{i \in [n]}$.

We first show the construction for $\{\bar{v}_i\}_{i \in [n]}$. For finite discrete signal spaces S_i, it is without loss of generality to assume that they are simply consecutive integers starting from zero. Our idea is to extend the original signal space of each bidder S_i to $\bar{S}_i = [\sum_{k=0}^{|S_i|-1} c^k]$, where $c = \lceil \frac{\max_{s \in S} v_{(1)}(s)}{\epsilon} \rceil + 1$ correlates to the maximum valuation of a single bidder over all signals. The signal space of all bidders is thus $\bar{S} = \prod_{i \in [n]} \bar{S}_i$. Specifically, each signal $s_i \in S_i$ is mapped to an exponential signal $c_{s_i} = \sum_{k=1}^{s_i - 1} c^k \in \bar{S}$. We define the set $\{c_s = \sum_{k=1}^{s} c^k\}_{s \in [\max_{i \in [n]} |S_i|]}$ as $\mathcal{C}$. Signal $\bar{s}_i \notin \mathcal{C}$ is a convex combination of its two closest signals in $\mathcal{C}$. Formally, for each signal $\bar{s}_i \in \bar{\mathcal{S}}_i$, let $\ell(\bar{s}_i) = \max_{s_i \in \mathcal{S}, c_{s_i} \leq \bar{s}_i} s_i$ and $(\bar{s}_i) = \min_{s_i \in \mathcal{S}, c_{s_i} \geq \bar{s}_i} s_i$ be the two closest integers smaller and larger than $\bar{s}_i$ separately. Notice that $\ell(\bar{s}_i) = (\bar{s}_i)$ when $\bar{s}_i \in \mathcal{C}$, and $\ell(\bar{s}_i) + 1 = (\bar{s}_i)$ otherwise. For the convenience of notation, we define this convex decomposition $\mu(\bar{s}_i) : \bar{\mathcal{S}}_i \to \Delta_{\mathcal{S}_i}$ as a distribution where $P(\ell(\bar{s}_i)) = \frac{c_{\ell(\bar{s}_i)+1} - \bar{s}_i}{c_{\ell(\bar{s}_i)+1} - c_{\ell(\bar{s}_i)}}$ and $P(\ell(\bar{s}_i)+1) = \frac{\bar{s}_i - c_{\ell(\bar{s}_i)}}{c_{\ell(\bar{s}_i)+1} - c_{\ell(\bar{s}_i)}}$. The decomposition of a signal profile $\mu(\bar{s})$ is the joint distribution over the decomposition of each bidder's signal $\bar{s}_i$, i.e., $\mu(\bar{s}) = \prod_{i=1}^{n} \mu(\bar{s}_i)$.

We construct valuation function $\bar{v}_i(\bar{s})$ as the expectation over $\mu(\bar{s})$ plus a small number, defined as the following

$$\bar{v}_i(\bar{s}) = \mathbb{E}_{s \sim \mu(\bar{s})}\left[v_i(s) + \epsilon \cdot \|s\|_1\right]$$

Lemma 5. *For any $i \in [n]$, the constructed valuation function $\bar{v}_i : \bar{\mathcal{S}} \to \mathbb{R}$ is strong-SOS.*

Proof. We will first show that $\bar{v}_i$ is an SOS valuation function. Next, we will show that for any $\bar{s}_{-j}$, $\bar{v}_i(\bar{s}_{-j}, \bar{s}_j)$ is a convex sequence in $\bar{s}_j$, i.e., $\bar{v}_i(\bar{s}_{-j}, \bar{s}_j + 1) - \bar{v}_i(\bar{s}_{-j}, \bar{s}_j) \leq \bar{v}_i(\bar{s}_{-j}, \bar{s}_j) - \bar{v}_i(\bar{s}_{-j}, \bar{s}_j - 1)$. This implies that for any $s'_j \leq s_j$ and $\delta \geq 0$, $\bar{v}_i(\bar{s}_{-j}, \bar{s}_j + \delta) - \bar{v}_i(\bar{s}_{-j}, \bar{s}_j) \leq \bar{v}_i(\bar{s}_{-j}, \bar{s}'_j + \delta) - \bar{v}_i(\bar{s}_{-j}, \bar{s}'_j)$. By combining these two results, we can conclude that $\bar{v}_i$ is strong-SOS.

We start by proving a useful claim.

Claim. For any $\ell \in [\mathcal{S} - 1]$ and $\bar{s}'_j \leq \bar{s}_j$,

$$\bar{v}_i(\bar{s}_{-jk}, \bar{s}_j, c_{\ell+1}) - \bar{v}_i(\bar{s}_{-jk}, \bar{s}_j, c_\ell) \leq \bar{v}_i(\bar{s}_{-jk}, \bar{s}'_j, c_{\ell+1}) - \bar{v}_i(\bar{s}_{-jk}, \bar{s}'_j, c_\ell)$$

Proof. If $r(\bar{s}_j) - 1 = \ell(\bar{s}'_j)$,

$$\left(\bar{v}_i(\bar{s}_{-jk}, \bar{s}_j, c_{\ell+1}) - \bar{v}_i(\bar{s}_{-jk}, \bar{s}_j, c_\ell)\right) - \left(\bar{v}_i(\bar{s}_{-jk}, \bar{s}'_j, c_{\ell+1}) - \bar{v}_i(\bar{s}_{-jk}, \bar{s}'_j, c_\ell)\right)$$

$$= \frac{\bar{s}_j - \bar{s}'_j}{c_{\ell(\bar{s}'_j)+1} - c_{\ell(\bar{s}'_j)}}\left(\bar{v}_i(\bar{s}_{-jk}, c_{\ell(\bar{s}_j)+1}, c_{\ell+1}) - \bar{v}_i(\bar{s}_{-jk}, c_{\ell(\bar{s}_j)+1}, c_\ell)\right) +$$

$$\frac{\bar{s}'_j - \bar{s}_j}{c_{\ell(\bar{s}'_j)+1} - c_{\ell(\bar{s}'_j)}}\left(\bar{v}_i(\bar{s}_{-jk}, c_{\ell(\bar{s}_j)}, c_{\ell+1}) - \bar{v}_i(\bar{s}_{-jk}, c_{\ell(\bar{s}_j)}, c_\ell)\right)$$

$$= \frac{\bar{s}_j - \bar{s}'_j}{c_{\ell(\bar{s}'_j)+1} - c_{\ell(\bar{s}'_j)}}\left[\left(\bar{v}_i(\bar{s}_{-jk}, c_{\ell(\bar{s}_j)+1}, c_{\ell+1}) - \bar{v}_i(\bar{s}_{-jk}, c_{\ell(\bar{s}_j)+1}, c_\ell)\right) - \left(\bar{v}_i(\bar{s}_{-jk}, c_{\ell(\bar{s}_j)}, c_{\ell+1}) - \bar{v}_i(\bar{s}_{-jk}, c_{\ell(\bar{s}_j)}, c_\ell)\right)\right]$$

$$= \frac{\bar{s}_j - \bar{s}'_j}{c_{\ell(\bar{s}'_j)+1} - c_{\ell(\bar{s}'_j)}}\mathbb{E}_{s_{-jk} \sim \mu(\bar{s}_{-jk})}\left[(v_i(s_{-jk}, \ell(\bar{s}_j) + 1, \ell + 1) - v_i(s_{-jk}, \ell(\bar{s}_j) + 1, \ell) + \epsilon) - (v_i(s_{-jk}, \ell(\bar{s}_j), \ell + 1) - v_i(s_{-jk}, \ell(\bar{s}_j), \ell) + \epsilon)\right]$$

$$\leq 0$$

$\frac{\bar{s}_j - \bar{s}'_j}{c_{\ell(\bar{s}_j)+1} - c_{\ell(\bar{s}_j)}} \geq 0$ since $\bar{s}'_j \leq \bar{s}_j$. For each $\mathbf{s}_{-jk} \sim \mu(\bar{\mathbf{s}}_{-jk})$, since v_i is SOS, $v_i(\mathbf{s}_{-jk}, \ell(\bar{s}_j) + 1, \ell + 1) - v_i(\mathbf{s}_{-jk}, \ell(\bar{s}_j) + 1, \ell) \leq v_i(\mathbf{s}_{-jk}, \ell(\bar{s}_j), \ell + 1) - v_i(\mathbf{s}_{-jk}, \ell(\bar{s}_j), \ell)$. By linearity of expectation, the second part of equation is negative.

If $r(\bar{s}_j) - 1 > \ell(\bar{s}'_j)$,

$$\bar{v}_i(\bar{\mathbf{s}}_{-jk}, \bar{s}_j, c_{\ell+1}) - \bar{v}_i(\bar{\mathbf{s}}_{-jk}, \bar{s}_j, c_\ell)$$
$$\leq \bar{v}_i(\bar{\mathbf{s}}_{-jk}, c_{\ell(\bar{s}_j)}, c_{\ell+1}) - \bar{v}_i(\bar{\mathbf{s}}_{-jk}, c_{\ell(\bar{s}_j)}, c_\ell)$$
$$\leq \bar{v}_i(\bar{\mathbf{s}}_{-jk}, c_{\ell(\bar{s}_j)-1}, c_{\ell+1}) - \bar{v}_i(\bar{\mathbf{s}}_{-jk}, c_{\ell(\bar{s}_j)-1}, c_\ell)$$
$$\leq \dots$$
$$\leq \bar{v}_i(\bar{\mathbf{s}}_{-jk}, c_{\ell(\bar{s}'_j)+1}, c_{\ell+1}) - \bar{v}_i(\bar{\mathbf{s}}_{-jk}, c_{\ell(\bar{s}'_j)+1}, c_\ell)$$
$$\leq \bar{v}_i(\bar{\mathbf{s}}_{-jk}, \bar{s}'_j, c_{\ell+1}) - \bar{v}_i(\bar{\mathbf{s}}_{-jk}, , \bar{s}'_j, c_\ell)$$

With the result from Claim 6, we are ready to show that $\bar{v}_i$ is SOS.

Lemma 6. *For any* $\bar{\mathbf{s}}_{-j} \in \bar{\mathcal{S}}_{-j}$, $s'_j \leq s_j$ *and* $\delta \geq 0$,

$$\bar{v}_i(\bar{\mathbf{s}}_{-j}, \bar{s}_j + \delta) - \bar{v}_i(\bar{\mathbf{s}}_{-j}, \bar{s}_j) \leq \bar{v}_i(\bar{\mathbf{s}}'_{-j}, \bar{s}_j + \delta) - \bar{v}_i(\bar{\mathbf{s}}'_{-j}, \bar{s}_j)$$

Proof. We will first show that $\bar{v}_i$ is SOS when $\ell(\bar{s}_i) = \ell(\bar{s}_i + \delta)$, and then prove the more generalized result. Specifically, let us define $\ell = \ell(\bar{s}_i) = \ell(\bar{s}_i + \delta)$.

$$\bar{v}_i(\bar{\mathbf{s}}_{-j}, \bar{s}_j + \delta) - \bar{v}_i(\bar{\mathbf{s}}_{-j}, \bar{s}_j)$$
$$= \frac{\bar{s}_j + \delta - c_\ell}{c_{\ell+1} - c_\ell} \bar{v}_i(\bar{\mathbf{s}}_{-j}, c_{\ell+1}) + \frac{c_{\ell+1} - (\bar{s}_j + \delta)}{c_{\ell+1} - c_\ell} \bar{v}_i(\bar{\mathbf{s}}_{-j}, c_\ell) -$$
$$\frac{\bar{s}_j - c_\ell}{c_{\ell+1} - c_\ell} \bar{v}_i(\bar{\mathbf{s}}_{-j}, c_{\ell+1}) - \frac{c_{\ell+1} - \bar{s}_j}{c_{\ell+1} - c_\ell} \bar{v}_i(\bar{\mathbf{s}}_{-j}, c_\ell)$$
$$= \frac{\delta}{c_{\ell+1} - c_\ell} \left(\bar{v}_i(\bar{\mathbf{s}}_{-j}, c_{\ell+1}) - \bar{v}_i(\bar{\mathbf{s}}_{-j}, c_\ell) \right)$$
$$\leq \frac{\delta}{c_{\ell+1} - c_\ell} \left(\bar{v}_i(\bar{\mathbf{s}}_{-\{1,j\}}, \bar{s}'_1, c_{\ell+1}) - \bar{v}_i(\bar{\mathbf{s}}_{-\{1,j\}}, \bar{s}'_1, c_\ell) \right)$$
$$\leq \frac{\delta}{c_{\ell+1} - c_\ell} \left(\bar{v}_i(\bar{\mathbf{s}}_{-\{1,2,j\}}, \bar{s}'_1, \bar{s}'_2, c_{\ell+1}) - \bar{v}_i(\bar{\mathbf{s}}_{-\{1,2,j\}}, \bar{s}'_1, \bar{s}'_2, c_\ell) \right)$$
$$\leq \dots$$
$$\leq \frac{\delta}{c_{\ell+1} - c_\ell} \left(\bar{v}_i(\bar{\mathbf{s}}'_{-j}, c_{\ell+1}) - \bar{v}_i(\bar{\mathbf{s}}'_{-j}, c_\ell) \right)$$
$$= \frac{\bar{s}_j + \delta - c_\ell}{c_{\ell+1} - c_\ell} \bar{v}_i(\bar{\mathbf{s}}'_{-j}, c_{\ell+1}) + \frac{c_{\ell+1} - (\bar{s}_j + \delta)}{c_{\ell+1} - c_\ell} \bar{v}_i(\bar{\mathbf{s}}'_{-j}, c_\ell) -$$
$$\frac{\bar{s}_j - c_\ell}{c_{\ell+1} - c_\ell} \bar{v}_i(\bar{\mathbf{s}}'_{-j}, c_{\ell+1}) - \frac{c_{\ell+1} - \bar{s}_j}{c_{\ell+1} - c_\ell} \bar{v}_i(\bar{\mathbf{s}}'_{-j}, c_\ell)$$
$$= \bar{v}_i(\bar{\mathbf{s}}'_{-j}, \bar{s}_j + \delta) - \bar{v}_i(\bar{\mathbf{s}}'_{-j}, \bar{s}_j)$$

The inequality above is from Claim 6.

We will next show that $\bar{v}_i$ is SOS when $\ell(\bar{s}_j)$ and $\ell(\bar{s}_j + \delta)$ could possibly be different, i.e., $\ell(\bar{s}_j) \leq \ell(\bar{s}_j + \delta)$.

$$\bar{v}_i(\bar{\mathbf{s}}_{-j}, \bar{s}_j + \delta) - \bar{v}_i(\bar{\mathbf{s}}_{-j}, \bar{s}_j)$$
$$= \left(\bar{v}_i(\bar{\mathbf{s}}_{-j}, c_{\ell(\bar{s}_j)+1}) - \bar{v}_i(\bar{\mathbf{s}}_{-j}, \bar{s}_j)\right) + \left(\bar{v}_i(\bar{\mathbf{s}}_{-j}, c_{\ell(\bar{s}_j)+2}) - \bar{v}_i(\bar{\mathbf{s}}_{-j}, c_{\ell(\bar{s}_j)+1})\right) + \cdots +$$
$$\left(\bar{v}_i(\bar{\mathbf{s}}_{-j}, \bar{s}_j + \delta) - \bar{v}_i(\bar{\mathbf{s}}_{-j}, c_{\ell(\bar{s}_j+\delta)})\right)$$
$$\leq \left(\bar{v}_i(\bar{\mathbf{s}}'_{-j}, c_{\ell(\bar{s}_j)+1}) - \bar{v}_i(\bar{\mathbf{s}}_{-j}, \bar{s}_j)\right) + \left(\bar{v}_i(\bar{\mathbf{s}}'_{-j}, c_{\ell(\bar{s}_j)+2}) - \bar{v}_i(\bar{\mathbf{s}}'_{-j}, c_{\ell(\bar{s}_j)+1})\right) + \cdots +$$
$$\left(\bar{v}_i(\bar{\mathbf{s}}'_{-j}, \bar{s}_j + \delta) - \bar{v}_i(\bar{\mathbf{s}}'_{-j}, c_{\ell(\bar{s}_j+\delta)})\right)$$
$$= \bar{v}_i(\bar{\mathbf{s}}'_{-j}, \bar{s}_j + \delta) - \bar{v}_i(\bar{\mathbf{s}}'_{-j}, \bar{s}_j)$$

Next, we will show that for any signal profile $\bar{\mathbf{s}}$, $\bar{v}_i(\bar{\mathbf{s}}_{-j}, \bar{s}_j + 1) - \bar{v}_i(\bar{\mathbf{s}}_{-j}, \bar{s}_j) \leq \bar{v}_i(\bar{\mathbf{s}}_{-j}, \bar{s}_j) - \bar{v}_i(\bar{\mathbf{s}}_{-j}, \bar{s}_j - 1)$.

When $\ell(\bar{s}_j - 1) = \ell(\bar{s}_j) = \ell$, $\bar{v}_i(\bar{\mathbf{s}}_{-j}, \bar{s}_j + 1) - \bar{v}_i(\bar{\mathbf{s}}_{-j}, \bar{s}_j)$ and $\bar{v}_i(\bar{\mathbf{s}}_{-j}, \bar{s}_j) - \bar{v}_i(\bar{\mathbf{s}}_{-j}, \bar{s}_j - 1)$ both can be rewritten as $\frac{1}{c_{\ell+1} - c_\ell}\left(\bar{v}_i(\bar{\mathbf{s}}_{-j}, c_{\ell+1}) - \bar{v}_i(\bar{\mathbf{s}}_{-j}, c_\ell)\right)$ and thus,

$$\bar{v}_i(\bar{\mathbf{s}}_{-j}, \bar{s}_j + 1) - \bar{v}_i(\bar{\mathbf{s}}_{-j}, \bar{s}_j) = \bar{v}_i(\bar{\mathbf{s}}_{-j}, \bar{s}_j) - \bar{v}_i(\bar{\mathbf{s}}_{-j}, \bar{s}_j - 1)$$

Otherwise, $\ell(\bar{s}_j - 1) < \ell(\bar{s}_j)$ implies that $\ell(\bar{s}_j - 1) = \ell(\bar{s}_j) - 1$. $\bar{s}_j \in \mathcal{C}$.

$$\bar{v}_i(\bar{\mathbf{s}}_{-j}, \bar{s}_j + 1) - \bar{v}_i(\bar{\mathbf{s}}_{-j}, \bar{s}_j)$$
$$= \frac{1}{c_{\ell(\bar{s}_j)+1} - c_{\ell(\bar{s}_j)}}\left(\bar{v}_i(\bar{\mathbf{s}}_{-j}, c_{\ell(\bar{s}_j)+1}) - \bar{v}_i(\bar{\mathbf{s}}_{-j}, c_{\ell(\bar{s}_j)})+\epsilon\right)$$
$$= \frac{1}{c^{\ell(\bar{s}_j)}}\left(\mathbb{E}_{\mathbf{s}_{-j} \sim \mu(\bar{\mathbf{s}}_{-j})}\left(v_i(\mathbf{s}_{-j}, \ell(\bar{s}_j) + 1) - v_i(\mathbf{s}_{-j}, \ell(\bar{s}_j))\right) + \epsilon\right)$$
$$\leq \frac{1}{c^{\ell(\bar{s}_j)}}\left(\max_{s \in \mathcal{S}} v_{(1)}(\mathbf{s}) + \epsilon\right)$$
$$\leq \frac{\epsilon}{c^{\ell(\bar{s}_j)-1}}$$
$$\leq \frac{1}{c^{\ell(\bar{s}_j)-1}}\left(\mathbb{E}_{\mathbf{s}_{-j} \sim \mu(\bar{\mathbf{s}}_{-j})}\left(v_i(\mathbf{s}_{-j}, \ell(\bar{s}_j)) - v_i(\mathbf{s}_{-j}, \ell(\bar{s}_j) - 1)\right) + \epsilon\right)$$
$$= \frac{1}{c_{\ell(\bar{s}_j-1)+1} - c_{\ell(\bar{s}_j-1)}}\left(\bar{v}_i(\bar{\mathbf{s}}_{-j}, c_{\ell(\bar{s}_j)}) - \bar{v}_i(\bar{\mathbf{s}}_{-j}, c_{\ell(\bar{s}_j)-1})+\epsilon\right)$$
$$= \bar{v}_i(\bar{\mathbf{s}}_{-j}, \bar{s}_j) - \bar{v}_i(\bar{\mathbf{s}}_{-j}, \bar{s}_j - 1)$$

Therefore, $v_i(\mathbf{s}_{-j}, \mathbf{s}_j)$ forms a convex sequence on $\mathbf{s}_j$. We can then conclude that for any $\delta > 0$,

$$\bar{v}_i(\bar{\mathbf{s}}_{-j}, \bar{s}_j + \delta) - \bar{v}_i(\bar{\mathbf{s}}_{-j}, \bar{s}_j) \leq \bar{v}_i(\bar{\mathbf{s}}_{-j}, \bar{s}'_j + \delta) - \bar{v}_i(\bar{\mathbf{s}}_{-j}, \bar{s}'_j)$$

Combining the two results we conclude our proof.

$$\bar{v}_i(\bar{\mathbf{s}}_{-j}, \bar{s}_j + \delta) - \bar{v}_i(\bar{\mathbf{s}}_{-j}, \bar{s}_j) \leq \bar{v}_i(\bar{\mathbf{s}}_{-j}, \bar{s}'_j + \delta) - \bar{v}_i(\bar{\mathbf{s}}_{-j}, \bar{s}'_j) \leq \bar{v}_i(\bar{\mathbf{s}}'_{-j}, \bar{s}'_j + \delta) - \bar{v}_i(\bar{\mathbf{s}}'_{-j}, \bar{s}'_j)$$

Finally, we show our reduction and prove the approximation result. Suppose mechanism $\bar{\mathcal{M}} = (\bar{x}, \bar{p})$ is ex-post IC & IR and achieves α-approximation on any strong-SOS valuation function. We will show that there exists a monotone allocation rule x such that x achieves $\alpha(1 - O(\epsilon))$-approximation on any SOS valuation setting. $x(\mathbf{s})$ simply takes the allocation rule of signal $c_{\mathbf{s}} \in \bar{\mathcal{S}}$, i.e., $\mathbf{x}(\mathbf{s}) = \bar{\mathbf{x}}(c_{\mathbf{s}})$. The monotonicity of $\bar{\mathbf{x}}(\bar{\mathbf{s}})$ directly implies that $x(\mathbf{s})$ is monotone: for any bidder $i \in [n]$, signals $\mathbf{s}_{-i}$ and $s_i \geq s_i'$,

$$x_i(\mathbf{s}_{-i}, s_i) = \bar{x}(c_{s_1}, c_{s_2}, \cdots, c_{s_i}, \cdots, c_{s_n}) \geq \bar{x}(c_{s_1}, c_{s_2}, \cdots, c_{s_i'}, \cdots, c_{s_n}) = x_i(\mathbf{s}_{-i}, s_i')$$

Since $\bar{\mathbf{x}}$ achieves α-approximation on strong-SOS valuation functions $\{\bar{v}_i\}_{i \in [n]}$, we have

$$\min_{\mathbf{s} \in \mathcal{S}} \frac{\sum_i x_i(\mathbf{s}) \cdot (v_i(\mathbf{s}) + \epsilon \cdot \|\mathbf{s}\|_1)}{v_{(1)}(\mathbf{s}) + \epsilon \cdot \|\mathbf{s}\|_1} = \min_{\mathbf{s} \in \mathcal{S}} \frac{\sum_i \bar{x}_i(c_{\mathbf{s}}) \cdot \bar{v}_i(c_{\mathbf{s}})}{\bar{v}_{(1)}(c_{\mathbf{s}})} \geq \min_{\bar{\mathbf{s}} \in \bar{\mathcal{S}}} \frac{\sum_i \bar{x}_i(\bar{\mathbf{s}}) \cdot \bar{v}_i(\bar{\mathbf{s}})}{\bar{v}_{(1)}(\bar{\mathbf{s}})} \geq \alpha$$

Based on our construction that $\bar{v}_i(\bar{\mathbf{s}}) = \mathbb{E}_{\mathbf{s} \sim \mu(\bar{\mathbf{s}})} [v_i(\mathbf{s}) + \epsilon \cdot \|\mathbf{s}\|_1]$,

$$\min_{\mathbf{s} \in \mathcal{S}} \frac{\sum_i x_i(\mathbf{s}) \cdot v_i(\mathbf{s})}{v_{(1)}(\mathbf{s})}$$

$$\geq \min_{\mathbf{s} \in \mathcal{S}} \frac{\sum_i x_i(\mathbf{s}) \cdot v_i(\mathbf{s})}{v_{(1)}(\mathbf{s}) + \epsilon \cdot \|\mathbf{s}\|_1}$$

$$\geq \min_{\mathbf{s} \in \mathcal{S}} \frac{\sum_i x_i(\mathbf{s}) \cdot (v_i(\mathbf{s}) + \epsilon \cdot \|\mathbf{s}\|_1)}{v_{(1)}(\mathbf{s}) + \epsilon \cdot \|\mathbf{s}\|_1} \cdot \min_{\mathbf{s} \in \mathcal{S}} \frac{\sum_i x_i(\mathbf{s}) \cdot v_i(\mathbf{s})}{\sum_i x_i(\mathbf{s}) \cdot (v_i(\mathbf{s}) + \epsilon \cdot \|\mathbf{s}\|_1)}$$

$$\geq \alpha \cdot \min_{\mathbf{s} \in \mathcal{S}, i \in [n]} \frac{v_i(\mathbf{s})}{v_i(\mathbf{s}) + \epsilon \cdot \|\mathbf{s}\|_1}$$

$$\geq \alpha \cdot \frac{1}{1 + \epsilon \cdot \frac{\max_{\mathbf{s} \in \mathcal{S}} \|\mathbf{s}\|_1}{\min_{\mathbf{s} \in \mathcal{S}} v_{(1)}(\mathbf{s})}}$$

$$= \alpha \cdot \left(1 - \frac{\max_{\mathbf{s} \in \mathcal{S}}}{\min_{\mathbf{s} \in \mathcal{S}} v_{(1)}(\mathbf{s})} \cdot \epsilon\right)$$

Therefore, x has $(1 - O(\epsilon))\alpha$-approximation on $\{v_i(\mathbf{s})\}_{i \in [n]}$

References

1. Amer, A., Talgam-Cohen, I.: Auctions with interdependence and sos: improved approximation. In: SAGT (2021)
2. Athey, S.: Single crossing properties and the existence of pure strategy equilibria in games of incomplete information. Econometrica **69**(4), 861–889 (2001)
3. Ausubel, L.: A Generalized Vickrey Auction. Econometric Society World Congress 2000 Contributed Papers 1257, Econometric Society, August 2000
4. Chawla, S., Fu, H., Karlin, A.: Approximate revenue maximization in interdependent value settings. In: EC 2014 (2014)
5. Che, Y.K., Kim, J., Kojima, F.: Efficient assignment with interdependent values. J. Econ. Theory **158**, 54–86 (2015)

6. Clarke, E.H.: Multipart pricing of public goods. Public Choice **11**, 17–33 (1971)
7. Dasgupta, P., Maskin, E.: Efficient Auctions. Q. J. Econ. **115**(2), 341–388 (2000)
8. Eden, A., Feldman, M., Fiat, A., Goldner, K.: Interdependent values without single-crossing. In: EC 2018 (2018)
9. Eden, A., Feldman, M., Fiat, A., Goldner, K., Karlin, A.R.: Combinatorial auctions with interdependent valuations: Sos to the rescue. In: EC 2019 (2019)
10. Eden, A., Feldman, M., Talgam-Cohen, I., Zviran, O.: Poa of simple auctions with interdependent values. In: Proceedings of the AAAI Conference on Artificial Intelligence, vol. 35(6), pp. 5321–5329 (2021)
11. Eden, A., Goldner, K., Zheng, S.: Private interdependent valuations. In: Proceedings of the 2022 Annual ACM-SIAM Symposium on Discrete Algorithms (SODA), pp. 2920–2939
12. Gkatzelis, V., Patel, R., Pountourakis, E., Schoepflin, D.: Prior-free clock auctions for bidders with interdependent values (2021)
13. Groves, T.: Incentives in teams. Econometrica **41**(4), 617–631 (1973)
14. Ito, T., Parkes, D.C.: Instantiating the contingent bids model of truthful interdependent value auctions. In: AAMAS 2006, pp. 1151–1158 (2006)
15. Jehiel, P., Moldovanu, B.: Efficient design with interdependent valuations. Econometrica **69**(5), 1237–1259 (2001)
16. Li, Y.: Approximation in mechanism design with interdependent values. In: EC 2013 (2013)
17. Maskin, E.: Auctions and Privatization, pp. 115–136. J.C.B. Mohr Publisher (1992)
18. McLean, R., Postlewaite, A.: Implementation with interdependent valuations. Theoret. Econ. **10**(3), September 2015
19. Milgrom, P.R., Weber, R.J.: A theory of auctions and competitive bidding. Econometrica **50**(5), 1089–1122 (1982)
20. Roughgarden, T., Talgam-Cohen, I.: Optimal and robust mechanism design with interdependent values. ACM Trans. Econ. Comput. **4**(3), June 2016
21. Vickrey, W.: Counterspeculation, auctions, and competitive sealed tenders. J. Financ. **16**(1), 8–37 (1961)
22. Vohra, R.V.: Mechanism design: a linear programming approach (2011)

Social Choice

On Best-of-Both-Worlds Fair-Share Allocations

Moshe Babaioff[1], Tomer Ezra[2(✉)], and Uriel Feige[1,3]

[1] Microsoft Research, Herzliya, Israel
moshe@microsoft.com
[2] Sapienza University of Rome, Rome, Italy
tomer.ezra@gmail.com
[3] Weizmann Institute, Rehovot, Israel
uriel.feige@weizmann.ac.il

Abstract. We consider the problem of fair allocation of indivisible items among n agents with additive valuations, when agents have equal entitlements to the goods, and there are no transfers. Best-of-Both-Worlds (BoBW) fairness mechanisms aim to give all agents both an ex-ante guarantee (such as getting the proportional share in expectation) and an ex-post guarantee. Prior BoBW results have focused on ex-post guarantees that are based on the "up to one item" paradigm, such as envy-free up to one item (EF1). In this work we attempt to give every agent a high *value* ex-post, and specifically, a constant fraction of her maximin share (MMS). There are simple examples in which previous BoBW mechanisms give some agent only a $\frac{1}{n}$ fraction of her MMS.

Our main result is a deterministic polynomial-time algorithm that computes a distribution over allocations that is ex-ante proportional, and ex-post, every allocation gives every agent at least half of her MMS. Moreover, the ex-post guarantee holds even with respect to a more demanding notion of a share, introduced in this paper, that we refer to as the *truncated proportional share* (TPS). Our guarantees are nearly best possible, in the sense that one cannot guarantee agents more than their proportional share ex-ante, and one cannot guarantee all agents value larger than a $\frac{n}{2n-1}$-fraction of their TPS ex-post.

Keywords: Fair division · Best-of-both-worlds · Maximin share · Truncated proportional share

1 Introduction

In this paper we consider fair allocation of indivisible items to agents with additive valuations. An *instance* $\mathcal{I} = (v, \mathcal{M}, \mathcal{N})$ of the fair allocation problem consists of a set $\mathcal{M}$ of m indivisible items, a set $\mathcal{N}$ of n agents, and vector $v = (v_1, v_2, \ldots, v_n)$ of non-negative additive valuations, with the valuation of agent $i \in \mathcal{N}$ for set $S \subseteq \mathcal{M}$ being $v_i(S) = \sum_{j \in S} v_i(j)$, where $v_i(j)$ denotes the value of agent i for item $j \in \mathcal{M}$. We assume that the valuation functions of the

© The Author(s), under exclusive license to Springer Nature Switzerland AG 2022
K. A. Hansen et al. (Eds.): WINE 2022, LNCS 13778, pp. 237–255, 2022.
https://doi.org/10.1007/978-3-031-22832-2_14

agents are known to the social planer, and that there are no transfers (no money involved). We further assume that all agents have equal entitlement to the items. An allocation A is a collection of n disjoint bundles $A_1, \ldots, A_n$ (some of which might be empty), where $A_i \subseteq \mathcal{M}$ for every $i \in \mathcal{N}$. A randomized allocation is a distribution over deterministic allocations. We wish to design randomized allocations that enjoy certain fairness properties.

Before discussing some standard fairness properties, we briefly motivate the *best of both worlds* (BoBW) framework, that considers both ex-ante and ex-post properties of randomized allocations. Consider an instance $\mathcal{I}_1$ with two agents and one indivisible item. Giving the item to one of the agents arbitrarily may be considered to be ex-post fair (because whatever allocation is chosen, it is unavoidable that some agent receives no item), but not ex-ante fair (as there was no a-priori reason to discriminate against the agent not receiving the item). Ex-ante fairness can be accomplished (while still maintaining ex-post fairness) by having an unbiased lottery to decide which agent gets the item. Consider now an allocation instance $\mathcal{I}_2$ with two agents and two equally valued items. We can have a lottery for $\mathcal{I}_2$, and have the winner receive both items. This would be ex-ante fair, but ex-post (with respect to the final allocation) it would not be fair (as there are allocations that give every agent one item). For $\mathcal{I}_2$, giving each agent one item is fair both ex-ante and ex-post. Examples such as those above illustrate why we want our allocation mechanism to concurrently enjoy *both* ex-ante and ex-post fairness guarantees, as each guarantee by itself seems not to be sufficiently fair.

For the purpose of defining ex-ante fairness properties of randomized allocations, we assume that agents are risk neutral. That is, the ex-ante value that an agent derives from a distribution over bundles is the same as the expected value of a bundle selected at random from this distribution. Consequently, when considering a distribution D over allocations (of $\mathcal{M}$ to $\mathcal{N}$), we also consider the expectation of this distribution, which can be interpreted as a *fractional allocation*. In this fractional allocation, the fraction of item i given to agent j exactly equals the probability with which agent i receives item j under D. We naturally extend the additive valuation functions of agents to fractional allocations, by considering the expected valuation, that is, an additive valuation where the value of a fraction q_j of item j to agent i is $q_j \cdot v_i(j)$.

1.1 Brief Review of Terminology and Notation

We briefly review some properties of allocations from the literature, properties that are most relevant to the current work and to prior related work.

We start with standard share definitions. The *proportional share* of agent i is $PS_i = \frac{v_i(\mathcal{M})}{n}$. We say that an allocation $A = (A_1, \ldots, A_n)$ is *proportional* if every agent i gets value at least PS_i (that is, $v_i(A_i) \geq \frac{v_i(\mathcal{M})}{n} = PS_i$), and a fractional (randomized) allocation is *ex-ante proportional* if she gets her proportional share in expectation. We say that an allocation A is *proportional up to one item (Prop1)* if for every agent i it holds that $v_i(A_i) \geq PS_i - \max_{j \in \mathcal{M} \setminus A_i} v_i(j)$. The *maximin share* MMS_i of agent i is the maximum value that i could secure if she was to partition $\mathcal{M}$ into n bundles, and receive the bundle with the lowest value under v_i.

We next discuss envy. An allocation is *envy free* (EF) if every agent (weakly) prefers her own bundle over that of any other agent, and a fractional (randomized) allocation is *ex-ante envy free* if for every agent, the expected value of her own allocation is at least as high as the expected value of the allocation of any other agent. Note that an allocation that is ex-ante envy free is ex-ante proportional. An allocation is *envy-free up to one good (EF1)* (*envy-free up to any good (EFX)*, respectively) if every agent weakly prefers her own bundle over that of any other agent, up to the most (least, respectively) valuable item in the other agent's bundle. Note that EF1 implies Prop1. Finally, an allocation is *envy-free up to one good more-and-less (EF_1^1)* if no agent i envies another agent j after removing one item from the set j gets, and adding one item (not necessarily the same item) to i. Note that EF_1^1 is weaker than $EF1$.

Finally, we consider notions of efficiency. An (fractional) allocation *Pareto dominates* another (fractional) allocation if it is weakly preferred by all agents, and strictly so by at least one. An integral allocation is *Pareto optimal (PO)* if no integral allocation Pareto dominates it. An allocation (integral or fractional) is *fractionally Pareto optimal (fPO)* if it is Pareto optimal, and moreover, no fractional allocation Pareto dominates it. Another notion of efficiency is that of *Nash Social Welfare maximization*. The *Nash Social Welfare (NSW)* of allocation $A = (A_1, \ldots, A_n)$ is $\left(\prod_{i \in \mathcal{N}} v_i(A_i) \right)^{\frac{1}{n}}$. In case of fractional allocations, we use the notation fNSW.

1.2 Previous BoBW Results for Additive Valuations

The state of the art BoBW results for additive valuations are presented in the two recent papers of [18], and [3]. Both of these works are based on the well known paradigm that we call here "faithful implementation of a fractional allocation": a distribution over deterministic allocations is a *faithful implementation of the fractional allocation* if the ex-ante (expected) value of every agent under the distribution is the same as it is in the fractional allocation, and ex-post (for any realization) it is the same as the expectation, up to the value of one item. Both papers use versions of the result of [13] showing that any fractional allocation can be faithfully implemented. Various versions of these results were presented in the past, and in the full version [5] we survey those results. In Sect. 2 we formally present a version of "faithful implementation" that summarizes the prior results, stated as Lemma 1.

By "faithful implementing" the fractional allocation that is the outcome of multiple executions of the probabilistic serial mechanism (a.k.a. *eating* mechanism) of [11] till there are no more items, the following BoBW result was proved in [3]. (The same theorem was established earlier in [18], but with a different proof.)

Theorem 1 ([3,18]). *There is a deterministic polynomial-time faithful implementation of a fractional allocation that is ex-ante envy free (and thus ex-ante* *proportional), and the implementation is supported on allocations that are (ex-post) EF1.*

By "faithful implementing" the fractional allocation that maximizes the fractional Nash Social Welfare, the following BoBW result was proved in [18].

Theorem 2 ([18]).   *There is a deterministic polynomial-time faithful implementation of a fractional allocation that is ex-ante fPO and ex-ante proportional, and the implementation is supported on allocations that are (ex-post) fPO, Prop1, and EF_1^1.*

The "up to one item" paradigm used in the ex-post guarantees of Theorems 1 and 2 is most useful when a difference of one item does not make a big difference in value. However, when items do have large values, it does not guarantee agents a high ex-post value. In contrast, we aim to give each agent "high enough value" ex-post, where value is measure compared to "what the agent deserves", captured by her fair share. Specifically, we aim to give every agent a large fraction ("an approximation") of her "fair share", e.g. half the agent's MMS. The following allocation instance shows that neither Theorem 1 nor Theorem 2 provide a constant approximation for the MMS ex-post, and both are supported only on allocations that are intuitively very unfair. Moreover, in this instance the MMS equals the proportional share, and hence one cannot dismiss this example as one in which the MMS is too small for the agents to care about.

Consider an instance with n identical items, each of value n. In this case it is clear each agent should get one item ex-post. Now, suppose that one of those big items is split into n small items, each of value 1. In this case we want one agent to get all of these small items, and each other agent to get one of the big items. Our next example shows that once these small items are not completely identical, but rather each agent slightly prefers a different one of them, then in both prior BoBW results, in every realization, one of the agents ends up getting only a small fraction of her MMS.

Example 1. The instance has $2n - 1$ items $\{s_1, s_2, \ldots, s_n\} \cup \{b_1, b_2, \ldots, b_{n-1}\}$. For some small $\epsilon > 0$, for every agent $i \in \mathcal{N}$, the additive valuation function v_i is as follows:

- $v_i(s_i) = 1 + \epsilon$.
- $v_i(s_j) = 1 - \frac{\epsilon}{n-1}$ for every $1 \leq j \leq n$ such that $j \neq i$.
- $v_i(b_j) = n$ for every $1 \leq j \leq n - 1$.

The MMS of every agent is n: one bundle contains all small items $\{s_1, \ldots, s_n\}$, and the remaining bundles each contain one of the remaining, big, items. The proportional share of every agent is also n.

In every allocation, at least one agent does not receive a big item, as there are fewer big items than agents. The algorithm of Theorem 1 gives every agent at most $\lceil \frac{|\mathcal{M}|}{n} \rceil$ items. Hence the agent that does not receive a big item receives a value of at most $2 - \epsilon$, whereas her MMS is n.

The algorithm of Theorem 2 starts with a fractional allocation that maximizes the fractional Nash Social Welfare. This fractional allocation necessarily allocates the small item s_i integrally to agent i, for every $i \leq n$. Consequently, also ex-post, every agent i gets the respective item s_i. By the pigeon-hole principle, in an ex-post allocation there is an agent that receives no item among the big items $\{b_1, b_2, \ldots, b_{n-1}\}$. This agent i receives only the small item s_i, and hence only a $\frac{1+\epsilon}{n}$ fraction of her MMS.

1.3   Our Contributions

In this paper we aim for a Best-of-Both-Worlds fairness result: a randomized allocation that gives every agent at least her proportional share ex-ante, and some guaranteed value ex-post. The ex-post guarantee we give is at least half the MMS, and in fact, stronger. We introduce a new notion of share that we refer to as the *truncated proportional share (TPS)*, which we believe might be of independent interest. We show that the TPS is at least as large as the MMS, and our BoBW result guarantees half of the TPS ex-post (and thus half the MMS ex post), while also giving each agent her proportional share ex-ante.

The Truncated Proportional Share. We next define the Truncated Proportional Share of an agent with an additive valuation. It equals the proportional share after the values of items of exceptionally high value have been truncated (hence the name truncated proportional share). As we will see later, this share has two advantages over MMS: it is at least as high as the MMS, and while the MMS is NP-hard to compute, the TPS is easy to compute. We alert the reader that in this paper we define TPS only with respect to additive valuation functions (while the definition of MMS extends without change beyond additive valuations).

Definition 1. *For a setting with n agents and a set of items $\mathcal{M}$, the* truncated proportional share $TPS_i = TPS_i(n, \mathcal{M}, v_i)$ *of agent i with additive valuation function v_i is the largest value t such that $\frac{1}{n} \sum_{j \in \mathcal{M}} \min[v_i(j), t] = t$.*

We note that the TPS is well defined, as $t = 0$ satisfies the equality, and the maximum is obtained as the RHS is linear, while the LHS is piece-wise-linear with finitely many segments (at most m). From the definition of TPS it is immediate to see that $TPS_i \leq PS_i$. The TPS_i is smaller than PS_i if and only if there is at least one *over-proportional item*, an item that by itself gives agent i value larger than PS_i. For example, whenever there are less items than agents (e.g. a single item and two agents that desire it) then for every agent $TPS_i = MMS_i = 0$ while $PS_i > 0$. Moreover, regardless of the presence of over-proportional items, $TPS_i \geq MMS_i$. This is because taking $t = MMS_i$ satisfies $\frac{1}{n} \sum_{j \in \mathcal{M}} \min[v_i(j), t] \geq t$ (as every one of the n bundles in the partition that determines MMS_i contributes at least t to the sum), which implies that t in Definition 1 is at least as large as MMS_i. Hence $MMS_i \leq TPS_i \leq PS_i$. In particular, guarantees with respect to the TPS imply at least the same guarantees with respect to MMS, and sometimes better.

The following example illustrates the TPS definition:

Example 2. There are $n = 4$ agents and $m = 5$ items. The values of the items for agent i are $2, 3, 4, 5, 6$. Her proportional share PS_i equals $\frac{2+3+4+5+6}{4} = 5$, and her truncated proportional share TPS_i can be seen to be 4.5. Her TPS is at least 4.5 since $\frac{2+3+4+4.5+4.5}{4} = 4.5$. Her TPS is at most 4.5 since for every $t > 4.5$, it holds that $\frac{2+3+4+\min(5,t)+\min(6,t)}{4} \leq \frac{9+2t}{4} < t$.

The TPS is a more tractable object than the MMS. The TPS can be computed by the following recursive procedure (see proof in the full version [5]): when $n = 1$ then $TPS_i = TPS_i(1, \mathcal{M}, v_i) = v_i(\mathcal{M})$, and when $n \geq 2$ then TPS_i is the minimum among $\frac{v_i(\mathcal{M})}{n}$, the proportional share of agent i, and her TPS in a *reduced instance* in which an item j of highest value is removed as well as one of the agents, that is, TPS_i in this case is $TPS_i(n - 1, \mathcal{M} \setminus \{j\}, v_i)$. This procedure provides a simple polynomial time algorithm for computing the TPS: if the proportional share of the reduced instance is smaller than that of the original instance, compute TPS_i for the reduced instance. If not, then TPS_i is the proportional share of the original instance. (In contrast, computing the MMS is NP-hard.)

Moreover, consider ρ_{TPS}, the highest fraction such that in every instance, there is an allocation giving every agent a ρ_{TPS} fraction of her TPS. It is easy to determine the exact value of ρ_{TPS}, which turns out to be $\frac{n}{2n-1}$. (In contrast, the exact value of the corresponding ρ_{MMS} is unknown [20,23].) To see that $\rho_{TPS} \geq \frac{n}{2n-1}$, we observe that a polynomial time allocation algorithm of [25] gives every agent a $\frac{n}{2n-1}$ fraction of her TPS. (For more details in the full version [5].) To see that $\rho_{TPS} \leq \frac{n}{2n-1}$, consider an instance with $2n - 1$ items, each of value 1. The TPS of every agent is $\frac{2n-1}{n}$, but in every allocation, at least one of the agents gets at most one item, and hence value at most 1.

The example above also shows that the TPS of an agent can be factor $\frac{2n-1}{n}$ larger than her MMS. This ratio is tight, because $MMS_i \geq \frac{n}{2n-1} TPS_i$ for every agent i. This follows by considering n agents with the same valuation function v_i, and recalling that there is an allocation that gives every agent at least a $\frac{n}{2n-1}$ fraction of her TPS. The n bundles of this allocation each have a value of at least $\frac{n}{2n-1} TPS_i$, and hence they form a partition of $\mathcal{M}$ that shows that $MMS_i \geq \frac{n}{2n-1} TPS_i$. We summarize the above discussion in the following proposition:

Proposition 1. *For any setting with n agents and any additive valuation v_i it holds that*

$$PS_i \geq TPS_i \geq MMS_i \geq \tfrac{n}{2n-1} \cdot TPS_i$$

Moreover, each of the above inequalities is strict for some instance, and holds as equality for some other instance.

Our Best-of-Both-Worlds Result. We now return to present our main result. Example 1 and Proposition 2 (to follow shortly) illustrate the difficulties of following the paradigm of starting with a simple to describe fractional allocation, and then faithfully implementing it (using Lemma 1). Consequently, instead of using this paradigm, we design an algorithm that generates a distribution over allocations that each gives every agent at least half of her TPS, with the additional property that every agent gets at least her proportional share in expectation. Along the way, we do use Lemma 1, but we apply it on fractional allocations that involve only carefully selected subsets of $\mathcal{M}$, rather than a fractional allocation that involves all of $\mathcal{M}$. Our main result is the following.

Theorem 3. *For every allocation instance with additive valuations, there is a randomized allocation that is ex-ante proportional, and gives each agent at least half of her TPS ex-post (and hence also at least half of her MMS), as well as being Prop1 ex-post. Moreover, there is a deterministic polynomial time algorithm that, given the valuation functions of the agents, computes such a randomized allocation, supported on at most n allocations.*

Theorem 3 is nearly the best possible in the following senses. First, it is not possible to guarantee every agent value that is strictly larger than her proportional share ex-ante (e.g., if all agents have the same valuation function). Second, the highest possible fraction of the truncated proportional share that can be guaranteed ex-post is at most $\frac{n}{2n-1} = \frac{1}{2} + \frac{1}{4n-2}$ (recall the example above with the $2n - 1$ identical items), which tends to half as n grows large, and the theorem indeed ensures a fraction of half. We also remark that for the instance in Example 1, while in the BoBW results from prior work [3,18] there is always an agent that gets only a small fraction of her MMS, the algorithm of Theorem 3 gives every agent her TPS (and her MMS) ex-post.

Another aspect in which Theorem 3 cannot be improved is with respect to its Pareto properties. While the prior result of [18] present a BoBW result (Theorem 2) with a distribution over allocations that is ex-ante fPO, our result does not give ex-ante fPO. We next show that if we want every agent to receive ex-post at least a constant fraction of her maximin share, getting the guarantee of ex-ante fPO is impossible. Moreover, this conflict between ex-ante fPO and half the MMS concerns every ex-post allocation that might potentially be in the support, not just one of them.

Proposition 2. *For every $n \geq 2$ and every $\epsilon > 0$ there are allocation instances with additive valuations, with the following property: for every ex-ante Pareto optimal (fPO) randomized allocation (whether ex-ante proportional or not), every allocation in its support gives some agent at most a $\frac{1+\epsilon}{n}$ fraction of her maximin share.*

The proof of Proposition 2 is based on Example 1, and can be found in the full version [5].

We thus see that we cannot hope to improve our result to also guarantee ex-ante fPO. How about the weaker condition of ex-post PO? The polynomial time algorithm referred to in Theorem 3 does not necessarily produce Pareto efficient allocations. However, the existential result in the theorem does hold simultaneously with a Pareto efficiency requirement, for the simple reason that ex-post replacement of an allocation by an allocation that Pareto dominates it cannot reduce the received fraction of the (ex-post) truncated proportional share (and ex-ante proportional share) of any of the agents. It is not clear whether this reallocation can be done in polynomial time. (For NP-hardness results associated with Pareto efficient reallocation, see [4,16].)

Corollary 1. *For every allocation instance with additive valuations, there is a randomized allocation that is supported on at most n allocations, is ex-ante proportional, and ex-post it gives every agent at least half of her TPS (and hence also at least half her MMS), as well as being ex-post PO.*[1]

1.4 Additional Related Work

The maximin share was introduced by [12]. In [23] it is shown that for agents with additive valuations, an allocation that gives each agent her MMS may not exist. A series of papers [2,9,19–21,23] considered the best fraction of the MMS that can be concurrently guaranteed to all agents, and the current state of the art (for additive valuations) is a $\frac{3}{4} + \Theta(\frac{1}{n})$-fraction of the MMS (whereas there are instances in which more than a $\frac{39}{40}$ fraction of the MMS cannot be achieved [17]). For the case of arbitrary (non-equal) entitlements, [7] define a share named the *AnyPrice share (APS)*, which is the value the agent can guarantee herself whenever her budget is set to her entitlement b_i (when $\sum_i b_i = 1$) and she buys her highest value affordable set when items are adversarially priced with a total price of 1. To approximate the APS, they extend our definition of the TPS to the case of unequal entitlements. For additive valuations they show that $TPS_i \geq APS_i$ and that the inequality is strict for some instances.

The fairness notion of Prop1 was introduced by [15]. The fairness notion of EF1 was implicitly used in [25], and was formally defined in [12]. EFX (envy-free up to any good) was introduced in [14] The notion of envy-free up to one good more-and-less (EF_1^1) was defined in [8], relaxing EF1.

For a subclass of additive valuations, that of additive *dichotomous* valuations, very strong BoBW results are known [3,6,22], which among other properties, are EF ex-ante, EFX ex-post, maximize welfare, and the underlying allocation mechanism is universally truthful. Such a strong combination of results is impossible to achieve for general additive valuations. In particular, the results of [1] imply that every universally truthful randomized allocation mechanism for two agents that allocates all items must sometimes not give an agent more than a $\frac{2}{m}$ fraction of her MMS ex-post. In the full version [5], we give a more extensive discussion on truthfulness.

In Sect. 1.2 we already discussed some previous BoBW results. We further remark that in [18] they present an instance for which there is no randomized

---

[1] The improvement to a PO allocation might not maintain the Prop1 property, yet each agent's value never decreases under that improvement.

allocation that is ex-ante proportional, ex-post EF1 and ex-post fPO. For the same instance, there is no randomized allocation that is ex-ante proportional, is ex-post fPO, and gives every agent a positive fraction of her MMS.

2   Preliminaries

The first paragraph of the introduction describes the basic setting and notation used in this paper.

As we shall be dealing with randomized allocations, let us introduce terminology that we shall use in this context. A *random allocation* is a distribution D over integral allocations $A^1, A^2, \ldots$. It induces an *expected* allocation A^*, where A^*_{ij} specifies for agent i and item j the probability that agent i receives item j, when an allocation is chosen at random from the underlying distribution D. These probabilities can be interpreted as fractions of the item that an agent receives ex-ante. Hence the expected allocation A^* can be viewed as a *fractional allocation*, in which items are divisible. Conversely, we say that the distribution D (namely, the random allocation) *implements* the fractional allocation A^* when the expectation of D is A^*. Finally, we note that an additive valuation function can be extended in a natural way from allocations to fractional allocations, by considering the expected valuation. That is, the value of a p_j fraction of item j to agent i is $p_j \cdot v_i(j)$, and the value of a fractional allocation A^* to agent i is $\sum_{j \in \mathcal{M}} A^*_{ij} \cdot v_i(j)$.

For the issue of computing randomized allocations there are two different notions of polynomial time computation. In a *random polynomial time implementation*, there is a randomized polynomial time algorithm that samples an allocation from the distribution D. In a *polynomial time implementation*, there is a deterministic polynomial time algorithm that lists all allocations in the support of D (implying in particular that the support contains at most polynomially many allocations), together with their associated probabilities.

2.1   Faithful Implementation

For general additive valuations, there is a very useful lemma that greatly simplifies the design of BoBW allocations. We refer to it here as the *faithful implementation lemma*. The lemma (sometimes with slight variations) was previously stated and used in BoBW results [3,13,18,22], and was used even earlier in approximation algorithms for maximizing welfare [26]. Restricted variants of it were introduced for scheduling problems [24], and were later used for allocation problems [10]. For an extensive discussion of the faithful implementation lemma, as well as its proof (presented for completeness), see the full version [5].

Lemma 1. *Let A^* be a fractional allocation of m items to n agents with additive valuations, and let f denote the number of strictly fractional variables in A^* (number of pairs (i, j) such that in A^*, the fraction of item j allocated to agent i is strictly between 0 and 1). Then there is a deterministic polynomial time* *implementation of A^*, supported only on allocations in which every agent gets value (ex-post) equal her ex-ante value (in the fractional allocation A^*), up to the value of one item. (For agent i, the corresponding one item is the item most valuable to i, among those items that are assigned to i under A^* in a strictly fractional fashion. Moreover, the values that the agent gets in any two allocations differ by at most the value of this single item.) The distribution of the implementation is supported over at most $f + 1$ allocations.*

Using Lemma 1, one trivially gets the following BoBW result (implicit in previous work), which is a baseline against which other BoBW results can be compared.

Proposition 3. *There is a deterministic polynomial time implementation of a fractional allocation that is ex-ante envy free, and the implementation is supported on allocations that are (ex-post) Prop1.*

Proof. Consider the *uniform fractional allocation*, that assigns a fraction of $\frac{1}{n}$ of every item to every agent. It is ex-ante envy free, as all agents get the same fractional allocation. Applying Lemma 1, it is implemented in deterministic polynomial time by allocations that are Prop1.   □

3   Main Result: The Best of both Worlds

In this section we prove our main result, Theorem 3. The main arguments appear below, missing details can be found in the full version [5]. We start with an overview of the proof.

Let $\mathcal{I} = (v, \mathcal{M}, \mathcal{N})$ be an input instance. For any instance $\mathcal{I}$ we denote the proportional share and the truncated proportional share of agent i by $PS_i(\mathcal{I})$ and $TPS_i(\mathcal{I})$ respectively. For the original instance, we omit the instance and denote the proportional share and the truncated proportional share of agent i by PS_i and TPS_i, respectively.

To prove the theorem we present a deterministic polynomial time algorithm that, given the input instance $\mathcal{I} = (v, \mathcal{M}, \mathcal{N})$, computes an implementation of a randomized allocation that gives every agent at least her proportional share ex-ante, and at least half of her truncated proportional share ex-post, and is supported on at most n allocations. Items that each by itself gives an agent her TPS will play a central role in our algorithm. We say that item j is *exceptional* for agent i if $v_i(j) \geq TPS_i$. Our algorithm has several phases:

1. Find a distribution over $4n$ matchings. Each of these matchings partitions the agents to two disjoint sets $\mathcal{N}_1$ and $\mathcal{N}_2$, and the items to three disjoint sets $\mathcal{M}(\mathcal{N}_1), \mathcal{M}(\mathcal{N}_2)$ and $\mathcal{M}_3$ ($\mathcal{M} = \mathcal{M}(\mathcal{N}_1) \cup \mathcal{M}(\mathcal{N}_2) \cup \mathcal{M}_3$, $|\mathcal{N}_1| = |\mathcal{M}(\mathcal{N}_1)|$ and $|\mathcal{N}_2| = |\mathcal{M}(\mathcal{N}_2)|$). Each agent in $\mathcal{N}_1$ is matched with an item in $\mathcal{M}(\mathcal{N}_1)$, and each agent in $\mathcal{N}_2$ is matched with an item in $\mathcal{M}(\mathcal{N}_2)$. The distribution over these matchings is computed in two steps:
   (a) Compute a distribution in which in every matching, every agent in $\mathcal{N}_1$ is matched to an item that is exceptional for him in $\mathcal{M}(\mathcal{N}_1)$ and such that:

    i. No unallocated item (an item in $\mathcal{M} \setminus \mathcal{M}(\mathcal{N}_1)$) is exceptional to any agent in $\mathcal{N}_2 = \mathcal{N} \setminus \mathcal{N}_1$.

    ii. The distribution over these (at most) $4n$ matchings gives each agent her proportional share conditioned on every agent in $\mathcal{N}_2$ eventually getting her TPS in expectation (as indeed is guaranteed by 1(b)ii below).

This step is accomplished using the linear program LP1 (see Sect. 3.1).

(b) Complete each partial matching to a complete matching by matching each agent in $\mathcal{N}_2$ to an item in $\mathcal{M}(\mathcal{N}_2)$, such that:

    i. Each agent prefers the item matched to him over any unmatched item (item in $\mathcal{M}_3$).

    ii. For the unmatched items, there still is a fractional allocation of $\mathcal{M}_3$ such that for each agent in $\mathcal{N}_2$, her expected value for the combination of her matched item and her fractional allocation is at least her TPS.

This step is accomplished by setting up and solving an appropriate maximum weight matching instance (see Sect. 3.1).

2. For each matching above, find a distribution over $m+1$ deterministic allocations that allocate $\mathcal{M}_3$, the unmatched items, with the following properties:

(a) in each allocation in the support, every agent in $\mathcal{N}_2$ gets (in total, over the matched item and the remaining allocation) at least half her TPS.

(b) In expectation, every agent in $\mathcal{N}_2$ gets (in total) her TPS.

This step is accomplished using the linear program LP2 (see Sect. 3.1).

3. From the distribution over (at most) $4n(m+1)$ allocations defined above, find a distribution over at most n of these allocations that is ex-ante proportional (all ex-post properties are preserved). This step is accomplished using the linear program LP3 (which can be found in the full version [5]).

Before elaborating on the above steps, we remark that there is some flexibility as to how they can be implemented. In particular, for the linear programs LP1 and LP2 described below, only their constraints matter, whereas their objective functions can be replaced by other objective functions without affecting the main claims of this paper.

3.1 The Algorithm and the Proof

We next move to formally describe all the steps of the algorithm and prove the theorem. In the full version [5] we give an example illustrating the steps of our algorithm.

Phase 1a: Maximal allocation of exceptional items.

We start by transforming the input instance $\mathcal{I}$ into a new instance $\mathcal{I}_1$. In $\mathcal{I}_1$, we add n auxiliary items to $\mathcal{M}$, and denote them by $a_1, \ldots, a_n$, thus obtaining a set $\mathcal{M}_1 = \mathcal{M} \cup \{a_1, \ldots, a_n\}$. For every $i \in \mathcal{N}$, we modify the original valuation function v_i to the following *unit demand* valuation function u_i.

- For every item $j \in \mathcal{M}$, if $v_i(j) \geq TPS_i(\mathcal{I})$ then $u_i(j) = v_i(j)$.
- For every item $j \in \mathcal{M}$, if $v_i(j) < TPS_i(\mathcal{I})$ then $u_i(j) = 0$.

- $u_i(a_i) = TPS_i(\mathcal{I})$.
- $u_i(a_j) = 0$ for $j \neq i$.
- u_i is *unit demand*. Namely, $u_i(S) = \max_{j \in S} u_i(j)$ for every $S \subseteq \mathcal{M}_1$.

We now set up a linear program that finds a fractional allocation that maximizes welfare in $\mathcal{I}_1$, subject to the constraint that the fractional value received by every agent i is at least $PS_i(\mathcal{I})$ (hence at least 1, due to our scaling). Variable x_{ij} denotes the fraction of item j received by agent i. Variable s_i denotes the value that agent i derives from the fractional allocation. We refer to the following linear program as LP1.

Maximize $\sum_{i \in \mathcal{N}} s_i$ subject to:

1. $\sum_{i \in \mathcal{N}} x_{ij} \leq 1$ for every item $j \in \mathcal{M}_1$. (Every item is fractionally allocated at most once.)
2. $\sum_{j \in \mathcal{M}_1} x_{ij} = 1$ for every agent $i \in \mathcal{N}$. (Agent i gets item fractions that sum to one item).
3. $s_i = \sum_{j \in \mathcal{M}_1} u_i(j) x_{ij}$ for every agent $i \in \mathcal{N}$. (Agent's i value is the sum of fraction of values that she receives from the fractional allocation.)
4. $s_i \geq PS_i(\mathcal{I})$ for every agent $i \in \mathcal{N}$. (Agent's i value is at least as high as $PS_i(\mathcal{I})$.)
5. $x_{ij} \geq 0$ for every agent $i \in \mathcal{N}$ and item $j \in \mathcal{M}_1$.

Proposition 4. *LP1 is feasible.*

Proof. Let E_i denote the set of items that are exceptional for agent i in the original instance $\mathcal{I}$. As there cannot be more than n items worth more than the proportional share, we have that $|E_i| \leq n$. Consider a solution for LP1 with $x_{ij} = \frac{1}{n}$ for every $i \in \mathcal{N}$ and $j \in E_i$, and $x_{ia_i} = 1 - \sum_{j \in E_i} x_{ij} = 1 - \frac{|E_i|}{n}$. It clearly satisfies constraints 1, 2, 3 and 5. In remains to establish that this solution satisfies constraint 4, that is, the constraint $s_i \geq PS_i(\mathcal{I})$.

We shall use the facts that $PS_i(\mathcal{I}) = \frac{1}{n} v_i(E_i) + \frac{1}{n} v_i(\mathcal{M} \setminus E_i)$ and $TPS_i(\mathcal{I}) = \frac{v_i(\mathcal{M} \setminus E_i)}{n - |E_i|}$. We can see that constraint 4 is satisfied by this solution:

$$s_i = \frac{1}{n} v_i(E_i) + \left(1 - \frac{|E_i|}{n}\right) u_i(a_i) = \frac{1}{n} v_i(E_i) + \frac{n - |E_i|}{n} TPS_i$$

$$= \frac{1}{n} \left(v_i(E_i) + v_i(\mathcal{M} \setminus E_i)\right) = PS_i(\mathcal{I})$$

We solve LP1, and find an optimal basic feasible solution for it, and denote it by A^*. It can be shown (see details in the full version [5]) that being a basic feasible solution, the number of positive x_{ij} variables in A^* is at most $4n - 1$.

We perform faithful randomized rounding on A^*. The following proposition follows immediately from the properties of A^* and Lemma 1, and hence its proof is omitted.

Proposition 5. *The faithful randomized rounding of A^* produces a distribution over allocations with the following properties:*

1. *In every allocation, every agent i gets exactly one item from $\mathcal{M}_1$. This item is either one of her exceptional items, or her auxiliary item a_i. In either case the u_i value of that item is least $TPS_i(\mathcal{I})$.*
2. *The distribution is supported on at most $4n$ allocations.*
3. *In expectation, every agent i gets value s_i with respect to u_i. Recall that $s_i \geq PS_i(\mathcal{I})$.*

Consider now an arbitrary allocation A' in the support of the faithful randomized rounding of A^*. With respect to A', let $\mathcal{N}_1 = \mathcal{N}_1(A')$ denote the set of agents that receive an item that was exceptional for them, and let $\mathcal{N}_2$ denote the set of agents that receive their auxiliary item. (Note that $\mathcal{N}_1 \cup \mathcal{N}_2 = \mathcal{N}$.)

The first phase ends by giving each agent of $\mathcal{N}_1$ the item that she receives under A', and not giving agents of $\mathcal{N}_2$ any item (as the auxiliary items do not really exist). Thus we have that $\mathcal{M}(\mathcal{N}_1)$ is the set of items matched to agents in $\mathcal{N}_1$. Observe that every agent $i \in \mathcal{N}_1$ gets at least TPS_i ex-post. The remaining phases will ensure that agents in $\mathcal{N}_2$ get at least half their TPS ex-post. They will also ensure that ex-ante, every agent gets at least her proportional share (this will make use of item 3 of Proposition 5).

Phase 1b: Completing the matching.

If $\mathcal{N}_2$ is empty, we go directly to Phase 2. Hence here we assume that $\mathcal{N}_2$ is non-empty.

Let $\mathcal{M}_2 \subset \mathcal{M}$ denote the subset of original items (not including the auxiliary items) that remain unallocated in A' (those items not allocated to $\mathcal{N}_1$). Let $\mathcal{I}_2$ denote the allocation instance that has $\mathcal{M}_2$ as its set of items, $\mathcal{N}_2$ as its set of agents, and the valuation function of every agent $i \in \mathcal{N}_2$ remains v_i (restricted to the items in $\mathcal{M}_2$). As $\mathcal{I}_2$ is obtained from $\mathcal{I}$ by removing $|\mathcal{N}_1|$ agents and $|\mathcal{N}_1|$ items, it holds that $TPS_i(\mathcal{I}_2) \geq TPS_i(\mathcal{I})$ for every agent $i \in \mathcal{N}_2$.

Importantly, recall that we may assume without loss of generality that $\mathcal{M}_2$ has no item that according to instance $\mathcal{I}$ was exceptional for an agent of $\mathcal{N}_2$. (If $\mathcal{M}_2$ contains an item j that is exceptional for $i \in \mathcal{N}_2$, then in A', give j instead of a_i to agent i, by this moving agent i out of $\mathcal{N}_2$ and into $\mathcal{N}_1$.) The fact that $TPS_i(\mathcal{I}_2) \geq TPS_i(\mathcal{I})$ (for $i \in \mathcal{N}_2$) implies that also in $\mathcal{I}_2$, $\mathcal{M}_2$ has no item that is exceptional for an agent of $\mathcal{N}_2$. Consequently, we infer that for every agent $i \in \mathcal{N}_2$:

- $TPS_i(\mathcal{I}_2) = \frac{v_i(\mathcal{M}_2)}{|\mathcal{N}_2|}$, and consequently also $\frac{v_i(\mathcal{M}_2)}{|\mathcal{N}_2|} \geq TPS_i(\mathcal{I})$.
- There are strictly more than $|\mathcal{N}_2|$ items $j \in \mathcal{M}_2$ with $v_i(j) > 0$. (This holds because $v_i(j) < TPS_i(\mathcal{I}_2)$ for every $j \in \mathcal{M}_2$.)

Let $B_i \subset \mathcal{M}_2$ denote the set of $|\mathcal{N}_2|$ items of highest value to agent $i \in \mathcal{N}_2$, breaking ties arbitrarily. Let $W_i = v_i(B_i)$. As $\mathcal{M}_2$ has more than $|\mathcal{N}_2|$ items of positive value for i, it follows that $W_i < v_i(\mathcal{M}_2)$.

We now transform the instance $\mathcal{I}_2$ into a new instance $\mathcal{I}_2'$. The set of items in $\mathcal{I}_2'$ is $\mathcal{M}_2$, and the set of agents is $\mathcal{N}_2$. Every agent $i \in \mathcal{N}_2$ has a *unit-demand* valuation function w_i, defined as follows. For $j \in B_i$ we have $w_i(j) = \frac{v_i(j)}{v_i(\mathcal{M}_2) - W_i}$ (observe that the denominator is positive), and for $j \notin B_i$ we have $w_i(j) = 0$.

In the matching completion phase, we find a welfare maximizing allocation B^* in $\mathcal{I}_2'$. Observe that this can be done in polynomial time, because agents are unit demand, and hence finding B^* amounts to solving an instance of maximum weight matching in a bipartite graph G, with $\mathcal{N}_2$ as the set of left side vertices, $\mathcal{M}_2$ as the set of right side vertices, and weight $w_i(j)$ for edge (i, j). In B^*, every agent $i \in \mathcal{N}_2$ receives an item from her respective set B_i (this follows because $|B_i| \geq |\mathcal{N}_2|$). We have that $\mathcal{M}(\mathcal{N}_2)$ is the set of items matched to agents in $\mathcal{N}_2$.

By the end of the matching phase, every agent holds one item. Agents in $\mathcal{N}_1$ received their item in Phase 1a (under A'), whereas agents in $\mathcal{N}_2$ received their item in Phase 1b (under B^*). Let e_i denote the item that has been allocated to agent i, and let $\mathcal{M}_3 = \mathcal{M} \setminus \{e_1, \ldots, e_n\}$ denote the set of items that are not yet allocated. A key property established by the first two phases is summarized in the following proposition.

Proposition 6. *For every agent $i \in \mathcal{N}$ it holds that $v_i(e_i) \geq \max_{j \in \mathcal{M}_3} v_i(j)$.*

Proof. For an agent $i \in \mathcal{N}_1$, the proposition follows from the optimality of the fractional allocation A^*. If there is an item $j \in \mathcal{M}_3$ with $v_i(j) > v_i(e_i)$, then in A' item j could replace item e_i for agent i, thus increasing the welfare of A'. This would imply that LP1 has a fractional solution of value higher than that of A^*, contradicting the optimality of A^*.

For an agent $i \in \mathcal{N}_2$, the proposition follows from the fact that B^* is a maximum weight matching: no unmatched item can be valued by an agent more than the item matched to her.   □

We are now ready to move to the next phase of our algorithm.

Phase 2: Allocating unmatched items.

In this phase, for each matching computed before, we allocate the items of $\mathcal{M}_3$, the items not in the matching. Every agent $i \in \mathcal{N}$ has her original valuation function v_i (with $v_i(\mathcal{M}) = n$).

We first compute a fractional allocation for the items of $\mathcal{M}_3$. This is done by solving a linear program that we refer to as LP2. In LP2, variable x_{ij} denotes the fraction of item $j \in \mathcal{M}_3$ allocated to agent i, and s_i denotes the value that agent i derives from the fractional allocation (under valuation function v_i). The parameters f_i are treated as constants in LP2. Their values are computed based on Phase 1b. Specifically, $f_i = \frac{\frac{v_i(\mathcal{M}_2)}{|\mathcal{N}_2|} - v_i(e_i)}{v_i(\mathcal{M}_2) - W_i}$ for $i \in \mathcal{N}_2$ (where e_i is the item allocated to agent i in Phase 1b, and $W_i = v_i(B_i)$, as defined in Phase 1b). We now present LP2.

Maximize $\sum_{i \in \mathcal{N}} s_i$ subject to:

1. $\sum_{i \in \mathcal{N}} x_{ij} \leq 1$ for every item $j \in \mathcal{M}_3$. (Every item is fractionally allocated at most once.)
2. $s_i = \sum_{j \in \mathcal{M}_3} v_i(j) x_{ij}$ for every agent $i \in \mathcal{N}$. (Agent's i value is the sum of the fractions of values that she receives from the fractional allocation.)
3. $s_i \geq f_i v_i(\mathcal{M}_3)$ for every agent $i \in \mathcal{N}_2$. (This is the key constraint that ties LP2 with the allocation B^* of Phase 1b. It applies only to agents in $\mathcal{N}_2$.)

4. $x_{ij} \geq 0$ for every agent $i \in \mathcal{N}$ and item $j \in \mathcal{M}_3$.

Note that LP2 may fractionally allocate items from $\mathcal{M}_3$ to agents in $\mathcal{N}_1$, but only after each agent in $\mathcal{N}_2$ receives items of sufficiently high value as dictated by Constraint 3.

Lemma 2. *LP2 is feasible.*

Proof. Constraints 2 and 4 are satisfied by every solution in which $x_{ij} \geq 0$ (for all i and j). It remains to show that constraints 1 and 3 can be satisfied simultaneously.

Recall the bipartite graph G from Phase 1b. In G, consider a fractional matching $F = \{y_{ij}\}$, where $y_{ij} = \frac{1}{|\mathcal{N}_2|}$ for every agent $i \in \mathcal{N}_2$ and item $j \in B_i$, and $y_{ij} = 0$ if $j \notin B_i$. Observe that for every agent $i \in \mathcal{N}_2$ we have $\sum_{j \in B_2} y_{ij} = 1$ and for every item $j \in \mathcal{M}_2$ we have $\sum_{i \in \mathcal{N}_2} y_{ij} \leq \frac{1}{|\mathcal{N}_2|}|\mathcal{N}_2| = 1$. Hence indeed F defines a fractional matching. In $\mathcal{I}_2$ the fractional matching F gives agent $i \in \mathcal{N}_2$ fractional value $\sum_{j \in B_i} y_{ij} v_i(j) = \frac{1}{|\mathcal{N}_2|} v_i(B_i) = \frac{W_i}{|\mathcal{N}_2|}$.

Being a fractional matching, F can be represented as a distribution D over integral matchings. In every one of these integral matchings, every agent $i \in \mathcal{N}_2$ is matched, because i is fully matched in F. Select a matching at random from the distribution D. Then in expectation, agent i gets an item of value $\sum_{j \in B_i} y_{ij} v_i(j) = \frac{W_i}{|\mathcal{N}_2|}$. Using E_D to denote expectation over choice from distribution D, and denoting by e_i the item received by i, we have that $E_D[v_i(e_i)] = \frac{W_i}{|\mathcal{N}_2|}$. Hence the expectation of f_i is $E_D \left[\frac{\frac{v_i(\mathcal{M}_2)}{|\mathcal{N}_2|} - v_i(e_i)}{v_i(\mathcal{M}_2) - W_i} \right] = \frac{\frac{v_i(\mathcal{M}_2)}{|\mathcal{N}_2|} - \frac{W_i}{|\mathcal{N}_2|}}{v_i(\mathcal{M}_2) - W_i} = \frac{1}{|\mathcal{N}_2|}$. By linearity of expectation, $E_D[\sum_{i \in \mathcal{N}_2} f_i] = 1$. This implies that there is a matching in G under which the sum of the respective f_i satisfies $\sum_{i \in \mathcal{N}_2} f_i \leq 1$. The matching that maximizes $\sum_{i \in [n]} \frac{v_i(e_i)}{v_i(\mathcal{M}_2) - W_i}$ (which is B^* that we use in the matching step, because we defined $w_i(j)$ to be $\frac{v_i(j)}{v_i(\mathcal{M}_2) - W_i}$) also minimizes $\sum_{i \in \mathcal{N}_2} f_i$, and hence has $\sum_{i \in \mathcal{N}_2} f_i \leq 1$. This implies that the solution with $x_{ij} = f_i$ for every $i \in \mathcal{N}_2$ and $j \in \mathcal{M}_3$, and $x_{ij} = 0$ for every $i \in \mathcal{N}_1$, is feasible for LP2. $\square$

Let C^* be a fractional allocation of $\mathcal{M}_3$ that is an optimal solution to LP2. Phase 2 ends by performing faithful randomized rounding of C^*. The following proposition follows immediately from the properties of C^* and Lemma 1, and hence its proof is omitted.

Proposition 7. *The faithful randomized rounding of C^* produces a distribution over allocations of the items of $\mathcal{M}_3$, with the following properties:*

1. *The distribution is supported on at most $m + 1$ allocations. (The number of constraints in LP2 is $|\mathcal{M}_3| + n + |\mathcal{N}_2|$. In a basic feasible solution, at least $|\mathcal{N}_2|$ of the s_i variables are positive, and so at most $|\mathcal{M}_3| + n = m$ of the x_{ij} variables are positive.)*
2. *Every agent $i \in \mathcal{N}_2$ gets ex-ante value $s_i \geq f_i \cdot v_i(\mathcal{M}_3)$.*
3. *Every agent $i \in \mathcal{N}_2$ gets ex-post value at least s_i, up to one item. That is, at least $s_i - \max_{j \in \mathcal{M}_3}[v_i(j)]$.*

The allocation algorithm above computes a distribution over $4n$ matchings in Phases 1a and 1b, and for each such matching, in Phase 2 it computes a distribution over $m + 1$ allocations of $\mathcal{M}_3$. We thus have a distribution over $4n(m + 1)$ allocations and we next prove that it satisfies the requirements of Theorem 3 (except the support reduction to n allocations, that is handled in Phase 3 in the full version [5]).

Every Agent Gets her Proportional Share Ex-ante. By item 3 of Proposition 5, with respect to A^*, every agent i gets value at least $PS_i(\mathcal{I})$ ex-ante. However, this value might have been attained by being allocated the respective auxiliary item a_i, of value $TPS_i(\mathcal{I})$. In this case, agent i does not actually get a_i, but is instead included in $\mathcal{N}_2$. Hence we need to show that for every agent $i \in \mathcal{N}_2$, her combined ex-ante value from Phases 1b and 2 is at least $TPS_i(\mathcal{I})$. This ex-ante value is at least $v_i(e_i) + f_i \cdot v_i(\mathcal{M}_3)$. We claim that indeed $v_i(e_i) + f_i \cdot v_i(\mathcal{M}_3) \geq TPS_i(\mathcal{I})$.

Recall that $f_i = \frac{\frac{v_i(\mathcal{M}_2)}{|\mathcal{N}_2|} - v_i(e_i)}{v_i(\mathcal{M}_2) - W_i}$. Observe also that $v_i(\mathcal{M}_2) - W_i \leq v_i(\mathcal{M}_3)$, because the total value of i for the $|\mathcal{N}_2|$ items allocated under B^* cannot be larger than $W_i = v_i(B_i)$ (as B_i contains the $|\mathcal{N}_2|$ items of highest value). Combining these observations we have that:

$$f_i = \frac{\frac{v_i(\mathcal{M}_2)}{|\mathcal{N}_2|} - v_i(e_i)}{v_i(\mathcal{M}_2) - W_i} \geq \frac{\frac{v_i(\mathcal{M}_2)}{|\mathcal{N}_2|} - v_i(e_i)}{v_i(\mathcal{M}_3)} = \frac{v_i(\mathcal{M}_2) - |\mathcal{N}_2| \cdot v_i(e_i)}{|\mathcal{N}_2| \cdot v_i(\mathcal{M}_3)}$$

We can now establish the claim.

$$v_i(e_i) + f_i v_i(\mathcal{M}_3) \geq v_i(e_i) + \frac{v_i(\mathcal{M}_2) - |\mathcal{N}_2| \cdot v_i(e_i)}{|\mathcal{N}_2| \cdot v_i(\mathcal{M}_3)} v_i(\mathcal{M}_3) = \frac{v_i(\mathcal{M}_2)}{|\mathcal{N}_2|} \geq TPS_i(\mathcal{I})$$

(for the last equality, see discussion in Phase 1b).

Every Agent Gets at Least Half her TPS Ex-post. For agents in $\mathcal{N}_1$, this holds by definition. For agents $i \in \mathcal{N}_2$, we have already shown that ex-ante they get at least $TPS_i(\mathcal{I})$. Item 3 of Proposition 7 implies that ex-post agent i gets a value of at least $TPS_i(\mathcal{I}) - \max_{j \in \mathcal{M}_3}[v_i(j)]$. If $\max_{j \in \mathcal{M}_3}[v_i(j)] \leq \frac{TPS_i(\mathcal{I})}{2}$, then at least a value of $\frac{TPS_i(\mathcal{I})}{2}$ remains. If $\max_{j \in \mathcal{M}_3}[v_i(j)] > \frac{TPS_i(\mathcal{I})}{2}$, then also $v_i(e_i) \geq \frac{TPS_i(\mathcal{I})}{2}$ (by Proposition 6), and hence i gets half her TPS already after Phase 1b.

The Allocation is Prop1 Ex-post. If there is an item that is exceptional for agent i, then an item that i values most, denoted as item j, necessarily satisfies $v_i(j) \geq PS_i$ (if $v_i(j) < PS_i$ then $TPS_i = PS_i$, and then j is not exceptional for i). In this case, every allocation gives i her proportional share, up to the item j. If there is no item that is exceptional for agent i, then $TPS_i(\mathcal{I}) = PS_i(\mathcal{I})$, and also, i ends up in $\mathcal{N}_2$. Item 3 of Proposition 7 ensures that she gets $TPS_i(\mathcal{I})$ up to one item, which in this case is equivalent to $PS_i(\mathcal{I})$ up to one item.

The Randomized Allocation is Supported on n Allocations. The combination of item 2 of Proposition 5 and item 1 of Proposition 7 implies that the randomized allocation is supported over at most $4n(m+1)$ allocations. In Phase 3 of our algorithm (described in the full version [5], for lack of space), we reduce this number to n.

The Randomized Allocation is Computed in Polynomial Time. This can be verified by inspection of the allocation algorithm (see details in the full version [5]).

4 Discussion

In the full version [5] we discuss directions in which it may be desirable to improve our results, presenting impossibilities of some natural extensions, as well as some open problems. In particular, we discuss incorporating additional considerations of economic efficiency in BoBW results, incentive compatibility issues, and we present BoBW results for allocation of indivisible chores (items of negative values). We also discuss other fairness guarantees that one may desire in BoBW results.

Acknowledgements. Tomer Ezra's research is Partially supported by the ERC Advanced Grant 788893 AMDROMA "Algorithmic and Mechanism Design Research in Online Markets" and MIUR PRIN project ALGADIMAR "Algorithms, Games, and Digital Markets".

References

1. Amanatidis, G., Birmpas, G., Christodoulou, G., Markakis, E.: Truthful allocation mechanisms without payments: Characterization and implications on fairness. In: Proceedings of the 2017 ACM Conference on Economics and Computation, EC 2017, pp. 545–562 (2017)
2. Amanatidis, G., Markakis, E., Nikzad, A., Saberi, A.: Approximation algorithms for computing maximin share allocations. ACM Trans. Algorithms (TALG) **13**(4), 1–28 (2017)
3. Aziz, H.: Simultaneously achieving ex-ante and ex-post fairness. In: Chen, X., Gravin, N., Hoefer, M., Mehta, R. (eds.) WINE 2020. LNCS, vol. 12495, pp. 341–355. Springer, Cham (2020). https://doi.org/10.1007/978-3-030-64946-3_24
4. Aziz, H., Biró, P., Lang, J., Lesca, J., Monnot, J.: Optimal reallocation under additive and ordinal preferences. In: Proceedings of the 2016 International Conference on Autonomous Agents & Multiagent Systems, pp. 402–410. ACM (2016)
5. Babaioff, M., Ezra, T., Feige, U.: Best-of-both-worlds fair-share allocations. arXiv preprint arXiv:2102.04909 (2021)

6. Babaioff, M., Ezra, T., Feige, U.: Fair and truthful mechanisms for dichotomous valuations. In: Thirty-Fifth AAAI Conference on Artificial Intelligence, pp. 5119–5126. AAAI Press (2021)
7. Babaioff, M., Ezra, T., Feige, U.: Fair-share allocations for agents with arbitrary entitlements. In: EC '21: The 22nd ACM Conference on Economics and Computation, p. 127. ACM (2021)
8. Barman, S., Krishnamurthy, S.K.: On the proximity of markets with integral equilibria. In: The Thirty-Third AAAI Conference on Artificial Intelligence, AAAI, pp. 1748–1755 (2019)
9. Barman, S., Krishnamurthy, S.K.: Approximation algorithms for maximin fair division. ACM Trans. Econ. Comput. (TEAC) **8**(1), 1–28 (2020)
10. Bezáková, I., Dani, V.: Allocating indivisible goods. ACM SIGecom Exchanges **5**(3), 11–18 (2005)
11. Bogomolnaia, A., Moulin, H.: A new solution to the random assignment problem. J. Econ. Theory **100**(2), 295–328 (2001)
12. Budish, E.: The combinatorial assignment problem: approximate competitive equilibrium from equal incomes. J. Polit. Econ. **119**(6), 1061–1103 (2011)
13. Budish, E., Che, Y.K., Kojima, F., Milgrom, P.: Designing random allocation mechanisms: theory and applications. Am. Econ. Rev. **103**(2), 585–623 (2013)
14. Caragiannis, I., Kurokawa, D., Moulin, H., Procaccia, A.D., Shah, N., Wang, J.: The unreasonable fairness of maximum nash welfare. ACM Trans. Econ. Comput. (TEAC) **7**(3), 1–32 (2019)
15. Conitzer, V., Freeman, R., Shah, N.: Fair public decision making. In: Proceedings of the 2017 ACM Conference on Economics and Computation, pp. 629–646 (2017)
16. De Keijzer, B., Bouveret, S., Klos, T., Zhang, Y.: On the complexity of efficiency and envy-freeness in fair division of indivisible goods with additive preferences. In: International Conference on Algorithmic Decision Theory, pp. 98–110 (2009)
17. Feige, U., Sapir, A., Tauber, L.: A tight negative example for MMS fair allocations. In: Feldman, M., Fu, H., Talgam-Cohen, I. (eds.) WINE 2021. LNCS, vol. 13112, pp. 355–372. Springer, Cham (2022). https://doi.org/10.1007/978-3-030-94676-0_20
18. Freeman, R., Shah, N., Vaish, R.: Best of both worlds: ex-ante and ex-post fairness in resource allocation. In: Proceedings of the 21st ACM Conference on Economics and Computation, pp. 21–22 (2020)
19. Garg, J., McGlaughlin, P., Taki, S.: Approximating maximin share allocations. Open access series in informatics 69 (2019)
20. Garg, J., Taki, S.: An improved approximation algorithm for maximin shares. In: Proceedings of the 21st ACM Conference on Economics and Computation, pp. 379–380 (2020)
21. Ghodsi, M., Hajiaghayi, M.T., Seddighin, M., Seddighin, S., Yami, H.: Fair allocation of indivisible goods: Improvements and generalizations. In: Proceedings of the 2018 ACM Conference on Economics and Computation, pp. 539–556. ACM (2018)
22. Halpern, D., Procaccia, A.D., Psomas, A., Shah, N.: Fair division with binary valuations: One rule to rule them all. In: Web and Internet Economics - 16th International Conference (2020)
23. Kurokawa, D., Procaccia, A.D., Wang, J.: Fair enough: Guaranteeing approximate maximin shares. J. ACM **65**(2), 8:1–8:27 (2018)
24. Lenstra, J.K., Shmoys, D.B., Tardos, É.: Approximation algorithms for scheduling unrelated parallel machines. Math. Program. **46**(1), 259–271 (1990)

25. Lipton, R.J., Markakis, E., Mossel, E., Saberi, A.: On approximately fair allocations of indivisible goods. In: Proceedings of the 5th ACM Conference on Electronic Commerce, pp. 125–131 (2004)
26. Srinivasan, A.: Budgeted allocations in the full-information setting. In: Approximation, Randomization and Combinatorial Optimization. Algorithms and Techniques, pp. 247–253. Springer (2008)

Nash Welfare Guarantees for Fair and Efficient Coverage

Siddharth Barman[1], Anand Krishna[1(✉)], Y. Narahari[1],
and Soumyarup Sadhukhan[2]

[1] Indian Institute of Science, Bengaluru, India
{barman,anandkrishna,narahari}@iisc.ac.in
[2] Indian Institute of Technology Kanpur, Kanpur, India

Abstract. We study coverage problems in which, for a set of agents and a given threshold T, the goal is to select T subsets (of the agents) that, while satisfying combinatorial constraints, achieve fair and efficient coverage among the agents. In this setting, the valuation of each agent is equated to the number of selected subsets that contain it, plus one. The current work utilizes the Nash social welfare function to quantify the extent of fairness and collective efficiency. We develop a polynomial-time $(18 + o(1))$-approximation algorithm for maximizing Nash social welfare in coverage instances. Our algorithm applies to all instances wherein, for the underlying combinatorial constraints, there exists an FPTAS for weight maximization. We complement the algorithmic result by proving that Nash social welfare maximization is APX-hard in coverage instances.

1 Introduction

Coverage problems, with a multitude of variants, are fundamental in theoretical computer science, combinatorics, and operations research. These problems capture numerous resource-allocation applications, such as electricity division [2,22], sensor allocation [20], program testing [18], and plant location [9].

Coverage problems entail identifying—for a given threshold $T \in \mathbb{Z}_+$ and a set of elements $[n]$—a collection of subsets, $F_1, F_2, \ldots, F_T \subseteq [n]$, that respect particular combinatorial constraints. Here, the problem objective is specified by considering, for each element $i \in [n]$, the number of selected subsets, F_t-s, that contain i. For instance, in the classic maximum coverage problem [16], the subsets, $F_1, \ldots, F_T$, are constrained to be from a given set family and the objective is to maximize the number of elements $i \in [n]$ that are contained in at least one of the F_t-s, i.e., maximize $|\cup_t F_t|$.

We study coverage problems where the ground set corresponds to a population of n agents and the cardinal valuation of each agent $i \in [n]$ depends on the number of selected subsets that contain i, i.e., the valuation of i depends on the coverage that i receives across the F_t-s. Our overarching goal is to select

Full version available at https://arxiv.org/abs/2207.01970.

K. A. Hansen et al. (Eds.): WINE 2022, LNCS 13778, pp. 256–272, 2022.
https://doi.org/10.1007/978-3-031-22832-2_15

subsets that, while satisfying combinatorial constraints, achieve fair and efficient coverage among the n agents.

Before detailing the model, we describe a stylized example that illustrates the applicability of the coverage framework. Consider an electricity grid operator tasked with apportioning electricity for T time periods among a set of n agents (consumers with varying electricity requirements). In a time period $t \in [T]$, the total demand of the n agents can exceed the available supply and, hence, the grid operator must select a subset of agents, $F_t \subseteq [n]$, whose electricity consumption can be fulfilled–agents in the subset F_t receive electricity during the tth time period and the remaining agents do not. An important desideratum in such load shedding scenarios is to achieve fairness along with economic efficiency; see the motivating work of Baghel et al. [2] for a thorough treatment of load shedding and its connections with the fair division literature. Indeed, the coverage framework provides an abstraction for this load shedding environment: for each $t \in T$, the selected subset F_t must satisfy a knapsack constraint[1] and the cardinal preference of each agent $i \in [n]$ is captured by the number of subsets that contain i, i.e., the number of time periods that i receives electricity.

Combinatorial Constraints. We study a coverage framework wherein, for each $t \in [T]$, the tth selected subset, $F_t \subseteq [n]$, must belong to a set family $\mathcal{I}_t$, i.e., each $\mathcal{I}_t \subseteq 2^{[n]}$ specifies the possible choices for the tth selection. Our results do not require the families $\mathcal{I}_t$-s to be given explicitly as input. Our results hold for any $\mathcal{I}_t$-s that admit a fully polynomial-time approximation scheme (FPTAS) for the weight maximization problem: given weights $w_1, \ldots, w_n \in \mathbb{R}_+$, for the n agents, find $\arg\max_{X \in \mathcal{I}_t} \sum_{i \in X} w_i$.

For instance, if each $\mathcal{I}_t$ contains the subsets that satisfy a knapsack constraint, then an FPTAS for weight maximization is known to exist [26]; in such a case weight maximization corresponds to the standard knapsack problem.[2] Furthermore, if the families $\mathcal{I}_t$-s are independent sets of matroids, then one can exactly solve the weight maximization problem in polynomial time [24]. It is relevant to note that matroids provide an expressive construct for numerous combinatorial constraints, e.g., cardinality and partition constraints. Hence, the coverage framework with matroids provides, by itself, an encompassing class of instances. Also, in instances wherein the sizes of the families $\mathcal{I}_t$-s are polynomially large, weight maximization can be efficiently solved by direct enumeration.

In addition, our result applies to settings that entail two-sided matchings: say, for each $t \in [T]$, we have a bipartite graph $G = (L \cup R, E)$, with $L \cup R = [n]$, and the goal is to select a matching, i.e., agents covered by the matching constitute the tth selected subset. We can express this matching setting in the current framework by including, in each $\mathcal{I}_t$, every subset of agents (i.e., subset of vertices in G) that is covered by some matching in G. Notably, such a formulation models

---

[1] In particular, the total demand of the agents in F_t should be at most the supply at time period t.

[2] Recall that in the electricity division example, the subsets F_t-s had to satisfy knapsack constraints.

two-sided markets [14,25] such as (i) ridesharing platforms, wherein the agent set consists of both the vehicle drivers and the passengers and (ii) recommendation engines, in which producers are recommended to consumers. Our result holds in such matching settings, since here weight maximization can be optimally solved in polynomial time via a maximum-weight matching algorithm.[3]

Agents' Valuations. As mentioned previously, we address settings in which each agent's valuation depends on the number of times it is covered among the selected subsets F_t-s. Specifically, for a solution $\mathcal{F} = (F_1, \ldots, F_T) \in \mathcal{I}_1 \times \ldots \times \mathcal{I}_T$, agent i's valuation is defined as $v_i(\mathcal{F}) := |\{t \in [T] : i \in F_t\}| + 1$. Note that the valuation of each agent is smoothed by adding 1. This smoothing enables us to achieve meaningful (multiplicative) approximation guarantees by shifting the valuations and, hence, the collective welfare away from zero. We also note that valuation smoothing has been considered in prior works in fair division; see, e.g., [10,12], and [17].

Nash Social Welfare. With the overarching aim of achieving fairness along with economic efficiency in coverage instances, we address the problem of maximizing Nash social welfare (NSW). This welfare function is defined as the geometric mean of agents' valuations and it achieves a balance between the extremes of social welfare (a well-studied objective for economic efficiency) and egalitarian welfare (a prominent fairness notion). NSW stands as a fundamental metric for quantifying the extent of fairness in numerous resource-allocation contexts; indeed, in recent years, NSW has been extensively studied in the fair division literature; see, e.g., [5,7,15,19,23] and many references therein.

Nash social welfare satisfies key fairness axioms, including scale freeness, symmetry, and the Pigou-Dalton transfer principle [21]. The Pigou-Dalton principle requires that the collective welfare should increase under a bounded transfer of value from a well-off agent i to a worse-off agent j. NSW satisfies this principle, since the geometric mean of a more balanced valuation profile (of the n agents) is higher than that of a skewed one. At the same time, if the increase in agent j's value is significantly less than the drop experienced by i, then NSW does not increase. That is, NSW prefers solutions[4] that have reduced inequality and, simultaneously, it accommodates for economic efficiency.

Furthermore, in various fair division contexts, prior works have shown that a solution that maximizes NSW satisfies additional fairness properties, e.g., [1,7,10,13,15]. Critically, the fact that Nash optimal solutions bear additional guarantees does not undermine the relevance of finding solutions with as high a Nash social welfare as possible. NSW cardinally ranks the solutions and, conforming to a welfarist perspective, one prefers solutions with higher NSW. Therefore, developing approximation guarantees for NSW maximization is a well-justified objective in and of itself.

---

[3] One can also address one-sided matching—with agents on one side and, say, indivisible slots on the other—as a transversal matroid.

[4] In the current context, a solution is a collection of T subsets $F_1, \ldots, F_T$ that are contained in the underlying set families $\mathcal{I}_1, \ldots, \mathcal{I}_T$, respectively.

1.1 Our Results and Techniques

We develop a constant-factor approximation algorithm for maximizing Nash social welfare in fair coverage instances. Given a set of n agents and threshold $T \in \mathbb{Z}_+$, our algorithm (Algorithm 1) computes in polynomial time a solution $\mathcal{F} = (F_1, \ldots, F_T) \in \mathcal{I}_1 \times \ldots \times \mathcal{I}_T$ whose Nash social welfare, $\mathrm{NSW}(\mathcal{F}) = (\prod_{i=1}^n v_i(\mathcal{F}))^{\frac{1}{n}}$, is at least $\frac{1}{18+o(1)}$ times the optimal (Theorem 1). As mentioned previously, the algorithm only requires blackbox access to an FPTAS for weight maximization over the set families $\mathcal{I}_1, \ldots, \mathcal{I}_T \subseteq 2^{[n]}$.

The algorithm starts with an arbitrary solution and iteratively performs updates till it essentially reaches a local maximum of the log social welfare $\varphi(\mathcal{F}) := \sum_{i=1}^n \log(v_i(\mathcal{F}))$. Here, for any solution $\mathcal{F} = (F_1, \ldots, F_T)$, a local update corresponds to replacing—for some $\tau \in [T]$—the subset F_τ with some other subset $A_\tau \in \mathcal{I}_\tau$. The algorithm performs the local updates by invoking, as a subroutine, the FPTAS for weight maximization.

It is relevant to note that while the algorithm is simple in design, its analysis entails novel insights. In particular, the domain of solutions, $\mathcal{I}_1 \times \ldots \times \mathcal{I}_T$, is combinatorial and, hence, it is not obvious if a local maximum solution of φ upholds any global approximation guarantees for φ, let alone for NSW. Furthermore, a multiplicative approximation bound for φ does not translate into a multiplicative guarantee for NSW: for any solution $\mathcal{F}$, we have $\frac{1}{n}\varphi(\mathcal{F}) = \log(\mathrm{NSW}(\mathcal{F}))$. Hence, even though a solution that (globally) maximizes φ also maximizes NSW, multiplicative approximation guarantees get exponentially worse when one moves from φ to NSW. This observation also implies that one cannot directly utilize the approximation guarantee known for the so-called concave coverage problem [3] to obtain a commensurate approximation ratio for NSW maximization.

Interestingly, in lieu of developing local-to-global approximation guarantees, we rely on counting arguments to establish the approximation ratio. We prove that, at a local maximum solution $\mathcal{F}$ (of the function φ) and for any integer $\alpha \geq 4$, the number of α-suboptimal agents is at most n/α; here, an agent i is said to be α-suboptimal iff i's current valuation $v_i(\mathcal{F})$ is (about) $1/\alpha$ times less than her optimal valuation. We complete the analysis by proving that these Markov-like bounds ensure that the computed solution $\mathcal{F}$ achieves an $(18 + o(1))$-approximation guarantee for NSW maximization.

In addition, we complement the algorithmic result by proving that, in fair coverage instances, NSW maximization is APX-hard (Theorem 2). This inapproximability result rules out a polynomial-time approximation scheme (PTAS) for NSW maximization in fair coverage instances.

1.2 Additional Related Work and Applications

The coverage framework generalizes the well-motivated setup of public decision making [8], albeit for agents that have binary additive valuations. The public decision making setup captures settings wherein decisions have to be made on T social issues, that can impact many of the n agents simultaneously. Specifically, each issue $t \in [T]$ is associated with a set of alternatives $A_t = \{a_t^1, a_t^2, \ldots, a_t^{\ell_t}\}$

and every agent $i \in [n]$ has an additive valuation over the issues. That is, for any outcome $\mathcal{A} = (a_1, a_2, \ldots, a_T) \in A_1 \times A_2 \times \ldots \times A_T$, agent i's utility is $u_i(\mathcal{A}) = \sum_{t=1}^{T} u_i^t(a_t)$; here $u_i^t(a_t) \in \mathbb{R}_+$ is the utility that i gains from the alternative $a_t \in A_t$.

Indeed, for agents $i \in [n]$ with binary additive valuations (i.e., $u_i^t(a) \in \{0, 1\}$ for all t and $a \in A_t$) the coverage framework generalizes public decision making: for every $t \in [T]$, define the set family $\mathcal{I}_t$ by including in it the set $F_a := \{i \in [n] : u_i^t(a) = 1\}$ for each $a \in A_t$. In particular, $\mathcal{I}_t$ contains a set F_a, for each alternative $a \in A_t$, where F_a is the set of agents that value alternative a. This reduction gives us set families of polynomial size ($|\mathcal{I}_t| = |A_t|$) and, hence, our results specialize to this case.

In the public decision making context, Conitzer et al. [8] obtain fairness guarantees in terms of relaxations of proportionality. They also show that Nash optimal solutions bear particular fairness properties. Complementing these results and for agents with (smoothed) binary additive valuations, the current work obtains approximation guarantees for NSW in public decision making.

The coverage framework also encompasses the standard fair division setting that entails allocation of m indivisible goods among n agents that have binary additive valuations. Multiple prior works have studied NSW in this discrete fair division setting; see, e.g., [6,15]. Here, each agent $i \in [n]$ prefers a subset of the goods $V_i \subseteq [m]$ and agent i's valuation $u_i(S) = |S \cap V_i|$, for any $S \subseteq [m]$. One can express this setting as a coverage instance by considering $T = m$ set families each comprised of singleton subsets. Specifically, for each good $g \in [m]$, we have a set family $\mathcal{I}_g$ that includes all singletons $\{i\}$ with the property that $g \in V_i$, i.e., subset $\{i\}$ is included in $\mathcal{I}_g$ iff agent i values good g. As in the public decision making setting, here we obtain a coverage instance with polynomially large $\mathcal{I}_t$-s.

With Nash welfare as a notion of fairness, Fluschnik et al. [12] study fair selection of indivisible goods under a knapsack constraint.[5] By contrast, the current work addresses combinatorial constraints over subsets of agents.

2 Notation and Preliminaries

An instance of a fair coverage problem is specified as a tuple $\langle [n], T, \{\mathcal{I}_t\}_{t=1}^T \rangle$, where $[n] = \{1, 2, \ldots, n\}$ denotes the set of agents and $T \in \mathbb{Z}_+$ denotes the number of subsets (of the agents) to be selected. Here, for each $t \in [T]$, the tth selected subset (say $F_t \subseteq [n]$) is constrained to be from the family $\mathcal{I}_t$, i.e., each $\mathcal{I}_t \subseteq 2^{[n]}$ specifies the possible choices for the tth selection. It is not necessary that the set families $\mathcal{I}_t$-s are given explicitly; our algorithmic result only requires a blackbox access to an FPTAS for weight maximization over $\mathcal{I}_t$-s.

For a fair coverage instance $\langle [n], T, \{\mathcal{I}_t\}_{t=1}^T \rangle$, a solution $\mathcal{F} = (F_1, F_2, \ldots, F_T)$ is a tuple with the property that $F_t \in \mathcal{I}_t$ for all $t \in [T]$. We address settings wherein the valuation of each agent depends on the number of times it is covered

---

[5] Fluschnik et al. [12] also highlight connections between NSW and proportional approval voting.

among the selected subsets. Specifically, for a solution $\mathcal{F} = (F_1, F_2, \ldots, F_T)$, the coverage value $v_i(\mathcal{F})$, of agent $i \in [n]$, is defined as $v_i(\mathcal{F}) := |\{t \in [T] : i \in F_t\}| + 1$. Note the coverage value of each agent is smoothed by adding 1. This smoothing ensures that the Nash social welfare of any solution is nonzero. We, in fact, show that if each agent's value is equated to exactly the number of times it is covered among the subsets, then one cannot achieve *any* multiplicative approximation guarantee for Nash social welfare maximization (refer to the full version [4]).

The Nash social welfare (NSW) of a solution $\mathcal{F}$ is defined as the geometric mean of the agents' coverage values, NSW $(\mathcal{F}) := \left(\prod_{i=1}^{n} v_i(\mathcal{F})\right)^{\frac{1}{n}}$. We will write $\mathcal{F}^* = (F_1^*, F_2^*, \ldots, F_T^*)$ to denote a solution that maximizes the Nash social welfare in a given fair coverage instance. Furthermore, a solution $\widehat{\mathcal{F}}$ is said to achieve a γ-approximation guarantee for the Nash social welfare maximization problem iff NSW$(\widehat{\mathcal{F}}) \geq \frac{1}{\gamma}$NSW $(\mathcal{F}^*)$. The current work develops a constant-factor approximation algorithm for NSW maximization in fair coverage instances.

As mentioned previously, the algorithm works with a blackbox access to an FPTAS for weight maximization over $\mathcal{I}_t$-s. Specifically, with parameter $\beta := \frac{1}{64nT^2}$, we will write APXMAXWT to denote a subroutine (blackbox) that takes as input weights $w_1, \ldots, w_n \in \mathbb{R}_+$, along with an index $t \in [T]$, and finds a $(1-\beta)$-approximation to $\max_{X \in \mathcal{I}_t} \sum_{i \in X} w_i$. The assumption that weight maximization over $\mathcal{I}_t$-s admits an FPTAS implies that a $(1 - \beta)$-approximation (with $\beta = \frac{1}{64nT^2}$) can be computed in polynomial time.

For any solution $\mathcal{F} = (F_1, \ldots, F_T)$, index $t \in [T]$, and subset $X \in \mathcal{I}_t$, write $(X, \mathcal{F}_{-t})$ to denote the solution obtained by replacing F_t with X, i.e., $(X, \mathcal{F}_{-t}) := (F_1, \ldots, F_{t-1}, X, F_{t+1}, \ldots, F_T)$. Finally, we will write $\varphi(\mathcal{F})$ to denote the log social welfare of the agents under solution $\mathcal{F}$, i.e., $\varphi(\mathcal{F}) := \sum_{i=1}^{n} \log(v_i(\mathcal{F}))$. Here, the logarithm is to the base e, i.e., we consider the natural logarithm of coverage values.

3   Approximation Algorithm for Nash Social Welfare

This section develops an $(18 + o(1))$-approximation algorithm for maximizing Nash social welfare in fair coverage instances. Given any instance $\langle [n], T, \{\mathcal{I}_t\}_{t=1}^T \rangle$, our algorithm ALG (Algorithm 1) starts with an arbitrary solution $\mathcal{F} = (F_1, \ldots, F_T) \in \mathcal{I}_1 \times \ldots \times \mathcal{I}_T$ and iteratively performs local updates as long as it experiences a sufficient (additive) increase in the log social welfare φ. Here, for any solution $\mathcal{F} = (F_1, \ldots, F_T)$, a local update corresponds to replacing—for some $\tau \in [T]$—the subset F_τ with some other subset $A_\tau \in \mathcal{I}_\tau$. For updating a solution $\mathcal{F}$ and with φ as a guiding objective, the algorithm addresses the problem of finding, for every $t \in [T]$, a subset $A_t \in \mathcal{I}_t$ that achieves $\max_{X \in \mathcal{I}_t} \varphi(X, \mathcal{F}_{-t}) - \varphi(\mathcal{F})$. Notably, we reduce this problem to that of weight maximization over $\mathcal{I}_t$-s, by setting appropriate weights w_i^t, for each agent $i \in [n]$ and each index $t \in [T]$. In particular, for a current solution $\mathcal{F} = (F_1, \ldots, F_T)$, the algorithm sets the weights as follows

$$
w_i^t = \begin{cases} \log{(v_i(\mathcal{F}))} \; - \; \log{(v_i(\mathcal{F}) - 1)} & \text{if } i \in F_t \\ \log{(v_i(\mathcal{F}) + 1)} \; - \; \log{(v_i(\mathcal{F}))} & \text{otherwise, if } i \in [n] \setminus F_t. \end{cases}
$$

We note that for each agent $i \in F_t$, the coverage value $v_i(\mathcal{F}) \geq 2$; this follows from the inclusion of '+1' in the definition of coverage value. Hence, the weights (specifically, the terms $\log{(v_i(\mathcal{F}) - 1)}$ for $i \in F_t$) are well defined. This is a relevant implication of smoothing the coverage values.

Moreover, this weight assignment ensures that, for every subset $X \subseteq [n]$, its weight $\sum_{i \in X} w_i^t = (\varphi(X, \mathcal{F}_{-t}) - \varphi(\mathcal{F})) + \sum_{j \in F_t} w_j^t$ (see Claim 3). Since the weight of the current subset F_t (i.e., $\sum_{j \in F_t} w_j^t$) is fixed, finding a subset $X \in \mathcal{I}_t$ with maximum possible weight is equivalent to finding a subset that maximizes $\varphi(X, \mathcal{F}_{-t}) - \varphi(\mathcal{F})$. In fact, we show that an FPTAS for this weight maximization suffices. As mentioned previously, we denote by $\text{ApxMaxWt}(t, w_1^t, \ldots, w_n^t)$ a subroutine (blackbox) that takes as input weights $w_1^t, \ldots, w_n^t \in \mathbb{R}_+$ and finds a $(1 - \beta)$-approximation to $\max_{X \in \mathcal{I}_t} \sum_{i \in X} w_i^t$; where the parameter $\beta = \frac{1}{64nT^2}$.

Hence, for updating the solution $\mathcal{F} = (F_1, \ldots, F_T)$, the algorithm invokes ApxMaxWt to obtain candidate subsets $A_1, A_2, \ldots, A_T$. If, for some index $\tau \in [T]$, replacing F_τ by A_τ leads to a sufficient additive increase φ, then Alg updates the solution to $(A_\tau, \mathcal{F}_{-\tau})$. Specifically, the algorithm sets parameter $\varepsilon := \frac{1}{16nT}$ and if $\varphi(A_\tau, \mathcal{F}_{-\tau}) - \varphi(\mathcal{F}) \geq \frac{\varepsilon n}{8T}$, then it updates the solution (see Lines 4 and 5 in Algorithm 1). Otherwise, if for all the candidate subsets the increase in φ is less than $\frac{\varepsilon n}{8T}$, the algorithm terminates.

Note that, for any solution $\widehat{\mathcal{F}}$, the log social welfare $\varphi(\widehat{\mathcal{F}})$ is at most $n \log(T + 1)$.[6] This observation, and the fact that in every iteration of Alg the log social welfare of the maintained solution increases by at least $\frac{\varepsilon n}{8T}$, imply that the algorithm terminates in polynomial time (Lemma 3). Overall, the algorithm efficiently finds a local maximum of φ.

We establish the approximation ratio via counting arguments. In the analysis, for each maintained solution $\mathcal{F}$, we consider the agents i whose current coverage value, $v_i(\mathcal{F})$, is sufficiently smaller than their optimal coverage value, $v_i(\mathcal{F}^*)$; recall that $\mathcal{F}^*$ denotes a Nash optimal solution. In particular, for a solution $\mathcal{F}$ and any integer $\alpha \in \mathbb{Z}_+$, we will write $S_\alpha^{\mathcal{F}}$ to denote the subset of agents whose coverage value is $\alpha(2.25 + \varepsilon)$ times less than their optimal, where, $\varepsilon = \frac{1}{16nT}$. Formally, for any $\alpha \in \mathbb{Z}_+$, the set of α-suboptimal agents is defined as[7]

$$
S_\alpha^{\mathcal{F}} := \left\{ i \in [n] : v_i(\mathcal{F}) < \frac{1}{\alpha(2.25 + \varepsilon)} v_i(\mathcal{F}^*) \right\}. \tag{1}
$$

First, we prove that, for any solution $\mathcal{F}$ and *any* integer $\alpha \geq 4$, if the number of α-suboptimal agents is more than $\frac{n}{\alpha}$, then there necessarily exists a local update

---

[6] Indeed, for any solution $\widehat{\mathcal{F}}$, we have $v_i(\widehat{\mathcal{F}}) \leq T + 1$, for all agents $i \in [n]$.

[7] Here, the constant 2.25 is selected to achieve the desired approximation ratio.

Algorithm 1. ALG

Input: Instance $\langle [n], T, \{\mathcal{I}_t\}_{t=1}^T \rangle$.
Output: A solution $\mathcal{F} = (F_1, \ldots, F_T)$.
1: Initialize $\mathcal{F} = (F_1, F_2, \ldots, F_T) \in \mathcal{I}_1 \times \mathcal{I}_2 \times \ldots \times \mathcal{I}_T$ to be an arbitrary solution and, for all agents $i \in [n]$, set coverage value $v_i = v_i(\mathcal{F})$. Set parameter $\varepsilon := \frac{1}{16nT}$.
2: For each $t \in [T]$ and all agents $i \in [n]$, set weight

$$w_i^t = \begin{cases} \log v_i - \log(v_i - 1) & \text{if } i \in F_t \\ \log(v_i + 1) - \log v_i & \text{if } i \in [n] \setminus F_t. \end{cases}$$

3: For each $t \in [T]$, set $A_t = \text{ApxMaxWt}(t, w_1^t, w_2^t, \ldots, w_n^t)$.
4: **while** there exists $\tau \in [T]$ such that $\varphi(A_\tau, \mathcal{F}_{-\tau}) - \varphi(\mathcal{F}) \geq \frac{\varepsilon n}{8T}$ **do**
5:     Update $\mathcal{F} \leftarrow (A_\tau, \mathcal{F}_{-\tau})$, i.e., update $F_\tau \leftarrow A_\tau$.
6:     For all agents $i \in [n]$, update coverage value $v_i = v_i(\mathcal{F})$.
7:     For each $t \in [T]$ and all agents $i \in [n]$, set weights w_i^t as in Line 2.
8:     Set $A_t = \text{ApxMaxWt}(t, w_1^t, w_2^t, \ldots, w_n^t)$ for all $t \in [T]$.
9: **end while**
10: **return** solution $\mathcal{F}$

that increases φ by a sufficient amount (Lemma 2). Contrapositively, we obtain that, for the solution finally obtained by ALG and for any $\alpha \geq 4$, the number of α-suboptimal agents is at most n/α. We complete the analysis by proving that this guarantee ensures that ALG achieves a constant-factor approximation ratio for NSW maximization; more formally, we will establish the following theorem (in Sect. 3.2).

Theorem 1 (Main Result). *Given any fair coverage instance $\langle [n], T, \{\mathcal{I}_t\}_{t=1}^T \rangle$, with blackbox access to an FPTAS for weight maximization over $\mathcal{I}_t$-s, ALG (Algorithm 1) computes—in polynomial time—an $\left(18 + \frac{1}{2nT}\right)$-approximate solution for the Nash social welfare maximization problem.*

3.1 Algorithm's Analysis

The following claim bounds the change in log social welfare φ when a solution is updated.

Claim 1. *For a solution $\mathcal{F} = (F_1, \ldots, F_T)$, let value $v_i := v_i(\mathcal{F})$ for all agents $i \in [n]$. Then, for any subset $X \subseteq [n]$ and any index $t \in [T]$, we have*

$$\varphi(X, \mathcal{F}_{-t}) - \varphi(\mathcal{F}) \geq \sum_{i \in X} \frac{1}{v_i + 1} - \sum_{j \in F_t} \frac{1}{v_j - 1}.$$

The proof of Claim 1 is deferred to the full version of the paper. Note that here, for each agent $j \in F_t$, the coverage value $v_j(\mathcal{F}) \geq 2$ and, hence, the subtracted terms, $\frac{1}{v_j - 1}$, in the claim are well defined.

Next, we bound the expected change in φ when—for any solution $\mathcal{F} = (F_1, \ldots, F_T)$—we replace F_t by F_t^*, for a $t \in [T]$ chosen uniformly at random.

Lemma 1. *For any solution* $\mathcal{F} = (F_1, \ldots, F_T)$ *and a Nash optimal solution* $\mathcal{F}^* = (F_1^*, \ldots, F_T^*)$, *let values* $v_i := v_i(\mathcal{F})$ *and* $v_i^* := v_i(\mathcal{F}^*)$, *for all agents* $i \in [n]$. *Then, uniformly sampling index* t *from the set* $[T]$, *we obtain*

$$\mathbb{E}_{t \in_R [T]} \left[\varphi(F_t^*, \mathcal{F}_{-t}) - \varphi(\mathcal{F}) \right] \geq \frac{1}{T} \sum_{i=1}^{n} \left(\frac{v_i^* - 1}{v_i + 1} \right) - \frac{n}{T}.$$

Proof. Invoking Claim 1, with $X = F_t^*$ for each $t \in [T]$, we obtain

$$\mathbb{E}_{t \in_R [T]} \left[\varphi(F_t^*, \mathcal{F}_{-t}) - \varphi(\mathcal{F}) \right] \geq \mathbb{E}_{t \in_R [T]} \left[\sum_{i \in F_t^*} \frac{1}{v_i + 1} - \sum_{j \in F_t} \frac{1}{v_j - 1} \right]$$

$$= \mathbb{E}_{t \in_R [T]} \left[\sum_{i \in [n]} \mathbb{1}\{i \in F_t^*\} \frac{1}{v_i + 1} \right.$$

$$\left. - \sum_{j \in [n] : v_j \geq 2} \mathbb{1}\{j \in F_t\} \frac{1}{v_j - 1} \right]$$

$$\text{(since } v_j \geq 2, \text{ for all } j \in F_t)$$

$$= \sum_{i \in [n]} \mathbb{P}\{i \in F_t^*\} \frac{1}{v_i + 1} - \sum_{j \in [n] : v_j \geq 2} \mathbb{P}\{j \in F_t\} \frac{1}{v_j - 1}. \quad (2)$$

Index t is selected uniformly at random from the set $[T]$. Also, by definition, v_i^* is equal to 1 plus the number of subsets that contain i in the Nash optimal solution $\mathcal{F}^* = (F_1^*, \ldots, F_T^*)$. Hence, the probability $\mathbb{P}\{i \in F_t^*\} = \frac{v_i^* - 1}{T}$, for all agents $i \in [n]$. Similarly, for the solution $\mathcal{F} = (F_1, \ldots, F_T)$, we have $\mathbb{P}\{j \in F_t\} = \frac{v_j - 1}{T}$, for all $j \in [n]$. These equations and inequality (2) give us

$$\mathbb{E}_{t \in_R [T]} \left[\varphi(F_t^*, \mathcal{F}_{-t}) - \varphi(\mathcal{F}) \right] \geq \sum_{i \in [n]} \frac{v_i^* - 1}{T} \cdot \frac{1}{v_i + 1} - \sum_{j \in [n] : v_j \geq 2} \frac{v_j - 1}{T} \cdot \frac{1}{v_j - 1}$$

$$\geq \frac{1}{T} \sum_{i \in [n]} \left(\frac{v_i^* - 1}{v_i + 1} \right) - \frac{n}{T}.$$

The lemma stands proved.

Next, we show that if, under a solution $\mathcal{F}$, the number of α-suboptimal agents is large, then the log social welfare can be sufficiently increased by replacing F_τ with F_τ^*, for some $\tau \in [T]$. Recall that $\mathcal{F}^* = (F_1^*, \ldots, F_T^*)$ denotes a Nash optimal allocation and $S_\alpha^{\mathcal{F}}$ denotes the set of α-suboptimal agents under solution $\mathcal{F}$; see Eq. (1).

Lemma 2. *For any solution* $\mathcal{F} = (F_1, \ldots, F_T)$ *and any* $\alpha \geq 4$, *if the number of* α-*suboptimal agents is at least* $\frac{n}{\alpha}$ *(i.e.,* $|S_\alpha^{\mathcal{F}}| > \frac{n}{\alpha}$*), then there exists an index* $\tau \in [T]$ *such that*

$$\varphi(F_\tau^*, \mathcal{F}_{-\tau}) - \varphi(\mathcal{F}) \geq \frac{\varepsilon n}{2T}.$$

Proof. Consider any solution $\mathcal{F}$ and integer $\alpha \geq 4$ such that $|S_\alpha^{\mathcal{F}}| > \frac{n}{\alpha}$. For each agent $i \in [n]$, write $v_i := v_i(\mathcal{F})$ and $v_i^* = v_i(\mathcal{F}^*)$. Now, Lemma 1 gives us

$$\mathbb{E}_{t \in R[T]}\left[\varphi(F_t^*, \mathcal{F}_{-t}) - \varphi(\mathcal{F})\right] \geq \frac{1}{T}\sum_{i=1}^{n}\left(\frac{v_i^* - 1}{v_i + 1}\right) - \frac{n}{T}$$

$$\geq \frac{1}{T}\sum_{i \in S_\alpha^{\mathcal{F}}}\left(\frac{v_i^* - 1}{v_i + 1}\right) - \frac{n}{T}$$

$$\geq \frac{1}{T}\sum_{i \in S_\alpha^{\mathcal{F}}}\left(\frac{\alpha(2.25 + \varepsilon)v_i - 1}{v_i + 1}\right) - \frac{n}{T}.$$

(by definition of $S_\alpha^{\mathcal{F}}$)

Claim 2. *For parameter $\varepsilon \in (0,1)$ along with any integers $\alpha \geq 4$ and $v \geq 1$, we have*

$$\frac{\alpha(2.25 + \varepsilon)v - 1}{v + 1} \geq \left(1 + \frac{\varepsilon}{2}\right)\alpha.$$

Claim 2 (proof appears in the full version [4]) shows that $\frac{\alpha(2.25+\varepsilon)v-1}{v+1} \geq \left(1 + \frac{\varepsilon}{2}\right)\alpha$, for all integers $\alpha \geq 4$ and $v \geq 1$. Therefore, the above-mentioned inequality simplifies to

$$\mathbb{E}_{t \in R[T]}\left[\varphi(F_t^*, \mathcal{F}_{-t}) - \varphi(\mathcal{F})\right] \geq \frac{1}{T}\sum_{i \in S_\alpha^{\mathcal{F}}}\left(1 + \frac{\varepsilon}{2}\right)\alpha - \frac{n}{T}$$

$$> \frac{1}{T}\frac{n}{\alpha}\left(1 + \frac{\varepsilon}{2}\right)\alpha - \frac{n}{T} \qquad (\text{since } |S_\alpha^{\mathcal{F}}| > \frac{n}{\alpha})$$

$$= \frac{\varepsilon n}{2T}.$$

Therefore, there exists a $\tau \in [T]$ such that

$$\varphi(F_\tau^*, \mathcal{F}_{-\tau}) - \varphi(\mathcal{F}) \geq \frac{\varepsilon n}{2T}.$$

This completes the proof of the lemma.

Using Lemma 2, we will establish in Corollary 1, below, that the algorithm continues to iterate as long as the number of α-suboptimal agents is more than n/α. The proof of the corollary also utilizes the following claim.

Claim 3. *Let $\mathcal{F} = (F_1, \ldots, F_T)$ be any solution considered in ALG (Algorithm 1) and, for all indices $t \in [T]$ and agents $i \in [n]$, let w_i^t-s be the corresponding weights set in Lines 2 or 7. Then, the weight of any subset $X \subseteq [n]$ satisfies*

$$\sum_{i \in X} w_i^t = (\varphi(X, \mathcal{F}_{-t}) - \varphi(\mathcal{F})) + \sum_{j \in F_t} w_j^t.$$

The proof of this claim appears in the full version of the paper.

Corollary 1. *For any solution* $\mathcal{F} = (F_1, \ldots, F_T)$ *considered in* ALG *(Algorithm 1) and any* $\alpha \geq 4$, *if the number of* α-*suboptimal agents is at least* $\frac{n}{\alpha}$ *(i.e.,* $|S_\alpha^{\mathcal{F}}| > \frac{n}{\alpha}$*), then the execution condition in the while-loop (Line 4) of* ALG *holds.*

Proof. Consider any solution $\mathcal{F}$ in ALG and integer $\alpha \geq 4$ such that $|S_\alpha^{\mathcal{F}}| > \frac{n}{\alpha}$. In such a case, we will show that there exists an index $\tau \in [T]$ for which the subset A_τ returned by the subroutine APXMAXWT$(\tau, w_1^\tau, \ldots, w_n^\tau)$ (in Line 8) satisfies $\varphi(A_\tau, \mathcal{F}_{-\tau}) - \varphi(\mathcal{F}) \geq \frac{\varepsilon n}{8T}$. Hence, the while-loop continues to iterate.

The desired index is in fact the one identified in Lemma 2. In particular, Lemma 2 ensures that for an index $\tau \in [T]$ we have

$$\varphi(F_\tau^*, \mathcal{F}_{-\tau}) - \varphi(\mathcal{F}) \geq \frac{\varepsilon n}{2T}. \tag{3}$$

Now, Claim 3 (with $X = F_\tau^*$) gives us

$$\sum_{i \in F_\tau^*} w_i^\tau = (\varphi(F_\tau^*, \mathcal{F}_{-\tau}) - \varphi(\mathcal{F})) + \sum_{i \in F_\tau} w_i^\tau$$

$$\geq \frac{\varepsilon n}{2T} + \sum_{i \in F_\tau} w_i^\tau. \qquad \text{(via inequality (3))}$$

Therefore,

$$\max_{X \in \mathcal{I}_\tau} \left\{ \sum_{i \in X} w_i^\tau \right\} \geq \sum_{i \in F_\tau} w_i^\tau + \frac{\varepsilon n}{2T}. \tag{4}$$

Recall that APXMAXWT$(\tau, w_1^\tau, \ldots, w_n^\tau)$ returns a set $A_\tau \in \mathcal{I}_\tau$ with the property that

$$\sum_{i \in A_\tau} w_i^\tau \geq (1 - \beta) \left(\max_{X \in \mathcal{I}_\tau} \sum_{i \in X} w_i^\tau \right). \tag{5}$$

Here, parameter $\beta = \frac{1}{64nT^2}$. Since $\varepsilon = \frac{1}{16nT}$, we have $\beta = \frac{\varepsilon}{4T}$. Inequalities (4) and (5) give us

$$\sum_{i \in A_\tau} w_i^\tau \geq (1 - \beta) \left(\sum_{i \in F_\tau} w_i^\tau + \frac{\varepsilon n}{2T} \right)$$

$$= \sum_{i \in F_\tau} w_i^\tau + \frac{\varepsilon n}{2T} - \beta \sum_{i \in F_\tau} w_i^\tau - \frac{\beta \varepsilon n}{2T}$$

$$\geq \sum_{i \in F_\tau} w_i^\tau + \frac{\varepsilon n}{2T} - \beta n - \frac{\beta \varepsilon n}{2T} \qquad \text{(since } \sum_{i \in F_\tau} w_i^\tau \leq n\text{)}$$

$$= \sum_{i \in F_\tau} w_i^\tau + \frac{\varepsilon n}{2T} - \frac{\varepsilon n}{4T} - \frac{\beta \varepsilon n}{2T} \qquad \text{(since } \beta = \frac{\varepsilon}{4T}\text{)}$$

$$\geq \sum_{i \in F_\tau} w_i^\tau + \frac{\varepsilon n}{2T} - \frac{\varepsilon n}{4T} - \frac{\varepsilon n}{8T} \qquad \text{(since } \beta \leq \frac{1}{4}\text{)}$$

$$= \sum_{i \in F_\tau} w_i^\tau + \frac{\varepsilon n}{8T}.$$

Applying Claim 3, with $X = A_\tau$, we get $\varphi(A_\tau, \mathcal{F}_{-\tau}) - \varphi(\mathcal{F}) \geq \frac{\varepsilon n}{8T}$. Therefore, the execution condition in the while-loop of ALG holds. This establishes the corollary.

We conclude the section by showing that the algorithm runs in polynomial time.

Lemma 3 (Runtime Analysis). *Given any fair coverage instance* $\langle [n], T,$ $\{\mathcal{I}_t\}_{t=1}^T\rangle$ *with blackbox access to an FPTAS for weight maximization over* $\mathcal{I}_t$*-s,* ALG *(Algorithm 1) terminates in time that is polynomial in* n *and* T.

Proof. For any solution $\mathcal{F}$, the coverage values $v_i(\mathcal{F}) \geq 1$, for agents $i \in [n]$. Hence, for the initial solution (arbitrarily) selected by the algorithm, we have $\varphi(\mathcal{F}) = \sum\limits_{i=1}^n \log(v_i(\mathcal{F})) \geq 0$. In addition, since the coverage values of the agents under any solution are at most $T+1$, the log social welfare φ across all solutions is upper bounded by $n\log(T + 1)$. Furthermore, note that in every iteration of ALG the log social welfare of the maintained solution increases additively by at least $\frac{\varepsilon n}{8T}$. These observations imply that the algorithm terminates after $O\left(nT^2 \log T\right)$ iterations; recall that $\varepsilon = \frac{1}{16nT}$. Since each iteration executes in polynomial time, the time complexity of the algorithm is polynomial in n and T. The lemma stands proved.

3.2   Proof of Theorem 1

This section establishes the approximation ratio of ALG. For the given fair coverage instance, let $\mathcal{F} = (F_1, \ldots, F_T)$ be the solution returned by ALG and $\mathcal{F}^* = (F_1^*, \ldots, F_T^*)$ be a Nash optimal allocation. Note that $v_i(\mathcal{F}) \geq 1$ and $v_i(\mathcal{F}^*) \leq T+1$, for all agents $i \in [n]$. Hence, for each agent $i \in [n]$, the following bound holds: $v_i(\mathcal{F}) \geq \frac{1}{T+1}v_i(\mathcal{F}^*)$.

We partition the set of agents $[n]$ considering the multiplicative gap between the coverage values under $\mathcal{F}$ and $\mathcal{F}^*$. Specifically, for each integer $d \in \{2, 3, \ldots, \lceil \log(T+1)\rceil\}$, define the set

$$X_{2^d} := \left\{ i \in [n] : \frac{1}{2^{d+1}}\frac{v_i(\mathcal{F}^*)}{(2.25 + \varepsilon)} \leq v_i(\mathcal{F}) < \frac{1}{2^d}\frac{v_i(\mathcal{F}^*)}{(2.25 + \varepsilon)} \right\}.$$

Furthermore, write $X' := [n] \setminus \left(\bigcup\limits_{d=2}^{\lceil \log(T+1)\rceil} X_{2^d} \right)$. Since all agents i satisfy $v_i(\mathcal{F}) \geq \frac{1}{T+1}v_i(\mathcal{F}^*)$, the subset X' only contains agents $j \in [n]$ with the property that $v_j(\mathcal{F}) \geq \frac{1}{4}\frac{v_j(\mathcal{F}^*)}{(2.25+\varepsilon)}$. Also, note that the subsets X_{2^d}-s and X' form a partition of the set of agents $[n]$; in particular, $|X'| + \sum_{d \geq 2} |X_{2^d}| = n$.

Recall that $S_\alpha^{\mathcal{F}}$ denotes the set of α-suboptimal agents (see Eq. (1)). Also, note that, with $\alpha = 2^d$, we have $X_\alpha \subseteq S_\alpha^{\mathcal{F}}$. Moreover, by the contrapositive of Corollary 1, for the solution $\mathcal{F} = (F_1, \ldots, F_T)$, returned by ALG, we have

$$|X_{2^d}| \leq |S_{2^d}^{\mathcal{F}}| \leq \frac{n}{2^d} \qquad \text{for all } 2 \leq d \leq \lceil \log(T+1)\rceil \tag{6}$$

For any subset of agents $Y \subseteq [n]$, write $\rho(Y) := \prod_{i \in Y} \frac{v_i(\mathcal{F})}{v_i(\mathcal{F}^*)}$, if subset $Y \neq \emptyset$. Otherwise, if $Y = \emptyset$, define $\rho(Y) := 1$. To bound the approximation ratio of the algorithm, we consider

$$\frac{\text{NSW}(\mathcal{F})}{\text{NSW}(\mathcal{F}^*)} = \left(\rho(X') \prod_{d=2}^{\lceil \log(T+1) \rceil} \rho(X_{2^d}) \right)^{\frac{1}{n}}$$

$$\geq \left(\left(\frac{1}{9+4\varepsilon} \right)^{|X'|} \prod_{d \geq 2} \rho(X_{2^d}) \right)^{\frac{1}{n}}$$

$$\left(v_j(\mathcal{F}) \geq \frac{1}{4(2.25+\varepsilon)} v_j(\mathcal{F}^*) \text{ for all } j \in X' \right)$$

$$\geq \left(\left(\frac{1}{9+4\varepsilon} \right)^{|X'|} \prod_{d \geq 2} \left(\frac{1}{2^{d+1}(2.25+\varepsilon)} \right)^{|X_{2^d}|} \right)^{\frac{1}{n}}$$

$$\left(v_i(\mathcal{F}) \geq \frac{1}{2^{d+1}(2.25+\varepsilon)} \text{ for all } i \in X_{2^d} \right)$$

$$= \frac{1}{9+4\varepsilon} \left(\prod_{d \geq 2} \left(\frac{1}{2^{d-1}} \right)^{\frac{|X_{2^d}|}{n}} \right) \quad \left(\text{since } |X'| + \sum_{d \geq 2} |X_{2^d}| = n \right)$$

$$\geq \frac{1}{9+4\varepsilon} \left(\prod_{d \geq 2} \left(\frac{1}{2^{d-1}} \right)^{\frac{1}{2^d}} \right). \quad \text{(via inequality (6))}$$

Claim 4. *For any integer $\ell \geq 2$, we have $\prod_{d=2}^{\ell} \left(\frac{1}{2^{d-1}} \right)^{\frac{1}{2^d}} \geq \frac{1}{2}$.*

The proof of Claim 4 appears in the full version of the paper. Hence, the stated approximation ratio follows

$$\frac{\text{NSW}(\mathcal{F})}{\text{NSW}(\mathcal{F}^*)} \geq \frac{1}{9+4\varepsilon} \left(\prod_{d=2}^{\lceil \log(T+1) \rceil} \left(\frac{1}{2^{d-1}} \right)^{\frac{1}{2^d}} \right) \geq \frac{1}{9+4\varepsilon} \cdot \frac{1}{2} = \frac{1}{18+8\varepsilon}.$$

4   APX-Hardness of Fair Coverage

This section shows that NSW maximization in fair coverage instances is APX-hard. In particular, we prove that there exists an absolute constant $\gamma > 1$ such that it is NP-hard to approximate the problem within factor γ. Hence, a constant-factor approximation is the best one can hope for NSW maximization in fair coverage instances, unless P = NP. The hardness result is obtained via an approximation preserving reduction from the following gap version of the maximum coverage problem.

Maximum k-Coverage [11]: Given a universe of elements $U = \{1, 2, \ldots, n\}$, a threshold $k \in \mathbb{Z}_+$, and a set family $\mathcal{S} = \{S_\ell \subseteq [n]\}_{\ell=1}^{N}$, it is NP-hard to distinguish between

- YES Instances: There exists a collection of k subsets in $\mathcal{S}$ that covers all the elements, i.e., the union of the k subsets is equal to $[n]$.
- NO Instances: Any collection of k subsets from $\mathcal{S}$ covers at most $\left(1 - \frac{1}{e}\right)n$ elements, i.e., the union of any k subsets from $\mathcal{S}$ has cardinality at most $\left(1 - \frac{1}{e}\right)n$.

This hardness result of Feige [11] holds even for instances that satisfy the following properties: (i) all the subsets in $\mathcal{S}$ have the same size τ, i.e., $|S_\ell| = \tau$ for all subsets $S_\ell \in \mathcal{S}$, and (ii) the threshold $k = n/\tau$. Properties (i) and (ii) will be utilized in our approximation preserving reduction.[8]

The APX-hardness result is established next. Notably, this negative result is applicable even for fair coverage instances in which the set families $\mathcal{I}_t$-s are explicitly given as input.

Theorem 2. *In fair coverage instances, it is NP-hard to approximate the maximum Nash social welfare within a factor of* 1.092.

Proof. Given an instance of the maximum k-coverage problem with universe $U = \{1, 2, \ldots, n\}$ and set family $\mathcal{S} = \{S_1, S_2, \ldots, S_N\}$ of τ-sized subsets of $[n]$, we construct a fair coverage instance with n agents and $T = k$. Since threshold $k = \frac{n}{\tau}$, we have $T = \frac{n}{\tau}$. To complete the construction and obtain an instance $\langle [n], T, \{\mathcal{I}_t\}_{t=1}^{T} \rangle$, we set the families $\mathcal{I}_t = \mathcal{S}$, for all $t \in [T]$.

First, we show that if the underlying maximum coverage instance is a YES instance, then the optimal NSW in the constructed fair coverage instance is at least 2. Note that in the YES case there exists a size-k collection $\mathcal{S}' = \{S_1', S_2', \ldots, S_k'\} \subseteq \mathcal{S}$ that covers all of $[n]$. Also, by construction, $T = k$ and $\mathcal{I}_t = \mathcal{S}$ for all $1 \leq t \leq k$. Hence, for each $t \in [T]$, we have $S_t' \in \mathcal{I}_t$. Therefore, the tuple $\mathcal{F}' = (S_1', S_2', \ldots, S_k')$ is a solution under which $v_i(\mathcal{F}') \geq 2$, for all agents $i \in [n]$.[9] This bound on the coverage value of the agents implies that in the current case, the optimal Nash social welfare is at least 2.

Now, we show that in the NO case the optimal NSW is at most c, for an absolute constant $c < 2$. Here, consider any solution $\mathcal{F} = (F_1, \ldots, F_T)$ in the constructed fair coverage instance. We have $T = k = \frac{n}{\tau}$ and, by construction, $F_t \in \mathcal{S}$. Furthermore, given that we are in the NO case, the collection of subsets $\{F_1, F_2, \ldots, F_T\} \subseteq \mathcal{S}$ covers at most $(1 - \frac{1}{e})n$ elements. Let L denote the set of agents not covered by the subsets F_t-s and write $\ell := |L| \geq \frac{n}{e}$. Since each agent $i \in L$ is not covered under $\mathcal{F}$, we have $v_i(\mathcal{F}) = 1$ for all $i \in L$. Furthermore, note

---

[8] The properties also ensure that in the YES case there is a collection of $k = \frac{n}{\tau}$ subsets that are pairwise disjoint and they cover all of $[n]$. That is, in the YES case there exists a perfect cover.

[9] In fact, for each agent i the coverage value $v_i(\mathcal{F}') = 2$, since i is contained in exactly one of the subsets S_t'-s. Recall that properties (i) and (ii) ensure that $\mathcal{S}'$ is a perfect cover.

that the agents in the set $L^c := [n] \setminus L$ are covered by the $T = k = \frac{n}{\tau}$ subsets $F_1, \ldots, F_T$, and each of these subsets is of size τ. Therefore,

$$
\begin{aligned}
\sum_{j \in L^c} v_j(\mathcal{F}) &= \sum_{t=1}^{n/\tau} |F_t| + |L^c| \\
&= \frac{n}{\tau}\tau + |L^c| \qquad &\text{(since } |F_t| = \tau \text{ for each } t) \\
&= n + (n - \ell). \qquad &(\ell = |L|)
\end{aligned}
$$

Hence, the average social welfare among agents in L^c satisfies $\frac{1}{|L^c|}\sum_{j \in L^c} v_j(\mathcal{F}) = \frac{2n-\ell}{n-\ell}$. This bound and the AM-GM inequality give us $\prod_{j \in L^c} v_i(\mathcal{F}) \le \left(\frac{2n-\ell}{n-\ell}\right)^{|L^c|}$. Therefore, we can bound the Nash social welfare of $\mathcal{F}$ as follows

$$
\text{NSW}(\mathcal{F}) = \left(\prod_{i \in L} v_i(\mathcal{F}) \prod_{j \in L^c} v_j(\mathcal{F})\right)^{\frac{1}{n}} \le 1^{\frac{\ell}{n}} \left(\frac{2n-\ell}{n-\ell}\right)^{\frac{n-\ell}{n}} = \left(\frac{2-\ell/n}{1-\ell/n}\right)^{\left(1-\frac{\ell}{n}\right)}
\tag{7}
$$

Note that the function $f(x) := \left(\frac{2-x}{1-x}\right)^{(1-x)}$ is decreasing in the interval $x \in [\frac{1}{e}, 1)$. Hence, using the fact that $\ell \ge \frac{n}{e}$ and inequality (7), we get

$$
\text{NSW}(\mathcal{F}) \le \left(\frac{2-1/e}{1-1/e}\right)^{1-\frac{1}{e}} \le 1.83
\tag{8}
$$

Since, in the NO case, inequality (8) holds for all solutions $\mathcal{F}$, we get that the optimal NSW is at most 1.83.

Overall, we get that in the YES case the optimal NSW is at least 2 and in the NO case it is at most 1.83. This multiplicative gap of $\frac{2}{1.83} > 1.092$ implies that a 1.092-approximation algorithm for NSW maximization can be used to distinguish between the two cases. Since this differentiation is NP-hard, a 1.092-approximation is NP-hard as well. The theorem stands proved.

5   Conclusion and Future Work

The current paper extends the scope of coverage problems from combinatorial optimization to fair division. In this setting, we develop algorithmic and hardness results for maximizing the Nash social welfare. The coverage framework considered in this work accommodates expressive combinatorial constraints and, hence, it models a range of applications. The framework also generalizes public decision making among agents that have binary additive valuations.

It would be interesting to extend the coverage framework to settings in which each agent i has value v_i^t for getting covered by the tth selected subset and her valuation is additive across the T selections. Online version of fair coverage is another interesting direction for future work.

Acknowledgements. Siddharth Barman gratefully acknowledges the support of a Microsoft Research Lab (India) grant and an SERB Core research grant (CRG/2021/006165).

References

1. Babaioff, M., Ezra, T., Feige, U.: Fair and truthful mechanisms for dichotomous valuations. In: Proceedings of the AAAI Conference on Artificial Intelligence, vol. 35, pp. 5119–5126 (2021)
2. Baghel, D.K., Levit, V.E., Segal-Halevi, E.: Fair division algorithms for electricity distribution. arXiv preprint arXiv:2205.14531 (2022)
3. Barman, S., Fawzi, O., Fermé, P.: Tight approximation guarantees for concave coverage problems. In: 38th International Symposium on Theoretical Aspects of Computer Science (STACS 2021). pp. 9:1–9:17 (2021)
4. Barman, S., Krishna, A., Narahari, Y., Sadhukhan, S.: Nash welfare guarantees for fair and efficient coverage (2022)
5. Barman, S., Krishnamurthy, S.K., Vaish, R.: Finding fair and efficient allocations. In: Proceedings of the 2018 ACM Conference on Economics and Computation, pp. 557–574 (2018)
6. Barman, S., Krishnamurthy, S.K., Vaish, R.: Greedy algorithms for maximizing nash social welfare. In: Proceedings of the 17th International Conference on Autonomous Agents and MultiAgent Systems, pp. 7–13 (2018)
7. Caragiannis, I., Kurokawa, D., Moulin, H., Procaccia, A.D., Shah, N., Wang, J.: The unreasonable fairness of maximum nash welfare. ACM Trans. Econ. Comput. **7**(3), 1–32 (2019)
8. Conitzer, V., Freeman, R., Shah, N.: Fair public decision making. In: Proceedings of the 2017 ACM Conference on Economics and Computation, EC 2017, pp. 629–646. Association for Computing Machinery, New York, NY, USA (2017). https://doi.org/10.1145/3033274.3085125, https://doi.org/10.1145/3033274.3085125
9. Cornuejols, G., Fisher, M.L., Nemhauser, G.L.: Exceptional paper - location of bank accounts to optimize float: an analytic study of exact and approximate algorithms. Manage. Sci. **23**(8), 789–810 (1977)
10. Fain, B., Munagala, K., Shah, N.: Fair allocation of indivisible public goods. In: Proceedings of the 2018 ACM Conference on Economics and Computation, pp. 575–592 (2018)
11. Feige, U.: A threshold of ln n for approximating set cover. J. ACM **45**(4), 634–652 (1998)
12. Fluschnik, T., Skowron, P., Triphaus, M., Wilker, K.: Fair knapsack. In: Proceedings of the AAAI Conference on Artificial Intelligence, vol. 33, pp. 1941–1948 (2019)
13. Garg, J., Kulkarni, P., Murhekar, A.: On fair and efficient allocations of indivisible public goods. In: 41st IARCS Annual Conference on Foundations of Software Technology and Theoretical Computer Science, pp. 22:1–22:19 (2021)
14. Gollapudi, S., Kollias, K., Plaut, B.: Almost envy-free repeated matching in two-sided markets. In: Chen, X., Gravin, N., Hoefer, M., Mehta, R. (eds.) WINE 2020. LNCS, vol. 12495, pp. 3–16. Springer, Cham (2020). https://doi.org/10.1007/978-3-030-64946-3_1

15. Halpern, D., Procaccia, A.D., Psomas, A., Shah, N.: Fair division with binary valuations: one rule to rule them all. In: Chen, X., Gravin, N., Hoefer, M., Mehta, R. (eds.) WINE 2020. LNCS, vol. 12495, pp. 370–383. Springer, Cham (2020). https://doi.org/10.1007/978-3-030-64946-3_26

16. Hochbaum, D.S.: Approximating covering and packing problems: set cover, vertex cover, independent set, and related problems. In: Hochbaum, D.S. (ed.) Approximation Algorithms for NP-Hard Problems, pp. 94–143. PWS Publishing Company, Boston (1996)

17. Kell, N., Sun, K.: Approximations for allocating indivisible items with concave additive valuations. arXiv preprint arxiv:2109.00081 (2021), https://arxiv.org/abs/2109.00081

18. Kicillof, N., Grieskamp, W., Tillmann, N., Braberman, V.: Achieving both model and code coverage with automated gray-box testing. In: Proceedings of the 3rd International Workshop on Advances in Model-based Testing, pp. 1–11 (2007)

19. Li, W., Vondrák, J.: A constant-factor approximation algorithm for nash social welfare with submodular valuations. In: 2021 IEEE 62nd Annual Symposium on Foundations of Computer Science (FOCS), pp. 25–36. IEEE (2022)

20. Marden, J.R., Wierman, A.: Distributed welfare games with applications to sensor coverage. In: Proceedings of the 47th IEEE Conference on Decision and Control, CDC 2008, 9-11 December 2008, Cancún, Mexico. pp. 1708–1713. IEEE (2008). https://doi.org/10.1109/CDC.2008.4738800, https://doi.org/10.1109/CDC.2008.4738800

21. Moulin, H.: Fair Division and Collective Welfare. MIT Press, London (2003)

22. Oluwasuji, O.I., Malik, O., Zhang, J., Ramchurn, S.D.: Solving the fair electric load shedding problem in developing countries. Auton. Agent. Multi-agent Syst. **34**(1), 1–35 (2020)

23. Rosenfeld, A., Talmon, N.: What should we optimize in participatory budgeting? an experimental study. arXiv preprint arXiv:2111.07308 (2021)

24. Schrijver, A.: Combinatorial Optimization: Polyhedra and Efficiency. Springer, Cham (2003). https://doi.org/10.1007/s10288-004-0035-9

25. Sühr, T., Biega, A.J., Zehlike, M., Gummadi, K.P., Chakraborty, A.: Two-sided fairness for repeated matchings in two-sided markets: a case study of a ride-hailing platform. In: Proceedings of the 25th ACM SIGKDD International Conference on Knowledge Discovery and Data Mining, KDD 2019, pp. 3082–3092. Association for Computing Machinery, New York, NY, USA (2019). https://doi.org/10.1145/3292500.3330793, https://doi.org/10.1145/3292500.3330793

26. Vazirani, V.V.: Approximation Algorithms, vol. 1. Springer, Cham (2001). https://doi.org/10.1007/978-3-662-04565-7

Tight Bounds on 3-Team Manipulations in Randomized Death Match

Atanas Dinev[1(✉)] and S. Matthew Weinberg[2]

[1] Massachusetts Institute of Technology, Cambridge, MA 02142, USA
adinev@mit.edu
[2] Princeton University, Princeton, NJ 08544, USA
smweinberg@princeton.edu

Abstract. Consider a round-robin tournament on n teams, where a winner must be (possibly randomly) selected as a function of the results from the $\binom{n}{2}$ pairwise matches. A tournament rule is said to be k-SNM-α if no set of k teams can ever manipulate the $\binom{k}{2}$ pairwise matches between them to improve the joint probability that one of these k teams wins by more than α. Prior work identifies multiple simple tournament rules that are 2-SNM-1/3 (Randomized Single Elimination Bracket [17], Randomized King of the Hill [18], Randomized Death Match [6]), which is optimal for $k = 2$ among all Condorcet-consistent rules (that is, rules that select an undefeated team with probability 1).

Our main result establishes that Randomized Death Match is 3-SNM-(31/60), which is tight (for Randomized Death Match). This is the first tight analysis of any Condorcet-consistent tournament rule and at least three manipulating teams. Our proof approach is novel in this domain: we explicitly find the most-manipulable tournament, and directly show that no other tournament can be more manipulable.

In addition to our main result, we establish that Randomized Death Match disincentivizes Sybil attacks (where a team enters multiple copies of themselves into the tournament, and arbitrarily manipulates the outcomes of matches between their copies). Specifically, for any tournament, and any team u that is not a Condorcet winner, the probability that u or one of its Sybils wins in Randomized Death Match approaches 0 as the number of Sybils approaches ∞.

1 Introduction

Consider a tournament on n teams competing to win a single prize via $\binom{n}{2}$ pairwise matches. A tournament rule is a (possibly randomized) map from these $\binom{n}{2}$ matches to a single winner. In line with several recent works [1,2,6,17,18], we study rules that satisfy some notion of fairness (that is, "better" teams should be more likely to win), and non-manipulability (that is, teams have no incentive to manipulate the matches).

More specifically, prior work identifies *Condorcet-consistence* (Definition 4) as one desirable property of tournament rules: whenever an undefeated team

© The Author(s), under exclusive license to Springer Nature Switzerland AG 2022
K. A. Hansen et al. (Eds.): WINE 2022, LNCS 13778, pp. 273–291, 2022.
https://doi.org/10.1007/978-3-031-22832-2_16

exists, a Condorcet-consistent rule selects that team as the winner with probability 1. Another desirable property is *monotonicity* (Definition 6): no team can unilaterally increase the probability that it wins by throwing a single match. Arguably, any sensible tournament rule should at minimum satisfy these two basic properties, and numerous such simple rules exist.

[1,2] further considered the following type of deviation: what if the same company sponsors multiple teams in an eSports tournament, and wants to maximize the probability that one of them wins the top prize?[1] In principle, these teams might manipulate the outcomes of the matches they play amongst themselves in order to achieve this outcome. Specifically, they call a tournament rule k-Strongly-Non-Manipulable (k-SNM, Definition 5), if no set of k teams can successfully manipulate the $\binom{k}{2}$ pairwise matches amongst themselves to improve the probability that one of these k teams wins the tournament. Unfortunately, even for $k = 2$, [1,2] establish that no tournament rule is both Condorcet-consistent and 2-SNM.

This motivated recent work in [6,17,18] to design tournament rules which are Condorcet-consistent *as non-manipulable as possible*. Specifically, [17] defines a tournament rule to be k-SNM-α if no set of k teams can manipulate the $\binom{k}{2}$ pairwise matches amongst themselves to increase total probability that any of these k teams wins *by more than* α (see Definition 5). These works design several simple Condorcet-consistent and 2-SNM-1/3 tournament rules, which is optimal for $k = 2$ (see [17]). In fact, the state of affairs is now fairly advanced for $k = 2$: each of [6,17,18] proposes a new 2-SNM-1/3 tournament rule. [18] considers a stronger fairness notion that they term Cover-consistent, and [6] considers probabilistic tournaments (see Sect. 1.3 for further discussion).

However, *significantly* less is known for $k > 2$. Indeed, only [18] analyzes manipulability for $k > 2$. They design a rule that is k-SNM-2/3 for all k, but that rule is non-monotone, and it is unknown whether their analysis of that rule is tight. Our main result provides a tight analysis of the manipulability of Randomized Death Match (first defined in [6]) when $k = 3$. We remark that this is: a) the first tight analysis of the manipulability of any Condorcet-consistent tournament rule for $k > 2$, b) the first analysis establishing a monotone tournament rule that is k-SNM-α for any $k > 2$ and $\alpha < 1$, and c) the strongest analysis to-date of any tournament rule (monotone or not) for $k = 3$. We overview our main result in more detail in Sect. 1.1 below.

Beyond our main result, we further consider manipulations through a *Sybil attack* (Definition 9). As a motivating example, imagine that a tournament rule is used as a proxy for a voting rule to select a proposal (voters compare each pair of proposals head-to-head, and this constitutes the pairwise matches input to a tournament rule). A proposer may attempt to manipulate the protocol with a Sybil attack, by submitting numerous nearly-identical clones of the same proposal. This manipulates the original tournament, with a single node u_1 corresponding to the proposal, into a new one with additional nodes $u_2, \ldots, u_m$ cor-

---

[1] Similarly, perhaps there are multiple athletes representing the same country or university in a traditional sports tournament.

responding to the Sybils. Each node $v \notin \{u_1, \ldots, u_m\}$ either beats all the Sybils, or none of them (because the Sybil proposals are essentially identical to the original). The questions then become: Can the proposer profitably manipulate the matches within the Sybils? Is it beneficial for a proposer to submit as many Sybils as possible? We first show that, when participating in Randomized Death Match, the Sybils can't gain anything by manipulating the matches between them. Perhaps more surprisingly, we show that Randomized Death Match is *Asymptotically Strongly Sybil-Proof*: as the number of Sybils approaches ∞, the collective probability that a Sybil wins RDM approaches *zero* (unless the original proposal is a Condorcet winner, in which case the probability that a Sybil wins is equal to 1, for any number of Sybils > 0).

1.1 Our Results

As previously noted, our main result is a tight analysis of the manipulability of Randomized Death Match (RDM) for coalitions of size 3. Randomized Death Match is the following simple rule: pick two uniformly random teams who have not yet been eliminated, and eliminate the loser of their head-to-head match.

Informal Theorem 1 *(See Theorem 6). RDM is 3-SNM-$\frac{31}{60}$. RDM is not 3-SNM-α for $\alpha < \frac{31}{60}$.*

Recall that this is the first tight analysis of any Condorcet-consistent tournament rule for any $k > 2$ and the first analysis establishing a monotone, Condorcet-consistent tournament rule that is k-SNM-α for any $k > 2$, $\alpha < 1$. Recall also that previously the smallest α for which a 3-SNM-α (non-monotone) Condorcet-consistent tournament rule is known is $2/3$.

Our second result concerns manipulation by Sybil attacks. A Sybil attack is where one team starts from a base tournament T, and adds some number $m-1$ of clones of their team to create a new tournament T' (they can arbitrarily control the matches within their Sybils, but each Sybil beats exactly the same set of teams as the cloned team) (See Definition 9). We say that a tournament rule r is *Asymptotically Strongly Sybil-Proof* (Definition 10) if for any tournament T and team $u_1 \in T$ that is not a Condorcet winner, the maximum collective probability that a Sybil wins (under r) over all of u_1's Sybil attacks with m Sybils goes to 0 as m goes to infinity. See Sect. 2 for a formal definition.

Informal Theorem 2 *(See Theorem 8). RDM is Asymptotically Strongly Sybil-Proof.*

1.2 Technical Highlight

All prior work establishing that a particular tournament rule is 2-SNM-1/3 follows a similar outline: for any T, cases where manipulating the $\{u, v\}$ match could potentially improve the chances of winning are coupled with two cases where manipulation cannot help. By using such a coupling argument, it is plausible that one can show that RDM is 3-SNM-$(\frac{1}{2} + c)$ for a small constant c.

However, given that Theorem 6 establishes that RDM is 3-SNM-31/60, it is hard to imagine that this coupling approach will be tractable to obtain the exact answer.

Our approach is instead drastically different: we find a particular 5-team tournament, and a manipulation by 3 teams that gains 31/60, and directly prove that this must be the worst case. We implement our approach using a first-step analysis, thinking of the first match played in RDM on an n-team tournament as producing a distribution over $(n-1)$-team tournaments.

The complete analysis inevitably requires some careful case analysis, but is tractable to execute fully by hand. Although this may no longer be the case for future work that considers larger k or more sophisticated tournament rules, our approach will still be useful to limit the space of potential worst-case examples.

1.3   Related Work

There is a vast literature on tournament rules, both within Social Choice Theory, and within the broad CS community [3,8–12,14,19]. The Handbook of Computational Social Choice provides an excellent survey of this broad field, which we cannot overview in its entirety [15]. Our work considers the model initially posed in [1,2], and continued in [6,17,18], which we overview in more detail below.

[1,2] were the first to consider Tournament rules that are both Condorcet-consistent and 2-SNM, and proved that no such rules exist. They further considered tournament rules that are 2-SNM and approximately Condorcet-consistent. [17] first proposed to consider tournament rules that are instead Condorcet-consistent and approximately 2-SNM. Their work establishes that Randomized Single Elimination Bracket is 2-SNM-1/3, and that this is tight.[2] [18] establish that Randomized King of the Hill (RKotH) is 2-SNM-1/3,[3] and [6] establish that Randomized Death Match is 2-SNM-1/3. [18] show further that RKotH satisfies a stronger fairness notion called Cover-consistence, and [6] extends their analysis to probabilistic tournaments. In summary, the state of affairs for $k = 2$ is quite established: multiple 2-SNM-1/3 tournament rules are known, and multiple different extensions beyond the initial model of [17] are known.

For $k > 2$, however, significantly less is known. [17] gives a simple example establishing that no rule is k-SNM-$\frac{k-1-\varepsilon}{2k-1}$ for any $\varepsilon > 0$, but no rules are known to match this bound for any $k > 2$. Indeed, [18] shows that this bound is not tight, and proves a stronger lower bound for $k \to \infty$. For example, a corollary of their main result is that no 939-SNM-1/2 tournament rule exists. They also design a non-monotone tournament rule that is k-SNM-2/3 for all k. Other than these results, there is no prior work for manipulating sets of size $k > 2$.

---

[2] Randomized Single Elimination Bracket iteratively places the teams, randomly, into a single-elimination bracket, and then 'plays' all matches that would occur in this bracket to determine a winner.

[3] Randomized King of the Hill iteratively picks a 'prince', and eliminates all teams beaten by the prince, until only one team remains.

In comparison, our work is the first to give a tight analysis of any Condorcet-consistent tournament rule for $k > 2$, and is the first proof that any monotone, Condorcet-consistent tournament rule is k-SNM-α for any $k > 2, \alpha < 1$.

Regarding our study of Sybil attacks, similar clone manipulations have been considered prior in Social Choice Theory under the name of *composition-consistency*. [13] introduces the notion of a *decomposition* of the teams in a tournament into components, where all the teams in a component are clones of each other with respect to the teams not in the component. [13] defines a deterministic tournament rule to be *composition-consistent* if it chooses the best teams from the best components[4]. In particular, composition-consistency implies that a losing team cannot win by introducing clones of itself or any other team. [13] shows that the tournament rules Banks, Uncovered Set, TEQ, and Minimal Covering Set are *composition-consistent*, while Top Cycle, the Slater, and the Copeland are not. Both computational and axiomatic aspects of *composition-consistency* have been explored ever since. [7] studies the structural properties of clone sets and their computational aspects in the context of voting preferences. In the context of probabilistic social choice, [4] gives probabilistic extensions of the axioms *composition-consistency* and *population-consistency* and uniquely characterize the probabilistic social choice rules, which satisfy both. In the context of scoring rules, [16] studies the incompatibility of *composition-consistency* and *reinforcement* (stronger than *population-consistency*) and decomposes composition-consistency into four weaker axioms. In this work, we consider Sybil attacks on Randomized Death Match. Our study of Sybil attacks differs from prior work on the relevant notion of *composition-consistency* in the following ways: (i) We focus on a randomized tournament rule (RDM), (ii) We study settings where the manipulator creates clones of themselves (i.e. not of other teams), (iii) We explore the asymptotic behavior of such manipulations (Definition 10, Theorem 8).

1.4 Roadmap

Section 2 follows with definitions and preliminaries, and formally defines Randomized Death Match (RDM). Section 3 introduces some basic properties and examples for the RDM rule as well as a recap of previous work for two manipulators. Section 4 consists of a proof that the manipulability of 3 teams in RDM is at most $\frac{31}{60}$ and that this bound is tight. Section 5 consists of our main results regarding Sybil attacks on a tournament. Section 6 concludes.

2 Preliminaries

In this section we introduce notation that we will use throughout the paper consistent with prior work in [6,17,18].

---

[4] For a full rigorous mathematical definition see Definition 10, [13].

Definition 1 (Tournament). *A (round robin) tournament T on n teams is a complete, directed graph on n vertices whose edges denote the outcome of a match between two teams. Team i beats team j if the edge between them points from i to j.*

Definition 2 (Tournament Rule). *A tournament rule r is a function that maps tournaments T to a distribution over teams, where $r_i(T) := \Pr(r(T) = i)$ denotes the probability that team i is declared the winner of tournament T under rule r. Let S be a set of teams. We use the shorthand $r_S(T) := \sum_{i \in S} r_i(T)$ to denote the probability that a team in S is declared the winner of tournament T under rule r.*

Definition 3 (S-adjacent). *Let S be a set of teams. Two tournaments T, T' are S-adjacent if for all i, j such that $\{i, j\} \not\subseteq S$, i beats j in T if and only if i beats j in T'.*

In other words, two tournaments T, T' are S-adjacent if the teams from S can manipulate the outcomes of the matches between them in order to obtain a new tournament T'.

Definition 4 (Condorcet-Consistent). *Team i is a Condorcet winner of a tournament T if i beats every other team (under T). A tournament rule r is Condorcet-consistent if for every tournament T with a Condorcet winner i, $r_i(T) = 1$ (whenever T has a Condorcet winner, that team wins with probability 1).*

Definition 5 (Manipulating a Tournament). *For a set S of teams, a tournament T and a tournament rule r, we define $\alpha_S^r(T)$ be the maximum winning probability that S can possibly gain by manipulating T to an S-adjacent T'. That is:*

$$\alpha_S^r(T) = \max_{T' : T' \text{ is } S\text{-adjacent to } T} \{r_S(T') - r_S(T)\}$$

For a tournament rule r, define $\alpha_{k,n}^r = \sup_{T, S : |S| = k, |T| = n} \{\alpha_S^r(T)\}$. Finally, define

$$\alpha_k^r = \sup_{n \in \mathbb{N}} \alpha_{k,n}^r = \sup_{T, S : |S| = k} \{\alpha_S^r(T)\}$$

If $\alpha_k^r \leq \alpha$, we say that r is k-Strongly-Non-Manipulable at probability α or k-SNM-α.

Intuitively, $\alpha_{k,n}^r$ is the maximum increase in collective winning probability that a group of k teams can achieve by manipulating the matches between them over tournaments with n teams. And α_k^r is the maximum increase in winning probability that a group of k teams can achieve by manipulating the matches between them over all tournaments.

Two other naturally desirable properties of a tournament rule are monotonicity and anonymity.

Definition 6 (Monotone). *A tournament rule r is monotone if T, T' are $\{u, v\}$-adjacent and u beats v in T, then $r_u(T) \geq r_u(T')$*

Definition 7 (Anonymous). *A tournament rule r is anonymous if for every tournament T, and every permutation σ, and all i, $r_{\sigma(i)}(\sigma(T)) = r_i(T)$*

Below we define the tournament rule that is the focus of this work.

Definition 8 (Randomized Death Match). *Given a tournament T on n teams the Randomized Death Match Rule (RDM) picks two uniformly random teams (without replacement) and plays their match. Then, eliminates the loser and recurses on the remaining teams for a total of $n - 1$ rounds until a single team remains, who is declared the winner.*

Below we define the notions of *Sybil Attack* on a tournament T, and the property of *Asymptotically Strongly Sybil-Proof* (ASSP) for a tournament rule r, both of which will be relevant in our discussion in Sect. 5.

Definition 9 (Sybil Attack). *Given a tournament T, a team $u_1 \in T$ and an integer m, define $Syb(T, u_1, m)$ to be the set of tournaments T' satisfying the following properties:*

1. *The set of teams in T' consists of $u_2 \ldots, u_m$ and all teams in T*
2. *If a, b are teams in T, then a beats b in T' if and only if a beats b in T.*
3. *If $a \neq u_1$ is a team in T and $i \in [m]$, then u_i beats a in T' if and only if u_1 beats a in T*
4. *The match between u_i and u_j can be arbitrary for each $i \neq j$*

Intuitively, $Syb(T, u_1, m)$ is the set of all Sybil attacks of u_1 at T with m Sybils. Each Sybil attack is a tournament $T' \in Syb(T, u_1, m)$ obtained by starting from T and creating m Sybils of u_1 (while counting u_1 as a Sybil of itself). Each Sybil beats the same set of teams from $T \backslash u_1$ and the matches between the Sybils $u_1, \ldots, u_m$ can be arbitrary. Every possible realization of the matches between the Sybils gives rise to new tournament $T' \in Syb(T, u_1, m)$ (implying $Syb(T, u_1, m)$ contains $2^{\binom{m}{2}}$ tournaments).

Definition 10 (Asymptotically Strongly Sybil-Proof). *A tournament rule r is Asymptotically Strongly Sybil-Proof (ASSP) if for any tournament T, team $u_1 \in T$ which is not a Condorcet winner,*

$$\lim_{m \to \infty} \max_{T' \in Syb(T, u_1, m)} r_{u_1, \ldots, u_m}(T') = 0$$

Informally speaking, Definition 10 claims that r is ASSP if the probability that a Sybil wins in the most profitable Sybil attack on T with m Sybils, goes to zero as m goes to ∞.

3   Basic Properties of RDM and Examples

In this section we consider a few basic properties of RDM and several examples on small tournaments. We will refer to those examples in our analysis later.

Throughout the paper we will denote RDM by r and it will be the only tournament rule we consider. We next state the first-step analysis observation that will be central to our analysis throughout the paper. For the remainder of the section let for a match e denote by $T|_e$ the tournament obtained from T by eliminating the loser in e. Let $S|_e = S \backslash x$, where x is the loser in e. Let d_x denote the number of teams x loses to and $T \backslash x$ the tournament obtained after removing team x from T.

Observation 3 (First-step analysis). *Let S be a subset of teams in a tournament T. Then*

$$r_S(T) = \frac{1}{\binom{n}{2}} \sum_e r_{S|_e}(T|_e) = \frac{1}{\binom{n}{2}} \sum_x d_x r_{S \backslash x}(T \backslash x)$$

(if $S = \{v\}$, then we define $r_{S \backslash v}(T \backslash v) = 0$, and if $x \notin S$, then $S \backslash x = S$)

Proof. The first equality follows from the fact that after we play e we are left with the tournament $T|_e$ and we sum over all possible e in the first round. To prove the second equality, notice that for any x the term $r_{S \backslash x}(T \backslash x)$ appears exactly d_x times in $\sum_e r_{S|_e}(T|_e)$ because x loses exactly d_x matches.

This first-step analysis observation can be used to show that adding teams that lose to every other team does not change the probability distribution of the winner.

Lemma 1. *Let T be a tournament and $u \in T$ loses to everyone. Then for all $v \neq u$, we have $r_v(T) = r_v(T \backslash u)$.*

Proof. See full version [5] for a proof.

As a natural consequence of Lemma 1 we show that the most manipulable tournament on $n + 1$ teams is at least as manipulable as the most manipulable tournament on n teams.

Lemma 2. $\alpha_{k,n}^r \leq \alpha_{k,n+1}^r$

Proof. See Appendix A.1 of [5] for a proof.

We now show another natural property of RDM, which is a generalization of Condorcet-consistent (Definition 4), namely that if a group of teams S wins all its matches against the rest of teams, then a team from S will always win.

Lemma 3. *Let T be a tournament and $S \subseteq T$ a group of teams such that every team in S beats every team in $T \backslash S$. Then, $r_S(T) = 1$.*

Proof. See Appendix A.1 of [5] for a proof.

As a result of Lemma 3 RDM is Condorcet-Consistent. As expected, RDM is also monotone (See Definition 6).

Lemma 4. *RDM is monotone.*

Proof. See Appendix A.1 of [5] for a proof.

Lemma 1 tells us that adding a team which loses to all other teams does not change the probability distribution of the other teams winning. Lemmas 1, 2, 3, 4 will be useful in our later analysis in Sects. 4 and 5. Now we consider a few examples of tournaments and illustrate the use of first-step analysis (Observation 3) to compute the probability distribution of the winner in them.

1. Let $T = \{a, b, c\}$, where a beats b, b beats c and c beats a. By symmetry of RDM, we have $r_a(T) = r_b(T) = r_c(T) = \frac{1}{3}$.
2. Let $T = \{a, b, c\}$ where a beats b and c. Then clearly, $r_a(T) = 1$ and $r_b(T) = r_c(T) = 0$.
3. By Lemma 1, it follows that the only tournament on 4 teams whose probability distribution cannot be reduced to a distribution on 3 teams is the following one $T = \{a_1, a_2, a_3, a_4\}$, where a_i beats a_{i+1} for $i = 1, 2, 3$, a_4 beats a_1, a_1 beats a_3 and a_2 beats a_4. By using what we computed in (1) and (2) combined with Lemma 1 we get by first step analysis

$$r_{a_1}(T) = \frac{1}{6}(r_{a_1}(T\backslash a_2) + 2r_{a_1}(T\backslash a_3) + 2r_{a_1}(T\backslash a_4)) = \frac{1}{6}(\frac{1}{3} + \frac{2}{3} + 2) = \frac{1}{2}$$

$$r_{a_2}(T) = \frac{1}{6}(r_{a_2}(T\backslash a_1) + 2r_{a_2}(T\backslash a_3) + 2r_{a_2}(T\backslash a_4)) = \frac{1}{6}(1 + \frac{2}{3}) = \frac{5}{18}$$

$$r_{a_3}(T) = \frac{1}{6}(r_{a_3}(T\backslash a_1) + r_{a_3}(T\backslash a_2) + 2r_{a_3}(T\backslash a_4)) = \frac{1}{6}(\frac{1}{3}) = \frac{1}{18}$$

$$r_{a_4}(T) = \frac{1}{6}(r_{a_4}(T\backslash a_1) + r_{a_4}(T\backslash a_2) + 2r_{a_4}(T\backslash a_3)) = \frac{1}{6}(\frac{1}{3} + \frac{2}{3}) = \frac{1}{6}$$

The above examples are really important in our analysis because: a) we will use them in later for our lower bound example in Sect. 4.1, and b) they are a short illustration of first-step analysis.

In the following subsection, we review prior results for 2-team manipulations in RDM, which will also be useful for our treatment of the main result in Sect. 4.

3.1   Recap: Tight Bounds on 2-Team Manipulations in RDM

[6] (Theorem 5.2) proves that RDM is 2-SNM-$\frac{1}{3}$ and that this bound is tight, namely $\alpha_2^{RDM} = \frac{1}{3}$. We will rely on this result in Sect. 4.

Theorem 4 *(Theorem 5.2 in [6]).* $\alpha_2^{RDM} = \frac{1}{3}$

[17] (Theorem 3.1) proves that the bound of $\frac{1}{3}$ is the best one can hope to achieve for a Condorcet-consistent rule.

Theorem 5 *(Theorem 3.1 in [17]). There is no Condorcet-consistent tournament rule on n players (for $n \geq 3$) that is 2-SNM-α for $\alpha < \frac{1}{3}$*

We prove the following useful corollary, which will be useful in Sect. 4.

Corollary 1. *Let T be a tournament and $u, v \in T$ two teams such that there is at most one match in which a team in $\{u, v\}$ loses to a team in $T \backslash \{u, v\}$. Then*

$$r_{u,v}(T) \geq \frac{2}{3}$$

Proof. See full version [5] for proof.

4 Main Result: $\alpha_3^{RDM} = 31/60$

The goal of this section is to prove that no 3 teams can improve their probability of winning by more than $\frac{31}{60}$ and that this bound is tight. We prove the following theorem

Theorem 6. $\alpha_3^{RDM} = \frac{31}{60}$

Our proof consists of two parts:

- Lower bound: $\alpha_3^{RDM} \geq \frac{31}{60}$, for which we provide a tournament T and a set S of size 3, which can manipulate to increase their probability by $\frac{31}{60}$
- Upper bound: $\alpha_3^{RDM} \leq \frac{31}{60}$, for which we provide a proof that for any tournament T no set S of size 3 can increase their probability of winning by more than $\frac{31}{60}$, i.e. RDM is 3-SNM-$\frac{31}{60}$

4.1 Lower Bound

Let r denote RDM. Denote by B_x the set of teams which team x beats. Consider the following tournament $T = \{u, v, w, a, b\}$:

$$B_a = \{u, v, b\}, B_b = \{u, v\}, B_u = \{v, w\}, B_v = \{w\}, B_w = \{a, b\}$$

Let $S = \{u, v, w\}$. By first-step analysis (Observation 3) and by using our knowledge in Sect. 3 for tournaments on 4 teams we can show that $r_{u,v,w}(T) = \frac{29}{60}$ (see full version [5] for full analysis) Let u and v throw their matches with w. i.e. T' is S-adjacent to T, where in T', w beats u and v and all other matches have the same outcomes as in T. Then, since w is a Condorcet winner, $r_{u,v,w}(T') = r_w(T') = 1$. Therefore,

$$\alpha_3^{RDM} \geq r_{u,v,w}(T') - r_{u,v,w}(T) = 1 - \frac{29}{60} = \frac{31}{60}$$

Thus, $\alpha_3^{RDM} \geq \frac{31}{60}$ as desired.

4.2 Upper Bound

Suppose we have a tournament T on $n \geq 3$ vertices and $S = \{u, v, w\}$ is a set of 3 (distinct) teams, where S will be the set of manipulator teams. Let I be the set of matches in which a team from S loses to a team from $T \backslash S$. Our proof for $\alpha_k^{RDM} \leq \frac{31}{60}$ will use the following strategy

- In the following section we introduce the first-step analysis framework by considering possible cases for the first played match. In each of these cases the loser of the match is eliminated and we are left with a tournament with one less team. We pair each match in T with its corresponding match in T' and we bound the gains of manipulation in each of the following cases separately (these correspond to each of the terms A, B, and C respectively in the analysis in the next section).
  - The first match is between two teams in S (there are 3 such matches).
  - The first match is between a team in S and a team in $T \backslash S$ and the team from S loses in the match (there are $|I|$ such matches).
  - The first match is any other match not covered by the above two cases
- We prove that if $|I| \leq 4$, then $\alpha_S^{RDM}(T) \leq \frac{31}{60}$ (i.e. the set of manipulators cannot gain more than $\frac{31}{60}$ by manipulating).
- We prove that if T is the most manipulable tournament on n vertices (i.e. $\alpha_S^{RDM}(T) = \alpha_{3,n}^{RDM}$), then $\alpha_S^{RDM}(T) \leq \frac{|I|+7}{3(|I|+3)}$
- We combine the above facts to finish the proof of Theorem 6

We first introduce some notation that we will use throughout this section. Suppose that T' and T are S-adjacent. Recall from Sect. 3 that for a match $e = (i, j)$, $T|_e$ is the tournament obtained after eliminating the loser in e. Also d_x is the number of teams that a team x loses to in T. For $x \in S$, let ℓ_x denote the number of matches x loses against a team in S when considered in T and let ℓ'_x denote the number of matches that x loses against a team in S when considered in T'. Let d_x^* denote the number of teams in $T \backslash S$ that x loses to. Notice that since T and T' are S-adjacent, $x \in S$ loses to exactly d_x^* teams in $T' \backslash S$ when considered in T'. Let $G = I \cup \{uv, vw, uw\}$ be the set of matches in which a team from S loses.

The First Step Analysis Framework. Notice that in the first round of RDM, a uniformly random match e from the $\binom{n}{2}$ matches is chosen. If $e \in G$ then we are left with $T \backslash x$ where x loses in e for some $x \in S$. If $e \notin G$, we are left with $T|_e$ and all teams in S are still in the tournament. For $x \in S$, there are ℓ_x matches in which they lose to a team from S and d_x^* matches in which they lose to a team from $T \backslash S$. We consider each of these cases and use first-step analysis (Observation 3) for both T and T' to obtain (see full version [5] for details)

$$r_{u,v,w}(T') - r_{u,v,w}(T) = \frac{1}{\binom{n}{2}}(A + B + C) \tag{1}$$

where

$$A = \sum_{x \in \{u,v,w\}} \ell'_x r_{\{u,v,w\} \backslash x}(T' \backslash x) - \ell_x r_{\{u,v,w\} \backslash x}(T \backslash x)$$

$$B = \sum_{x \in S} d_x^*(r_{\{u,v,w\} \backslash x}(T' \backslash x) - r_{\{u,v,w\} \backslash x}(T \backslash x))$$

$$C = \sum_{e \notin G} r_{u,v,w}(T'|_e) - r_{u,v,w}(T|_e)$$

Upper Bounds on A, B and C. We now prove some bounds on the terms A, B and C (defined in Sect. 4.2) which will be useful later. Recall that I denotes the set of matches in which a team from S loses from a team from $T \backslash S$. We begin with bounding A in the following lemma

Lemma 5. *For all S-adjacent T and T', we have $A \le \frac{7}{3}$. Moreover, if $|I| \le 1$, then $A \le 1$.*

Proof. See Appendix A.2 in [5] for a proof.

Next, we show the following bound on the term B.

Lemma 6. *For all S-adjacent T, T' we have*

$$B \le \frac{d_u^* + d_v^* + d_w^*}{3} = \frac{|I|}{3}$$

Moreover, if $|I| \le 1$, then $B = 0$

Proof. See Appendix A.2 in [5] for a proof.

We introduce some more notation. For $n \in \mathbb{N}$, define $M_n(a_1, a_2, a_3)$ as the maximum winning probability gain that three teams $\{u, v, w\}$ can achieve by manipulation in a tournament T of size n, in which there are exactly a_i teams in $T \backslash S$ each of which beats exactly i teams of S. Formally,

$$M_n(a_1, a_2, a_3) = \max \Big\{ r_S(T') - r_S(T) \big| T, T' \text{ are } S\text{-adjacent}, |T| = n, |S| = 3,$$

$$a_i \text{ teams in } T \backslash S \text{ beat exactly } i \text{ teams in } S \Big\}$$

Additionally, let L_i be the set of teams in $T \backslash S$ each of which beats exactly i teams in S. Let Q be the set of matches in which two teams from L_i play against each other or in which a team from L_i loses to a team from S for $i = 1, 2, 3$. Notice that $|Q| = 2a_1 + a_2 + \binom{a_1}{2} + \binom{a_2}{2} + \binom{a_3}{2}$ if there are a_i teams in $S \backslash T$ each of which beat i teams from S.

With the new notation, we are now ready to prove a bound on the term C. Recall that

$$C = \sum_{e \notin G} r_{u,v,w}(T'|e) - r_{u,v,w}(T|e)$$

where $G = I \cup \{uv, vw, uw\}$ is the set of matches in which a team from S loses. Then we have the following bound on C.

Lemma 7. *For all S-adjacent T and T' we have that C is at most*

$$\left(2a_1 + \binom{a_1}{2}\right) M_{n-1}(a_1 - 1, a_2, a_3) + \left(a_2 + \binom{a_2}{2}\right) M_{n-1}(a_1, a_2 - 1, a_3)$$

$$+ \binom{a_3}{2} M_n(a_1, a_2, a_3 - 1) + \sum_{e \notin G \cup Q} r_{u,v,w}(T'|e) - r_{u,v,w}(T|e)$$

Proof. See Appendix A.2 in [5] for a proof.

The Case $|I| \leq 4$. We summarize our claim when $|I| \leq 4$ in the following lemma

Lemma 8. *Let T be a tournament, and S a set of 3 teams. Suppose that there are at most 4 matches in which a team in S loses to a team in $T \backslash S$ (i.e. $|I| \leq 4$). Then $\alpha_S^{RDM}(T) \leq \frac{31}{60}$*

Proof. We will show that $M_n(a_1, a_2, a_3) \leq f(a_1, a_2, a_3)$ by induction on $n \in \mathbb{N}$ for the values of (a_1, a_2, a_3) and $f(a_1, a_2, a_3)$ given in Table 1 below. Notice that when there are at most 4 matches between a team in S and a team in $T \backslash S$, in which the teams from S loses, then we fall into one of the cases shown in the table for (a_1, a_2, a_3).

Table 1. Upper bounds on $M_n(a_1, a_2, a_3)$

(a_1, a_2, a_3)	$f(a_1, a_2, a_3)$
$(0, 0, 0)$	0
$(1, 0, 0)$	$\frac{1}{6}$
$(2, 0, 0)$	$\frac{23}{60}$
$(3, 0, 0)$	$\frac{407}{900}$
$(4, 0, 0)$	$\frac{4499}{9450}$
$(0, 1, 0)$	$\frac{1}{2}$
$(0, 2, 0)$	$\frac{31}{60}$
$(1, 1, 0)$	$\frac{1}{2}$
$(2, 1, 0)$	$\frac{131}{260}$
$(0, 0, 1)$	0
$(1, 0, 1)$	$\frac{11}{27}$

1. Base case. Our base case is when $n = 3$. If we are in the case of 3 teams then S wins with probability 1, so the maximum gain S can achieve by manipulation is clearly 0, which satisfies all of the bounds in the table.

2. Induction step. Assume that $M_k(a_1, a_2, a_3) \leq f(a_1, a_2, a_3)$ hold for all $k < n$ and a_1, a_2, a_3 as in Table 1. We will prove the statement for $k = n$. By using the upper bounds on A, B, C in Lemmas 5–7 and the inductive hypothesis we can show that (see [5] for full analysis)

$$M_n(a_1, a_2, a_3) \leq \frac{1}{\binom{n}{2}} \left[A' + B' + (2a_1 + \binom{a_1}{2}) f(a_1 - 1, a_2, a_3) \right.$$

$$+ (a_2 + \binom{a_2}{2}) f(a_1, a_2 - 1, a_3) + \binom{a_3}{2} f(a_1, a_2, a_3 - 1)$$

$$\left. + (\binom{n}{2} - 3(1 + a_1 + a_2 + a_3) - \binom{a_1}{2} - \binom{a_2}{2} - \binom{a_3}{2}) f(a_1, a_2, a_3) \right] \quad (\Delta)$$

where $A' = 1$ and $B' = 0$ if $|I| \leq 1$ and $A' = \frac{7}{3}$ and $B' = \frac{|I|}{3}$ if $|I| \geq 2$. Next, we apply the formula (Δ) to each of the cases in Table 1, to obtain $M_n(a_1, a_2, a_3) \leq f(a_1, a_2, a_3)$. We defer the reader to [5] for full details.

This finishes the induction and the proof for the bounds in Table 1. Note that $f(a_1, a_2, a_3) \leq \frac{31}{60}$ for all a_1, a_2, a_3 in Table 1 and this bounds is achieved when $(a_1, a_2, a_3) = (0, 2, 0)$ i.e. there are 2 teams that beat exactly two of S as is the case in the optimal example in Sect. 4.1. Thus, we get that if there are at most 4 matches that a team from S loses from a team in $T \backslash S$, then $\alpha_S^{RDM}(T) \leq \frac{31}{60}$. This finishes the proof of the lemma.

General Upper Bound for the Most Manipulable Tournament

Lemma 9. *Suppose that $\alpha_S^{RDM}(T) = \alpha_{3,n}^{RDM}$. Let I be the set of matches a team of S loses to a team from $T \backslash S$. Then*

$$\alpha_{3,n}^{RDM} = \alpha_S^{RDM}(T) \leq \frac{|I| + 7}{3(|I| + 3)}$$

Proof. Let T and T' be S-adjacent tournaments on n vertices such that $S = \{u, v, w\}$ and

$$\alpha_{3,n}^{RDM} = \alpha_S^{RDM}(T) = r_S(T') - r_s(T)$$

I.e. T is the "worst" example on n vertices. By (1) we have

$$\alpha_{3,n}^{RDM} = \frac{1}{\binom{n}{2}}(A + B + C)$$

where A, B and C were defined in Sect. 4.2. By Lemma 5 we have

$$A \leq \frac{7}{3}$$

and by Lemma 6

$$B \leq \frac{d_u^* + d_v^* + d_w^*}{3} = \frac{|I|}{3}$$

Let $e \notin G$. Notice that both $T'|_e$ and $T|_e$ are tournaments on $n-1$ vertices and by definition of G, u, v, w are not eliminated in both $T'|_e$ and $T|_e$. Moreover, $T'|_e$ and $T|_e$ are S-adjacent. Therefore, for every $e \notin G$, we have by Lemma 2

$$r_{u,v,w}(T'|_e) - r_{u,v,w}(T|_e) \leq \alpha_{3,n-1}^{RDM} \leq \alpha_{3,n}^{RDM}$$

By using the above on each term in C and the fact that $|G| = 3 + |I|$, we get that

$$C \leq (\binom{n}{2} - (3 + |I|))\alpha_{3,n}^{RDM}$$

By using the above 3 bounds we get

$$\alpha_{3,n}^{RDM} \leq \frac{1}{\binom{n}{2}}(\frac{7}{3} + \frac{|I|}{3} + (\binom{n}{2} - (3 + |I|))\alpha_{3,n}^{RDM})$$

$$\iff (|I| + 3)\alpha_{3,n}^{RDM} \leq \frac{|I| + 7}{3}$$

$$\iff \alpha_{3,n}^{RDM} = \alpha_S^{RDM}(T) \leq \frac{|I| + 7}{3(|I| + 3)}$$

as desired.

Proof of Theorem 6. Suppose that T is the most manipulable tournament on n vertices i.e. it satisfies $\alpha_S^{RDM}(T) = \alpha_{3,n}^{RDM}$. If $|I| \leq 4$, then by Lemma 8, we have that

$$\alpha_{3,n}^{RDM} = \alpha_S^{RDM}(T) \leq \frac{31}{60}$$

If $|I| \geq 5$, then by Lemma 9

$$\alpha_{3,n}^{RDM} = \alpha_S^{RDM}(T) \leq \frac{|I| + 7}{3(|I| + 3)} \leq \frac{5 + 7}{3(5 + 3)} = \frac{1}{2}$$

where above we used that $\frac{x+7}{3(x+3)}$ is decreasing for $x \geq 5$. Combining the above bounds, we obtain $\alpha_{3,n}^{RDM} \leq \frac{31}{60}$ for all $n \in \mathbb{N}$. Therefore,

$$\alpha_k^{RDM} = \max_{n \in \mathbb{N}} \alpha_{k,n}^{RDM} \leq \frac{31}{60}$$

which proves the upper bound and finishes the proof of Theorem 6.

5 Sybil Attacks on Tournaments

5.1 Main Results on Sybil Attacks on Tournaments

Recall our motivation from the Introduction. Imagine that a tournament rule is used as a proxy for a voting rule to select a proposal. The proposals are compared head-to-head, and this constitutes the pairwise matches in the resulting tournament. A proposer can try to manipulate the protocol with a Sybil attack and submit many nearly identical proposals with nearly equal strength relative to the other proposals. The proposer can choose to manipulate the outcomes of the head-to-head comparisons between two of his proposals in a way which maximizes the probability that a proposal of his is selected. In the tournament T his proposal corresponds to a team u_1, and the tournament T' resulting from the Sybil attack is a member of $Syb(T, u_1, m)$ (Recall Definition 9). The questions that we want to answer in this section are: (1) Can the Sybils manipulate their matches to successfully increase their collective probability of winning? and (2) Is it beneficial for the proposer to create as many Sybils as possible?

The first question we are interested in is whether any group of Sybils can manipulate successfully to increase their probability of winning. It turns out that the answer is No. We first prove that the probability that a team that is not a Sybil wins does not depend on the matches between the Sybils.

Lemma 10. *There exists a function q that takes in as input integer m, tournament T, team $u_1 \in T$, and team $v \in T \backslash u_1$ with the following property. For all $T' \in Syb(T, u_1, m)$, we have*

$$r_v(T') = q(m, T, u_1, v)$$

where the dependence on u_1 is encoded as the outcomes of its matches with the rest of T.

Proof. See Appendix A.3 in [5] for a full proof.

Note that by Lemma 10 $r_v(T') = q(m, T, u_1, v)$ does not depend on which tournament $T' \in Syb(T, u_1, m)$ is chosen. Now, we prove our first promised result. Namely, that no number of Sybils in a Sybil attack can manipulate the matches between them to increase their probability of winning.

Theorem 7. *Let T be a tournament, $u_1 \in T$ a team, and m and integer. Let $T'_1 \in Syb(T, u_1, m)$. Let $S = \{u_1, \ldots, u_m\}$. Then*

$$\alpha_S^{RDM}(T'_1) = 0$$

Proof. Let T'_1 and T'_2 be S-adjacent. By Definition 9, $T'_2 \in Syb(T, u_1, m)$. Therefore by Lemma 10, $r_v(T'_1) = r_v(T'_2) = q(m, T, u_1, v)$ for all $v \in T \backslash u_1$. Using this we obtain

$$r_S(T'_1) = 1 - \sum_{v \in T \backslash u_1} r_v(T'_1) = 1 - \sum_{v \in T \backslash u_1} r_v(T'_2) = r_S(T'_2)$$

Therefore, $r_S(T'_1) = r_S(T'_2)$ for all S-adjacent T'_1, T'_2, which implies the desired result.

We are now ready to prove our second main result. Namely, that RDM is Asymptotically Strongly Sybil-Proof (Definition 10). Before we present the result (Theorem 8) we will try to convey some intuition for why RDM should be ASSP.

Let u_1 be a team to create Sybils of (u_1 is not Condorcet Winner). Let A be the set of teams which u_1 beats and B the set of teams which u_1 loses to in T. Observe that the only way a Sybil can win is when all the teams from B are eliminated before all the Sybils. The teams from B can only be eliminated by teams from A. However, as m increases there are more Sybils and, thus, the teams from A are intuitively more likely to all lose the tournament before the teams from B. When there are no teams from A left and at least one team from B left, no Sybil can win. In fact, this intuition implies something stronger than RDM being ASSP: the collective winning probability of the Sybils and the teams from A converges to 0 as m converges to ∞ (or, equivalently, the probability that a team from B wins goes to 1). This intuition indirectly lies behind the technical details of the proof of Theorem 8. Let $p(m, T, u_1)$ be the collective probability that any Sybil or a team in A wins. It is not hard to see that by Lemma 10 it doesn't depend on the matches between the Sybils (see [5] for full details). Then we have the following theorem.

Theorem 8. *Randomized Death Match is Asymptotically Strongly Sybil-Proof. In fact a stronger statement holds, namely if $u_1 \in T$ is not a Condorcet winner, then*

$$\lim_{m \to \infty} p(m, T, u_1) = 0$$

Proof. See Appendix A.3 in [5] for a full proof.

5.2   On a Counterexample to an Intuitive Claim

We will use Theorem 8 to prove that RDM does not satisfy a stronger version of the monotonicity property in Definition 6. First, we give a generalization of the definition for monotonicity given in Sect. 3

Definition 11 (Strongly monotone). *Let r be a tournament rule. Let T and let $C \cup D$ be any partition of the teams in T into two disjoint sets. A tournament rule r is strongly monotone for every $(u, v) \in C \times D$, if T' is $\{u, v\}$-adjacent to T such that u beats v in T' we have $r_C(T') \geq r_C(T)$*

Intuitively, r is Strongly monotone if whenever flipping a match between a team from C and a team from D in favor of the team from C makes C better off. Notice that if $|C| = 1$ this is the usual definition of monotonicity (Definition 6), which is satisfied by RDM by Lemma 4. However, RDM is not strongly monotone, even though strong monotonicity may seem like an intuitive property to have.

Theorem 9. *RDM is not strongly monotone*

Proof. See [5] for a proof.

6   Conclusion and Future Work

We use a novel first-step analysis to nail down the minimal α such that RDM is 3-SNM-α. Specifically, our main result shows that $\alpha_3^{RDM} = \frac{31}{60}$. Recall that this is the first tight analysis of any Condorcet-consistent tournament rule for any $k > 2$, and also the first monotone, Condorcet-consistent tournament rule that is k-SNM-α for any $k > 2, \alpha < 1$. We also initiate the study of manipulability via Sybil attacks, and prove that RDM is Asymptotically Strongly Sybil-Proof.

Our technical approach opens up the possibility of analyzing the manipulability of RDM (or other tournament rules) whose worst-case examples are complicated-but-tractable. For example, it is unlikely that the elegant coupling arguments that work for $k = 2$ will result in a tight bound of 31/60, but our approach is able to drastically reduce the search space for a worst-case example, and a tractable case analysis confirms that a specific 5-team tournament is tight. Our approach can similarly be used to analyze the manipulability of RDM for $k > 3$, or other tournament rules. However, there are still significant technical barriers for future work to overcome in order to keep analysis tractable for large k, or for tournament rules with a more complicated recursive step. Still, our techniques provide a clear approach to such analyses that was previously non-existent.

References

1. Altman, A., Kleinberg, R.: Nonmanipulable randomized tournament selections. In: Proceedings of the AAAI Conference on Artificial Intelligence, vol. 24 (2010)
2. Altman, A., Procaccia, A.D., Tennenholtz, M.: Nonmanipulable selections from a tournament. In: Computational Foundations of Social Choice. Dagstuhl Seminar Proceedings, 07 March–12 March 2010, vol. 10101. Schloss Dagstuhl - Leibniz-Zentrum für Informatik, Germany (2010). http://drops.dagstuhl.de/opus/volltexte/2010/2560/
3. Banks, J.S.: Sophisticated voting outcomes and agenda control. Soc. Choice Welfare **1**(4), 295–306 (1985). https://doi.org/10.1007/BF00649265. http://www.jstor.org/stable/41105790
4. Brandl, F., Brandt, F., Seedig, H.G.: Consistent probabilistic social choice. Econometrica **84**(5), 1839–1880 (2016)
5. Dinev, A., Weinberg, S.M.: Tight bounds on 3-team manipulations in randomized death match. In: Hansen, K.A., et al. (eds.) WINE 2022. LNCS, vol. 13778, pp. 273–291. Springer, Cham (2022)
6. Ding, K., Weinberg, S.M.: Approximately strategyproof tournament rules in the probabilistic setting. In: 12th Innovations in Theoretical Computer Science Conference, ITCS (2021)
7. Elkind, E., Faliszewski, P., Slinko, A.: Clone structures in voters' preferences. In: Proceedings of the 13th ACM Conference on Electronic Commerce, pp. 496–513 (2012)
8. Fishburn, P.C.: Condorcet social choice functions. SIAM J. Appl. Math. **33**(3), 469–489 (1977). http://www.jstor.org/stable/2100704
9. Fisher, D.C., Ryan, J.: Optimal strategies for a generalized "scissors, paper, and stone" game. Am. Math. Monthly **99**(10), 935–942 (1992). http://www.jstor.org/stable/2324486

10. Laffond, G., Laslier, J.F., Le Breton, M.: The bipartisan set of a tournament game. Games Econ. Behav. **5**(1), 182–201 (1993). https://EconPapers.repec.org/RePEc: eee:gamebe:v:5:y:1993:i:1:p:182-201
11. Kim, M.P., Suksompong, W., Williams, V.V.: Who can win a single-elimination tournament? SIAM J. Discrete Math. **31**(3), 1751–1764 (2017)
12. Kim, M.P., Williams, V.V.: Fixing tournaments for kings, chokers, and more. In: Twenty-Fourth International Joint Conference on Artificial Intelligence (2015)
13. Laffond, G., Lainé, J., Laslier, J.F.: Composition-consistent tournament solutions and social choice functions. Soc. Choice Welfare **13**(1), 75–93 (1996). https://doi. org/10.1007/BF00179100
14. Miller, N.R.: A new solution set for tournaments and majority voting: further graph-theoretical approaches to the theory of voting. Am. J. Polit. Sci. **24**(1), 68–96 (1980). http://www.jstor.org/stable/2110925
15. Moulin, H.: Handbook of Computational Social Choice. Cambridge University Press, Cambridge (2016). https://doi.org/10.1017/CBO9781107446984
16. Öztürk, Z.E.: Consistency of scoring rules: a reinvestigation of composition-consistency. Int. J. Game Theory **49**(3), 801–831 (2020). https://doi.org/10.1007/s00182-020-00711-7
17. Schneider, J., Schvartzman, A., Weinberg, S.M.: Condorcet-consistent and approximately strategyproof tournament rules. In: Proceedings of the 8th Innovations in Theoretical Computer Science Conference (ITCS) (2017)
18. Schvartzman, A., Weinberg, S.M., Zlatin, E., Zuo, A.: Approximately strategyproof tournament rules: on large manipulating sets and cover-consistence. In: 11th Innovations in Theoretical Computer Science Conference, ITCS 2020, 12–14 January 2020
19. Stanton, I., Williams, V.V.: Rigging tournament brackets for weaker players. In: IJCAI Proceedings-International Joint Conference on Artificial Intelligence, vol. 22, p. 357 (2011)

Auditing for Core Stability in Participatory Budgeting

Kamesh Munagala[1]([⊠])(iD), Yiheng Shen[1], and Kangning Wang[2]

[1] Duke University, Durham, NC 27708-0129, USA
{kamesh,ys341}@cs.duke.edu
[2] Simons Institute for the Theory of Computing, Berkeley, CA 94720-2190, USA
kangning@berkeley.edu

Abstract. We consider the *participatory budgeting* problem where each of n voters specifies additive utilities over m candidate projects with given sizes, and the goal is to choose a subset of projects (i.e., a *committee*) with total size at most k. Participatory budgeting mathematically generalizes multiwinner elections, and both have received great attention in computational social choice recently. A well-studied notion of group fairness in this setting is *core stability*: Each voter is assigned an "entitlement" of $\frac{k}{n}$, so that a subset S of voters can pay for a committee of size at most $|S| \cdot \frac{k}{n}$. A given committee is in the core if no subset of voters can pay for another committee that provides each of them strictly larger utility. This provides proportional representation to all voters in a strong sense. In this paper, we study the following auditing question: Given a committee computed by some preference aggregation method, how close is it to the core? Concretely, how much does the entitlement of each voter need to be scaled down by, so that the core property subsequently holds? As our main contribution, we present computational hardness results for this problem, as well as a logarithmic approximation algorithm via linear program rounding. We show that our analysis is tight against the linear programming bound. Additionally, we consider two related notions of group fairness that have similar audit properties. The first is *Lindahl priceability*, which audits the closeness of a committee to a market clearing solution. We show that this is related to the linear programming relaxation of auditing the core, leading to efficient exact and approximation algorithms for auditing. The second is a novel weakening of the core that we term the *sub-core*, and we present computational results for auditing this notion as well.

Keywords: Computational social choice · Participatory budgeting · Core stability

Supported by NSF grant CCF-2113798.

K. Wang—This work was done while Kangning Wang was at Duke University.

K. A. Hansen et al. (Eds.): WINE 2022, LNCS 13778, pp. 292–310, 2022.
https://doi.org/10.1007/978-3-031-22832-2_17

1   Introduction

The *participatory budgeting* problem [1,6,12,19,24] is motivated by real-world elections where voters decide which projects a city should fund subject to a budget constraint on the total cost of these projects. In this problem, there are m candidate projects forming a set C, and n voters. Each candidate j is associated with a size/cost s_j.

The *multiwinner election* problem [2,9,14,18,36] is commonly seen in practice, and has received significant research attention recently. Mathematically, it is a specialization of the participatory budgeting problem, where each candidate is of the same unit size.

In both settings, our goal is to pick a subset $T \subseteq C$ of candidates – which we call a *committee* – with total size at most a given value k, that is, $\sum_{j \in T} s_j \le k$. Each voter i has a utility function $U_i(T)$ over subsets $T \subseteq C$ of candidates. In this paper, we assume the utility functions $\{U_i\}$ are *additive* across candidates. For some of our results, we also look at the more restricted case of multiwinner elections with *approval* (i.e. 0/1-additive) utilities: Each candidate is of unit size; each voter i "approves" a subset $A_i \subseteq C$ of candidates, and for any committee T, the utility function of voter i is simply $U_i(T) = |T \cap A_i|$, the number of approved candidates in the committee. We call this the APPROVAL ELECTION setting.

Core Stability. In both multiwinner elections and participatory budgeting, the methods used to aggregate preferences of voters are typically very simple, for instance, choosing the candidates who receive the most approval votes. This leads to a tension of such rules with *fairness* of the resulting outcome in terms of proportional representation of minority opinions, and a social planner may want to quantify this tension for any given election.

A notion of fairness in this context, which has been studied for over a century, is that of *core stability* [17,20,29,35,36]. This captures a strong notion of proportional representation. Given a committee W of size k, think of k as a budget, and split it equally among the n voters, so that each voter is entitled to a budget of $\frac{k}{n}$. For any subset $S \subseteq [n]$ of voters, their total entitlement is $|S| \cdot \frac{k}{n}$. If there is another committee T of size at most the entitlement $|S| \cdot \frac{k}{n}$, such that each voter $i \in S$ strictly prefers T to W, i.e., $U_i(T) > U_i(W)$ for all $i \in S$, then these voters would have a justified complaint with W. A committee W where no subset $S \subseteq [n]$ of voters have a justified complaint is termed *core stable*.

The core has a "fair taxation" interpretation [23,29]. The quantity $\frac{k}{n}$ can be thought of as the tax contribution of a voter, and a committee in the core has the property that no sub-group of voters could have spent their share of tax money in a way that *all* of them were better off. It subsumes notions of fairness such as Pareto-optimality, proportionality, and various forms of justified representation [3,4,22] that have been extensively studied in multiwinner election and fairness literature. Note that the core is *oblivious* to how the demographic slices are defined – it attempts to be fair to *all* subsets of voters. This is a desirable feature in practice, since demographic slices are often not known upfront, and there could be hidden sub-groups that can only be inferred from voter preferences.

Approximate Stability. The core is a very appealing group fairness notion; however, even in very simple settings, the core could be empty [20]. This motivates approximation, where the entitlement $\frac{k}{n}$ of each voter is scaled by a factor of θ.

Definition 1. *For $\theta \leq 1$, a committee W of size at most k lies in the θ-approximate core if for all $S \subseteq [n]$, there is no deviating committee T with size at most $\theta \cdot |S| \cdot \frac{k}{n}$, such that for all $i \in S$, we have $U_i(T) > U_i(W)$.*

It is known [25] that a $\frac{1}{32}$-approximate core solution always exists for very general utility functions of the voters.

Auditing for Approximate Stability. Though the existence of approximate core solutions is a strong positive result, the algorithms for finding such solutions are often complex. Indeed, even in settings where the core is known to be always non-empty, for instance when candidates can be chosen fractionally [23], the non-emptiness is an existence result that needs an expensive fixed point computation. On the other hand, in practice, what are implemented are typically the simplest and most explainable social choice methods such as Single Transferable Vote (STV). Therefore, from the perspective of a societal decision maker, such as a civic body running a participatory budgeting election, it becomes important to answer the following *auditing* question for any given election:

> Given a committee W of size at most k found by some implemented preference aggregation method, how close is it to being core stable, i.e., what is the smallest value of θ_c such that W *does not* lie in the θ_c-approximate core for that instance?

Note that if a committee W lies in the core, then $\theta_c > 1$, else $\theta_c \leq 1$. Such an auditing question is useful even if the decision maker themselves is not sensitive for fairness because it allows for review of implemented decision rules via a third party or government agency. Further, the set of deviating voters that correspond to the θ_c-approximation yield a demographic that are unhappy with the current outcome, and this can be analyzed further by policy makers.

We term the above question as the *core auditing* problem. In this paper, we study the computational complexity of core auditing. In that process, we define both stronger and weaker notions of fairness and audit these notions as well.

1.1   Our Results

Hardness and Approximation Algorithm. We show in Sect. 3 that for APPROVAL ELECTIONS, the value of θ_c in the core auditing problem is NP-HARD to approximate to a factor better than $1 + \frac{1}{e} > 1.367$. We further show that this APX-HARDNESS persists even when voters are allowed to choose a fractional deviating committee. We also show that the problem remains NP-HARD when each voter approves a constant number of candidates, and each candidate is approved by a constant number of voters. These results significantly strengthen the NP-HARDNESS result presented in [11].

On the positive side, in Sect. 4, we design an $O(\min(\log m, \log n))$ approximation algorithm for the value θ_c, where m and n are the number of candidates and voters respectively. We do this via linear program rounding. Our program (and indeed, our auditing question itself) is an interesting generalization of the densest subgraph problem [15], where the goal is to choose a subgraph with maximum average degree. Given a graph, treat voters as edges and candidates as vertices that are approved by the incident edges; further assume any voter needs utility 2 (that is, both end-points) in a feasible deviation. Then, the value of θ_c is precisely the density of the densest subgraph (to scaling). We combine ideas from the rounding for densest subgraph (where the rounding produces the integer optimum without approximation) with that from maximum coverage to design our rounding scheme. We further show that our linear program has an integrality gap of $\Omega(\min(\log m, \log n))$, showing that we cannot do any better against an LP lower bound. Our proof in Sect. 4 applies to the APPROVAL ELECTION setting. We extend this to general candidate sizes and arbitrary additive utilities via knapsack cover inequalities in Sect. 5, leading to an $O(\min(\log m, \log n))$ approximation factor. In the full paper [31], we show that both our hardness and approximation results easily extend to settings where candidates can be fractionally chosen in the committees.

It is an interesting question to close the gap between our hardness result (constant factor) and our approximation ratio. The difficulty lies with density problems in general, where hardness of approximation results have been hard to come by; see for instance, the k-densest subgraph problem [27].

Lindahl Priceability. A closely related notion of fairness, considered in [23,29, 32,33] is that of committees that can be supported by market clearing prices. The notion of *Lindahl equilibrium* is a pricing scheme that strengthens the core, meaning that if the former exists, it lies in the core. In this scheme, each voter i is assigned price p_{ij} for candidate j, and these prices are such that for any candidate, the total price is equal to its size. If a voter buys their optimal set of candidates subject to the total price paid being at most their entitlement, k/n, then all voters choose the *same* committee. This is therefore a market clearing notion with per-voter prices such that the optimal voter action given these prices and equal entitlements results in a common committee being chosen. If committees could be chosen fractionally, it is known via a fixed point argument that the Lindahl equilibrium always exists [23]. However, these need not exist when considering integer committees.

In this paper, we consider an integer version of this concept that we term *Lindahl priceability*. We show that this notion implies the core. As with the core, in Sect. 6, we define the approximation factor θ_ℓ to which a given committee satisfies Lindahl priceability, via scaling the entitlement k/n of each voter by that factor. We show via LP duality not only that the quantity θ_ℓ can be audited in polynomial time for APPROVAL ELECTIONS, but also that this computation coincides with the LP relaxation to the core auditing program. This results in a novel and somewhat surprising connection between the Lindahl priceability and the core for APPROVAL ELECTIONS, where the approximation factor θ_ℓ for

Lindahl priceability is found via the LP relaxation to the program that computes the approximation factor θ_c for the core. Further, our approach easily extends to show computational results for general utilities and sizes.

Our notion is related to the *cost efficient Lindahl equilibrium* proposed recently in [33] for APPROVAL ELECTIONS. However, there is a crucial difference: While they translate the fractional Lindahl equilibrium to the integer case, we translate the gradient optimality conditions implied by the fractional equilibrium to the integer case. To illustrate that our definition is different, note that while there are simple instances of APPROVAL ELECTIONS on which the former notion does not exist, we do not know such an instance for our definition.

Weak Priceability and Sub-core. In Sect. 7, we finally connect our work to another notion of priceability first studied in [34]. This notion is a relaxation of Lindahl priceability for APPROVAL ELECTIONS, where voters cannot greedily augment the current committee given the prices and their entitlement. We term this "weak priceability" and use this to define a new relaxation of the core, termed *sub-core*, which only allows voters to deviate and gain utility from super-sets. We show that weakly priceable committees lie in the sub-core. Further, though the sub-core appears like a weak notion of fairness, we show that it remains NP-HARD to audit. We finally present an $O(\min(\log m, \log n))$ approximation to the auditing question using same techniques as for auditing the core.

In practice, committees found by social choice rules are likely to be much better approximations to the sub-core compared to the core. Hence, it is desirable to show a practitioner closeness to weaker notions of fairness such as the sub-core in addition to closeness to the core.

Omitted Proofs. For lack of space, all omitted proofs are in the full paper [31].

1.2 Related Work

Proportionality in Social Choice. The earliest work that considers proportional representation dates back to the late 1800's [17], and several voting rules attempting to achieve it, such as PAV [36] and Phragmén [10] rules also date back to then. There has been resurgence of interest in axiomatizing proportionality [3,4,8,14,22,30] partly driven by real-world applications of such elections to areas such as participatory budgeting [1,6,24], and partly due to local bodies and countries implementing rules such as ranked choice voting that attempt to achieve proportionality, in their elections. These advances have made auditing fairness notions such as closeness to the core and weaker group fairness notions imperative in these settings.

Notions of Approximate Core. In addition to the notion of approximation presented in Definition 1, a different notion allows deviating voters to use their entire entitlement, but requires them to extract at least a factor $\theta > 1$ larger utility on deviation. Under this notion, it is shown in [34] that a classic voting rule called

PAV [36] achieves a 2-approximation. This result was generalized to show a constant approximate core for arbitrary submodular utility functions and general candidate sizes in [32]. An analogous result for clustering was presented in [16]. Our work directly shows that this notion of approximation can be audited in a *bicriteria* fashion as follows: If the given committee is a c-approximation without violating entitlements, we can determine if it is a c-approximation had entitlements been violated by a factor of $O(\log m)$. It is an interesting open question to remove the bicriteria nature of this result.

Auditing for Fairness. The question of auditing has become salient given the increasing democratization of societal decision making, for instance via processes like participatory budgeting. In the context of social choice, there are natural properties that are easy to achieve algorithmically but hard to audit. For instance, checking if an arbitrary outcome is Pareto-optimal is computationally hard [5], while achieving it via some algorithm is easy. We take a further step in this direction by studying the approximate audit of arguably the strongest possible group fairness notion, the core, as well as related fairness properties.

Going beyond social choice, the notion of auditing for group fairness has gained prevalence in machine learning. Here, the "voters" are data points, and the "committee" is a classifier. We wish to audit if the classifier provides comparable accuracy for various demographic slices. The work of [26] formulates and presents algorithms for this problem.

2   Mathematical Program for θ_c

For most of this paper, we consider the APPROVAL ELECTION setting. Recall that in this setting, each voter i "approves" a set $A_i \subseteq C$ of unit-sized candidates, and its utility for a committee $T \subseteq C$ is simply $U_i(T) = |A_i \cap T|$. Our hardness results hold even for this simple setting, while our approximation algorithms hold for general additive utilities and sizes (see Sect. 5).

We first present a mathematical program that computes θ_c given a committee $W \subseteq C$ of size at most k, as in Definition 1. In this program, there is a variable $z_i \in \{0, 1\}$ that captures whether voter i deviates, and a variable $x_j \in \{0, 1\}$ that captures whether candidate j is present in the deviating committee. If this is a feasible deviation, then the utility of each voter for which $z_i = 1$ must strictly increase, which means $\forall i \in [n]$, $\sum_{j \in A_i} x_j \geq (U_i(W) + 1) \cdot z_i$.

Next, let $R = \frac{n}{k}$. Then, the budget available to the deviating voters is $\frac{1}{R} \sum_i z_i$, while the size of the committee to which they deviate is $\sum_{j \in C} x_j$. This means the entitlement k/n of each voter must be scaled by a factor of $R \cdot \frac{\sum_{j \in C} x_j}{\sum_i z_i}$, so that the voters with $z_i = 1$ do not have enough entitlement to pay for this deviating committee. Since the goal is to have no deviations at all, the value θ_c is simply the solution to the following mathematical program:

$$\text{Minimize } R \cdot \frac{\sum_{j \in C} x_j}{\sum_i z_i}, \text{ s.t.}$$

$$\forall i \in [n], \quad \sum_{j \in C \cap A_i} x_j \geq z_i \cdot (U_i(W) + 1);$$

$$\forall i \in [n], \ \forall j \in [m], \ x_j, z_i \in \{0, 1\}.$$

The above program attempts to maximize the ratio of the number of constraints satisfied via setting z_i to 1, to the number of x_j variables set to 1.

3 Hardness of Auditing the Core

As mentioned before, all hardness results in this section apply to the APPROVAL ELECTION setting, where the utilities are binary, and candidate sizes are unit. We first show that the core auditing problem, that is, the problem of computing θ_c for a given committee W, is NP-HARD even in a "constant degree" setting. This strengthens an NP-HARDNESS result for the core in [11].

Theorem 1. *Deciding whether a committee W does not lie in the core (that is, deciding whether its $\theta_c \leq 1$) is NP-HARD when each voter approves at most 6 candidates (that is, $|A_i| \leq 6$ for all voters $i \in [n]$), and each candidate lies in at most 2 of the sets A_i.*

We now show that the core auditing problem is in fact APX-HARD.

Theorem 2. *For any constant $\gamma > 0$, approximating θ_c to within a factor of $1 + \frac{1}{e} - \gamma$ is NP-HARD.*

We will reduce from the maximum set coverage problem on regular instances.

Lemma 1 (Regular Maximum Coverage [21]). *The universe contains qd elements. There are ξ sets, each with d elements. It is NP-HARD to distinguish between the following two cases:*

1. *"YES" instances: There exist q sets that cover the universe.*
2. *"NO" instances: No collection of q sets can cover $(1 - 1/e + \varepsilon) \cdot qd$ elements.*

Proof (Proof of Theorem 2). For each instance of the regular Max Covering Problem, there are qd elements and ξ sets. We construct the following instance for auditing the core:

- There are ξ main candidates. Each candidate corresponds to a set. There are $\frac{1}{e}(q-1)qd$ dummy candidates.
- There are two group of voters. The first group contains $\frac{1}{e} \cdot qd$ voters. They each approve $q - 1$ disjoint dummy candidates, and all the main candidates.

- The second group contains qd voters. Each of these voters corresponds to an element of the covering instance. She approves the main candidates whose corresponding set contains her corresponding element. Therefore, there are $(1 + 1/e)qd$ voters. Add dummy voters who do not approve any candidates, so that the total number of voters is $n = q(q - 1)d^2$.
- The budget for committee selection is $k = (q - 1)qd$. The current committee W contains all the dummy candidates. All voters in the first group have utility $q - 1$ while all voters in the second group have utility 0 in W.
- Note that each voter is assigned a budget of $\frac{1}{R} = \frac{k}{n} = \frac{(q-1)qd}{q(q-1)d^2} = \frac{1}{d}$.

If the maximum coverage instance is a "YES" instance, choose as the deviating committee the q main candidates whose corresponding sets cover the universe. We call a voter "satisfied" if her utility has strictly increased compared to the current committee W. From the program in Sect. 2, θ_c is $R = d$ times the minimum ratio of the total number of selected candidates to the number of satisfied voters. Since we have selected q candidates, the voters in the first group receive utility q and are therefore satisfied. Moreover, since the chosen candidates' corresponding sets cover the universe of qd elements, the voters in the second group receive utility at least one, and are therefore satisfied. Therefore,

$$\theta_c \le R \cdot \frac{q}{qd \cdot (1 + 1/e)} = \frac{1}{1 + 1/e}.$$

Suppose the maximum coverage instance is a "NO" instance. We will show that $\theta_c \ge 1 - o(1)$. First suppose a deviating committee is composed of $s < q$ main candidates. These candidates can cover at most ds voters from the second group. For the first group, they provide utility s to each voter. If t of these voters are satisfied, we must have chosen $(q - s)t$ dummy candidates. This means the scaling factor needed is at least

$$R \cdot \frac{s + (q - s)t}{ds + t} = \frac{s + (q - s)t}{s + t/d} > 1.$$

If the number of main candidates in the deviating committee is at least q, the voters in the first group are all satisfied and we don't need to choose dummy candidates. Consider an arbitrary q-candidate subset of these selected candidates. All voters in the first group are satisfied by these candidates, since they receive utility q from them. Since the coverage instance is a "NO" instance, no more than $(1 - 1/e + \varepsilon) \cdot kd$ voters in the second group are satisfied by this subset. Suppose there are r remaining candidates in the deviation, each candidate can only increase the number of satisfied voters by at most $r \cdot d$. Therefore,

$$\theta_c \ge R \cdot \frac{q + r}{r \cdot d + (1 - 1/e + \varepsilon) \cdot q \cdot d + (1/e) \cdot q \cdot d} = \frac{R}{d} \cdot \frac{q + r}{r \cdot d + q \cdot d \cdot (1 + \varepsilon)} \ge \frac{1}{1 + \varepsilon}.$$

Since the gap of θ_c between the constructed auditing instance from "YES" instances and from "NO" instances is at least $\frac{1+1/e}{1+\varepsilon}$, approximating θ_c to within this factor is NP-HARD.

Hardness of Auditing Fractional Committees. One natural question is whether the above hardness stems from the integrality requirement on the committee (the x_j variables in the program in Sect. 2) or the voters (the z_i variables). In the full paper [31], we show that the auditing problem remains hard to approximate to constant factors even when the committees can be chosen *fractionally*. This corresponds to allowing the variables $\{x_j\}$ to be fractional in $[0, 1]$. This shows that the hardness of the problem stems mainly from insisting $\{z_i\}$ be integral. The proof of this result is similar to the previous proof.

4 A Logarithmic Approximation for Auditing the Core

Our main result in this section is the following theorem, which we prove for the APPROVAL ELECTION setting. The proof for general candidate sizes and general additive utilities is presented in Sect. 5.

Theorem 3. *Given a committee W of size at most k, its θ_c value can be computed within $O(\min(\log m, \log n))$ factor in polynomial time, where m, n are the total number of candidates and voters respectively.*

4.1 LP Relaxation

Given a committee W, we start with the mathematical program from Sect. 2 and relax the variables to be fractional. This yields the following program. To see that this is a relaxation, if $z_i = 0$ for some i, then the first constraint is trivially satisfied. On the other hand, if $z_i = 1$, then we can increase all y_{ij} so that $y_{ij} = x_j$, thereby recovering the constraint in the integer program from Sect. 2. Therefore, any solution to the integer program is a feasible solution to the program below.

$$\text{Minimize } R \cdot \frac{\sum_{j=1}^{m} x_j}{\sum_{i=1}^{n} z_i}, \text{ s.t.}$$

$$\forall i \in [n], \sum_{j \in A_i} y_{ij} \geq z_i \cdot (U_i(W) + 1);$$

$$\forall i \in [n], \forall j \in A_i, y_{ij} \leq x_j;$$

$$\forall i \in [n], \forall j \in A_i, y_{ij} \leq z_i;$$

$$\forall i \in [n], j \in [m], x_j, z_i, y_{ij} \geq 0.$$

This can be written as a LP if we omit the denominator from the objective and add the constraint $\sum_i z_i \geq 1$, and hence can be solved in polynomial time.

Denote $u_i = U_i(W)$. For the committee W, we further denote

$$\theta_p = R \cdot \frac{\sum_{j=1}^{m} x_j}{\sum_{i=1}^{n} z_i} \tag{1}$$

where the variables are set based on the optimal solution to the linear relaxation. Therefore, $\theta_p \leq \theta_c$.

4.2   Proof of Theorem 3

We will show that θ_p is an $O(\log m)$ approximation to θ_c. The proof of the $O(\log n)$ approximation is similar and presented in the full paper [31].

By scaling the LP, we can assume that $\max_i\{z_i\} = 1$. Therefore, all $y_{ij} \leq z_i \leq 1$ and $x_j = \min_{i:j\in A_i}\{y_{ij}\} \leq 1$: all the variables are in the range $[0, 1]$.

$O(\log m)$ *Approximation.* Given the fractional solution, we note that $y_{ij} = \min(x_j, z_i)$. We now construct an integral solution by the following steps:

1. Pick $\alpha \in [0, 1]$ uniformly at random. If $z_i > \alpha$, set $\hat{z}_i = 1$; else $\hat{z}_i = 0$.
2. Let $x'_j = \max\{\frac{1}{2m^2}, x_j\}$.
3. If $2x'_j > \alpha$, then set $\hat{x}_j = 1$; else set $\hat{x}_j = 1$ with probability $2x'_j/\alpha$. We round each $\hat{x}_j$ independently.
4. If $\hat{z}_i = 1$, check if $\sum_{j\in A_i} \min\{\hat{x}_j, \hat{z}_i\} \geq u_i$. If so, set $\hat{z}_i = 1$; else set $\hat{z}_i = 0$.

Suppose the largest z_i is $z_{i^*} = 1$, we have $\sum_{j\in A_{i^*}} y_{ij} \geq 1$. Therefore, for some j, $y_{i^*j} \geq 1/m$. Therefore $\sum_{j=1}^m x_j \geq \frac{1}{m}$. Since Step 2 increases $\sum_j x_j$ by at most $\frac{1}{2m}$, we have $\frac{\sum_j x'_j}{\sum_j x_j} \leq 3/2$.

We first bound the expectation of $\hat{x}_j$. If $x'_j < 1/2$, since $x'_j \geq \frac{1}{2m^2}$, we have:

$$\mathbb{E}[\hat{x}_j] = \int_{\alpha=0}^{2x'_j} 1 \, d\alpha + \int_{\alpha=2x'_j}^1 2x'_j/\alpha \, d\alpha = 2x'_j + 2x'_j \cdot \ln\alpha \Big|_{2x'_j}^1 \leq 2x'_j \cdot (1 + 2\ln m).$$

Therefore, we have

$$\mathbb{E}\Big[\sum_j \hat{x}_j\Big] \leq 2(1 + 2\ln m)\sum_j x'_j \leq 3(1 + 2\ln m)\sum_j x_j.$$

We now bound $\mathbb{E}\Big[\sum_i \hat{z}_i\Big]$. Let $P_i \triangleq \{j \in A_i : 2x'_j < \alpha\}$, $Q_i \triangleq \{j \in A_i : 2x'_j \geq \alpha\}$. Since $\hat{x}_j = 1$ for $j \in Q_i$, conditioned on $\hat{z}_i = 1$, we have:

$$\Pr\left(\hat{z}_i = 0\right) = \Pr\left(\sum_{j\in A_i} \min\{\hat{x}_j, \hat{z}_i\} < u_i + 1\right) = \Pr\left(\sum_{j\in P_i} \hat{x}_j < u_i + 1 - |Q_i|\right).$$

By the constraints in the optimization and since $y_{ij} = \min(x_j, z_i)$, we have

$$\sum_{j\in P_i} \min\{x_j, z_i\} + \sum_{j\in Q_i} \min\{x_j, z_i\} \geq z_i \cdot (u_i + 1).$$

Since the second term is capped by $z_i \cdot |Q_i|$, we have $\sum_{j\in P_i} x_j \geq z_i \cdot ((u_i + 1) - |Q_i|)$. When $\hat{z}_i = 1$, we have $\alpha < z_i$, and thus

$$\mathbb{E}\left[\sum_{j\in P_i} \hat{x}_j\right] \geq 2 \cdot \left(\sum_{j\in P_i} x'_j\right)/\alpha \geq 2 \cdot ((u_i+1) - |Q_i|) \cdot z_i/\alpha \geq 2 \cdot ((u_i+1) - |Q_i|).$$

By Chernoff Bounds on the independent binary random variables $\{\hat{x}_j\}$, we have

$$\Pr\left(\sum_{j \in P_i} \hat{x}_j < u_i + 1 - |Q_i| \,\Big|\, \hat{z}_i = 1\right) < \left(\frac{e^{-1/2}}{\sqrt{1/2}}\right)^2 = 2/e.$$

Therefore, we have

$$\mathbb{E}\left[\sum_i \hat{\hat{z}}_i\right] = \sum_i \mathbb{E}\left[\hat{z}_i \cdot (1 - \Pr(\hat{\hat{z}}_i = 0))\right] \geq \sum_i \mathbb{E}\left[\hat{z}_i \cdot (1 - \frac{2}{e})\right] \geq (1 - \frac{2}{e}) \cdot \sum_i z_i.$$

Since $\{\hat{x}_j\}$ and $\{\hat{\hat{z}}_i\}$ form a valid solution to the program in Sect. 2, there exists a setting of these variables such that

$$\frac{\sum_j \hat{x}_j}{\sum_i \hat{\hat{z}}_i} \leq \frac{\mathbb{E}[\sum_j \hat{x}_j]}{\mathbb{E}[\sum_i \hat{\hat{z}}_i]} \leq \frac{3(1 + 2\ln m)}{1 - 2/e} \cdot \frac{\sum_j x_j}{\sum_i z_i} = \frac{3(1 + 2\ln m)}{1 - 2/e} \cdot \theta_p.$$

Therefore, we have $\theta_p \leq \theta_c \leq \frac{3(1+2\ln m)}{1-2/e} \cdot \theta_p$, completing the proof of the $O(\log m)$ approximation. The proof of the $O(\log n)$ approximation is in the full paper [31].

4.3   Integrality Gap Instance

In the full paper [31], we prove the following theorem, which shows the analysis in Sect. 4.2 is tight.

Theorem 4. *There exists a committee s.t.* $\theta_p = O\left(\frac{1}{\log \min(m,n)}\right)$ *and* $\theta_c = \Theta(1)$.

5   Extension to Arbitrary Utilities and Sizes

We now extend the result in the previous section to the setting where the candidates have general sizes s_j, and voters have arbitrary additive utilities over candidates. We assume voter i has utility $u_{ij} \in \mathbb{Z}^+ \cup \{0\}$ for candidate j. Given a committee W of size at most k, the utility of voter i for the committee is $U_i(W) = \sum_{j \in W} u_{ij}$. We restrict the utilities to be integral, so that if $U_i(T) > U_i(W)$, then $U_i(T) \geq U_i(W) + 1$. Let $A_i = \{j \in C \mid u_{ij} > 0\}$.

LP Formulation. A natural modification to the program in Sect. 2 for θ_c has unbounded integrality gap. We make two modifications to the linear program. First, in the optimal integer solution, we guess the candidate j^* with largest size. This means we set $x_j = 0$ for all j such that $s_j > s_{j^*}$, and delete these items. Since the numerator in the objective is at least s_{j^*}, we can set $x_j = 1$ for all j with $s_j < \frac{s_{j^*}}{m}$, and this only increases the numerator by a constant factor. Let S denote the set of these "small" items; we ignore these items, and set $U_i(W)$ to be $U_i(W) - U_i(S \cap A_i)$. If the latter quantity is smaller than zero, then we can set $z_i = 1$ and delete this voter from further consideration; this only lowers the objective. We let m denote the number of candidates and n denote the number of voters in the residual instance. We now scale the sizes so that the remaining items have sizes in $\left[\frac{1}{m}, 1\right]$. Let $R = \frac{n}{k}$.

Next, we add knapsack cover constraints [13, 28]. Let $\hat{U}_i(S) = \max(0, U_i(W) + 1 - U_i(S))$, and let $u_{ijS} = \min(u_{ij}, \hat{U}_i(S))$ The resulting LP is presented below. In this LP, first set of constraints can be interpreted as follows: Even if the x_j for $j \in S$ are all set to 1, so that voter i already has utility $U_i(S)$, if voter i is chosen by the integer program, the remaining $\{y_{ij}\}$ must push the total utility above $U_i(W)$. Further, any utility value u_{ij} on the LHS can be truncated at $\hat{U}_i(S)$ and the constraint should still hold. This constraint is clearly true for any S in the integer program; the LP encodes the fractional version of all of them.

$$\text{Minimize } R \cdot \frac{\sum_{j=1}^{m} s_j x_j}{\sum_{i=1}^{n} z_i}, \text{ s.t.}$$

$$\forall i \in [n], S \subseteq [m], \sum_{j \in A_i \setminus S} u_{ijS} y_{ij} \geq z_i \cdot \hat{U}_i(S);$$

$$\forall i \in [n], \ \forall j \in A_i, \ y_{ij} \leq \min(x_j, z_i);$$

$$\forall i \in [n], j \in [m], \ x_j, z_i, y_{ij} \geq 0.$$

This LP has exponentially many constraints. For any given solution (x, y, z) and fixed voter i, we divide the first set of constraints by z_i and use the polynomial-time dynamic programming procedure exactly as in [13] to find the most violated constraint to a $(1+\epsilon)$ approximation, for constant $\epsilon > 0$. Omitting standard details, this implies the LP can be solved to a $(1 + \epsilon)$ approximation in polynomial time via the Ellipsoid algorithm.

Rounding. The rounding is similar to Sect. 4.2, leading to the following theorem, whose proof is presented in the full paper [31].

Theorem 5. *For the setting with arbitrary additive utilities and sizes, θ_c can be approximated to an $O(\min(\log m, \log n))$ factor in polynomial time.*

6   Auditing Lindahl Priceability

In this section, we study fairness of a committee in terms of closeness to market clearing. The concept is motivated by *Lindahl equilibrium* [23, 29], a market clearing concept for public goods. Such market clearing notions have been widely studied as fairness concepts in Economics [7, 37]. Our main result is the following novel connection to the core – auditing the approximation of a committee to Lindahl priceability reduces to the LP relaxation for auditing for core stability, hence leading to a polynomial time auditing algorithm.

We consider the APPROVAL ELECTION setting below. The extension to arbitrary utilities and sizes is presented in the full paper [31].

6.1 Lindahl Priceability

As in the definition of core stability, we first scale the entitlements so that the entitlement of each voter is set to 1 instead of k/n. Each candidate now requires $R = n/k$ entitlement to be paid for. A feasible committee of size k corresponds to a total entitlement of n in this scaling.

A committee W of size at most k is Lindahl priceable if there exists a price system $\{p_{ij}\}$ from voters to candidates, such that the following hold:

1. $\forall j \in [m]$, $\sum_i p_{ij} \leq R$, and
2. $\forall i \in [n]$, $T \subseteq C$, if $|T \cap A_i| \geq |W \cap A_i| + 1$, then $\sum_{j \in T} p_{ij} > 1$.

The first condition above means that for each candidate, the prices from all voters sum up to at most $R = n/k$, so that each candidate is not "over-paid". Note that the first set of constraints can be made equalities by raising the prices $\{p_{ij}\}$, so the candidates are exactly paid for. The second condition means a voter cannot afford any committee that she strictly prefers to W.

Lindahl priceability can be viewed as an integral version of the *gradient optimality conditions* in the fractional Lindahl equilibrium [23]. As mentioned before, this makes our definition subtly different from a related concept in [33]. Analogous to the fractional Lindahl equilibrium, the following proposition holds, and we present a proof later in this section.

Proposition 1. *If a committee is Lindahl priceable, it lies in the core.*

6.2 Auditing via Duality

As with core stability, we now define the best approximation to Lindahl priceability achievable by a committee W. Formally, we only allow a voter to use $\theta_p < 1$ endowment if they want to deviate to a committee with larger utility.

Definition 2 (θ-Approximate Lindahl Priceability). *A committee W of size at most k is θ-approximate Lindahl priceable if there exists a price system $\{p_{ij}\}$ from voters to candidates, such that the following conditions hold:*

1. $\forall j \in [m]$, $\sum_i p_{ij} \leq R$, and
2. $\forall i \in [n]$, $T \subseteq C$, if $|T \cap A_i| \geq |W \cap A_i| + 1$, then $\sum_{j \in T} p_{ij} > \theta$.

The Lindahl priceability ratio of a committee W is the smallest θ for which the committee is not θ-approximate Lindahl priceable. Our main result is the following theorem that ties Lindahl priceability ratio to the fractional relaxation of θ_c. As a corollary, this shows that determining if a committee W is Lindahl priceable is polynomial time solvable.

Theorem 6. *For a committee W, its Lindahl priceability ratio is θ_p from Eq. (1).*

Proof. For simplicity, let $u_i = U_i(W)$. Let the Lindahl priceability ratio of the instance be θ_ℓ. Fix the prices $\{p_{ij}\}$ achieving this. Then the minimum entitlement needed for a voter i to deviate to a committee of utility larger than u_i is captured by the following linear program:

$$\text{Minimize} \sum_{j \in A_i} p_{ij}\gamma_{ij}, \text{ s.t.}$$

$$\sum_{j \in A_i} \gamma_{ij} \geq u_i + 1;$$

$$\forall j \in A_i, \ \gamma_{ij} \leq 1;$$

$$\forall j \in A_i, \ \gamma_{ij} \geq 0.$$

Here, the variable γ_{ij} corresponds to the fraction to which this voter chooses candidate j. In the optimal solution, these variables will be integers. Since the Lindahl priceability ratio is θ_ℓ, Condition (2) of Definition 2 implies objective of the above LP is at least θ_ℓ for any $i \in [n]$.

Now take the dual of the above, where the dual variable for the first constraint is λ_i and the dual variable for the second constraint is α_{ij}. We obtain:

$$\text{Maximize } \theta_i, \text{ s.t.}$$

$$\forall j \in A_i, \lambda_i - \alpha_{ij} \leq p_{ij};$$

$$(u_i + 1)\lambda_i - \sum_{j \in A_i} \alpha_{ij} \geq \theta_i;$$

$$\forall j \in [m], \ \lambda_i, \alpha_{ij} \geq 0.$$

Since the optimal $\theta_i \geq \theta_\ell$, this solution satisfies $(u_i + 1)\lambda_i - \sum_{j \in A_i} \alpha_{ij} \geq \theta_\ell$. Since $\{p_{ij}\}$ satisfy Condition (1) in Definition 2, $\{p_{ij}\}, \{\alpha_{ij}\}, \{\lambda_i\}$ and $\theta = \theta_\ell$ are feasible for the following program:

$$\text{Maximize } \theta, \text{ s.t.}$$

$$\forall i \in [n], j \in A_i, \lambda_i - \alpha_{ij} \leq p_{ij};$$

$$\forall i \in [n], \ (u_i + 1)\lambda_i - \sum_{j \in A_i} \alpha_{ij} \geq \theta;$$

$$\forall j \in [m], \ \sum_{i \in T_j} p_{ij} \leq R;$$

$$\forall i \in [n], j \in [m], \ \lambda_i, \alpha_{ij}, p_{ij} \geq 0.$$

We now claim that the optimal solution to the above program must be exactly θ_ℓ. If it is larger, this larger value θ' must be feasible for the per-voter duals, which means the per-voter primals have value at least θ'. Then the Lindahl priceability is at least θ', contradicting the definition of θ_ℓ.

Finally, take the dual for the LP above, let y_{ij}, z_i, x_j respectively be the dual variable of the three constraints. The dual is the following:

$$\text{Minimize} R \cdot \sum_{j=1}^{m} x_j, \text{ s.t.}$$

$$\forall i \in [n], \sum_{j \in A_i} y_{ij} \geq z_i \cdot (u_i + 1);$$

$$\forall i \in [n], \forall j \in A_i, \ y_{ij} \leq x_j;$$

$$\forall i \in [n], \forall j \in A_i, \ y_{ij} \leq z_i;$$

$$\sum_i z_i \geq 1;$$

$$\forall i \in [n], j \in [m], \ z_i, x_j, y_{ij} \geq 0.$$

This optimal value (which is θ_ℓ) is also the definition of θ_p, completing the proof.

Note that if $\theta_\ell > 1$, then since $\theta_c \geq \theta_p = \theta_\ell > 1$, we have $\theta_c > 1$. Therefore, if a committee is Lindahl priceable, it lies in the core, showing Proposition 1.

7   Sub-core for APPROVAL ELECTIONS

Given our approximation results for auditing the core, we can ask if such results can also be derived for weaker fairness notions. Such an auditing notion would be interesting to a practitioner in addition to auditing the core, since it is quite likely an implemented rule and resulting committee would be closer to satisfying a weaker but still reasonable notion of fairness compared to the core. We present a new weakening of the core for APPROVAL ELECTIONS, that we term the sub-core, that we show also admits approximate auditing. Note that this result is not implied by our results for the core; indeed, there are weakenings of the core, such as EJR, that we do not know how to efficiently audit.

7.1   Weak Priceability

In the multiwinner election setting, suppose the final condition in Lindahl priceability is relaxed so that each voter is only allowed to add candidates to its deviating committee, we get the following relaxed version of priceability. Recall that $R = n/k$, where n is the total number of voters.

Definition 3 (Weak Priceability). *A committee W of size at most k is weakly priceable if there exists a set of prices $\{p_{ij}\}$ from each voter v_i to each candidate c_j, such that the following two conditions hold:*

1. $\forall j \in [m], \ \sum_i p_{ij} \leq R.$
2. $\forall i \in [n], \ d \in A_i \setminus W, p_{id} + \sum_{j \in A_i \cap W} p_{ij} > 1.$

This notion is equivalent to "priceability" as proposed in [34]. Unlike Lindahl priceability, there are many natural and greedy voting rules, such as the Phragmén rule [10], that satisfy weak priceability, making it a desirable property to study in practice.

7.2   Sub-core

If we proceed as in the proof of Theorem 6 and take the dual of the weak priceability ratio, we obtain a new concept of fairness that we call the *sub-core*.

Definition 4 (Sub-core). *A committee W lies in the* sub-core *if there is no $S \subseteq V$ and committee T with $|T| \leq \frac{|S|}{n} \cdot k$, s.t. $A_i \cap W \subsetneq A_i \cap T$ for all $i \in S$.*

The sub-core prevents any group of voters from deviating to a new committee in which each voter's approved candidates forms a proper superset of the approved candidates in the original committee.

Clearly, any committee that lies in the core also lies in the sub-core. The following proposition shows the sub-core is a weakening of weak priceability.

Proposition 2. *If a committee is weakly priceable, then it lies in the sub-core.*

Since weakly priceable committees can be easily found by greedy procedures [34], this shows that the sub-core is always non-empty.

Hardness of Auditing Sub-core. Though the sub-core seems like a weak and satisfiable fairness condition (it insists voters have no greedy deviation to a better committee), we show that deciding if a given committee W lies in the sub-core is actually NP-HARD. Towards this end, we observe that the core and sub-core coincide when each voter approves at most 2 candidates (i.e., for all voters i, we have $|A_i| \leq 2$). To see this, suppose the original committee was W and a subset of voters deviate to T. If a deviating voter had original utility zero, then $A_i \cap W = \varnothing$, so that $A_i \cap T \supsetneq A_i \cap W$. Similarly, if $|A_i \cap W| = 1$ and $|A_i \cap T| = 2$, then $A_i \cap T = A_i \supsetneq A_i \cap W$. This shows any deviation satisfies the sub-core property, so that the core coincides with the sub-core.

Theorem 7. *If each voter only approves at most two candidates, deciding whether a committee W does not lie in the sub-core (or core) is NP-COMPLETE.*

Approximately Auditing Sub-core Property. Similar to θ_c, we can now define a parameter θ_{sc} showing how close a committee is to the sub-core.

Definition 5. *For $\theta \leq 1$, a committee W of size k lies in the θ-approximate sub-core if for all subsets of voters $T \subseteq [n]$, there is no deviating committee T' with size at most $\theta \cdot |T| \cdot \frac{k}{n}$, such that for all $i \in T$, we have $A_i \cap W \subsetneq A_i \cap T'$.*

Given a committee W, θ_{sc} is defined as the smallest θ such that W is not in the θ-approximate sub-core. The following theorem shows the sub-core can be approximately audited. This positive result on auditing makes the sub-core a desirable weakening of the core property.

Theorem 8. *Given any committee W, θ_{sc} has an $O(\min(\log m, \log n))$ approximation in polynomial time, where m, n are the total number of candidates and voters respectively.*

8   Conclusion

Note that our theoretical approximation results for auditing are worst case guarantees. In practice, the linear program value θ_p will provide a lower bound on θ_c, and if this can be rounded so that the integer solution has value $\alpha\theta_p$ for some small $\alpha \geq 1$, then this sandwiches $\theta_c \in [\theta_p, \alpha\theta_p]$. Further, the rounding outputs a deviating set of voters and their chosen committee, which will be of interest as a demographic that is not well-represented by the current committee.

The main open question arising from this work is closing the gap between the positive and hardness results for auditing the core. As mentioned before, showing such results for density objectives is challenging [27]. A related question is existence results: A major open question in social choice is whether there is a committee in the core for APPROVAL ELECTIONS. Though there are voting rules that find committees in the approximate core [16,25,34], these results do not translate to the exact core. Even more specifically, it is an open question whether there is always a committee that is Lindahl priceable.

Finally, it would be interesting to use the techniques in this paper to approximately audit other notions of fairness or efficiency in social choice. For instance, consider the notion of *extended justified representation* (or EJR, [5]), where a group of $t \cdot n/k$ voters can only deviate if they all approve at least t candidates in common. Since this notion is weaker than the core, it is easier to show existence – indeed the PAV rule [36] satisfies EJR but fails the core. However, imposing restrictions on the deviation does not necessarily make it easier to audit such notions [33], and we do not know how to audit EJR approximately. We showed a particular weakening of the core, the sub-core, that can be approximately audited, and it would be interesting to study the landscape of efficient auditing more systematically.

References

1. The Stanford Participatory Budgeting Platform. https://pbstanford.org
2. Aziz, H., Brandt, F., Elkind, E., Skowron, P.: Computational social choice: the first ten years and beyond. In: Steffen, B., Woeginger, G. (eds.) Computing and Software Science. LNCS, vol. 10000, pp. 48–65. Springer, Cham (2019). https://doi.org/10.1007/978-3-319-91908-9_4
3. Aziz, H., Brill, M., Conitzer, V., Elkind, E., Freeman, R., Walsh, T.: Justified representation in approval-based committee voting. Soc. Choice Welfare **48**(2), 461–485 (2017). https://doi.org/10.1007/s00355-016-1019-3
4. Aziz, H., Elkind, E., Huang, S., Lackner, M., Fernández, L.S., Skowron, P.: On the complexity of extended and proportional justified representation. In: AAAI, pp. 902–909 (2018)
5. Aziz, H., Lang, J., Monnot, J.: Computing Pareto optimal committees. In: Kambhampati, S. (ed.) IJCAI, pp. 60–66 (2016)
6. Aziz, H., Shah, N.: Participatory budgeting: models and approaches. In: Rudas, T., Péli, G. (eds.) Pathways Between Social Science and Computational Social Science. CSS, pp. 215–236. Springer, Cham (2021). https://doi.org/10.1007/978-3-030-54936-7_10

7. Brainard, W.C., Scarf, H.E.: How to compute equilibrium prices in 1891. Am. J. Econ. Sociol. **64**(1), 57–83 (2005)
8. Brams, S.J., Kilgour, D.M., Sanver, M.R.: A minimax procedure for electing committees. Public Choice **132**(3), 401–420 (2007). https://doi.org/10.1007/s11127-007-9165-x
9. Brandt, F., Conitzer, V., Endriss, U., Lang, J., Procaccia, A.D.: Handbook of Computational Social Choice, 1st edn. Cambridge University Press, Cambridge (2016)
10. Brill, M., Freeman, R., Janson, S., Lackner, M.: Phragmén's voting methods and justified representation. In: AAAI, pp. 406–413 (2017)
11. Brill, M., Gölz, P., Peters, D., Schmidt-Kraepelin, U., Wilker, K.: Approval-based apportionment. In: AAAI, pp. 1854–1861 (2020)
12. Cabannes, Y.: Participatory budgeting: a significant contribution to participatory democracy. Environ. Urban. **16**(1), 27–46 (2004)
13. Carr, R.D., Fleischer, L.K., Leung, V.J., Phillips, C.A.: Strengthening integrality gaps for capacitated network design and covering problems. In: SODA, pp. 106–115 (2000)
14. Chamberlin, J.R., Courant, P.N.: Representative deliberations and representative decisions: proportional representation and the Borda rule. Am. Polit. Sci. Rev. **77**(3), 718–733 (1983)
15. Charikar, M.: Greedy approximation algorithms for finding dense components in a graph. In: Jansen, K., Khuller, S. (eds.) APPROX 2000. LNCS, vol. 1913, pp. 84–95. Springer, Heidelberg (2000). https://doi.org/10.1007/3-540-44436-X_10
16. Chen, X., Fain, B., Lyu, L., Munagala, K.: Proportionally fair clustering. In: ICML, pp. 1032–1041 (2019)
17. Droop, H.R.: On methods of electing representatives. J. Stat. Soc. Lond. **44**(2), 141–202 (1881)
18. Endriss, U.: Trends in Computational Social Choice. Lulu.com (2017)
19. Fain, B., Goel, A., Munagala, K.: The core of the participatory budgeting problem. In: Cai, Y., Vetta, A. (eds.) WINE 2016. LNCS, vol. 10123, pp. 384–399. Springer, Heidelberg (2016). https://doi.org/10.1007/978-3-662-54110-4_27
20. Fain, B., Munagala, K., Shah, N.: Fair allocation of indivisible public goods. In: EC (2018)
21. Feige, U.: A threshold of ln n for approximating set cover. J. ACM (JACM) **45**(4), 634–652 (1998)
22. Fernández, L.S., et al.: Proportional justified representation. In: AAAI, pp. 670–676 (2017)
23. Foley, D.K.: Lindahl's solution and the core of an economy with public goods. Econometrica J. Econometric Soc. **38**(1), 66–72 (1970)
24. Goel, A., Krishnaswamy, A.K., Sakshuwong, S., Aitamurto, T.: Knapsack voting for participatory budgeting. ACM Trans. Econ. Comput. **7**(2), 1–27 (2019)
25. Jiang, Z., Munagala, K., Wang, K.: Approximately stable committee selection. In: STOC, pp. 463–472 (2020)
26. Kearns, M., Neel, S., Roth, A., Wu, Z.S.: Preventing fairness gerrymandering: auditing and learning for subgroup fairness. In: ICML, pp. 2564–2572 (2018)
27. Khot, S.: Ruling out PTAS for graph min-bisection, dense k-subgraph, and bipartite clique. SIAM J. Comput. **36**(4), 1025–1071 (2006)
28. Kolliopoulos, S.G., Young, N.E.: Approximation algorithms for covering/packing integer programs. J. Comput. Syst. Sci. **71**(4), 495–505 (2005)
29. Lindahl, E.: Just taxation-a positive solution. In: Classics in the Theory of Public Finance, pp. 168–176 (1958)

30. Monroe, B.L.: Fully proportional representation. Am. Polit. Sci. Rev. **89**(4), 925–940 (1995)

31. Munagala, K., Shen, Y., Wang, K.: Auditing for core stability in participatory budgeting (2022). https://doi.org/10.48550/ARXIV.2209.14468. https://arxiv.org/abs/2209.14468

32. Munagala, K., Shen, Y., Wang, K., Wang, Z.: Approximate core for committee selection via multilinear extension and market clearing. In: SODA, pp. 2229–2252 (2022)

33. Peters, D., Pierczyński, G., Shah, N., Skowron, P.: Market-based explanations of collective decisions. In: AAAI, vol. 35, no. 6, pp. 5656–5663 (2021)

34. Peters, D., Skowron, P.: Proportionality and the limits of welfarism. In: EC, pp. 793–794 (2020)

35. Scarf, H.E.: The core of an N person game. Econometrica **35**(1), 50–69 (1967)

36. Thiele, T.N.: Om flerfoldsvalg. Oversigt over det Kongelige Danske Videnskabernes Selskabs Forhandlinger **1895**, 415–441 (1895)

37. Varian, H.R.: Two problems in the theory of fairness. J. Public Econ. **5**(3), 249–260 (1976)

Core-Stable Committees Under Restricted Domains

Grzegorz Pierczyński[✉] and Piotr Skowron

University of Warsaw, Warsaw, Poland
g.pierczynski@mimuw.edu.pl

Abstract. We study the setting of committee elections, where a group of individuals needs to collectively select a given-size subset of available objects. This model is relevant for a number of real-life scenarios including political elections, participatory budgeting, and facility-location. We focus on the core—the classic notion of proportionality, stability and fairness. We show that for a number of restricted domains including voter-interval, candidate-interval, single-peaked, and single-crossing preferences the core is non-empty and can be found in polynomial time. We show that the core might be empty for strict top-monotonic preferences, yet we introduce a relaxation of this class, which guarantees non-emptiness of the core. Our algorithms work both in the randomized and discrete models. We also show that the classic known proportional rules do not return committees from the core even for the most restrictive domains among those we consider (in particular for 1D-Euclidean preferences). We additionally prove a number of structural results that give better insights into the nature of some of the restricted domains, and which in particular give a better intuitive understanding of the class of top-monotonic preferences.

1 Introduction

We consider the model of committee elections, where the goal is to select a fixed-size subset of objects based on the preferences of a group of individuals. The objects and the individuals are typically referred to as the *candidates* and the *voters*, respectively, a convention that we follow in our paper. Yet, the candidates do not need to represent humans. For example, the model of committee elections describes the problems of (1) locating public facilities—there the candidates correspond to possible physical locations in which the facilities can be built [21,31], (2) presenting results by a search engine in response to a user query—there, the candidates are web-pages, and voters are potential users searching for a given query [32], (3) selecting validators in the blockchain, where the candidates are the users of the protocol [9,10]. For more examples that fall into the category of committee elections see the recent book chapter [20] and survey [23].

In numerous applications that fit the model of committee elections it is critical to select a subset of candidates, hereinafter called a *committee*, in a fair and proportional manner. Proportionality is one of the fundamental requirements of methods for selecting representative bodies, such as parliaments, faculty boards, etc. Yet, even in the context of facility location the properties that corresponds to proportionality are desirable:

© The Author(s), under exclusive license to Springer Nature Switzerland AG 2022
K. A. Hansen et al. (Eds.): WINE 2022, LNCS 13778, pp. 311–329, 2022.
https://doi.org/10.1007/978-3-031-22832-2_18

more objects should be built in densely populated areas, ideally ensuring that the distribution of the locations of the built facilities resembles the distribution of the locations of the potential users. In case of searching, the returned results should contain items that are interesting to different types of users, or—in other words—the preferences of each minority of users should be represented in the returned set of results. Finally, the validators in the blockchain should proportionally represent the protocol users in order to make validation robust against coordinated attacks of malicious users. In all these examples it is important to select a proportional committee, yet it is not entirely clear what it means that the committee proportionally reflects the opinions of the voters, let alone how to find such a committee.

The problem of formalizing the intuitive idea of proportionality has been often addressed in the literature and a plethora of axioms have been proposed (see [23][Section 5] and [20][Section 2.3.3]). Among them, the notion of the core is particularly interesting. This concept, borrowed from cooperative game theory [11,26] can be intuitively described as follows: assume our goal is to select a committee of k candidates based on the ballots submitted by n voters. Then, a group of n/k voters should intuitively have the right to select one committee member, and—analogously—a group of $\ell \cdot n/k$ voters should be able to decide about ℓ members of the elected committee. This intuition is formalized as follows: we say that a committee W is in the core if no group S of $\ell \cdot n/k$ voters can propose a set of ℓ candidates T such that each voter from S prefers T over W.

The notion of the core is intuitively appealing, universal, and strong. It applies to different types of voters' ballots, in particular to the *ordinal* and *approval* ones. In the ordinal model the voters rank the candidates from the most to the least preferred one, while in the approval model, each voter only marks the candidates that she find acceptable— we say that the voter approves such candidates. Being in the core implies numerous other fairness-related properties, among them properties which are rather strong on their own. For example, in the approval model the core implies the properties of extended justified representation (EJR) [2], proportional justified representation (PJR) [30], and justified representation (JR) [2]. For ordinal ballots, being in the core implies the properties of unanimity, consensus committee, and solid coalitions [15], as well as Dummett's proportionality [13] and proportionality for solid coalitions (PSC) [4]. In the full version of our paper [28] we also explain that for ordinal ballots being in the core is equivalent to satisfying full local stability [3].

While core-stability—the property of a voting rule that requires that each elected committee should belong to the core—is highly desired, it is also very demanding. For ordinal ballots there exists no core-stable rule, and for approval ballots it is one of the major open problems in computational social choice to determine whether the property is satisfiable. Given the property is so demanding, so far the literature focussed on its relaxed versions—either the weaker properties which we mentioned before, or the approximate [22,27] and the randomized [12] variants of the core have been extensively studied.

In our work we explore a different, yet related approach. Our point is that before we look at how a voting rule works in the general case, at the very minimum we shall ensure that it behaves well on well-structured preferences. Thus, the main question that we state in this paper is whether core-stability can be satisfied for certain natural restricted domains of voters' preferences, and what is the computational complexity of finding committees that belong to the core given elections where the voters' preferences come from restricted domains. The idea to restrict the scope only to instances in which the preferences are somehow well-structured is not new [17], yet to the best of our knowledge it has never been considered in the context of the core.

Our Contribution

Our work contributes to two areas of computational social choice. First, we prove a number of structural theorems that describe existing domain restrictions. In particular, our results give a more intuitive explanation of the class of top-monotonic preferences. The original definition of this class is somewhat cumbersome. We show two independent conditions that provide alternative characterisations of top-monotonic preferences provided the voters' preference rankings have no ties. We also introduce two new domain restrictions which are natural, and which provide sufficient conditions for the existence of core-stable rules. One of our new classes generalizes voter-interval and candidate-interval domains [16], and the other class is a weakening of the domain of top-monotonic preferences; yet our class still includes single-peaked [6] and single-crossing preferences [25,29].

Second, we prove the existence of core-stable rules under the assumption that the voters' preferences come from certain restricted domains, in particular from domains of voter-interval, candidate-interval, single-peaked, and single-crossing preferences. Interestingly, we show a single algorithm that is core-stable for all four aforementioned domains. At the same time, we show that if we restrict our attention to top-monotonic elections no core-stable rule exists.

The idea of our algorithm is the following. We first find a fractional (randomized) committee that is in the core. We pick those candidates that have been selected with probability equal to one. We choose the remaining candidates using a variant of the median rule applied to the truncated instance of the original election. Thus, our results hold both in the discrete and in the probabilistic case.

2 Preliminaries

For each $t \in \mathbb{N}$, we set $[t] = \{1, 2, \ldots, t\}$.

2.1 Elections, Preferences, and Committtees

An *election* is a tuple $E = (N, C, k)$, where $N = [n]$ is a set of n *voters*, C is a set of m *candidates*, and k is the desired *committee size*. Each voter $i \in N$ submits her weak ranking $\succsim_i$ over the candidates—for each $i \in N$ and $a, b \in C$, we say that voter i weakly prefers candidate a over candidate b if $a \succsim_i b$. We set $a \sim_i b$ if $a \succsim_i b$ and $b \succsim_i a$, and we write $a \succ_i b$ if $a \succsim_i b$ and $a \not\sim_i b$. For a voter $i \in N$ and $j \in [m]$, by $\mathrm{pos}_i(j)$ we denote the equivalence class of candidates ranked on the j-th position by voter i. Formally, a candidate c belongs to $\mathrm{pos}_i(j)$ if there are $(j - 1)$ candidates

$a_1, \ldots, a_{j-1}$ such that $a_1 \succ_i a_2 \succ_i \ldots \succ_i a_{j-1} \succ_i c$ and if there exist no j candidates $a_1, \ldots, a_j$ for which $a_1 \succ_i a_2 \succ_i \ldots \succ_i a_j \succ_i c$. By d_i we denote the number of the nonempty positions in the i-th voter's preference list. For each $j \in [d_i]$, by $\mathrm{pos}_i([j])$ we denote $\bigcup_{q \leqslant j} \mathrm{pos}_i(q)$. For each $i \in N$, by top_i and bot_i we denote the sets of candidates ranked respectively at the highest and the lowest position (note that $\mathrm{top}_i = \mathrm{pos}_i(1)$ and $\mathrm{bot}_i = \mathrm{pos}_i(d_i)$).

We distinguish two specific types voters' preferences.

Approval preferences. The preferences are *approval*, if for each candidate $c \in C$ and each voter $i \in N$ either $c \in \mathrm{top}_i$ or $c \in \mathrm{bot}_i$. We say that i *approves* c if $c \in \mathrm{top}_i$.

Strict preferences. The preferences are *strict*, if for all $a, b \in C$ ($a \neq b$) and each $i \in N$ it holds that $a \nsim_i b$.

We call k-element subsets of C size-k *committees*, or in short committees, if the size k is clear from the context. We extend this notion to the continuous model as follows: a *fractional committee* is a function $p \colon C \to [0, 1]$ that assigns to each candidate from $c \in C$ a value $p(c)$ such that $0 \leqslant p(c) \leqslant 1$; intuitively $p(c)$ can be thought of as the probability that candidate c is a member of the selected committee. We extend this notation to sets, defining $p(T) = \sum_{c \in T} p(c)$ for each $T \subseteq C$. The value of $p(C)$ is the *size* of the fractional committee. If for a candidate c it holds that $p(c) = 1$, then we say that c is *elected*, otherwise she is *unelected*. If for an unelected candidate c it holds that $p(c) > 0$, then c is *partially elected*. If there are no partially elected candidates in p, then we say that p is a *discrete committee* (or simply a committee) and associate it with the set $\{c \in C \colon p(c) = 1\}$.

A *voting rule* is an algorithm that takes as input an election, and returns a nonempty set of committees, hereinafter called winning committees.[1] A fractional voting rule is an algorithm that given an election returns a fractional committee.

The notion of a fractional committee is similar to several probabilistic concepts considered in the literature. For instance, in probabilistic social choice (see the book chapter [7]) we also assign fractional values to candidates. The main difference is that in probabilistic social choice, the whole value that we want to divide among the candidates can be assigned to fewer than k candidates; in particular it is feasible to set $p(c) = k$ for one candidate and $p(c') = 0$ for all c', $c' \neq c$. Thus, intuitively, in probabilistic social choice each candidate is divisible and appears in an unlimited quantity. Viewed from this perspective, probabilistic social choice extends the discrete model of approval-based apportionment [8]. Several works have considered axioms of proportionality for probabilistic social choice [1, 18], yet unfortunately their results do not apply to fractional committees.

Another concept related to fractional committees is where we assign probabilities to committees instead of individual candidates. The notions of proportionality in this setting have been considered, e.g., in [12]. It is worth noting that fractional committees can induce probability distributions over committees, e.g., by applying sampling techniques, such as dependent rounding [33], that ensure we always select k candidates. Yet, there is no one-to-one equivalence between the two settings, thus the results of [12] do not apply to fractional committees.

---

[1] Typically a voting rule would return a single winning committee, but ties are possible.

2.2   The Core as a Concept of Proportionality

There are numerous axioms that aim at formalizing the intuitive idea of proportionality. In this paper we focus on one of the strongest such properties, the *core* [2]. The idea behind the definition of the core is the following: a group of voters S should be allowed to decide about a subset of candidates that is proportional to the size of S; for example a group consisting of 70% of voters should have the right to decide about 70% of the elected candidates. The core prohibits situations where a group S can propose a proportionally smaller set of candidates T such that each voter from S would prefer T to the committee at hand.

Definition 1 (The core). *Given an election instance $E = (N, C, k)$, we say that a committee W is in the* core, *if for each $S \subseteq N$ and each subset of candidates T with $|T| \leqslant k \cdot |S|/n$ there is a voter $v \in S$ weakly preferring W to T.*

In the above definition we still need to specify how the voters compare committees, i.e., how their preferences over individual candidates can be extended to the preferences over committees. For each voter $i \in N$ by $\rhd_i$ we denote the partial order over 2^C being the result of extending the preference relation of i. Throughout the whole paper we use the *lexicographic* extension, both in case of discrete and fractional committees, defined formally as follows for discrete ones:

$$W \rhd_i T \iff \exists \sigma \in [d_i].\ |\text{pos}_i(\sigma) \cap W| > |\text{pos}_i(\sigma) \cap T|$$
$$\text{and } \forall \varrho < \sigma.\ |\text{pos}_i(\varrho) \cap W| = |\text{pos}_i(\varrho) \cap T|.$$

and for fractional ones:

$$p \rhd_i p' \iff \exists \sigma \in [d_i].\ p(\text{pos}_i(\sigma)) > p'(\text{pos}_i(\sigma))$$
$$\text{and } \forall \varrho < \sigma.\ p(\text{pos}_i(\varrho)) = p'(\text{pos}_i(\varrho)).$$

Note that for approval preferences it boils down to counting approved candidates in T and T' (since in Definition 1 $|W| \geqslant |T|$, a voter weakly prefers W over T whenever she approves at least as many candidates in W as in T). An alternative preference extension is considered in the full version of our paper [28].

Definition 1 naturally extends to fractional committees.

Definition 2 (The core (for fractional committees)). *Given an election instance $E = (N, C, k)$, we say that a fractional committee p is in the core, if for each $S \subseteq N$ and each fractional committee p' with $p'(C) \leqslant k \cdot |S|/n$, there exists a voter $i \in S$ such that i weakly prefers p over p'.*

We say that a voting rule is *core-stable* if it always returns committees in the core.

3   Restricted Domains

A voting rule specifies an outcome of an election independently of what the voters' preferences look like. Similarly, core-stability puts certain structural requirements on the selected committees that should be satisfied in every possible election. However, the space of all elections is rich and it might be too demanding to expect a voting rule to satisfy a strong property in each possible case. (This is the case for the core: there are elections with strict rankings where no non-fractional committee belongs to the core [12, 19]; the question whether the core is always non-empty assuming approval preferences is still open.) Instead, what is often desired is that a voting rule should satisfy strong notions of proportionality when the voters' preferences are in some sense logically consistent. This motivates focusing primarily on election instances where the voters' preferences are well-structured, or—in other words—come from certain restricted domains.

To the best of our knowledge, none of the known voting rules is core-stable, even for more restricted domains than considered in our work (see the full version of our paper [28] for a few examples).

In this section we describe several known and introduce one new preference domain. We also provide alternative conditions characterising some of the considered domains. These results will help us in our further analysis of voting methods, but are also interesting on their own.

3.1 Strict Preferences

For strict ordinal preferences we start by recalling the definitions of the following two classes.

Definition 3 (Single-crossing preferences). *Given an election instance* $E = (N, C, k)$, *we say that* E *has single-crossing preferences if there exists a linear order* $\sqsupseteq$ *over voters such that for each voters* $x \sqsupseteq y \sqsupseteq z$ *and candidates* $a, b \in C$ *such that* $a \succ_y b$ *we have that* $b \succ_x a \implies a \succ_z b$.

Intuitively, we say that preferences are single-crossing if the voters can be ordered in such a way that for each pair of candidates, $a, b \in C$, the relative order between a and b changes at most once while we move along the voters.

Definition 4 (Single-peaked preferences). *Given an election instance* $E = (N, C, k)$, *we say that* E *has single-peaked preferences if there exists a linear order* $\sqsupseteq$ *over candidates such that for each voter* $i \in N$ *and candidates* $a \sqsupseteq b \sqsupseteq c$ *we have that* $\text{top}_i = a \implies b \succ_i c$ *and* $\text{top}_i = c \implies b \succ_i a$.

We will now recall the definition of the *top monotonic* domain [5]. This domain is defined assuming the voters submit their preferences as weak orders. For strict preferences it generalizes both the single-peaked and single-crossing domain. We call all candidates that are ranked in the top position by at least one voter top-candidates.

Definition 5 (Top monotonicity (TM)). *Given an election* $E = (N, C, k)$, *we say that* E *has* top monotonic *preferences if there exists a linear order* $\sqsupseteq$ *over the candidates such that the two following conditions hold:*

1. *for all candidates a, b, c and voters i, j such that $a \in \text{top}_i$ and $b \in \text{top}_j$ it holds that:*

$$\begin{matrix} a \sqsupset b \sqsupset c \ or \\ c \sqsupset b \sqsupset a \end{matrix} \implies \begin{cases} b \succsim_i c & \text{if } c \in \text{top}_i \cup \text{top}_j \\ b \succ_i c & \text{otherwise} \end{cases}$$

2. *the same implication holds also for all top-candidates a, b, c and voters i, j such that $a \succsim_i b, c$ and $b \succsim_j a, c$.*

The definition of top-monotonic preferences is complex and somewhat counterintuitive. We will first show that for strict orders this definition can be equivalently characterized by two much simpler and more intuitive conditions.

Definition 6 (Single-top-peaked preferences). *Given an election $E = (N, C, k)$, we say that E has* single-top-peaked *(STP) preferences if there exists a linear order $\sqsupset$ over the candidates such that for all candidates $a \sqsupset b \sqsupset c$ such that b is a top-candidate, and each voter i it holds that $\text{top}_i = a \implies b \succ_i c$ and $\text{top}_i = c \implies b \succ_i a$.*

Proposition 1. *For strict rankings, single-top-peakedness is equivalent to top-monotonicity.*

Proof. Observe that the first condition in the definition of TM implies STP. Now, we will show the reverse implication. Consider an STP election. We will show that it satisfies the two conditions specified in Definition 5.

Note that in the strict model if the premise of the first condition is satisfied, then $\text{top}_i = \{a\}$ and $\text{top}_j = \{b\}$ and $c \notin \text{top}_i \cup \text{top}_j$. Hence, the first condition follows from the condition for STP.

Consider now the second condition. If $a \succsim_i b, c$ and $b \succsim_j a, c$, then in the strict model it holds that $a \succ_i b, c$ and $b \succ_j a, c$. Let us consider two cases: first assume that $a \sqsupset b \sqsupset c$. We know that $\text{top}_i = \{d\}$ for some $d \in C \setminus \{b, c\}$. If $d \sqsupset b$, then $b \succ_i c$ follows from the definition of STC (for voter i and candidates d, b, c). Suppose now that $b \sqsupset d$. But then from the definition of STP (for voter i and candidates a, b, d) we obtain that $b \succ_i a$, a contradiction. The reasoning for the case when $c \sqsupset b \sqsupset a$ is analogous.

It is clear that the definition of STP is closely connected to the definition of single-peaked preferences (only the condition is partially weakened to the candidates that are ranked top by some voter). One could also consider the analogous weakening for single-crossing preferences.

Definition 7 (Single-top-crossing preferences). *Given an election $E = (N, C, k)$, we say that E has* single-top-crossing *(STC) preferences if there exists a linear order $\sqsupset$ over voters such that for all voters $x \sqsupset y \sqsupset z$ and a candidate $a \in C$, we have that $a \succ_x \text{top}_y \implies \text{top}_y \succ_z a$.*

Although the definitions of STC and STP look different, they are in fact equivalent.

Proposition 2. *For strict rankings, single-top-peakedness is equivalent to single-top-crossingness.*

Proof. Consider an STC election E and a linear order $\sqsupset$ over voters given by the definition of STC. We say that i precedes j if $j \sqsupset i$. We construct the linear order over candidates as follows: consider some $a, b \in C$ such that a is the top preference for some voter $i \in N$. From the definition of STC, we know that voters preferring b to a can all either succeed or precede i. If they succeed i, then we add constraint $b \sqsupset a$, otherwise we add constraint $a \sqsupset b$. If there are no voters preferring b to a, we add no constraint. We repeat this step for each pairs $a, b \in C$. Finally, if after the previous step some pairs are still uncomparable, we complete the order in any transitive way.

We will show that the constraints placed during the first step of the procedure are transitive. Indeed, consider (for the sake of contradiction) three candidates a, b, c such that the procedure placed constraints $a \sqsupset b$, $b \sqsupset c$ and $c \sqsupset a$. Hence, we know that at least two out of these three candidates are top-candidates. Assume without the loss of generality that a and b are top-candidates. Let i_a, i_b be voters ranking top respectively a and b (naturally, $i_a \sqsupset i_b$). We know that all the voters preceding i_a prefer a to c and all the voters preceding i_b prefer c to b. There exists at least one voter i preferring c to b (as otherwise constraint $b \sqsupset c$ would not be added) and $i_b \sqsupset i$. By transitivity of the preference relation, we know that i prefers a over b. Consequently, i_a, i_b and i together with candidate a witness STC violation. The obtained contradiction shows that the order $\sqsupset$ is indeed transitive.

We will now prove that such linear order $\sqsupset$ over candidates satisfies the conditions of STP. Indeed, consider any three candidates $a \sqsupset b \sqsupset c$ such that b is a top-candidate and a voter $i \in N$. Let $\text{top}_i = \{a\}$. As b is a top-candidate, there exists a voter j such that $\text{top}_j = \{b\}$. As $a \sqsupset b$, it holds that $i \sqsupset j$. Then if we had that $c \succ_i b$, our procedure would place constraint $c \sqsupset b$, a contradiction. Hence $b \succ_i c$. The proof for the case $\text{top}_i = \{c\}$ is analogous.

Now we will prove the reverse implication. Let E be an STP election with a linear order $\sqsupset$ over the candidates. Consider the following linear order $\sqsupset$ over the voters: for each $i, j \in N$ we have that if $\text{top}_i \sqsupset \text{top}_j$ then $i \sqsupset j$. Now consider three voters x, y, z and a candidate a such that $a \succ_x \text{top}_y$. Suppose that $a \sqsupset \text{top}_y$. Then from the properties of top monotonicity and the fact that $\text{top}_y \sqsupset \text{top}_z$, we have that z has preference ranking $\text{top}_z \succ_z \text{top}_y \succ_z a$. Suppose now that $\text{top}_y \sqsupset a$. But since $\text{top}_x \sqsupset \text{top}_y \sqsupset a$, the fact that $a \succ_x \text{top}_y$ leads to the contradiction with the definition of top monotonicity, which completes the proof. □

Recall that single-crossingness implies single-peakedness for narcissist domains, i.e., under the assumption that each candidate is ranked top at least once [14]. Since for narcissist domains a single-peaked profile is also single-top-peaked, we get a corollary: single-peakedness is equivalent to single-top-crossingness assuming narcissist preferences.

The class of top monotonic preferences (TM) puts a focus on the top positions in the voters' preference rankings. For example, an election in which the voters unanimously rank a single candidate as their most preferred choice is top-monotonic, independently of how the other candidates are ranked. This suggests that TM offers a combinatorial structure that might be useful in the analysis of single-winner elections, but which might not help to reason about committees. Indeed, below we define a new class which is a natural strengthening of TM. In Sect. 4 we show that the core is always nonempty for elections belonging to our newly defined class, and we show that this is not the case for the original class of TM.

Definition 8 (Recursive single-top-crossing (r-STC) preferences). *Given an election $E = (N, C, k)$, we say that E has recursive single-top-crossing preferences if every subinstance of E obtained by removing some candidates from E is STC.*

Although r-STC is stricter than STC, it still contains both single-peaked and single-crossing preferences. This follows from the fact that both single peaked and single-crossing preferences are top monotonic and that single-peakedness and single-crossingness is preserved under the operation of removing candidates from the election.

3.2 Approval Elections

In the approval model we first recall the definitions of two classic domain restrictions, the voter-interval and the candidate-interval models [16].

Definition 9 (Voter-interval (VI) preferences). *Given an approval election instance $E = (N, C, k)$, we say that E has voter-interval preferences if there exists a linear order $\sqsupset$ over N such that for all voters $v_1, v_2, v_3 \in N$ and for each candidate $c \in \text{top}_{v_1} \cap \text{top}_{v_3}$, we have that $v_1 \sqsupset v_2 \sqsupset v_3 \implies c \in \text{top}_{v_2}$. Intuitively, each candidate is approved by a consistent interval of voters.*

Definition 10 (Candidate-interval (CI) preferences). *Given an approval election instance $E = (N, C, k)$, we say that E has candidate-interval preferences if there exists a linear order $\sqsupset$ over C such that for each voter $i \in N$ and all candidates $a, c \in \text{top}_i, b \in C$ we have that $a \sqsupset b \sqsupset c \implies b \in \text{top}_i$. Intuitively, each voter approves a consistent interval of candidates.*

Below we introduce a new class that generalizes both CI and VI domains. In Sect. 4.2 we will prove that the core is always nonempty if preferences come from this new domain.

Definition 11 (Linearly consistent (LC) preferences). *Given an approval election instance $E = (N, C, k)$, we say that E has linearly consistent preferences, if there exists a linear order $\sqsupset$ over $N \cup C$ such that for each voters $i, j \in N$ ($i \sqsupset j$) and candidates $a, b \in C$ ($a \sqsupset b$), if $a \in \text{top}_j$, then $a \succsim_i b$. In words, if i approves b and j approves a, then i approves a.*

Proposition 3. *Each VI or CI election is LC.*

Proof. The case of voter-interval preferences. Let $\sqsupset$ be a linear order over N that witnesses that preferences are voter-interval. Let us sort N by this order. For each candidate c, by first_c we denote $\max\{i \in N : c \in \text{top}_i\}$. Let us now associate each candidate c to first_c (breaking the tie between c and first_c arbitrarily). If two candidates a, b are associated to the same point, we also break the tie between them arbitrarily. In such a way we obtained an order $\sqsupset$ over $N \cup C$. For simplicity, for each $x, y \in N \cup C$ by $x \sqsupseteq y$ we denote "$x \sqsupset y$ or $x = y$".

Consider two voters, i and j, with $i \sqsupset j$, and two candidates, a and b, with $a \sqsupset b$. Assume i approves b and j approves a. We will prove that i approves a. Since $a \sqsupset b$, by our definition $\text{first}_a \sqsupseteq \text{first}_b$. Since i approves b, $\text{first}_b \sqsupseteq i$, and so $\text{first}_a \sqsupseteq i$. If $i = \text{first}_a$, then i approves a. Otherwise, $\text{first}_a \sqsupset i$. Consequently, first_a, i, and j are three voters, such that $\text{first}_a \sqsupset i \sqsupset j$. Since the preferences are voter-interval we infer that i approves a.

The case of candidate-interval preferences. Let $\sqsupset$ be a linear order on C witnessing the candidate-interval property. Let us sort C by this order. We associate each voter $i \in N$ with $(\max \text{top}_i)$, again breaking all the ties arbitrarily. Consider two voters, i and j with $i \sqsupset j$, and two candidates a and b, with $a \sqsupset b$. Further, assume that i approves b and j approves a. Since $i \sqsupset j$, we get that $(\max \text{top}_i) \sqsupseteq (\max \text{top}_j)$, and since j approves a, we have $(\max \text{top}_j) \sqsupseteq a$. Consequently, $(\max \text{top}_i) \sqsupseteq a$. If $(\max \text{top}_i) = a$, then i approves a. Otherwise, $(\max \text{top}_i)$, a and b are three candidates, such that $(\max \text{top}_i) \sqsupset a \sqsupset b$. Given that preferences are candidate-interval, and that i approves b, we get that i approves a.

In the full version of our paper [28] we compare the domain of linearly consistent preferences with the one of *seemingly single-crossing (SSC)* preferences [17]—yet another known class that generalizes VI and CI domains.

4   Finding Core-Stable Committees for Restricted Domains

In this section, we describe an algorithm for finding committees that takes as input preferences represented as weak orders. We will show that if the preferences are approval linearly consistent (LC), or strict recursive single-top-crossing (r-STC), then the returned committee belongs to the core. Our algorithm works in polynomial time, assuming that we are given the linear order $\sqsupset$ over $N \cup C$, provided by the definitions of these preference classes. For approval preferences, such an order can be found in polynomial time for candidate-interval and voter-interval domains [17]. It is not known whether it is the case for general LC preferences—hence, for this class we show only the non-emptiness of the core. However, LC is mainly a technical domain, allowing us to present a coherent algorithm for both voter-interval and candidate-interval preferences. In case of r-STC preferences, the linear order witnessing this class is the same as the one witnessing top monotonicity which can be found in polynomial time [24].

Hereinafter we assume that the fraction n/k is integral. It is without loss of generality due to the following remark:

Remark 1. Consider an election E and the instance E' obtained from E by multiplying each voter k times. If a committee is not in the core for E, it is not in the core for E'.

The algorithm, which we call CORECOMMITTEE, consists of two phases: first we construct a fractional committee and then we discretize it. The first phase (further called the BESTREPRESENTATIVE algorithm) is the following: imagine that each voter has an equal probability portion k/n to distribute, and that we want to choose one candidate (her *representative*) who gets this portion. Initially, the fractional committee p is empty. We iterate over the set of voters, sorted according to the relation $\sqsupset$. Let P_i denote the set of unelected candidates at the moment of considering voter $i \in N$. The representative of i is defined as a candidate $r_i \in P_i$ such that for each $c \in P_i$ it holds that either $r_i \succ_i c$ or that $r_i \sim_i c$ and $r_i \sqsupset c$. Next, $p(r_i)$ is increased by k/n. Note that, as n/k is integral, the election probability of each candidate does not exceed 1.

In Sect. 4.1 we prove that after this phase the obtained fractional committee p is in the core for all strict elections and all LC approval elections. Denote by W_1 the set of candidates c such that $p(c) = 1$. Before the second phase of the algorithm, remove candidates from W_1 from the election together with the voters who are represented by them, obtaining a smaller election $E_2 = (N_2, C_2, k_2)$. By k_2 we denote $k - |W_1|$ (remaining seats in the committee) and by n_2 we denote $n - |W_1| \cdot n/k$ (remaining voters). Renumerate the voters so that they are numbers from $[n_2]$ (and in case of r-STC elections, resort them so that E_2 is still r-STC). Note that by definition $n_2/k_2 = (n - |W_1| \cdot n/k)/(k - |W_1|) = n \cdot (1 - |W_1|/k)/(k - |W_1|) = n/k$.

The second phase (further called the MEDIANRULE algorithm) is simple: for each $q \in [k_2]$ denote by m_q the $(q-1) \cdot n/k + 1$st voter. Further we will refer to these voters as *median voters*. Then elect committee $W_2 = \{r_{m_q} : q \in [k_2]\}$.

Finally, we return the committee $W = W_1 \cup W_2$. In Sect. 4.2 we show that the final committee W belongs to the core for LC and r-STC preferences.

4.1 Core Stability for Fractional Committees

We will now show that the committee elected by BESTREPRESENTATIVE is always in the core for LC approval elections and for all elections with strict preferences. The proof is the same for those two models; we only use the following property:

Definition 12. *Given an election $E = (N, C, k)$, we say that E is well-ordered, if there exists a linear order $\sqsupset$ over $N \cup C$ such that for all voters $i, j \in N$ ($i \sqsupset j$) and candidates $a, b \in C$ ($a \sqsupset b$), if $a \sim_j b$ and $a, b \notin \mathrm{bot}_j$, then $a \succsim_i b$.*

It is clear that every strict election is well-ordered for every order $\sqsupset$ (the premise is never satisfied). For approval elections this definition is a weakening of Definition 11 (because for approval elections $a \in \mathrm{top}_j \implies a \notin \mathrm{bot}_j$), hence every LC election is well-ordered.

For convenience, for $i \in N$ by p_i we denote the fractional committee p after considering voter i. Let $\sigma_i \in [m]$ be the number such that $r_i \in \mathrm{pos}_i(\sigma_i)$. From how the algorithm BESTREPRESENTATIVE works, we have that for every voter $i \in N$ and a candidate $c \in \mathrm{pos}_i([\sigma_i - 1])$ it holds that $p_{i-1}(c) = 1$ (and also $p_j(c) = 1$ for every $j \geqslant i$).

Theorem 1. *Each fractional committee elected by BESTREPRESENTATIVE belongs to the core for well-ordered elections.*

Proof. Let us start from the following remark, obtained naturally from the very construction of the algorithm.

Remark 2. For each $i \in N$ and $c \in C$, there exists $q \in [n/k]$ such that $p_i(c) = q \cdot k/n$. In particular, q is the number of voters for whom c is a representative.

We will prove the following invariant: for each $i \in N$, p_i satisfies the condition of the fractional core (see Definition 2) with the additional restriction that $S \subseteq [i]$. We will prove the invariant by induction.

For the first voter the invariant is clearly true. Assume, there exists $i \in N$ satisfying the invariant. We will prove that the invariant holds also for voter $(i + 1)$.

For the sake of contradiction suppose that there exists a group $S \subseteq [i+1]$ and a fractional committee p'_{i+1} such that for each $v \in S$ we have that v prefers lexicographically p'_{i+1} to p_{i+1}.

First, note that if $(i + 1) \notin S$, then the invariant does not hold also for i, a contradiction. This is the case because the election probability of no candidate is decreased during a loop iteration. Hence, $(i + 1) \in S$.

By the definition of BESTREPRESENTATIVE we have that for each $\varrho < \sigma_{i+1}$ and $c \in \mathrm{pos}_{i+1}(\varrho)$ it holds that $p_{i+1}(c) = 1$. From that, in particular we have the following equation:

$$\forall \varrho < \sigma_{i+1}. \; p'_{i+1}(\mathrm{pos}_{i+1}(\varrho)) \leqslant |\mathrm{pos}_{i+1}(\varrho)| = p_{i+1}(\mathrm{pos}_{i+1}(\varrho))$$

Hence, as $(i + 1)$ prefers lexicographically p' to p:

$$\forall \varrho < \sigma_{i+1}. \; p'_{i+1}(\mathrm{pos}_{i+1}(\varrho)) = p_{i+1}(\mathrm{pos}_{i+1}(\varrho)) \tag{1}$$

It also needs to hold that:

$$p'_{i+1}(\mathrm{pos}_{i+1}([\sigma_{i+1}])) > p_{i+1}(\mathrm{pos}_{i+1}([\sigma_{i+1}])) \tag{2}$$

We can conclude that $\sigma_{i+1} < d_{i+1}$, as otherwise voter $(i + 1)$ could not prefer p'_{i+1} over p_{i+1}. Consequently:

$$r_{i+1} \notin \mathrm{bot}_{i+1} \tag{3}$$

Suppose that $p'_{i+1}(r_{i+1}) = 0$. From (2) and the fact that for all $c \in \mathrm{pos}_{i+1}(\sigma_{i+1})$ with $c \sqsupset r_{i+1}$ we have $p(c) = 1$, we infer that there exists $a \in \mathrm{pos}_{i+1}(\sigma_{i+1})$ such that $r_{i+1} \sqsupset a$ and $p'_{i+1}(a) > 0$. From Remark 2 we have that $p'_{i+1}(a) \geqslant k/n$. Now we modify p'_{i+1} by moving the fraction of k/n from a to r_{i+1}. By Definition 12 and (3) we have that for every $v \in S$ (naturally, $v \sqsupset (i + 1)$) it holds that $r_{i+1} \succsim_v a$. Thus, after the change p'_{i+1} still witnesses core violation for S.

Now consider a fractional committee p'_i obtained from p'_{i+1} by decreasing the probability portion of r_{i+1} by k/n. We will show that p'_i together with $S \setminus \{(i + 1)\}$ witness the core violation for p_i. Indeed, the election probability of no candidate except r_{i+1} changed, and the election probability of r_{i+1} changed in the same way: in p_{i+1} and p'_{i+1} it is higher by k/n than in p_i and p'_i, respectively. Hence, if for a voter $v \in S$ it holds that $p'_{i+1} \rhd_v p_{i+1}$, then also $p'_i \rhd_v p_i$. Besides, we have that $p'_i(C) \leqslant k \cdot |S-1|/n$, so we obtain a contradiction with our inductive assumption.

4.2 Core Stability for Discrete Committees

In this section we present our main result: that the committee W elected by CORECOMMITTEE is in the core for LC and r-STC preferences. The algorithm for these two restricted domains is the same, but the proof techniques used for these models differ significantly.

The proof for LC elections heavily relies on the following two technical lemmas:

Lemma 1. *Given an LC election* CORECOMMITTEE *elects exactly k candidates.*

Proof. We will show that MEDIANRULE elects exactly k_2 candidates. Suppose for the sake of contradiction that there are two median voters i, j in E_2 such that $r_i = r_j$. Without loss of generality assume $i \sqsupset j$. Consider now any voter v between these median voters. If $r_v \sqsupset r_i$ then from the definition of LC, i approves r_v, and so r_v should be selected as i's representative, a contradiction. If $r_j = r_i \sqsupset r_v$, then from the definition of LC, v approves r_j, and so r_j should be selected as v's representative, a contradiction. Hence, $r_v = r_i$. But then we have that after running BESTREPRESENTATIVE, r_i was a representative for at least n/k voters and was not elected, a contradiction.

Lemma 2. *Let W be the committee elected by* CORECOMMITTEE. *For each $i \in N$, $|W \cap \mathrm{top}_i| + 1 > p(\mathrm{top}_i)$.*

Proof. Let us start with the following remark.

Remark 3. Consider an LC election E and two voters i, j who were not removed from the election after the first phase, such that $i \sqsupset j$. Then either $r_i = r_j$ or $r_i \sqsupset r_j$.

Indeed, towards a contradiction assume that $r_j \sqsupset r_i$. From LC we have that i approves r_j and r_j should be i's representative.

Consider a voter $i \in N$. Define part_i as $p(\mathrm{top}_i) - |W_1 \cap \mathrm{top}_i|$. As W_1 contains all candidates c such that $p(c) = 1$, then part_i is intuitively the joint sum of election probabilities of partially elected candidates in top_i. From Remark 2 we have that:

$$\mathrm{part}_i = q \cdot k/n \tag{4}$$

where q is the number of voters for whom a candidate from $\mathrm{top}_i \setminus W_1$ is a representative. Naturally, such voters could not be removed from the election after the execution of BESTREPRESENTATIVE.

We will prove that $\mathrm{part}_i < |W_2 \cap \mathrm{top}_i| + 1$. From the fact that $W = W_1 \cup W_2$ and $W_1 \cap W_2 = \emptyset$, it will imply the desired statement. We will now focus on upper-bounding q from (4).

Consider three voters v_1, v_2, v_3 such that $v_1 \sqsupset v_2 \sqsupset v_3$ and $r_{v_1}, r_{v_3} \in \mathrm{top}_i$. We will prove that then also $r_{v_2} \in \mathrm{top}_i$. Indeed, from Remark 3 we have that either $r_{v_2} \in \{r_{v_1}, r_{v_3}\}$ (and the statement is true) or $r_{v_1} \sqsupset r_{v_2} \sqsupset r_{v_3}$. First, consider the case, when $v_2 \sqsupset i$. Since i approves r_{v_1} by LC applied to voters v_2, i and candidates r_{v_1} and r_{v_2}, we get that also v_2 approves r_{v_1}, a contradiction with Remark 3. Second, we look at the case when $i \sqsupset v_2$. From LC applied to v_2, i and candidates r_{v_2} and r_{v_3} and by the fact that i approves r_{v_3} we get that i also approves r_{v_2}, which is what we wanted to prove.

Hence, these q voters from (4) need to form a consistent interval among all non-removed voters. Besides, we know that there is no more than $|W_2 \cap \mathrm{top}_i|$ median voters inside this interval and that between each two median voters there is $n/k - 1$ non-removed voters. Hence:

$$q \leqslant (|W_2 \cap \mathrm{top}_i| + 1) \cdot (n/k - 1) + |W_2 \cap \mathrm{top}_i| = (|W_2 \cap \mathrm{top}_i| + 1) \cdot n/k - 1$$

and:

$$\mathrm{part}_i = q \cdot k/n < (|W_2 \cap \mathrm{top}_i| + 1) \cdot n/k \cdot k/n = |W_2 \cap \mathrm{top}_i| + 1$$

which completes the proof.

Theorem 2. *For LC elections,* CORECOMMITTEE *elects committees from the core.*

Proof. We know that fractional committee p elected by BESTREPRESENTATIVE belongs to the core. Suppose now that W is not in the core. Hence, there exists a nonempty set $S \subseteq N$ and a committee T of size $|S| \cdot k/n$ such that $|W \cap \text{top}_i| < |T \cap \text{top}_i|$ for each $i \in S$—alternatively, $|W \cap \text{top}_i| + 1 \leqslant |T \cap \text{top}_i|$.

From Lemma 2 we know that for each $i \in S$ we have $p(\text{top}_i) < |W \cap \text{top}_i| + 1 \leqslant |T \cap \text{top}_i|$. Let us define a fractional committee p' such that $p'(c) = 1$ for $c \in T$ and $p'(c) = 0$ otherwise. Hence, S and p' witness the violation of the core for p, which is contradictory with Theorem 1.

For r-STC elections, we again start with proving two technical lemmas:

Lemma 3. CORECOMMITTEE *for strict r-STC election E elects exactly k candidates.*

Proof. We need to show that MEDIANRULE elects exactly k_2 candidates. Suppose for the sake of contradiction that there are two median voters i, j in E_2 such that $r_i = r_j$. From STC it follows that $r_v = r_i$. But this means that after running BESTREPRESEN-TATIVE, r_i was a representative for at least n/k voters and was not elected, a contradiction.

Lemma 4. *Consider an STC election $E = (N, C, k)$ and apply MEDIANRULE to E to obtain the committee W. If $|W| = k$, then W is in the core.*

Proof. Towards a contradiction suppose that the statement of the lemma is not true. Without loss of generality, assume that E is an election with the smallest k among those for which the statement of the lemma does not hold. Let S and T be subsets of voters and candidates, respectively, that witness that the committee returned by the median rule does not belong to the core.

Observe that there are at least two candidates from W that do not belong to T. Indeed, if there were only one such candidate, we would have that $|T| = |W|$ (as $T \backslash W$ is nonempty) and $|S| = n$. In particular, in such a case all median voters would belong to S. Consequently, the most preferred candidates of the median voters would belong to T, hence $W \subseteq T$, a contradiction.

Let us fix a candidate $a \in W \backslash T$ that is elected by the greatest median voter $(i \cdot n/k + 1)$. In particular, $i \neq 0$. For a candidate $b \in T$ by $S_b \subseteq S$ we denote the subset of voters in S preferring b to a. Since E is single-top-crossing, it holds that either $S_b \subseteq [i \cdot n/k]$ or $S_b \subseteq N \backslash [i \cdot n/k]$.

Now we split E into two smaller elections $E_{\text{low}} = ([i \cdot n/k], C, i)$ and $E_{\text{grt}} = (N \backslash [i \cdot n/k], C, k - i)$. By W_{low} and W_{grt} we denote the committees elected by the median rule for E_{low} and E_{grt}, respectively. Observe that $W_{\text{low}} \sqcup W_{\text{grt}} = W$.

Let us also split S and T into two parts, as follows:

$$S_{\text{low}} = S \cap [i \cdot n/k], \qquad S_{\text{grt}} = S \cap (N \backslash [i \cdot n/k]),$$
$$T_{\text{low}} = \{c \in T : S_c \subseteq [i \cdot n/k]\}, \qquad T_{\text{grt}} = \{c \in T : S_c \subseteq N \backslash [i \cdot n/k]\}.$$

Note that $S_{\text{low}} \cup S_{\text{grt}} = S$ and $T_{\text{low}} \cup T_{\text{grt}} = T$. Hence, if we had that both $|T_{\text{low}}| > |S_{\text{low}}| \cdot n/k$ and $|T_{\text{grt}}| > |S_{\text{grt}}| \cdot n/k = (|S| - |S_{\text{low}}|) \cdot n/k$, then we would have also $|T| >$

$|S| \cdot n/k$, a contradiction. Hence, for at least one of the pairs $(S_{\text{low}}, T_{\text{low}}), (S_{\text{grt}}, T_{\text{grt}})$ the opposite inequality holds. Without the loss of generality, assume that $|T_{\text{low}}| \leqslant |S_{\text{low}}| \cdot n/k$.

We claim that the pair $(S_{\text{low}}, T_{\text{low}})$ witnesses the core violation for E_{low} and committee W_{low}.

Consider a voter $j \in S_{\text{low}}$. We know that there exists a candidate $c \in T \backslash W$ such that $c \succ_j W \backslash T$. First observe that W_{low} and T_{grt} are disjoint—indeed, for every candidate $b \in T_{\text{grt}}$ we have that $S_b \subseteq N \backslash [i \cdot n/k]$. As a result, there is no median voter in $[i \cdot n/k]$ who prefers b to a, hence $b \notin W_{\text{low}}$. From this fact we conclude that $W_{\text{low}} \backslash T_{\text{low}} = W_{\text{low}} \backslash T \subseteq W \backslash T$. Consequently, $c \succ_j W_{\text{low}} \backslash T_{\text{low}}$.

Further, observe that $c \in T_{\text{low}}$. Indeed, voter j prefers c to $W \backslash T$, thus in particular j prefers c to a. Consequently, $j \in S_c$, and thus $S_c \subseteq S_{\text{low}}$, from which we get that $c \in T_{\text{low}}$. Since $c \in T_{\text{low}}$ and $c \succ_j W_{\text{low}} \backslash T_{\text{low}}$, we get that j prefers lexicographically T_{low} to W_{low}.

Finally, we obtain that if the core was violated for E, it also needs to be violated for E_{low}, which is contradictory to our assumption that E minimizes the value of k.

Theorem 3. *For r-STC elections,* CoreCommittee *elects committees from the core.*

Proof. From Lemma 4, we conclude that in CoreCommittee algorithm, W_2 is in the core for E_2.

For the sake of contradiction suppose that the statement of the theorem is not true. Then there exist a set $S \subseteq N$ and a set $T \subseteq C$ witnessing the violation of the condition of the core. For every candidate $c \in C$, by $R(c)$ we denote set $\{i \in N : r_i = c\}$. Note that for a candidate $c \in W_1$ and a voter $i \in S$ such that $i \in R(c)$, we have $c \in T$. Hence, $S \cap \bigcup_{c \in T \cap W_1} R(c) = S \cap \bigcup_{c \in W_1} R(c)$. Consider now sets $S \cap N_2$ and $T \cap C_2$. It holds that:

$$|T \cap C_2| = |T| - |T \cap W_1| \leqslant |S| \cdot k/n - \left| \bigcup_{c \in T \cap W_1} R(c) \right| \cdot k/n$$

$$\leqslant |S| \cdot k/n - \left| S \cap \bigcup_{c \in T \cap W_1} R(c) \right| \cdot k/n$$

$$\leqslant \left| S \backslash \bigcup_{c \in W_1} R(c) \right| \cdot k/n = |S \cap N_2| \cdot k/n = |S \cap N_2| \cdot k_2/n_2.$$

Further, for each voter $i \in S \cap N_2$ we have that $(T \rhd_i W) \wedge (W_1 \subseteq T) \implies (T \cap C_2) \rhd_i W_2$. Consequently, $S \cap N_2$ and $T \cap C_2$ witness the violation of the core condition for committee W_2, which is contradictory to Lemma 3 and Lemma 4.

Corollary 1. *The core is always nonempty and can be found in polynomial time for the following classes of voters' preferences: (1) voter-interval, (2) candidate-interval, and (3) recursive single-top-crossing. For linearly consistent voters' preferences, the core is always nonempty.*

In Theorem 4 below we show that the condition of recursiveness in the definition of the class of r-STC preferences is necessary for the nonemptyness of the core. Thus, in a way Theorem 3 gives a rather precise condition on the existence of the core-stable committees for strict voters' preferences. For approval preferences one cannot easily argue that the conditions are precise, since it is still a major open question whether a core-stable committee exists in each approval election.

Theorem 4. *There exists a top-monotonic election with strict preferences, where the core is empty.*

Proof. Let A be a Condorcet cycle of $r = 100$ candidates:

$$a_1 \succ a_2 \succ \ldots \succ a_r$$
$$a_2 \succ a_3 \succ \ldots \succ a_r \succ a_1$$
$$\ldots$$
$$a_r \succ a_1 \succ a_2 \succ \ldots \succ a_{r-1}$$

Now let B, C, D, E and F be five clones of A. Thus in $A \cup B \cup \ldots \cup F$ we have $6r = 600$ candidates. We add two more candidates, namely g and h.

Consider the following profile with 600 voters:

$$g \succ A \succ B \succ C \succ D \succ E \succ F \succ h$$
$$g \succ B \succ C \succ A \succ E \succ F \succ D \succ h$$
$$g \succ C \succ A \succ B \succ F \succ D \succ E \succ h$$
$$h \succ D \succ E \succ F \succ A \succ B \succ C \succ g$$
$$h \succ E \succ F \succ D \succ B \succ C \succ A \succ g$$
$$h \succ F \succ D \succ E \succ C \succ A \succ B \succ g$$

For example, the first two votes in this profile are:

$$g \succ a_1 \succ \ldots \succ a_r \succ b_1 \succ \ldots \succ b_r \succ \ldots \succ f_1 \succ \ldots \succ f_r \succ h$$
$$g \succ a_2 \succ \ldots \succ a_r \succ a_1 \succ b_2 \succ \ldots \succ b_r \succ b_1 \succ \ldots \succ f_2 \succ \ldots \succ f_r \succ f_1 \succ h$$

The above profile is single-top-crossing since there are only two top-candidates, g and h, and each of them crosses with each other candidate only once.

Let $k = 7$, and consider a committee W. We will show that W does not belong to the core. Without loss of generality, we can assume that $g, h \in W$, as there exists more than $600/7$ voters who rank each of these candidates as their favourite one. Further, since the profile is symmetric, without loss of generality we can also assume that it contains at most two candidates from $A \cup B \cup C$. If the two candidates belong to the same clone, say A, then we take a candidate $c \in C$, and observe that 200 voters (the second and the third group) prefer $\{c, g\}$ over W. Otherwise, if the two candidates are from two different clones, say A and B (the situation is symmetric), then we take the clone which is preferred by the majority (in this context A) and select the candidate $a \in A$ that is preferred by $r - 1$ voters to the member of $W \cap A$. There are $2r - 2 = 198$ voters who prefer $\{g, a\}$ to W. Thus, W does not belong to the core.

5   Conclusions and Open Questions

In this work we have determined the existence of core-stable committees for a number of restricted domains both in the approval and in the ordinal models of voters' preferences. We have additionally presented a number of results that give better insights into the structures of the known domains. In particular, our results give a better understanding of the class of top-monotonic preferences. Let us conclude with two open questions that we find particularly important.

In the full version of our paper [28] we show that classic committee election rules that are commonly considered proportional are not core-stable even if the voters' preferences come from certain restricted domains. Since these domains are natural and can be intuitively explained, one would expect a good rule to behave well for such well-structured elections. This leads us to the following important open question.

Question 1. Is there a natural voting rule that satisfies the strongest axioms of proportionality, and which at the same time satisfies the core for restricted domains?

The requirement that a rule should be "natural" says in particular that its definition cannot conditionally depend on whether the election at hand comes from a restricted domain. Question 1 is valid for both approval and ordinal preferences.

Additionally, it would be interesting to check how often the classic rules violate the core, especially in the case of restricted domains. One can make such a quantitative comparison via experiments. This however raises the algorithmic questions of how hard it is to verify if a given committee (in our case the committee returned by the particular rule) belongs to the core.

Question 2. What is the computational complexity of deciding whether a given committee belongs to the core?

This question is interesting in the general case, and as well as for each preference domain studied in this work.

References

1. Aziz, H., Bogomolnaia, A., Moulin, H.: Fair mixing: the case of dichotomous preferences. In: EC-2019, pp. 753–781 (2019)
2. Aziz, H., Brill, M., Conitzer, V., Elkind, E., Freeman, R., Walsh, T.: Justified representation in approval-based committee voting. Soc. Choice Welfare **48**(2), 461–485 (2017). https://doi.org/10.1007/s00355-016-1019-3
3. Aziz, H., Elkind, E., Faliszewski, P., Lackner, M., Skowron, P.: The Condorcet principle for multiwinner elections: from shortlisting to proportionality. In: IJCAI-2017, pp. 84–90, August 2017
4. Aziz, H., Lee, B.: The expanding approvals rule: improving proportional representation and monotonicity. Soc. Choice Welfare **54**(1), 1–45 (2020). https://doi.org/10.1007/s00355-019-01208-3
5. Barberà, S., Moreno, B.: Top monotonicity: a common root for single peakedness, single crossing and the median voter result. Games Econom. Behav. **73**(2), 345–359 (2011)
6. Black, D.: On the rationale of group decision-making. J. Polit. Econ. **56**(1), 23–34 (1948)

7. Brandt, F.: Rolling the dice: recent results in probabilistic social choice. In: Endriss, U. (ed.) Trends in Computational Social Choice, pp. 3–26. AI Access (2017)

8. Brill, M., Gölz, P., Peters, D., Schmidt-Kraepelin, U., Wilker, K.: Approval-based apportionment. In: AAAI-2020, pp. 1854–1861 (2020)

9. Burdges, J., et al.: Overview of Polkadot and its design considerations. arXiv preprint arXiv:2005.13456 (2020)

10. Cevallos, A., Stewart, A.: A verifiably secure and proportional committee election rule. arXiv preprint arXiv:2004.12990 (2020)

11. Chalkiadakis, G., Elkind, E., Wooldridge, M.: Computational aspects of cooperative game theory. In: Synthesis Lectures on Artificial Intelligence and Machine Learning, vol. 5, no. 6, pp. 1–168 (2011)

12. Cheng, Y., Jiang, Z., Munagala, K., Wang, K.: Group fairness in committee selection. In: EC-2019, pp. 263–279 (2019)

13. Dummett, M.: Voting Procedures. Oxford University Press, Oxford (1984)

14. Elkind, E., Faliszewski, P., Skowron, P.: A characterization of the single-peaked single-crossing domain. Soc. Choice Welfare **54**, 167–181 (2020). https://doi.org/10.1007/s00355-019-01216-3

15. Elkind, E., Faliszewski, P., Skowron, P., Slinko, A.: Properties of multiwinner voting rules. Soc. Choice Welfare **48**(3), 599–632 (2017). https://doi.org/10.1007/s00355-017-1026-z

16. Elkind, E., Lackner, M.: Structure in dichotomous preferences. In: IJCAI-2015, pp. 2019–2025 (2015)

17. Elkind, E., Lackner, M., Peters, D.: Structured preferences. In: Trends in Computational Social Choice, pp. 187–207 (2017)

18. Fain, B., Goel, A., Munagala, K.: The core of the participatory budgeting problem. In: Cai, Y., Vetta, A. (eds.) WINE 2016. LNCS, vol. 10123, pp. 384–399. Springer, Heidelberg (2016). https://doi.org/10.1007/978-3-662-54110-4_27

19. Fain, B., Munagala, K., Shah, N.: Fair allocation of indivisible public goods. In: Proceedings of the 2018 ACM Conference on Economics and Computation, pp. 575–592 (2018). Extended version. arXiv:1805.03164

20. Faliszewski, P., Skowron, P., Slinko, A., Talmon, N.: Multiwinner voting: a new challenge for social choice theory. In: Endriss, U. (ed.) Trends in Computational Social Choice, pp. 27–47. AI Access (2017)

21. Farahani, R.Z., Hekmatfar, M. (eds.): Facility Location: Concepts, Models, Algorithms and Case Studies. Springer, Heidelberg (2009). https://doi.org/10.1007/978-3-7908-2151-2

22. Jiang, Z., Munagala, K., Wang, K.: Approximately stable committee selection. In: STOC-2020, pp. 463–472 (2020)

23. Lackner, M., Skowron, P.: Approval-based committee voting: axioms, algorithms, and applications. Technical report. arXiv:2007.01795 [cs.GT], arXiv.org (2020)

24. Magiera, K., Faliszewski, P.: Recognizing top-monotonic preference profiles in polynomial time. J. Artif. Intell. Res. **66**, 57–84 (2019)

25. Mirrlees, J.: An exploration in the theory of optimal income taxation. Rev. Econ. Stud. **38**, 175–208 (1971)

26. Osborne, J.M., Rubinstein, A.: A Course in Game Theory, vol. 1. The MIT Press, Cambridge (1994)

27. Peters, D., Skowron, P.: Proportionality and the limits of welfarism. In: EC-2020, pp. 793–794 (2020). Extended version. arXiv:1911.11747

28. Pierczyński, G., Skowron, P.: Core-stable committees under restricted domains (2021). https://doi.org/10.48550/ARXIV.2108.01987. https://arxiv.org/abs/2108.01987

29. Roberts, K.W.S.: Voting over income tax schedules. J. Public Econ. **8**(3), 329–340 (1977)

30. Sánchez-Fernández, L., et al.: Proportional justified representation. In: Proceedings of the 31st AAAI Conference on Artificial Intelligence (AAAI-2017), pp. 670–676 (2017)

31. Skowron, P., Faliszewski, P., Lang, J.: Finding a collective set of items: from proportional multirepresentation to group recommendation. Artif. Intell. **241**, 191–216 (2016)
32. Skowron, P., Lackner, M., Brill, M., Peters, D., Elkind, E.: Proportional rankings. In: IJCAI-2017, pp. 409–415 (2017)
33. Srinivasan, A.: Distributions on level-sets with applications to approximation algorithms. In: FOCS-2001, pp. 588–597 (2001)

Beyond the Worst Case: Semi-random Complexity Analysis of Winner Determination

Lirong Xia[1] and Weiqiang Zheng[2($\boxtimes$)]

[1] Rensselaer Polytechnic Institute, Troy, USA
[2] Yale University, New Haven, USA
weiqiang.zheng@yale.edu

Abstract. The computational complexity of winner determination is a classical and important problem in computational social choice. Previous work based on worst-case analysis has established NP-hardness of winner determination for some classic voting rules, such as Kemeny, Dodgson, and Young.

In this paper, we revisit the classical problem of winner determination through the lens of *semi-random analysis*, which is a worst average-case analysis where the preferences are generated from a distribution chosen by the adversary. Under a natural class of semi-random models that are inspired by recommender systems, we prove that winner determination remains hard for Dodgson, Young, and some multi-winner rules such as the Chamberlin-Courant rule and the Monroe rule. Under another natural class of semi-random models that are extensions of the Impartial Culture, we show that winner determination is hard for Kemeny, but is easy for Dodgson. This illustrates an interesting separation between Kemeny and Dodgson.

Keywords: Computational social choice · Winner determination · Semi-random complexity

1 Introduction

Voting is one of the most popular methods for group decision-making. In large-scale, high-frequency group decision-making scenarios, it is highly desirable that the winner can be computed in a short amount of time. The complexity of *winner determination* under common voting rules is thus not only a classic theoretical problem in computational social choice [15, chapter 4, 5], but also an important consideration in practice.

In this paper, we focus on several classic voting rules: the Kemeny rule, the Dodgson rule, and the Young rule, whose winner determination problems are denoted as KEMENYSCORE, DODGSONSCORE, and YOUNGSCORE, respectively. The Kemeny rule, which is closely related to the Feedback Arc Set problem [1,2], is a classical method for recommender systems and information retrieval [19].

K. A. Hansen et al. (Eds.): WINE 2022, LNCS 13778, pp. 330–347, 2022.
https://doi.org/10.1007/978-3-031-22832-2_19

The Dodgson rule and the Young rule have also been extensively studied in the literature [15,16,23,34].

Previous work has established the (worst-case) NP-hardness of winner determination under the Kemeny rule, the Dodgson rule, and the Young rule [5,34]. Using average-case analysis, McCabe-Dansted et al. [28] and Homan and Hemaspaandra [25] showed that DODGSONSCORE admits an efficient algorithm that succeeds with high probability, where each ranking is generated i.i.d. uniformly, known as the *Impartial Culture (IC)* assumption in social choice. Unfortunately, IC or generally any i.i.d. distribution has been widely criticized of being unrealistic (see, e.g., [30, p. 30], [22, p. 104], and [26]). It remains unknown whether there exists an efficient algorithm for DODGSONSCORE beyond IC. This motivates us to ask the following question:

What is the complexity of winner determination beyond worst-case analysis and IC?

One promising idea is to tackle this question through the lens of *smoothed complexity analysis* [6,37], a beautiful and powerful framework for analyzing the performance of algorithms in practice. Smoothed analysis can be seen as a worst average-case analysis, where the adversary first arbitrarily chooses an instance, and then Nature adds random noise (perturbation) to it, based on which the expected runtime of an algorithm is evaluated. Smoothed analysis explains why the simplex method is fast despite its worst-case exponential time complexity [36]. It has been successfully applied to many fields to understand the practical performance of algorithms, see the survey by Spielman and Teng [37].

Smoothed analysis belongs to the more general approach of complexity analysis under *semi-random models* [9,20], where the problem instance contains an adversarial component and a random component. In this paper, we adopt the semi-random model called the *single-agent preference model* [38], where the adversary chooses a preference distribution for each agent from a set Π of distributions. Note that if Π consists of only the uniform distribution, then the model is equivalent to IC. By varying Π, the model can provide a smooth transition from average-case analysis to worst-case analysis. Under this model, Xia and Zheng [41] proved the semi-random hardness of computing Kemeny ranking and Slater ranking with mild assumptions. However, their hardness results do not imply hardness of KEMENYSCORE under the same model, because KEMENYSCORE is easier than computing the Kemeny ranking (see Definition 2). The semi-random complexity of the Dodgson rule and the Young rule were also left as open questions [41].

Our Contributions. We provide the first set of results on the computational complexity of winner determination under the following two classes of semi-random models.

The first class of models are inspired by recommender systems and information retrieval, where the number of alternatives m can be very large and it is inefficient for an intelligent system to learn the total ranking. In such cases, one often uses top-K ranking algorithms that only recover the top-K ranking with high accuracy for $K = o(m)$ [17,29]. Similarly in social choice, the collected preference from an agent is more robust over her few top-ranked alternatives and may be much more noisy over the remaining alternatives (see Example 1). Formally, we capture such features in Assumption 1. Then, we prove in Theorem 1 and Theorem 2 that DODGSONSCORE and YOUNGSCORE remain hard under Assumption 1 unless NP = ZPP. Similar semi-random hardness results also hold for some *multi-winner* rules, i.e., the Chamberlin-Courant rule and the Monroe rule (Theorem 3).

The second class of semi-random models are called α-Impartial Culture (α-IC for short, see Definition 4) where $\alpha \in [0, 1]$. They are a relaxation of IC such that a single ranking receives probability $1 - \alpha$ and the other rankings are uniformly distributed. When α is $\frac{1}{\mathcal{O}(\text{poly}(m))}$ away from 1, we illustrate an interesting separation between the complexity of KEMENYSCORE and that of DODGSONSCORE: winner determination is hard for KEMENYSCORE (Theorem 4) while being easy for DODGSONSCORE (Theorem 5).

1.1 Related Works and Discussions

Smoothed and Semi-random Analysis. Semi-random models have been widely adopted to analyse the performance of algorithms in practice and to circumvent worst-case computational hardness in the field of combinatorial optimization [11,21], mathematical programming [36], and recently in algorithmic game theory [3,10,13,14,33]. We refer the readers to recent surveys of semi-random models [20] and beyond worst-case analysis [35] for a comprehensive literature review. We mention here that the partial alternative randomization model in Example 1 is inspired by the partial bit randomization model which has been applied to smoothed complexity analysis [4] and smoothed competitive ratio analysis [7].

Recently, semi-random analysis has also been proposed in the field of social choice [6,38]. The smoothed probability of paradoxes and ties, and strategyproofness in voting are studied [18,38–40]. As mentioned above, Xia and Zheng [41] studied complexity of computing Kemeny and Slater rankings under semi-random models. We are not aware of other semi-random complexity results in computational social choice, which motivates this work.

Beier and Vöcking [8] studied the case of the integer linear programs (ILPs) over the unit cube and showed that a problem has polynomial smoothed complexity if and only if it admits a pseudo-polynomial algorithm. Since winner determination under voting rules studied in this paper can also be formulated as ILPs, one might be tempted to think that the results in [8] also apply to the single-agent preference model. However, this is not true because they only considered continuous perturbation for *real numbers*, while the set of rankings is *discrete*. Their conclusion works for discrete combinatorial optimization problems only if the continuous noise is added to the so-called *stochastic parameters* that are real numbers, so that the problem's combinatorial structure remains unchanged, which is not the case of our setting.

Complexity of Winner Determination. There is a large body of literature on worst-case computational complexity of winner determination under various voting rules. Bartholdi et al. [5] proved that computing DODGSON-SCORE and YOUNGSCORE are NP-hard, respectively. They also provided the NP-completeness of KEMENYSCORE, which holds even for only four voters [19]. The problem of computing Dodgson winner, Young ranking, and Kemeny ranking were proved to be Θ_2^P complete [23,24,34].

2   Model and Preliminaries

Basics of Voting. Let $\mathcal{A}_m = \{a_1, \ldots, a_m\}$ denote the set of m alternatives and $\mathcal{L}(\mathcal{A}_m)$ the set of rankings (linear orders) over $\mathcal{A}_m$. A *(preference) profile* $P \in \mathcal{L}(\mathcal{A}_m)^n$ is a collection of n agents' rankings, which is also called their *preferences*. Throughout the paper, we assume without loss of generality that $m \geq 3$ since winner determination is easy for 2 alternatives. For any ranking $R \in \mathcal{L}(\mathcal{A}_m)$, we denote $\mathrm{Top}_K(R)$ the top-K ranking of R. For a permutation σ over $\mathcal{A}_m$ and any distribution π over $\mathcal{L}(\mathcal{A}_m)$, we denote $\sigma(\pi)$ the permuted distribution where $\mathrm{Pr}_{\sigma(\pi)}(\sigma(R)) = \mathrm{Pr}_\pi(R)$ for all $R \in \mathcal{L}(\mathcal{A}_m)$.

The Dodgson Rule, the Young Rule and the Kemeny Rule. The Condorcet winner of preference profile P is defined as the alternative $a \in \mathcal{A}_m$ who is preferred to every $b \in \mathcal{A}_m$ by strictly more than half of the agents. The *Dodgson score* of a in P is defined as the smallest number of sequential exchanges of adjacent alternatives in rankings of P to make a the Condorcet winner. The *Young score* of a in P is defined as the size of the largest subset of preferences where a is the Condorcet winner. The Dodgson rule chooses the alternatives with the lowest Dodgson score as winners, and the Young rule chooses the alternatives with the highest Young score as winners. The winner determination problems of the Dodgson rule and the Young rule are defined as follows.

Definition 1 (DODGSONSCORE **and** YOUNGSCORE). *Given $P \in \mathcal{L}(\mathcal{A}_m)^n$, $a \in \mathcal{A}_m$, and $t \in \mathbb{N}$, in* DODGSONSCORE *(respectively,* YOUNGSCORE*), we are asked to decide whether the Dodgson score (respectively, Young score) of a in P is at most (respectively, at least) t.*

The *Kendall's Tau distance (KT distance)* between two linear orders $R, R' \in \mathcal{L}(\mathcal{A})$, denoted by $\mathrm{KT}(R, R')$, is the number of pairwise disagreements between R and R'. Given a profile P and a linear order R, the *KT distance* between R and P is defined to be $\mathrm{KT}(P, R) = \sum_{R' \in P} \mathrm{KT}(R, R')$. The *Kemeny score* of an alternative a in P is defined as the minimum KT distance between any linear order that ranks a at the top. The Kemeny rule chooses the alternatives with the lowest Kemeny score. Besides, the Kemeny ranking is defined as the ranking with minimum KT distance to P. The winner determination problem of the Kemeny rule is defined as follows:

Definition 2 (KEMENYSCORE). *Given $P \in \mathcal{L}(\mathcal{A}_m)^n$ and $t \in \mathbb{N}$, in* KEMENYSCORE, *we are asked to decide if there exists an alternative $a \in \mathcal{A}_m$ whose Kemeny score is at most t.*

If we can compute the Kemeny ranking, then we can compute its KT distance to P in polynomial time and then decides KEMENYSCORE. Thus KEMENYSCORE is easier than computing the Kemeny ranking.

Semi-random Complexity Analysis. We use the following semi-random model, proposed in [38] and used for semi-radom complexity analysis in [41].

Definition 3 (Single-agent preference model [38]). *A single-agent prefer-ence model for m alternatives is denoted by $\mathcal{M}_m = (\Theta_m, \mathcal{L}(\mathcal{A}_m), \Pi_m)$. Π_m is a set of distributions over $\mathcal{L}(\mathcal{A}_m)$ indexed by a parameter space Θ_m such that for each parameter $\theta \in \Theta_m$, $\pi_\theta \in \Pi_m$ is its corresponding distribution.*

We say $\mathcal{M}_m$ is *P-samplable* if there exists a poly-time sampling algorithm for each distribution in Π_m. It is the "most natural restriction" on general dis-tributions, which is less restrictive than the commonly-studied *P-computable* distributions [12, p. 17,18]. We say $\mathcal{M}_m$ is *neutral* if for any $\pi \in \Pi_m$ and any permutation σ over $\mathcal{A}_m$, we have $\sigma(\pi) \in \Pi_m$. Note that winner determina-tion under all the above voting rules is in P when m is bounded above by a constant. Therefore, we are given a sequence of single-agent preference models $\vec{\mathcal{M}} = \{\mathcal{M}_m = (\Theta_m, \mathcal{L}(\mathcal{A}_m), \Pi_m) : m \geq 3\}$. We say $\vec{\mathcal{M}}$ is P-samplable (respec-tively, neutral) if Π_m is P-samplable (respectively, neutral) for any $m \geq 3$.

We introduce the following generalization of the Impartial Culture model, which is P-samplable and neutral.

Definition 4 (α-Impartial Culture). *Fix $\alpha \in [0, 1]$. α-Impartial Culture (α-IC) is a single-agent preference model $\mathcal{M}_m = (\Theta_m, \mathcal{L}(\mathcal{A}_m), \Pi_m)$ such that $\Theta_m = \mathcal{L}(\mathcal{A}_m)$ and for each $R \in \mathcal{L}(\mathcal{A}_m)$, distribution π_R is defined as*

$$\Pr_{R' \sim \pi_R}[R'] = \frac{\alpha}{m!} + (1-\alpha)\mathbf{1}[R' = R],$$

where $\mathbf{1}[R' = R] = 1$ if $R = R'$ and $\mathbf{1}[R' = R] = 0$ otherwise. Fix $\vec{\alpha} = (\alpha_m)_{m \geq 3}$ such that $\alpha_m \in [0, 1]$ for all $m \geq 3$. Denote $\vec{\alpha}$-IC the sequence of models $\{\alpha_m\text{-}IC : m \geq 3\}$. It is easy to see that $\vec{\alpha}$-IC is P-samplable and neutral.

The semi-random profile P according to $\mathcal{M}_m$ is generated as follows. First, the adversary chooses $\vec{\pi} = (\pi_1, \dots, \pi_n) \in \Pi_m^n$. Then agent j's ranking will be independently (but not necessarily identically) generated from π_j for any $j \in [n]$. The semi-random version of winner determination under $\vec{\mathcal{M}}$ is defined as follows, which is similar to the definition in a recent paper on smoothed hardness of two-player Nash equilibrium [13].

Definition 5 (SEMI-RANDOM-DODGSONSCORE). *Fix a sequence of single-agent preference models $\vec{\mathcal{M}}$. Given alternative $a \in \mathcal{A}_m$, $t \in \mathbb{N}$ and a semi-random profile P drawn from $\mathcal{M}_m$, we are asked to decide whether the Dodgson score of a is at most t, with probability at least $1 - \frac{1}{m}$.[1]*

---

[1] The algorithm is allowed to return "Failure" with probability at most $\frac{1}{m}$. However, when it returns YES or NO, the answer must be correct. Our hardness results hold even for algorithms that are only required to succeed with probability $o(1)$.

Definition 6 (Semi-Random-KemenyScore). *Fix a sequence of single-agent preference models $\vec{\mathcal{M}}$. Given $t \in \mathbb{N}$ and a semi-random profile P drawn from $\mathcal{M}_m$, we are asked to decide whether there exists an alternative whose Kemeny score of a is at most t, with probability at least $1 - \frac{1}{m}$.*

The definition of Semi-Random-YoungScore is similar.

3   Semi-random Hardness of DodgsonScore and YoungScore

In many applications, such as recommender systems and information retrieval, the number of alternatives m can be very large and it is inefficient for an intelligent system to learn the total ranking. In such cases, one often uses Top-K ranking algorithms which only recover the top-K ranking with high accuracy for $K = o(m)$ [17,29]. Similarly, the collected preference from an agent is more robust over her few top-ranked alternatives and can be much more noisy over the remaining alternatives. Such features are captured by Assumption 1 below. Informally, Assumption 1 states that there *exists* a distribution in Π_m that does not significantly "perturb" one top-K ranking for $K = \Theta(m^{\frac{1}{d}})$ where $d \geq 1$.

Assumption 1 (Top-K concentration). *A series of single-agent preference models $\vec{\mathcal{M}}$ is P-samplable, neutral, and satisfies the following condition: there exists a constant $d > 1$ such that for any sufficiently large m and $K = \lceil m^{\frac{1}{d}} \rceil$, there exists $\mathcal{A}' \subseteq \mathcal{A}_m$, $R' \in \mathcal{L}(\mathcal{A}')$, and $\pi \in \Pi_m$, such that $|\mathcal{A}'| = K$ and*

$$\Pr_{R \sim \pi} \left(Top_K(R) = R' \right) \geq 1 - \frac{1}{K}.$$

The following *partial alternative randomization model*, in the spirit of partial bit randomization model [4,7], satisfies Assumption 1. The partial bit randomization model applies to m-bits non-negative integer by randomly flipping its $m - K$ least significant bits while keeping its K most significant bits unchanged.

Example 1. The *partial alternative randomization model* is denoted by $\mathcal{M}_m(K)$ and has parameter space $\mathcal{L}(\mathcal{A}_m)$. For any $R \in \mathcal{L}(\mathcal{A}_m)$, the distribution π_R is obtained by uniformly at random perturbing the order of the $m - K$ least preferred alternatives in R and keeping the top-K ranking unchanged. For any constant d and $K \geq m^{\frac{1}{d}}$, the model is P-samplable, neutral and satisfies Assumption 1. Note that in such model, each ranking receives probability at most $\frac{1}{(m-K)!} = \frac{1}{\Omega(\exp m)}$.

We show that for models that satisfying Assumption 1, winner determination under the Dodgson rule and the Young rule is hard unless NP = ZPP. Note that NP $\neq$ ZPP is widely believed to hold in complexity theory. The high-level idea is to combine the existence of a top-K concentration distribution guaranteed by Assumption 1 and neutrality, to show that for any possible input of a NP-complete problem, the adversary is able to construct a distribution of voting profile such that efficient semi-random winner determination implies a coRP algorithm for the NP-complete problem. Thus NP $\subseteq$ coRP and it implies NP = ZPP by the following reasoning. Recall that RP $\subseteq$ NP. Therefore, RP $\subseteq$ NP $\subseteq$ coRP, which means that RP = RP $\cap$ coRP. Recall that RP $\cap$ coRP =ZPP. We have RP = ZPP, which means that coRP = coZPP. Since coZPP = ZPP, it follows that NP $\subseteq$ coRP = coZPP = ZPP.

Theorem 1 (Semi-random hardness of DODGSONSCORE). *For any serie of single-agent preference models $\vec{\mathcal{M}}$ that satisfies Assumption 1, there exists no polynomial-time algorithm for* SEMI-RANDOM-DODGSONSCORE *under $\vec{\mathcal{M}}$ unless* NP = ZPP.

Proof. **Overview of the proof.** We leverage the reduction in [5] that reduces the NP-complete problem EXACT COVER BY 3-SETS (X3C) to DODGSON-SCORE. An instance of X3C is denoted by (U, S) including a q-element set U such that q is divisible by 3 and a collection S of 3-element subsets of U. We are asked to decide whether S contains an exact cover for U, i.e., a subcollection S' of S such that every element of U occurs in exactly one member of S'.

Suppose that SEMI-RANDOM-DODGSONSCORE has a polynomial-time algorithm, denoted as Alg. We will use Alg to construct a coRP algorithm for X3C. Formally, the proof proceeds in two steps. For any instance of X3C, in Step 1, we follow the original reduction to construct a profile P_1. Then we construct a parameter profile P^Θ using the semi-random model $\vec{\mathcal{M}}$ based on P_1. Note that a parameter profile corresponds to distribution over profiles. In Step 2, we show that Alg can be leveraged to Algorithm 1 to prove that X3C is in coRP, which implies NP = ZPP as shown above.

Let (U, S) be any instance of X3C such that $U = \{u_1, u_2, \cdots, u_q\}$ and $S = \{S_1, S_2, \cdots, S_s\}$ is a collection of s distinct 3-element subsets of U. We assume without loss of generality that $q/3 \le s \le q^3/6$ because (U, S) must be a NO instance if $s < q/3$ and there are at most $\binom{q}{3} \le q^3/6$ distinct 3-element subsets of U.

Step 1. Construct Profile P_1 and Parameter Profile P^Θ. We first use the reduction by Bartholdi et al. [5] to construct a voting profile $P_1 \in \mathcal{L}(\mathcal{A}_{m_1})^n$ of polynomial-size in q. The proof of Lemma 1 can be found in the full version.

Lemma 1. *We can construct a profile $P_1 \in \mathcal{L}(\mathcal{A}_{m_1})^n$ with $m_1 = 2q + s + 1 = \mathcal{O}(q^3)$, $n \le 2(q + 1)s + 1 = \mathcal{O}(q^4)$, and an alternative c such that $(P_1, c, \frac{4q}{3})$ is a YES instance of* DODGSONSCORE *if and only if (U, S) is a YES instance of* X3C. *The construction can be done in polynomial time in q.*

The following observation of the Dodgson rule is crucial for the proof. We introduce one more notation here. For any profile $P \in \mathcal{L}(\mathcal{A}_m)^n$, we denote **AppLast**$(P, m')$ the set of profiles obtained from P by appending m' extra alternatives to the bottom of each agent's preferences in any order. It follows that $|\mathbf{AppLast}(P, m')| = (m'!)^n$.

Lemma 2. *For any profile* $P_1 \in \mathcal{L}(\mathcal{A}_m)^n$, *any integer* $m' \geq 1$ *and profile* $P_2 \in$ **AppLast**(P_1, m'), *the following holds for any alternative* $a \in \mathcal{A}_m$:

– *If* a *is Dodgson winner in* P_1, *then* a *is also Dodgson winner in* P_2.
– *The Dodgson score of* a *in* P_1 *is equal to that in* P_2.

The proof of Lemma 2 follows by definition and can be found the full version. Informally, Lemma 2 states that by appending alternatives at the bottom of each agent's preference order, the winner and score under the Dodgson rule are robust.

Let $m = (2m_1 n)^d = \text{poly}(q)$, where d is the constant defined in Assumption 1. We create a set of $m - m_1$ dummy alternatives called D. The total alternative set is set as $\mathcal{A}_m = \mathcal{A}_{m_1} \cup D$. Denote $P_1 = (R_i^1)_{i \in [n]}$. We define $P := \textbf{AppLast}(P_1, m - m_1) = (R_i)_{i \in [n]}$ by appending the dummy alternatives in D. We remark that by the definition of **AppLast**, each ranking R_i in P is of the form

$$R_i = \mathcal{A}_{m_1} \succ_{R_i} D$$

where the order of $\mathcal{A}_{m_1}$ in R_i is the same as R_i^1.

Now we construct the parameter profile P^Θ based on P such that each parameter corresponds to a preference order in P. According to Assumption 1, there exists $K = \lceil m^{\frac{1}{d}} \rceil \geq m_1$, $\mathcal{A}' \subseteq \mathcal{A}_m$, $R' \in \mathcal{L}(\mathcal{A}')$, and $\pi \in \Pi_m$, such that $|\mathcal{A}'| = K$ and $\Pr_{R \sim \pi}(\text{Top}_K(R) = R') \geq 1 - \frac{1}{K}$. Let

$$R^* := \mathcal{A}' \succ_{R^*} (\mathcal{A}_m \setminus \mathcal{A}')$$

where the order in $\mathcal{A}'$ is the same as R' and the order in $\mathcal{A}_m \setminus \mathcal{A}'$ is arbitrary. Denote the parameter corresponding to this specific distribution $\pi \in \Pi_m$ as θ. For every $i \in [n]$, we can find a permutation σ_i over $\mathcal{L}(\mathcal{A}_m)$ such that $\sigma_i(R^*) = R_i$. We then apply permutation σ_i to the predefined distribution π and get a new distribution $\sigma_i(\pi)$ which is also in Π_m since $\mathcal{M}_m$ is neutral. Now we define the parameter profile $P^\Theta := (\theta_i)_{i \in [n]}$, where θ_i is the parameter corresponding to $\sigma_i(\pi)$. Since $K \geq m_1$, we have

$$\Pr_{R \sim \pi_{\theta_i}} (\text{Top}_{m_1}(R) = R_i^1) \geq 1 - \frac{1}{K}$$

and the construction of P^Θ can be done in polynomial time of q.

Step 2. Use Alg to solve X3C **.** For a profile $P \in \mathcal{L}(\mathcal{A}_m)^n$, we denote $\text{Top}_K(P)$ the collection of top-K ranking of each preference order in P. We now prove that we can construct a coRP Algorithm for X3C based on Alg.

Claim 1. *If* $\text{Top}_{m_1}(P') = P_1$, *then* $(P', c, 4q/3)$ *is a YES instance for* DODGSONSCORE *if and only if* (U, S) *is a YES instance for* X3C.

Proof. We know that $P' \in \textbf{AppLast}(P_1, m - m_1)$ by definition. According to Lemma 2, we know that the Dodgson score of c in P' is the same as the Dodgson score of c in P_1. Therefore, $(P', c, \frac{4q}{3})$ is a YES instance of DODGSONSCORE if and only if $(P_1, c, \frac{4q}{3})$ is a YES instance of DODGSONSCORE, which is also equivalent to (U, S) is a YES instance by Lemma 1. □

Algorithm 1. Randomized Algorithm for X3C

Input: An X3C instance (U, S) and Alg for DODGSONSCORE

1: Construct profile P_1 and parameter profile P^Θ according to Step 1.
2: Sample a profile P' from $\vec{\mathcal{M}}_m$ given P^Θ.
3: **if** $\text{Top}_{m_1}(P') \neq P_1$ **then**
4:     Return YES.
5: **end if**
6: Run Alg on $(P', c, 4q/3)$.
7: **if** Alg returns YES **then**
8:     Return YES.
9: **else**
10:     Return NO.
11: **end if**

Notice that sampling P' from P^Θ takes polynomial time because $\vec{\mathcal{M}}$ is P-samplable (Assumption 1). It follows that Algorithm 1 is a polynomial-time algorithm. Recall that A coRP algorithm always returns YES to YES instances, and returns NO with constant probability to NO instances. Since Algorithm 1 returns NO only if $\text{Top}_{m_1}(P') = P_1$ and $(P', c, 4q/3)$ is a NO instance, by Claim 1 it is clear that if (U, S) is a YES instance then Algorithm 1 returns YES. Therefore, to prove that Algorithm 1 is an coRP algorithm it suffices to prove that if (U, S) is a NO instance then Algorithm 1 returns NO with constant probability.

Claim 2. $\Pr\left(\text{Top}_{m_1}(P') = P_1\right) \geq 1/2$.

Proof. $P' = (R_i')_{i \in [n]}$ is sampled from $P^\Theta = (\theta_i)_{i \in [n]}$. Recall that $m = (2m_1 n)^d$ and $K = m^{\frac{1}{d}}$. Thus $K \geq m_1$ and we know that by construction in Step 1 and Assumption 1 that for all $i \in [n]$,

$$\Pr_{R_i' \sim \pi_{\theta_i}} \left(\text{Top}_{m_1}(R_i') = R_i^1\right) \geq \Pr_{R_i' \sim \pi_{\theta_i}} \left(\text{Top}_K(R_i') = R_i^1\right) \geq 1 - \frac{1}{K} = 1 - \frac{1}{2m_1 n}.$$

Thus we can derive

$$\Pr_{P' \sim P^\Theta} \left(\text{Top}_{m_1}(P') = P_1\right) \geq \prod_{i=1}^n \left(\Pr_{R_i' \sim \pi_{\theta_i}} \left(\text{Top}_K(R_i') = R_i^1\right)\right)$$

$$\geq (1 - \frac{1}{2m_1 n})^n \geq 1 - \frac{1}{2m_1} \geq \frac{1}{2}. \qquad \square$$

When (U, S) is a NO instance of X3C, $\text{Alg}(P', c, \frac{4q}{3})$ returns NO with probability at least $1 - \frac{1}{m}$ by definition 5. Note that Algorithm 1 returns YES when $\text{Top}_{m_1}(P') \neq P_1$ which happens with probability at most $\frac{1}{2}$ by Claim 2. Therefore, for any profile $P' \in \mathcal{L}(\mathcal{A}_m)^n$ such that $\text{Top}_{m_1}(P') = P_1\}$, $\text{Alg}(P', c, \frac{4q}{3})$ succeeds with probability at least $1 - \frac{2}{m} \geq \frac{1}{3}$. According to Claim 1 and 2, we know that Algorithm 1 returns NO for any NO instance with probability at least $\frac{1}{2} \times \frac{1}{3} = \frac{1}{6}$. This completes the proof.

We prove a similar result for YOUNGSCORE with proof in the full version. $\square$

Theorem 2 (Semi-random hardness of YOUNGSCORE). *For any single-agent preference model $\vec{\mathcal{M}}$ that satisfies Assumption 1, there exists no polynomial-time algorithm for* SEMI-RANDOM-YOUNGSCORE *under $\vec{\mathcal{M}}$ unless* NP=ZPP.

Proof sketch. We first extend Lemma 2 for the Dodgson rule to the Young rule. With that in hand, the proof is then very similar to that of Theorem 1. The main difference is now that we use the reduction in [16] to construct the profile in Step 1 and then a coRP algorithm for the NP-complete problem X3C, which leads to NP = ZPP.   □

3.1 Extension to Multi-winner Voting Rules

A multi-winner voting rule selects a winning *k-committee*, which is a k-size subset of alternatives. We consider the Chamberlin-Courant (CC) rule and the Monroe rule that assign each k-committee a score and choose the k-committee with the highest (respectively, lowest) score as the winner. Definitions of the two voting rules and their corresponding winner determination problems and the proof of the following theorem can be found in the full version. We remark that winner determination under the CC rule and the Monroe rule are both NP-hard [27,32].

Theorem 3 (Semi-random hardness of CC and Monroe). *For any single-agent preference model $\vec{\mathcal{M}}$ that satisfies Assumption 1, there exists no polynomial-time algorithm for the semi-random version of the winner determination problems of the CC rule and the Monroe rule under $\vec{\mathcal{M}}$ unless* NP = ZPP.

Proof sketch. We first prove counter parts of Lemma 2 for the CC rule and the Monroe rule. Then the proof follows the same idea in the proof of Theorem 1, except that we use different reductions to construct the profile in Step 1.   □

4 KEMENYSCORE V.S. DODGSONSCORE

In this section, we present to two results regarding the Kemeny and Dodgson rule under the $\vec{\alpha}$-IC model. In Theorem 4, we show that SEMI-RANDOM-KEMENYSCORE has no polynomial time algorithm under $(1 - \frac{1}{m})$-IC unless NP = ZPP. In contrast, we provide an efficient algorithm for SEMI-RANDOM-DODGSONSCORE under $(1 - \frac{1}{m})$-IC when $n = \Omega(m^2 \log^2 m)$ (Theorem 5). The two results together provide an interesting separation of the semi-random complexity of winner determination under different NP-hard rules.

4.1 Semi-random Hardness of KEMENYSCORE

KEMENYSCORE is NP-complete and is easier than computing the Kemeny ranking, which is Θ_2^P-complete [24]. Thus hardness result for computing the Kemeny ranking [41] does not imply the semi-random hardness of KEMENYSCORE. Nevertheless, under the same assumption made in [41], we can prove the semi-random hardness of KEMENYSCORE. To better illustrate the separation of semi-random complexity between Kemeny and Dodgson, we state the result in a special case under the $\vec{\alpha}$-IC model first. The formal statement of the general assumption and theorem as well as its proof are defered to Sect. 4.3.

Theorem 4. *For any constant $d \geq 0$ and $\vec{\alpha} = (\alpha_m)_{m \geq 3}$ such that $\alpha_m \in [0, 1 - \frac{1}{m^d}]$ for any sufficiently large m, there exists no polynomial-time algorithm for* SEMI-RANDOM-KEMENYSCORE *under $\vec{\alpha}$-IC unless* NP = ZPP.

Note that for $d \geq 0$, $(1 - \frac{1}{m^d})$-IC is close to the average case in the sense that any distribution in Π_m is only $O(\frac{1}{m^d})$ away from the uniform distribution in total variation distance. Therefore, Theorem 4 shows that KEMENYSCORE remains hard even for models that are close to the average case.

4.2 Semi-random Easiness of DODGSONSCORE

In contrast to the Kemeny rule, we prove that winner determination under the Dodgson rule is tractable under models close to the average case, i.e., $(1 - \frac{1}{m})$-IC. We remark here that although $(1 - \frac{1}{m})$-IC is close to IC, $(1 - \frac{1}{m})$-IC may concentrate on a single ranking with probability as large as $\Theta(\frac{1}{m})$, while every ranking in IC has probability exactly $\frac{1}{m!} = o(\frac{1}{\exp(m)})$.

Since 1-IC is equivalent to IC, the following theorem that works for any $\alpha \in [1 - \frac{1}{m}, 1]$ thus generalizes previous results that only work for IC [25, 28].

Theorem 5 (Semi-random easiness of DODGSONSCORE). *For any $\vec{\alpha} = (\alpha_m)_{m \geq 3}$ such that $\alpha_m \in [1 - \frac{1}{m}, 1]$ for sufficiently large m, there exists a polynomial-time algorithm for* SEMI-RANDOM-DODGSONSCORE *under $\vec{\alpha}$-IC that succeeds with probability at least $1 - 2(m - 1) \exp\left(-\frac{n}{72m^2}\right)$.*

Proof. The algorithm runs the polynomial-time greedy algorithm, denoted as GREEDY in [25] as a subroutine. Given (P, a), the output of GREEDY(P, a) belongs to $\mathbb{Z} \times$ ("definitely", "maybe") such that if GREEDY(P, a) outputs $(s, \text{"definitely"})$, then s is the Dodgson score of a in P. Given DODGSONSCORE instance (P, a, t), the algorithm runs GREEDY(P, a) first. Then if GREEDY(P, a) outputs $(s, \text{"definitely"})$, the algorithm outputs YES or NO based on whether $s \leq t$. Otherwise the algorithm declares failure. Therefore, it suffices to prove that when P is generated from $\vec{\alpha}$-IC, GREEDY(P, a) outputs with "definitely" with high probability.

The following lemma, a simple extension of [25, Theorem 4.1.1], gives a sufficient condition under which GREEDY(P, a) outputs with "definitely". We introduce some new notations here. For two distinct alternatives a, b and voter i, by $a \prec_i b$ we mean voter i prefers b to a. By $a \ll_i b$ we mean that not only voter i prefers b to a, but also there is no other alternative c such that voter i prefers b to c and prefers c to a i.e., $a \prec_i c \prec_i b$.

Lemma 3. *Given $P = (\prec_i)_{i \in [n]}$. For each alternative $a \in \mathcal{A}_m$, if for all $b \in \mathcal{A}_m \setminus \{a\}$ there exists $\beta > 0$ such that $|\{i \in [n] : a \prec_i b\}| \leq \frac{n}{2} + \beta$ and $|\{i \in [n] : a <_i b\}| \geq \beta$ then* GREEDY(P, a) *outputs with "definitely".*

We give a sketch of the proof for Lemma 3 here. Recall that the Dodgson score of an alternative a is the smallest number of exchanges between adjacent alternatives that makes a a Condorcet winner. Now consider alternative $b \neq a$ such that a needs extra β votes to defeat b. If $|\{i \in [n] : a <_i b\}| \geq \beta$, then a defeats b after exactly β exchanges, which is also necessary. If this is the case for any alternative $b \neq a$, then we can decide in polynomial time the Dodgson score of a with certainty.

Claim 3. *For any profile $P = (\succ_i)_{i \in [n]}$ generated from α_m-IC, alternatives $a, b \in \mathcal{A}_m$, and $\beta = (\frac{3}{4} - \frac{1}{2m})\frac{n}{m} > 0$, We have*

- $\Pr\left[|\{i \in [n]|a \prec_i b\}| > \frac{n}{2} + \beta\right] < \exp\left(-\frac{n}{72m^2}\right);$
- $\Pr\left[|\{i \in [n]|a <_i b\}| < \beta\right] < \exp\left(-\frac{n}{72m^2}\right).$

Proof. Due to the space limit, we only prove the first inequality and leave the proof of the second inequality in the full version. We need the following technical lemma, which is a straightforward application of Hoeffding's inequality for bounded random variables, hence we omit the proof.

Lemma 4. *Let $X_1, \cdots, X_n$ be a sequence of mutually independent random variables. If there exist $q, p \in [0, 1]$ such that $q \leq p$ and for each $i \in \{1, \cdots, n\}$,*

$$\Pr[X_i = 1 - p] = q \text{ and } \Pr[X_i = -p] = 1 - q,$$

then for all $d > 0$, we have $\Pr[\sum_{i=1}^n X_i > d] < e^{-2d^2/n}$.

Fix any $i \in [n]$. Denote $\pi_i \in \Pi_m$ the preference distribution of agent i. Since $\alpha_m \geq 1 - \frac{1}{m}$, we know $\Pr_{\pi_i}[R] \geq \frac{m-1}{m \cdot m!}$ for any preference order $R \in \mathcal{L}(\mathcal{A}_m)$. Note that there are exactly $\frac{m!}{2}$ rankings in $\mathcal{L}(\mathcal{A}_m)$ such that a is ranked above b. Therefore, we have

$$\Pr[a \prec_i b] = 1 - \Pr[b \prec_i a] \leq 1 - \frac{m!}{2} \cdot \frac{m-1}{m \cdot m!} = \frac{m+1}{2m}$$

For each $i \in [n]$, define X_i as

$$X_i = \begin{cases} \frac{m-1}{2m} & \text{if } a \prec_i b \\ -\frac{m+1}{2m} & \text{otherwise} \end{cases}$$

It follows that $|\{i \in [n]|a \prec_i b\}| > \frac{n}{2} + \beta$ only if

$$\sum_{i=1}^n X_i > \frac{m-1}{2m}\left(\frac{n}{2} + \beta\right) - \frac{m+1}{2m}\left(\frac{n}{2} - \beta\right) = \left(\frac{1}{4} - \frac{1}{2m}\right)\frac{n}{m} \geq \frac{n}{12m} \quad (m \geq 3)$$

Note that $\Pr[X_i = \frac{m-1}{2m}] = \Pr[a \prec_i b] \leq \frac{m+1}{2m}$. The claim follows by setting $d = \frac{n}{12m}$ and $p = \frac{m+1}{2m}$ in Lemma 4. □

Applying union bound for all $m - 1$ alternatives in $\mathcal{A}_m - \{a\}$ to Claim 3, we have

$$\Pr\left[\forall b \neq a, |\{i \in [n] : a \prec_i b\}| > \frac{n}{2} + \beta \text{ or } |\{i \in [n] : a <_i b\}| < \beta\right]$$

$$\leq 2(m-1)\exp\left(-\frac{n}{72m^2}\right)$$

According to Lemma 3, with probability at least $1 - 2(m - 1)\exp\left(-\frac{n}{72m^2}\right)$, GREEDY$(P, a)$ outputs with "definitely". This completes the proof. □

According to Theorem 5, we know that under $(1 - \frac{1}{m})$-IC, SEMI-RANDOM-DODGSONSCORE is in P when $n = \Omega(m^2 \log^2 m)$. By Theorem 4, SEMI-RANDOM-KEMENYSCORE has no polynomial time algorithm under $(1 - \frac{1}{m})$-IC unless NP = ZPP. The two results together provide an interesting separation of the semi-random complexity of winner determination under different NP-hard rules.

4.3   Proof of Theorem 4

We introduce some notations before the statement of assumption and the proof. For a profile $P \in \mathcal{L}(\mathcal{A}_m)^n$, its *weighted majority graph WMG(P)* is a weighted directed graph, and its vertices are represented by $\mathcal{A}_m$. For any pair of alternatives $a, b \in \mathcal{A}_m$, the weight on edge $a \to b$ is the number of agents that prefer a to b minus the number of agents that prefer b to a. For a distribution π over rankings, we define its weighted majority graph WMG(π) similarly: For any pair of alternatives $a, b \in \mathcal{A}_m$, the weight on edge $a \to b$ is the probability that a ranking prefers a to b minus the probability that a ranking prefers b to a. For each 3-cycle $a \to b \to c \to a$, its weight is defined as the sum of the weights on its three edges $a \to b$, $b \to c$, and $c \to a$.

Assumption 2 ([41]). $\vec{\mathcal{M}}$ *is P-samplable, neutral, and satisfies the following condition: there exist constants $k \geq 0$ and $A > 0$ such that for any $m \geq 3$, there exist $\pi_{3c} \in \Pi_m$ such that WMG(π_{3c}) has a 3-cycle G_{3c} with weight at least $\frac{A}{m^k}$*

Assumption 2 is weaker than Assumption 1. That's because in the distribution π guaranteed by Assumption 1, the top-K ranking remains unchanged with probability at least $1 - \frac{1}{K}$, which implies that the 3-cycle formed by the top-3 alternatives has weight $\geq 1 - \frac{2}{K}$ with $K = m^{\frac{1}{d}}$ for constant d. For $\alpha_m \in [0, 1 - \frac{1}{m^d}]$, the model α_m-IC has a 3-cycle with weight at least $\mathcal{O}(\frac{1}{m^d})$ and thus also satisfies Assumption 2. We prove in Theorem 6 the smoothed hardness of Kemeny under Assumption 2 which implies Theorem 4.

Theorem 6 (Smoothed Hardness of Kemeny). *For any single-agent preference model $\vec{\mathcal{M}}$ that satisfies Assumption 2, there exists no polynomial-time algorithm for SEMI-RANDOM-KEMENYSCORE unless NP=ZPP.*

Proof. Suppose that SEMI-RANDOM-KEMENYSCORE has a polynomial-time algorithm, denoted as Alg. We use it to construct a coRP algorithm for the NP-complete problem EULERIAN FEEDBACK ARC SET (EFAS) [31], which implies NP = ZPP as discussed in the proof of Theorem 1. An instance of EFAS is denoted by (G, t), where $t \in \mathbb{N}$ and G is a directed unweighted Eulerian graph, which means that there exists a closed Eulerian walk that passes each edge exactly once. We are asked to decide whether G can be made acyclic by removing no more than t edges.

Given a single-agent preference model, a *(fractional) parameter profile* $P^\Theta \in \Theta_m^n$ is a collection of $n > 0$ parameters, where n may not be an integer. Note that P^Θ naturally leads to a fractional preference profile, where the weight on each ranking represents its total weighted "probability" under all parameters in P^Θ. We include an illustrating example of fractional parameter profile and fractional preference profile in the full version.

Let $(G = (V, E), t)$ be any EFAS instance, where $|V| = m$.

Claim 4 ([41]). *We can construct a fractional preference profile P_G^Θ in polynomial time in m such that there exists a constant k*

- $|P_G^\Theta| = \mathcal{O}(m^{k+2})$,
- P_G^Θ consists of $\mathcal{O}(m^5)$ types of parameters,
- $WMG(P_G^\Theta) = G$.

Let $K = 13 + 2k$, which means that $K > 12$. We first define a parameter profile $P_G^{\Theta*}$ of $n = \Theta(m^K)$ parameters that is approximately $\frac{m^K}{|P_G^\Theta|}$ copies of P_G^Θ up to $\mathcal{O}(m^5)$ in L_∞ error. Formally, let

$$P_G^{\Theta*} = \left\lfloor P_G^\Theta \cdot \frac{m^K}{|P_G^\Theta|} \right\rfloor \tag{1}$$

Let $n = |P_G^{\Theta*}|$. Because the number of different types of parameters in $P_G^{\Theta*}$ is $\mathcal{O}(m^5)$, we have $n = m^K - \mathcal{O}(m^5)$, $\|WMG(P_G^{\Theta*}) - WMG(P_G^\Theta \cdot \frac{m^K}{|P_G^\Theta|})\|_\infty = \mathcal{O}(m^5)$, and $\|WMG(P_G^{\Theta*}) - G \cdot \frac{m^K}{|P_G^\Theta|})\|_\infty = \mathcal{O}(m^5)$. Let $f(G, R)$ denote the number of backward arcs of linear order R in a directed graph G. The following useful claim calculates the KT distance between R and the parameter profile $P_G^\Theta \cdot \frac{m^K}{|P_G^\Theta|}$. The proof of Claim 5 can be found in the full version.

Claim 5. *For any linear order $R \in \mathcal{L}(\mathcal{A}_m)$, the KT distance between R and the fractional parameter profile $P_G^\Theta \cdot \frac{m^K}{|P_G^\Theta|}$ is $\mathrm{KT}\left(P_G^\Theta \cdot \frac{m^K}{|P_G^\Theta|}, R\right) = M + \frac{m^K}{|P_G^\Theta|} \cdot f(G, R)$, where $M = \frac{m^K}{2}\left(\binom{m}{2} - \frac{|E|}{|P_G^\Theta|}\right)$.*

We now prove that Alg returns the correct answer to (G, t) with probability at least $1 - \exp(-\Omega(m))$. Let $G_n = G \cdot \frac{m^K}{|P_G^\Theta|}$. The following claim bounds the probability that $WMG(P')$ is different from G_n by more than $\Omega(m^{\frac{K+1}{2}})$.

Algorithm 2. Algorithm for EFAS.

Input: EFAS Instance (G, t), Alg

1: Compute a parameter profile $P_G^{\Theta *}$ according to (1).
2: Sample a profile P' from $\vec{\mathcal{M}}_m$ given $P_G^{\Theta *}$.
3: **if** $\|\mathrm{WMG}(P') - G_n\|_1 > \binom{m}{2} \cdot m^{\frac{K+1}{2}}$ **then**
4:     Return YES.
5: **end if**
6: Run Alg on $\left(P', M + t \cdot \frac{m^K}{|P_G^{\Theta}|} + m^{k+10}\right)$.
7: **if** Alg returns NO **then**
8:     Return NO.
9: **else**
10:     Return YES.
11: **end if**

Claim 6 ([41]). $\Pr\left[\|WMG(P') - G_n\|_1 > \binom{m}{2} \cdot m^{\frac{K+1}{2}}\right] < \exp(-\Omega(m))$.

Claim 7. *If* $\|WMG(P') - G_n\|_1 \leq \binom{m}{2} \cdot m^{\frac{K+1}{2}}$, *then* $\left(P', M + t \cdot \frac{m^K}{2|P_G^{\Theta}|} + m^{k+10}\right)$ *is a YES instance of* KEMENYSCORE *if and only if* (G, t) *is a YES instance of* EFAS.

Proof. If (G, t) is a YES instance of EFAS, then there exists a linear order R such that there are at most t backward arcs in G according to R. Considering R as a ranking over alternatives, we have $\mathrm{KT}\left(P_G^{\Theta} \cdot \frac{m^K}{|P_G^{\Theta}|}, R\right) \leq M + t \cdot \frac{m^K}{|P_G^{\Theta}|}$. By assumption we know $|\mathrm{KT}(P', R) - \mathrm{KT}(P_G^{\Theta} \cdot \frac{m^K}{|P_G^{\Theta}|}, R)| = \mathcal{O}(m^{\frac{K+5}{2}})$. Therefore, the kemeny score of ranking R is at most

$$\mathrm{KT}(P', R) \leq \mathrm{KT}\left(P_G^{\Theta} \cdot \frac{m^K}{|P_G^{\Theta}|}, R\right) + \mathcal{O}(m^{\frac{K+5}{2}}) < M + t \cdot \frac{m^K}{|P_G^{\Theta}|} + m^{k+10},$$

which means $\left(P', M + t \cdot \frac{m^K}{|P_G^{\Theta}|} + m^{k+10}\right)$ is a YES instance.

If (G, t) is a NO instance of EFAS, then for any linear order R of $|V|$, there are at least $t+1$ backward arcs in G according to R. We have for any $R \in \mathcal{L}(\mathcal{A}_m)$, $\mathrm{KT}\left(P_G^{\Theta} \cdot \frac{m^K}{|P_G^{\Theta}|}, R\right) \geq M + (t+1) \cdot \frac{m^K}{|P_G^{\Theta}|}$. Therefore, for any $R \in \mathcal{L}(\mathcal{A}_m)$, we have

$$\mathrm{KT}(P', R) \geq \mathrm{KT}\left(P_G^{\Theta} \cdot \frac{m^K}{|P_G^{\Theta}|}, R\right) - \mathcal{O}(m^{\frac{K+5}{2}})$$

$$\geq M + t \cdot \frac{m^K}{|P_G^{\Theta}|} + \frac{m^K}{|P_G^{\Theta}|} - \mathcal{O}(m^{\frac{K+5}{2}})$$

$$= M + t \cdot \frac{m^K}{|P_G^{\Theta}|} + \Theta(m^{k+11}) - \mathcal{O}(m^{k+9})$$

$$> M + t \cdot \frac{m^K}{|P_G^{\Theta}|} + m^{k+10},$$

which means $\left(P', M + t \cdot \frac{m^K}{|P_G^{\ominus}|} + m^{k+10}\right)$ is a NO instance of KEMENYSCORE.

Note that Algorithm 2 only returns NO in line 8, when $\|\text{WMG}(P') - G_n\|_1 > \binom{m}{2} \cdot m^{\frac{K+1}{2}}$ and Alg returns NO. By Claim 7, we know that Algorithm 2 never returns NO for any YES instance of EFAS, or equivalently, it always returns YES for YES instance. Since $\|\text{WMG}(P') - G_n\|_1 \le \binom{m}{2} \cdot m^{\frac{K+1}{2}}$ holds with probability at least $1 - \exp(-\Omega(m))$ and Alg returns with probability at least $1 - \frac{1}{m}$, we know that Algorithm 2 returns NO for NO instance of EFAS with at least constant probability. This proves that EFAS is in coRP and completes the proof. □

5   Conclusion

In this paper, we conduct semi-random complexity analysis of winner determination under various voting rules. We give the first semi-random complexity results for the Dodgson rule, the Young rule, the Chamberlin-Courant rule, and the Monroe rule. We also prove a hardness result for the Kemeny rule and a semi-random easiness result for the Dodgson rule, illustrating an interesting separation between the semi-random complexity of winner determination under different NP-hard voting rules.

As for future direction, an ambitious goal is to develop a dichotomy theorem for the semi-random complexity of winner determination: winner determination is efficient if and only if the semi-random model satisfies certain conditions. The semi-random complexity of winner determination under models beyond Assumption 1 is a natural and interesting problem. We also conjecture that under the average-case analysis, YOUNGSCORE is easy to decide with high probability but KEMENYSCORE remains hard.

Acknowledgements. We thank anonymous reviewers for helpful feedback and suggestions. LX acknowledges NSF #1453542 and a gift fund from Google for support.

References

1. Ailon, N., Charikar, M., Newman, A.: Aggregating inconsistent information: ranking and clustering. J. ACM **55**(5) (2008). Article No. 23
2. Alon, N.: Ranking tournaments. SIAM J. Discrete Math. **20**, 137–142 (2006)
3. Bai, Y., Feige, U., Gölz, P., Procaccia, A.D.: Fair allocations for smoothed utilities. In: Proceedings of EC (2022)
4. Banderier, C., Beier, R., Mehlhorn, K.: Smoothed analysis of three combinatorial problems. In: Proceedings of MFCS (2003)
5. Bartholdi, J., III., Tovey, C., Trick, M.: Voting schemes for which it can be difficult to tell who won the election. Soc. Choice Welfare **6**, 157–165 (1989)
6. Baumeister, D., Hogrebe, T., Rothe, J.: Towards reality: smoothed analysis in computational social choice. In: Proceedings of AAMAS (2020)

7. Becchetti, L., Leonardi, S., Marchetti-Spaccamela, A., Schäfer, G., Vredeveld, T.: Average-case and smoothed competitive analysis of the multilevel feedback algorithm. Math. Oper. Res. **31**(1), 85–108 (2006)
8. Beier, R., Vöcking, B.: Typical properties of winners and Losersin discrete optimization. SIAM J. Comput. **35**(4), 855–881 (2006)
9. Blum, A.: Some tools for approximate 3-coloring. In: Proceedings of FOCS (1990)
10. Blum, A., Gölz, P.: Incentive-compatible kidney exchange in a slightly semi-random model. In: Proceedings of EC (2021)
11. Blum, A., Spencer, J.: Coloring random and semi-random k-colorable graphs. J. Algorithms **19**(2), 204–234 (1995)
12. Bogdanov, A., Trevisan, L.: Average-case complexity. Found. Trends Theor. Comput. Sci. **2**(1), 1–106 (2006)
13. Boodaghians, S., Brakensiek, J., Hopkins, S.B., Rubinstein, A.: Smoothed complexity of 2-player Nash equilibria. In: Proceedings of FOCS (2020)
14. Boodaghians, S., Kulkarni, R., Mehta, R.: Smoothed efficient algorithms and reductions for network coordination games. In: Proceedings of ITCS (2020)
15. Brandt, F., Conitzer, V., Endriss, U., Lang, J., Procaccia, A.D. (eds.): Handbook of Computational Social Choice. Cambridge University Press, Cambridge (2016)
16. Caragiannis, I., et al.: On the approximability of Dodgson and Young elections. In: Proceedings of SODA (2009)
17. Chen, X., Li, Y., Mao, J.: A nearly instance optimal algorithm for top-k ranking under the multinomial logit model. In: Proceedings of SODA (2018)
18. Ding, K., Weinberg, S.M.: Approximately strategyproof tournament rules in the probabilistic setting. In: Proceedings of ITCS (2021)
19. Dwork, C., Kumar, R., Naor, M., Sivakumar, D.: Rank aggregation methods for the web. In: Proceedings WWW (2001)
20. Feige, U.: Introduction to semi-random models. In: Roughgarden, T. (ed.) Beyond the Worst-Case Analysis of Algorithms. Cambridge University Press, Cambridge (2021)
21. Feige, U., Kilian, J.: Heuristics for finding large independent sets, with applications to coloring semi-random graphs. In: Proceedings of FOCS (1998)
22. Gehrlein, W.V.: Condorcet's Paradox. Springer, Heidelberg (2006). https://doi.org/10.1007/3-540-33799-7
23. Hemaspaandra, E., Hemaspaandra, L.A., Rothe, J.: Exact analysis of Dodgson elections: Lewis Carroll's 1876 voting system is complete for parallel access to np. J. ACM **44**, 806–825 (1997)
24. Hemaspaandra, E., Spakowski, H., Vogel, J.: The complexity of Kemeny elections. Theoret. Comput. Sci. **349**(3), 382–391 (2005)
25. Homan, C.M., Hemaspaandra, L.A.: Guarantees for the success frequency of an algorithm for finding Dodgson-election winners. J. Heuristics **15**, 403–423 (2009)
26. Lehtinen, A., Kuorikoski, J.: Unrealistic assumptions in rational choice theory. Philos. Soc. Sci. **37**(2), 115–138 (2007)
27. Lu, T., Boutilier, C.: Budgeted social choice: From consensus to personalized decision making. In: Proceeding of IJCAI (2011)
28. McCabe-Dansted, J.C., Pritchard, G., Slinko, A.: Approximability of Dodgson's rule. Soc. Choice Welfare **31**, 311–330 (2008)
29. Mohajer, S., Suh, C., Elmahdy, A.: Active learning for top-k rank aggregation from noisy comparisons. In: Proceedings of ICML (2017)
30. Nurmi, H.: Voting Paradoxes and How to Deal with Them. Springer, Heidelberg (1999). https://doi.org/10.1007/978-3-662-03782-9

31. Perrot, K., Pham, T.V.: Feedback arc set problem and np-hardness of minimum recurrent configuration problem of chip-firing game on directed graphs. Ann. Comb, 373–396 (2015)
32. Procaccia, A.D., Rosenschein, J.S.R., Zohar, A.Z.: On the complexity of achieving proportional representation. Soc. Choice Welfare **30**(3), 353–362 (2008)
33. Psomas, A., Schvartzman, A., Weinberg, S.M.: Smoothed analysis of multi-item auctions with correlated values. In: Proceedings of EC (2019)
34. Rothe, J., Spakowski, H., Vogel, J.: Exact complexity of the winner problem for Young elections. Theory Comput. Syst. **36**(4), 375–386 (2003)
35. Roughgarden, T.: Beyond the Worst-Case Analysis of Algorithms. Cambridge University Press, Cambridge (2021)
36. Spielman, D.A., Teng, S.H.: Smoothed analysis of algorithms: why the simplex algorithm usually takes polynomial time. J. ACM **51**(3) (2004)
37. Spielman, D.A., Teng, S.H.: Smoothed analysis: an attempt to explain the behavior of algorithms in practice. Commun. ACM **52**(10), 76–84 (2009)
38. Xia, L.: The smoothed possibility of social choice. In: Proceedings of NeurIPS (2020)
39. Xia, L.: How likely are large elections tied? In: Proceedings of EC (2021)
40. Xia, L.: The semi-random satisfaction of voting axioms. In: Proceedings of NeurIPS (2021)
41. Xia, L., Zheng, W.: The smoothed complexity of computing Kemeny and slater rankings. In: Proceedings of AAAI (2021)

Abstracts

Strategyproof and Proportionally Fair Facility Location

Haris Aziz[1], Alexander Lam[1(✉)], Barton E. Lee[2], and Toby Walsh[1,3]

[1] UNSW Sydney, Sydney, Australia
{haris.aziz,alexander.lam1,t.walsh}@unsw.edu.au
[2] ETH Zürich, Zürich, Switzerland
bartonlee@ethz.ch
[3] Data61 CSIRO, Eveleigh, Australia

Abstract. We focus on a simple, one-dimensional collective decision problem (often referred to as the facility location problem) and explore issues of strategyproofness and proportional fairness. We present several characterization results for mechanisms that satisfy strategyproofness and varying levels of proportional fairness. We also characterize one of the mechanisms as the unique equilibrium outcome for any mechanism that satisfies natural fairness and monotonicity properties. Finally, we identify strategyproof and proportionally fair mechanisms that provide the best welfare-optimal approximation among all mechanisms that satisfy the corresponding fairness axiom.
ArXiV link: https://arxiv.org/pdf/2111.01566.pdf

Keywords: Facility location · Mechanism design · Social choice · Fairness · Strategyproofness

© The Author(s), under exclusive license to Springer Nature Switzerland AG 2022
K. A. Hansen et al. (Eds.): WINE 2022, LNCS 13778, p. 351, 2022.
https://doi.org/10.1007/978-3-031-22832-2

Robust Misspecified Models and Paradigm Shifts

Cuimin Ba[(✉)]

University of Pennsylvania, Philadelphia, PA 19104, USA
cuiminba@sas.upenn.edu

Abstract. This paper studies which misspecified models are likely to persist when the decision maker compares her model with competing models. I present a framework where the agent learns about an action-dependent data-generating process and makes decisions repeatedly. Aware of potential model misspecification, she uses a Bayes factor criterion to switch between models according to how well they fit the data. The main result provides a characterization of persistent models based on the model-induced equilibrium, properties of the learning process such as priors and the switching threshold, and the set of competing models that may arise. I show that misspecified models can be robust against a wide range of competing models—including the true data-generating process, despite the agent having an infinite amount of data. Moreover, simple misspecified models with entrenched priors may have even better robustness properties than correctly specified models. I use these results to provide learning foundations for the persistence of systemic biases in two canonical applications: first, in a natural class of effort-choice problems, overconfidence in one's ability is more robust than underconfidence; second, an oversimplified binary view in politics trumps a correct view and leads to polarization when individuals consume media without fully recognizing the media bias.

Keywords: Robust misspecified models · Learning with misspecified models · Self-confirming equilibrium · Berk-Nash equilibrium

The full paper is available at https://cuiminba.github.io/Papers/Robust_Models.pdf.

K. A. Hansen et al. (Eds.): WINE 2022, LNCS 13778, p. 352, 2022.
https://doi.org/10.1007/978-3-031-22832-2

Information Design in Allocation with Costly Verification

Yi-Chun Chen[1], Gaoji Hu[2]([⊠]), and Xiangqian Yang[3]

[1] Department of Economics and Risk Management Institute,
National University of Singapore, Singapore 117570, Singapore
ecsycc@nus.edu.sg
[2] School of Economics, Shanghai University of Finance and Economics,
Shanghai 200433, China
hugaoji@sufe.edu.cn
[3] Department of Economics, Hunan University, Changsha 410082, China
yangxiangqian@hnu.edu.cn

Abstract. This paper studies information design in the context of allocation with costly verification à la [1]. Particularly, a principal who values an object allocates it to one or more agents. Agents learn private information (signals) from an information designer about the allocation payoff to the principal. Monetary transfer is not available but the principal can costly verify agents' private signals. The information designer can influence the agents' signal distributions, based upon which the principal maximizes the allocation *surplus*. An agent's utility is simply the *probability* of obtaining the good.

With a single agent, we characterize (i) the agent-optimal information, (ii) the principal-worst information, and (iii) the principal-optimal information. For concrete examples, making the signal distribution the least informative is principal-worst and the most informative being principal-optimal. An agent-optimal information pools information above a cutoff signal and fully reveals information below the cutoff. Even though the objectives of the principal and the agent are not directly comparable, any agent-optimal information is principal-worst, but not the converse.

With multiple agents, agent-optimal information maximizes the total probability of agents' obtaining the good. Compared with the prior distribution, under some agent-optimal information, all agents can be better off; while under some other agent-optimal information, some agents get worse off. Moreover, agent-optimal informations may deliver different payoffs to the principal, which implies that an agent-optimal information need *not* be principal-worst.

The principal's payoff under the principal-worst information provides an upper bound for the payoff that can be achieved by a "robust" mechanism which does not depend on details of the agent's type distribution. We find a robust mechanism that does achieve such an upper bound payoff, which is therefore an optimal robust mechanism. Moreover, allowing for correlated distributions does not affect the result.

https://ssrn.com/abstract=4245445.

K. A. Hansen et al. (Eds.): WINE 2022, LNCS 13778, pp. 353–354, 2022.
https://doi.org/10.1007/978-3-031-22832-2

Reference

1. Ben-Porath, E., Dekel, E., Lipman, B.L.: Optimal allocation with costly verification. Am. Econ. Rev. **104**(12), 3779–3813 (2014)

Optimal Private Payoff Manipulation Against Commitment in Extensive-form Games

Yurong Chen[1]($\boxtimes$) (iD), Xiaotie Deng[1]($\boxtimes$) (iD), and Yuhao Li[2]($\boxtimes$) (iD)

[1] Center on Frontiers of Computing Studies, School of Computer Science,
Peking University, Beijing 100871, China
{chenyurong,xiaotie}@pku.edu.cn
[2] Columbia University, New York, NY 10027, USA
yuhaoli@cs.columbia.edu

Abstract. To take advantage of strategy commitment, a useful tactic of playing games, a leader must learn enough information about the follower's payoff function. However, this leaves the follower a chance to provide fake information and influence the final game outcome. Through a carefully contrived payoff function misreported to the learning leader, the follower may induce an outcome that benefits him more, compared to the ones when he truthfully behaves.

We study the follower's optimal manipulation via such strategic behaviors in extensive-form games. Followers' different attitudes are taken into account. An optimistic follower maximizes his true utility among all game outcomes that can be induced by some payoff function. A pessimistic follower only considers misreporting payoff functions that induce a unique game outcome. For all the settings considered in this paper, we characterize all the possible game outcomes that can be induced successfully. We show that it is polynomial-time tractable for the follower to find the optimal way of misreporting his private payoff information. Our work completely resolves this follower's optimal manipulation problem on an extensive-form game tree.

Full version of the paper can be found at https://arxiv.org/abs/2206.13119.

Keywords: Stackelberg equilibrium · Strategic behavior · Private information manipulation · Extensive-form games

This work is supported by Science and Technology Innovation 2030 - "The Next Generation of Artificial Intelligence" Major Project No. (2018AAA0100901).
Y. Li—Supported by NSF grants CCF-1563155, CCF-1703925, IIS-1838154, CCF-2106429 and CCF-2107187.

K. A. Hansen et al. (Eds.): WINE 2022, LNCS 13778, p. 355, 2022.
https://doi.org/10.1007/978-3-031-22832-2

Optimal Feature-Based Market Segmentation and Pricing

Titing Cui[1]([⊠]) [ID] and Michael L. Hamilton[2] [ID]

[1] Katz Graduate School of Business, University of Pittsburgh,
Pittsburgh, PA 15260, USA
tic54@pitt.edu
[2] Katz Graduate School of Business, University of Pittsburgh, Pittsburgh,
PA 15260, USA
mhamilton@katz.pitt.edu

Abstract. In this work, we study semi-personalized pricing strategies where a seller uses features about their customers to segment the market, and customers are offered segment-specific prices. In general, finding jointly optimal market segmentation and pricing policies is computationally intractable, with practitioners often resorting to heuristic segment-then-price strategies. In response, we study how to optimize and analyze feature-based market segmentation and pricing under the assumption that the seller has a trained (noisy) regression model mapping features to valuations. First, we establish novel hardness and approximation results in the case when model noise is independent. Second, in the common cases when the noise in the model is log-concave, we show the joint segmentation and pricing problem can be efficiently solved, and characterize a number of attractive structural properties of the optimal feature-based market segmentation and pricing. Finally, we conduct a case study using home mortgage data, and show that compared to heuristic approaches, our optimal feature-based market segmentation and pricing policies can achieve nearly all of the available revenue with only a few segments.

The full paper can be found at https://papers.ssrn.com/sol3/papers.cfm?abstract_id=4151103.

Keywords: Market segmentation · Personalized pricing · Third degree price discrimination · Regression

K. A. Hansen et al. (Eds.): WINE 2022, LNCS 13778, p. 356, 2022.
https://doi.org/10.1007/978-3-031-22832-2

Competition Among Parallel Contests

Xiaotie Deng[1], Ningyuan Li[1], Weian Li[1], and Qi Qi[2(✉)]

[1] Center on Frontiers of Computing Studies, School of Computer Science,
Peking University, Beijing, China
{xiaotie,liningyuan,weian_li}@pku.edu.cn
[2] Gaoling School of Artificial Intelligence, Renmin University of China,
Beijing, China
qi.qi@ruc.edu.cn

Abstract. Contest depicts a scene in which many players compete for several designed prizes, capturing many realistic game-theoretical settings involving competition, and is an important part of mechanism design theory, which has attracted the attention of many researchers from the past to the present. So far, most of the research literature in contest theory has focused on the setting of a single contest and aimed to design the rewarding policy to achieve some specific goals. With the emergence of crowdsourcing competitions, contests are becoming increasingly popular. More and more contests are run in parallel nowadays.

In this paper, we investigate the model of multiple contests held in parallel, where each contestant selects one contest to join and each contest designer decides the prize structure to compete for the participation of contestants. We first analyze the strategic behaviors of contestants and completely characterize the symmetric Bayesian Nash equilibrium. As for the strategies of contest designers, when other designers' strategies are known, we show that computing the best response is NP-hard and propose a fully polynomial time approximation scheme (FPTAS) to output the ϵ-approximation best response. When other designers' strategies are unknown, we perform a worst-case analysis of one designer's strategy. An upper bound on the worst-case utility of any strategy is derived and taken as a benchmark. We propose a method to construct a strategy whose utility can guarantee a constant ratio of the benchmark.

The full version is available at https://arxiv.org/abs/2210.06866.

Keywords: Competition · Parallel contests · Equilibrium behavior · Best response · Safety level

This research was partially supported by the National Natural Science Foundation of China (NSFC) (No.62172012), and the Research Funds of Renmin University of China (22XNKJ07) and No.297522503709, Beijing Outstanding Young Scientist Program No.BJJWZYJH012019100020098, and Intelligent Social Governance Platform, Major Innovation & Planning Interdisciplinary Platform for the "Double-First Class" Initiative, Renmin University of China.

K. A. Hansen et al. (Eds.): WINE 2022, LNCS 13778, p. 357, 2022.
https://doi.org/10.1007/978-3-031-22832-2

Revenue Management with Product Retirement and Customer Selection

Adam N. Elmachtoub[1], Vineet Goyal[1], Roger Lederman[2], and Harsh Sheth[1(✉)]

[1] Department of Industrial Engineering and Operations Research and Data Science Institute, Columbia University, New York, USA
{ae2516,vg2277,hts2112}@columbia.edu
[2] Amazon, Seattle, USA
rllederm@amazon.com

Abstract. We consider a multi-period, multi-product revenue management problem where in each period the seller has a fixed inventory of multiple substitutable products to sell over a fixed time horizon. In each time period, the seller chooses which subset of products to retire and also selects a customer to visit. When a product is retired, it becomes unavailable to all future customers. When a customer is selected, all available products – non-retired products with positive remaining inventory – are offered for the customer to choose from. The objective of the seller is to dynamically retire products and select customers in order to maximize the total expected revenue over a fixed time horizon. Such product retirement decisions are essential when the seller is not able to personalize the set of products offered to each customer.

When customers choose according to the same multinomial logit model, we show that a deterministic product retirement policy is asymptotically optimal as the inventories grow large. For multiple customer types, we give an asymptotically optimal policy for product retirement and customer selection when the upper bound linear program has an optimal solution with specific structure. We show that such solution can always be found when there are only two products. In the general case with multiple customer types and products, we design a linear programming-based policy that guarantees a constant fraction of the optimal dynamic retirement-selection policy. Finally, we show that our policies perform well in numerical experiments calibrated with real data, compared to natural benchmarks.

Full paper: https://papers.ssrn.com/sol3/papers.cfm?abstract_id=4033922.

Work of this author was conducted while employed by IBM Research.

K. A. Hansen et al. (Eds.): WINE 2022, LNCS 13778, p. 358, 2022.
https://doi.org/10.1007/978-3-031-22832-2

Order Selection Problems in Hiring Pipelines

Boris Epstein$^{(\boxtimes)}$ and Will Ma

Graduate School of Business, Columbia University, New York, NY 10027, USA
{bepstein25,wm2428}@gsb.columbia.edu

Abstract. Motivated by hiring pipelines, we study two order selection problems in which applicants for a finite set of positions must be interviewed or made offers sequentially. There is a finite time budget for interviewing or making offers, and a stochastic realization after each decision, leading to computationally-challenging problems.

In the first problem we study sequential interviewing. In this setting, a firm must interview candidates to fill out k job positions. There is a pool of n candidates, and the values obtained by the firm from hiring each one of them are non-negative random variables sampled from known, independent distributions. The firm can interview candidates to learn the realization of their values. Up to T interviews can be sequentially carried out, after which k out of the T interviewed candidates are chosen for hire. We show that a computationally tractable, non-adaptive policy that must make offers immediately after interviewing is approximately optimal, assuming offerees always accept their offers.

In the second problem, there are again k positions but we assume that the n applicants have already been interviewed. They accept offers independently according to known probabilities. Moreover, offers can be sent in parallel, under the constraint that at each of the T time periods the amount of offers sent does not exceed the amount of positions remaining. We develop a computationally tractable policy that makes offers for the different positions in parallel, which is approximately optimal even relative to a policy that can make kT offers sequentially.

Our two results both generalize and improve the guarantees in the work of Purohit et al. [1] on hiring algorithms, from 1/2 and 1/4 to approximation factors that are at least $1 - 1/e \approx 63.2\%$. Our algorithms work by solving LP relaxations of the corresponding problems, and then rounding the optimal solutions to decide which candidates to interview and send offers to.

Keywords: Hiring · Order selection · Stochastic probing · Adaptivity gap

© The Author(s), under exclusive license to Springer Nature Switzerland AG 2022
K. A. Hansen et al. (Eds.): WINE 2022, LNCS 13778, pp. 359–360, 2022.
https://doi.org/10.1007/978-3-031-22832-2

A full version of this paper can be found in https://arxiv.org/abs/2210.04059.

Reference

1. Purohit, M., Gollapudi, S., Raghavan, M.: Hiring under uncertainty. In: International Conference on Machine Learning, pp. 5181–5189. PMLR (2019)

Strategyproofness in Kidney Exchange with Cancellations

Itai Feigenbaum[1,2]([✉])

[1] Lehman College, City University of New York, Bronx, New York 10468, USA
itai.feigenbaum@lehman.cuny.edu
[2] The Graduate Center, City University of New York, New York, NY 10016, USA

Patients requiring kidney transplant may have proxy donors: people who want to donate a kidney to the patient, but can't due to medical incompatibility. p is called a proxy patient of d if d is a proxy donor of p. Some patients, called overloaded, have multiple proxy donors. The pool of a patient is the set of her proxy donors. Patients can swap proxy donors, so that each swapping patient ends up with a compatible donor. A matching is a set of planned transplants resulting from swaps, altruistic donations from donors without proxy patients, and donations to the waiting list of patients without proxy donors. $C_{\max}$ is the maximum cycle size allowed, and $\Lambda_{\max}$ is the maximum number of donations allowed from a pool.

In practice, many planned transplants get canceled. Cancellation of a transplant from donor d to patient p can be direct—due to reasons involving d and p, or indirect—due to cancellation of a transplant to d's proxy patient. We assume that donors in directly-but-not-indirectly canceled transplants are redirected to donate to the waiting list. We want matchings to maximize the objective of the expected number of actually executed transplants. Exact maximization introduces perverse incentives for overloaded patients, who can increase the probability they receive a kidney by hiding some of their proxy donors.

We design the SuperGreedy Algorithm, which incentivizes patients to fully reveal their pools: each patient maximizes the probability of receiving a kidney by full revelation. We then assume a uniformly constant direct cancellation probability $1 - \alpha$ for all transplants. When $\alpha < \frac{1}{\Lambda_{\max}}$, we prove a bound of $\frac{\Lambda_{\max}}{1 - \Lambda_{\max}\alpha} \max\{(1 + 2\alpha - 2\alpha^2), (1 - \alpha)(1 + C_{\max}\alpha)\}$ on SuperGreedy's approximation ratio. Next, we implement SuperGreedy via integer programming, and simulate it on realistic data. Our results show much better performance than our theoretical bound. Specifically, we get an upper bound on the average approximation ratio of 1.142 when $\Lambda_{\max} = 1$ and $\alpha = 0.3$; as $\Lambda_{\max}$ increases and α decreases, the bound decreases, down to 1.016 when $\Lambda_{\max} = 4$ and $\alpha = 0.1$. A full version is available at http://www.itaifeigenbaum.com/WINEStrategicCancellations.pdf.

© The Author(s), under exclusive license to Springer Nature Switzerland AG 2022
K. A. Hansen et al. (Eds.): WINE 2022, LNCS 13778, p. 361, 2022.
https://doi.org/10.1007/978-3-031-22832-2

Tractable Fragments of the Maximum Nash Welfare Problem

Jugal Garg[1], Edin Husić[2], Aniket Murhekar[1(✉)], and László Végh[2]

[1] University of Illinois, Urbana-Champaign, USA
{jugal,aniket2}@illinois.edu
[2] London School of Economics and Political Science, London, UK
{e.husic,l.vegh}@lse.ac.uk

We study the problem of maximizing Nash welfare (MNW) while allocating indivisible goods to asymmetric agents. The Nash welfare of an allocation is the weighted geometric mean of agents' utilities, and the allocation with maximum Nash welfare is known to satisfy several desirable fairness and efficiency properties. However, computing such an MNW allocation is NP-hard, even for two agents with identical, additive valuations. Hence, we aim to identify tractable classes that either admit a PTAS, an FPTAS, or an exact polynomial-time algorithm. To this end, we design a PTAS for finding an MNW allocation for the case of asymmetric agents with identical, additive valuations, thus generalizing a similar result for symmetric agents [2]. We also extend our PTAS to compute a nearly Nash-optimal allocation which also satisfies the best fairness guarantee offered by the optimal MNW allocation (a weighted relaxation of envy-freeness); showing we do not need to compromise fairness for tractability. Our techniques can also be adapted to give (i) a PTAS for the problem of computing the optimal p-mean welfare, and (ii) a polynomial time algorithm for computing an MNW allocation for identical agents with k-ary valuations when k is a constant, where every agent has at most k different values for the goods. Next, we consider the special case where every agent finds at most two goods valuable, and show that this class admits an efficient algorithm, even for general monotone valuations. In contrast, we note that when agents can value three or more goods, maximizing Nash welfare is NP-hard, even when agents are symmetric and have additive valuations, showing our algorithmic result is essentially tight. Finally, we show that for constantly many asymmetric agents with additive valuations, the MNW problem admits an FPTAS. The full version of the paper is available at [1].

Work supported by the NSF Grant CCF-1942321 and ERC Starting Grant ScaleOpt.

K. A. Hansen et al. (Eds.): WINE 2022, LNCS 13778, pp. 362–363, 2022.
https://doi.org/10.1007/978-3-031-22832-2

References

1. Garg, J., Husic, E., Murhekar, A., Végh, L.A.: Tractable fragments of the maximum Nash welfare problem. CoRR abs/2112.10199 (2021). arxiv.org/abs/2112.10199
2. Nguyen, T.T., Rothe, J.: Minimizing envy and maximizing average Nash social welfare in the allocation of indivisible goods. Discrete Appl. Math. **179**(C), 54–68 (2015)

Project Selection with Partially Verifiable Information

Sumit Goel[iD] and Wade Hann-Caruthers[(✉)][iD]

California Institute of Technology, Pasadena, CA 91125, USA
{sgoel,whanncar}@caltech.edu

Abstract. We consider a principal agent project selection problem with asymmetric information. There are N projects and the principal must select exactly one of them. Each project provides some profit to the principal and some payoff to the agent and these profits and payoffs are the agent's private information. If the principal could use transfers, it could essentially sell the firm to the agent and extract the entire surplus. If transfers are not feasible and the agent is unconstrained in its reporting, the principal can do no better than to choose the ex-ante optimal project. However, the agent's ability to manipulate may be constrained due to environmental factors, or because it may be required to furnish evidence in support of its claims. Motivated by such considerations, we consider the problem under a natural partial verifiability constraint of no-overselling wherein the agent cannot report a project to be more profitable than it actually is.

To study this problem, we first characterize the set of implementable mechanisms. As we show, every implementable mechanism can be decomposed into two functions. The first maps each vector of reported profits π to a subset of projects, and the second maps each subset of projects T and vector of reported agent payoffs α to the agent-preferred project in T. The first function can be understood as determining the set of projects the principal makes available for the agent to choose from. We show that such a function corresponds to an implementable mechanism if and only if it is increasing: if every project's reported profit under π' is (weakly) higher than under π, then every project made available when π is reported must also be made available when π' is reported.

Using this characterization, we study the principal's problem of finding an optimal mechanism for two different objectives: maximizing expected profit and maximizing the probability of choosing the most profitable project. For both objectives, we find that in the case of two projects, the optimal mechanism takes the form of a simple cutoff mechanism. The simple structure of the optimal mechanism also allows us to find evidence in support of the well-known ally-principle which says

---

The full version of the paper is available at https://arxiv.org/abs/2007.00907.

K. A. Hansen et al. (Eds.): WINE 2022, LNCS 13778, pp. 364–365, 2022.
https://doi.org/10.1007/978-3-031-22832-2

that the principal delegates more authority to an agent who shares their preferences. In particular, we find the optimal cutoff for the case where principal agent payoffs are distributed bivariate normal and show that it decreases as the payoffs become more correlated.

Keywords: Mechanism design · Partial verifiability · Cutoff mechanisms

Revenue Management Under a Price Alert Mechanism

Bo Jiang[1], Zizhuo Wang[2], and Nanxi Zhang[1,3](✉)

[1] Research Institute of Interdisciplinary Science, Shanghai University of Finance and Economics, Shanghai, China
`jiang.bo@mail.shufe.edu.cn`
[2] School of Data Science, The Chinese University of Hong Kong, Shenzhen, Shenzhen 518172, China
`wangzizhuo@cuhk.edu.cn`
[3] Sauder School of Business, University of British Columbia, Vancouver, BC, Canada
`nanxi.zhang@sauder.ubc.ca`

Abstract. Many online platforms adopt a price alert mechanism to facilitate customers tracking the price changes. This mechanism allows customers to register their valuation to the system when they find the price is larger than the valuation on their arrival period. Once the price drops below the customers' registered price, a message will be sent to notify them. In this paper, we study the optimal pricing problem under this mechanism. First, when the customer's waiting time is one period, we show that it is optimal for the seller to use a threshold to decide whether to accept or reject a registered price, and the price trajectory under the optimal policy has a stochastic cyclic decreasing structure. When the customer's valuation is a uniform distribution, the analytical form of the optimal policy is further obtained. When the customer's patience level is two periods, we obtain the structure of the optimal policy by showing the asymmetric role each registered price plays in the optimal policy. Then we consider the case when the customer can stay in the system for an infinite number of periods. We derive an asymptotic optimal policy for this case. We find that adopting the price alert mechanism always increases social welfare; however, it may hurt the customer surplus when the seller has a large discount factor. Finally, we consider the case when the customers can strategically react to the price alert mechanism by timing their purchases and reporting false valuations. Using a Stackelberg's game model, we obtain the seller's optimal threshold type of policy. We

Link to the full paper: https://papers.ssrn.com/sol3/papers.cfm?abstract_id=4154861. The first author's research is supported by the National Natural Science Foundation of China (Grants 72171141, 72150001 and 11831002), and Program for Innovative Research Team of Shanghai University of Finance and Economics. The second author's research is partly supported by the National Science Foundation of China (NSFC) Grant 72150002.

K. A. Hansen et al. (Eds.): WINE 2022, LNCS 13778, pp. 366–367, 2022.
https://doi.org/10.1007/978-3-031-22832-2

show that the price alert mechanism can still be helpful to the seller, although the advantage diminishes when customers are very strategic.

Keywords: Price alert mechanism · Threshold property · Stochastic cyclic decreasing price · Strategic customer

Bicriteria Nash Flows over Time

Tim Oosterwijk[1]([✉])(iD), Daniel Schmand[2](iD), and Marc Schröder[3](iD)

[1] Vrije Universiteit Amsterdam, De Boelelaan 1105, 1081, HV Amsterdam,
The Netherlands
t.oosterwijk@vu.nl
[2] University of Bremen, Bibliothekstraße 5, 28359 Bremen, Germany
schmand@uni-bremen.de
[3] Maastricht University, Tongersestraat 53, 6211, LM Maastricht, The Netherlands
m.schroder@maastrichtuniversity.nl

Traffic congestion imposes a huge economic loss to the economy. As such, there has been a huge effort to understand congestion using theoretical models. The dynamic model that gained most attention for modelling traffic is the deterministic fluid queuing model, already introduced by Vickrey [4]. A common drawback of most of the models is the simplified assumption that road network users only aim for minimizing their arrival time [1,2]. However, in traffic networks in particular, users are not always that single-minded. In this paper we extend the state-of-the-art game theoretic traffic models with a multi-criteria objective function. We assume that users try to minimize costs subject to arriving at the sink before a given deadline. Here, costs could be thought of as an intrinsic preference a user has regarding the different route choices and queuing dynamics only play a role for the arrival time of a user.

We determine the existence and the structure of Nash flows over time and fully characterize the price of anarchy for this model, which measures the ratio of the quality of the Nash flow and the optimal flow. We evaluate the quality both with respect to the throughput for a given deadline and the makespan for a given amount of flow. We prove the following three results. (i) In series-parallel graphs, both prices of anarchy are unbounded. (ii) In parallel path graphs the throughput-PoA is at most 2, or at most $e/(e-1)$ if all transit times are 0. Both bounds are tight. (iii) In parallel path graphs the makespan-PoA is at most $e/(e-1)$, independent of transit times, and this is tight. All our upper bounds are also valid for dynamic equilibria in the deterministic fluid queuing model.

The full version of the paper can be found in [3].

References

1. Cominetti, R., Correa, J., Larré, O.: Dynamic equilibria in fluid queueing networks. Oper. Res. **63**(1), 21–34 (2015)
2. Koch, R., Skutella, M.: Nash equilibria and the price of anarchy for flows over time. Theor. Comput. Syst. **49**(1), 71–97 (2011)
3. Oosterwijk, T., Schmand, D., Schörder, M.: Bicriteria nash flows over time. arXiv (2021). arxiv.org/abs/2111.08589
4. Vickrey, W.S.: Congestion theory and transport investment. Am. Econ. Rev. **59**(2), 251–260 (1969)

K. A. Hansen et al. (Eds.): WINE 2022, LNCS 13778, p. 368, 2022.
https://doi.org/10.1007/978-3-031-22832-2

Truthful Generalized Linear Models

Yuan Qiu[1(✉)], Jinyan Liu[2], and Di Wang[3]

[1] Georgia Institute of Technology, Atlanta, GA 30332, USA
yuan.qiu@gatech.edu
[2] Beijing Institute of Technology, Beijing 100081, China
jyliu@bit.edu.cn
[3] King Abdullah University of Science and Technology, Thuwal 23955, Saudi Arabia
di.wang@kaust.edu.sa

Abstract. In this paper we study estimating Generalized Linear Models (GLMs) in the case where the agents (individuals) are strategic or self-interested and they concern about their privacy when reporting data. Compared with the classical setting, here we aim to design mechanisms that can both incentivize most agents to truthfully report their data and preserve the privacy of individuals' reports, while their outputs should also close to the underlying parameter. In the first part of the paper, we consider the case where the covariates are sub-Gaussian and the responses are heavy-tailed where they only have the finite fourth moments. First, motivated by the stationary condition of the maximizer of the likelihood function, we derive a novel private and closed form estimator. Based on the estimator, we propose a mechanism which has the following properties via some appropriate design of the computation and payment scheme for several canonical models such as linear regression, logistic regression and Poisson regression: (1) the mechanism is $o(1)$-jointly differentially private (with probability at least $1 - o(1)$); (2) it is an $o(\frac{1}{n})$-approximate Bayes Nash equilibrium for a $(1 - o(1))$-fraction of agents to truthfully report their data, where n is the number of agents; (3) the output could achieve an error of $o(1)$ to the underlying parameter; (4) it is individually rational for a $(1-o(1))$ fraction of agents in the mechanism; (5) the payment budget required from the analyst to run the mechanism is $o(1)$. In the second part, we consider the linear regression model under more general setting where both covariates and responses are heavy-tailed and only have finite fourth moments. By using an ℓ_4-norm shrinkage operator, we propose a private estimator and payment scheme that have similar properties as in the sub-Gaussian case.

Di Wang is supported in part by King Abdullah University of Science and Technology URF/1/4663-01-01, FCC/1/1976-49-01, and a funding from SDAIA-KAUST AI center. Jinyan Liu is partially supported by National Natural Science Foundation of China (NSFC Grant No.62102026).

K. A. Hansen et al. (Eds.): WINE 2022, LNCS 13778, pp. 369–370, 2022.
https://doi.org/10.1007/978-3-031-22832-2

The full version of the paper is available at https://arxiv.org/pdf/2209.07815.pdf.

Keywords: Generalized linear models · Bayesian game · Differential privacy · Sub-gaussian and heavy-tailed data · Truthful mechanism design

Algorithmic Challenges in Ensuring Fairness at the Time of Decision

Jad Salem[1(✉)], Swati Gupta[1], and Vijay Kamble[2]

[1] Georgia Institute of Technology, Atlanta, GA 30332, USA
{jsalem7,swatig}@gatech.edu
[2] University of Illinois Chicago, Chicago, IL 60607, USA
kamble@uic.edu

Abstract. Algorithmic decision-making in societal contexts such as retail pricing, loan administration, recommendations on online platforms, etc., often involves experimentation with decisions for the sake of learning, which results in perceptions of unfairness amongst people impacted by these decisions. It is hence necessary to embed appropriate notions of fairness in such decision-making processes. The goal of this paper is to highlight the rich interface between temporal notions of fairness and online decision-making through a novel meta-objective of ensuring *fairness at the time of decision*. Given some arbitrary comparative fairness notion for static decision-making (e.g., students should pay at most 90% of the general adult price), a corresponding online decision-making algorithm satisfies fairness at the time of decision if the said notion of fairness is satisfied for any entity receiving a decision *in comparison to all the past decisions*. We show that this basic requirement introduces new methodological challenges in online decision-making. We illustrate the novel approaches necessary to address these challenges in the context of stochastic convex optimization with bandit feedback under a comparative fairness constraint that imposes lower bounds on the decisions received by entities depending on the decisions received by everyone in the past. The paper showcases novel research opportunities in online decision-making stemming from temporal fairness concerns.

Keywords: Fairness · Online learning · Bandit convex optimization

---

This work was partially supported by NSF grant 2112533.
A complete version of this work can be found at https://arxiv.org/abs/2103.09287.

K. A. Hansen et al. (Eds.): WINE 2022, LNCS 13778, p. 371, 2022.
https://doi.org/10.1007/978-3-031-22832-2

Eliminating Waste in Cadaveric Organ Allocation

Peng Shi[✉][iD] and Junxiong Yin[iD]

USC Marshall School of Business, Los Angeles, CA 90089, USA
{pengshi,junxiong.yin}@marshall.usc.edu

Abstract. There is a shortage in the supply of cadaveric organs in most countries, but many successfully procured and medically tenable organs are currently being discarded. This wastage of cadaveric organs exacerbates the shortage in organ supply and the financial strains on healthcare systems. Many reforms have been or are currently being implemented to address the wastage problem. However, we show that waste will still be a problem as long as the allocation mechanism continues to prioritize patients by their waiting times, which incentivizes patients to reject organs of reasonable quality now to wait for better offers in the future. Such waiting is risky, as the patients' health conditions may deteriorate while they wait, and they may no longer be fit to receive transplants when the ideal offers come. Through analyzing a theoretical model, we show that the necessary and sufficient conditions to eliminating waste are to disincentivize waiting by allocating over-demanded organ types only to the patients who recently signed up for transplantation, and to give the patients who are not allocated their ideal organs an opportunity to take another offer. However, such a policy may be contentious as it no longer prioritizes patients by waiting times. Moreover, it may reduce the welfare of the patients who are most willing to wait. The benefits of eliminating waste should be weighed against these costs when making policy decisions.

Keywords: Market design · Matching markets · Organ allocation · Wait-list

Link to the full paper:
https://papers.ssrn.com/sol3/papers.cfm?abstract_id=4069084

Matrix-Exact Covers
of Minimum-Cost-Spanning-Tree Games

Zhibin Tan[iD], Zhigang Cao[iD], and Zhengxing Zou[✉][iD]

School of Economics and Management, Beijing Jiaotong University,
No. 3 Shangyuancun, Beijing 100044, China
{tanzhibin,zgcao,zhxzou}@bjtu.edu.cn

Abstract. The minimum-cost-spanning-tree (m.c.s.t.) game is a classical cooperative game model that has been extensively studied. Many important solutions for m.c.s.t. games depend on certain pruning operations that construct a new m.c.s.t. game by reducing the costs of certain edges such that the new game is simpler, and its core is a subset of the core of the original game. Examples include the classical irreducible graph, based on which the irreducible core (Bird, 1976), the Folk rule (Feltkamp et al., 1994; Bergantiños and Vidal-Puga, 2007), and the DK rule (Dutta and Kar, 2004) are defined, and the cycle-complete graph (Trudeau, 2012). However, these operations often make the relevant solutions use very little information of the original cost matrix. As criticized by Bogomolnaia and Moulin (2010), "this drastic pruning of the cost data throws away much information relevant to the fairness of the eventual cost sharing".

To answer this criticism, we address the problem of decreasing the connection costs as much as possible in an m.c.s.t. game such that its core does not change. We define the desired m.c.s.t. game as the matrix-exact cover of the original m.c.s.t. game, and show its existence and uniqueness by providing an explicit formula. Our results also imply that the set of all m.c.s.t. games with the same core possesses a meet-semilattice structure: if two m.c.s.t. games have the same core, and we construct a new m.c.s.t. game by defining the cost of each edge as the minimum between the corresponding connection costs of the two games, then the new game has the same core too.

Keywords: Cooperative games · Core · Exact games · Matrix-exact games · Meet-semilattice

A full version is available at https://papers.ssrn.com/sol3/papers.cfm?abstract_id=4238837

Zhigang Cao was supported by the National Natural Science Foundation of China (Grants No. 71922003, 71871009, 72271016), and Beijing Natural Science Foundation (Grant No. Z220001). Zhengxing Zou was supported by the National Natural Science Foundation of China (Grant No. 72101015).

K. A. Hansen et al. (Eds.): WINE 2022, LNCS 13778, p. 373, 2022.
https://doi.org/10.1007/978-3-031-22832-2

Personalized Assortment Optimization Under Consumer Choice Models with Local Network Effects

Tong Xie[1] and Zizhuo Wang[2(✉)]

[1] Booth School of Business, University of Chicago, Chicago, IL 60637, USA
tong.xie@chicagobooth.edu
[2] School of Data Science, The Chinese University of Hong Kong, Shenzhen,
Shenzhen 518172, China
wangzizhuo@cuhk.edu.cn

Abstract. In this paper, we introduce a consumer choice model in which each consumer's utility is affected by the purchase probabilities of his/her neighbors in a network. Such a consumer choice model is a general model to characterize consumer choice under network effect. We first characterize the choice probabilities under such a choice model. Then we consider the associated personalized assortment optimization problem. Particularly, the seller is allowed to offer a personalized assortment to each consumer, and the consumer chooses among the products according to the proposed choice model. We show that the problem is NP-hard even if the consumers form a star network. Despite of the complexity of the problem, we show that if the consumers form a star network, then the optimal assortment to the central consumer cannot be strictly larger than that without network effects; and the optimal assortment to each peripheral consumer must be a revenue-ordered assortment that is a subset of the optimal assortment without network effect. We also present a condition when revenue-ordered assortments can achieve a provable performance. Then in view of the fact that each node in a network can represent a group of consumers, we propose a novel idea in which the sellers are allowed to offer "randomized assortments" to each node in the network. We show that allowing for randomized assortments may further increase the revenue, and under certain conditions, the optimal assortment for the central consumer must be a combination of two adjacent revenue-ordered assortments and thus efficient algorithm can be developed. Finally, we extend the results to directed acyclic graphs (DAGs), showing that a mixture of adjacent revenue-ordered assortments is optimal under certain conditions.

Keywords: Revenue management · Consumer choice models · Network effects · Assortment optimization

Link to the full paper: https://papers.ssrn.com/sol3/papers.cfm?abstract_id=3788880.
The second author's research is partly supported by the National Science Foundation of China (NSFC) Grant 72150002.

Exploring the Tradeoff Between Competitive Ratio and Variance in Online-Matching Markets

Pan Xu(✉)

New Jersey Institute of Technology, Newark, NJ 07102, USA
pxu@njit.edu

Abstract. In this paper, we propose an online-matching-based model to study the assignment problems arising in a wide range of online-matching markets, including online recommendations, ride-hailing platforms, and crowdsourcing markets. It features that each assignment can request a random set of resources and yield a random utility, and the two (cost and utility) can be arbitrarily correlated with each other. We present two linear-programming-based parameterized policies to study the tradeoff between the competitive ratio (CR) on the total utilities and the variance on the total number of matches (unweighted version). The first one (SAMP) is to sample an edge according to the distribution extracted from the clairvoyant optimal, while the second (ATT) features a time-adaptive attenuation framework that leads to an improvement over the state-of-the-art competitive-ratio result. We also consider the problem under a large-budget assumption and show that SAMP achieves asymptotically optimal performance in terms of competitive ratio.

Here is the arXiv link to the full version: http://arxiv.org/abs/2209.07580.

Keywords: Online matching · Competitive ratio · Variance analysis

---

Pan Xu is partially supported by NSF CRII Award IIS-1948157.

Author Index

Printed in the United States
by Baker & Taylor Publisher Services